Applications of Digital Wireless Technologies to Global Wireless Communications

Feher/Prentice Hall Digital and Wireless Communications Series

Carne, E. Bryan. *Telecommunications Primer: Signal, Building Blocks and Networks*

Feher, Kamilo. *Wireless Digital Communications: Modulation and Spread Spectrum Applications*

Garg, Vijay, Kenneth Smolick, and Joseph Wilkes. *Applications of CDMA in Wireless/Personal Communications*

Garg, Vijay and Joseph Wilkes. *Wireless and Personal Communications Systems*

Pelton, N. Joseph. *Wireless Satellite Telecommunications: The Technology, the Market & the Regulations*

Ricci, Fred. *Personal Communications Systems Applications*

Sampei, Seiichi, *Applications of Digital Wireless Technologies to Global Wireless Communications*

Other Books by Dr. Kamilo Feher

Advanced Digital Communications: Systems and Signal Processing Techniques

Telecommunications Measurements, Analysis and Instrumentation

Digital Communications: Satellite/Earth Station Engineering

Digital Communications: Microwave Applications

Available from CRESTONE Engineering Books, c/o G. Breed, 5910 S. University Blvd., Bldg. C-18 #360, Littleton, CO 80121, Tel. 303-770-4709, Fax 303-721-1021, or from DIGCOM, Inc., Dr. Feher and Associates, 44685 Country Club Drive, El Macero, CA 95618, Tel. 916-753-0738, Fax 916-753-1788.

Applications of Digital Wireless Technologies to Global Wireless Communications

Seiichi Sampei

To join a Prentice Hall PTR
mailing list, point to:
http://www.prenhall.com/register

Prentice Hall PTR
Upper Saddle River, NJ 07458
http://www.prenhall.com

Library of Congress Cataloging-in-Publication Data

Sampei, Seiichi.
    Application of digital wireless technologies to global
wireless communications / Seiichi Sampei.
    p. cm. — (Feher/Prentice Hall digital and wireless
communications series)
    Includes bibliographical references and index.
    ISBN 0-13-214272-4
    1. Global system for mobile communications. 2. Global system
for mobile communications—Japan. I. Title.
    TK5103.483.S26 1997
    621.3845'0952—dc21

TK
5103.483
.S26
1997

97-3635
CIP

Editorial/production supervision: *BooksCraft, Inc., Indianapolis, IN*
Cover design director: *Jerry Votta*
Cover design: *Amy Rosen*
Acquisitions editor: *Paul W. Becker*
Manufacturing manager: *Alexis R. Heydt*

© 1997 by Prentice Hall PTR
Prentice-Hall, Inc.
A Simon & Schuster Company
Upper Saddle River, NJ 07458

The publisher offers discounts on this book when ordered in bulk quantities.
For more information, contact:

Corporate Sales Department
Prentice Hall PTR
One Lake Street
Upper Saddle River, NJ 07458
Phone: 800-382-3419   Fax: 201-236-7141
E-mail: corpsales@prenhall.com.

All product names mentioned herein are the trademarks of their respective owners.

Printed in the United States of America

10 9 8 7 6 5 4 3 2

ISBN: 0-13-214272-4

Prentice-Hall International (UK) Limited, *London*
Prentice-Hall of Australia Pty. Limited, *Sydney*
Prentice-Hall Canada Inc., *Toronto*
Prentice-Hall Hispanoamericana, S.A., *Mexico*
Prentice-Hall of India Private Limited, *New Delhi*
Prentice-Hall of Japan, Inc., *Tokyo*
Simon & Schuster Asia Pte. Ltd., *Singapore*
Editora Prentice-Hall do Brasil, Ltda., *Rio de Janeiro*

Table of Contents

Preface

Since the early 1990s, the wireless world has been changing very rapidly, from analog to digital, from voice service to multimedia service, and from regional systems to global systems. This progress is due to recent development of key wireless communication technologies as well as to rapidly increasing demand for wireless access to the Internet that has the same transmission quality as wired Internet systems. To satisfy such requirements, we have to develop high-quality multitransmission rate capability, perhaps from several kbit/s for voice to 1–2 Mbit/s for multimedia services including video transmission, while saving radio spectrum resources as much as possible. At present, although multimedia transmission technology using the global system for mobile communications (GSM) and personal digital cellular (PDC) systems are being developed that can partially support voice and nonvoice integrated services, their maximum bit rate is limited to around 10 kbit/s, which is too low to support multimedia services. Thus, third-generation wireless communications systems are now being developed in the International Telecommunication Union Radio Section (ITU-R). The standardization process in the IRU-R has been delayed due to consideration of the migration from existing digital cellular or digital cordless systems to third-generation systems, prematurity of wireless and device technologies, and the other market issues. However, this delay has sparked various productive debates on topics such as time division multiple access (TDMA) versus code division multiple access (CDMA), a single standard versus plural standards, how to cope with various operation environments, and so on. Although migration is a political problem (and a political issue actually has a big impact on the specification of wireless communication systems), wireless communication engineers have to fully understand the available wireless communication technologies to fairly evaluate candidates for third-generation systems in the future. Actually, some debates tend to be misleading. For example, the TDMA versus CDMA debate is sometimes not an actual access technology comparison but just a comparison between existing digital cellular systems—IS-54 versus IS-95 or GSM versus IS-95. Therefore, this book focuses on key technologies for designing both TDMA-based and CDMA-based wireless communication systems. In wireless communication systems, it is most important to fully understand propagation path conditions and their impact on modulation/demodulation and access schemes as well as on synchronization performance. Therefore, there is a great deal of discussion in the book about modulation/demodulation, access schemes, and synchronization circuit designs, taking into consideration propagation path characteristics specific to wireless communication systems.

Although there are many standardized digital wireless communication systems in the world, this book intentionally discusses standardized Japanese systems, such as the PDC system and the personal handyphone system (PHS). This is because there are few English-language publications about Japanese standardized systems, while there are many books and documents about European and U.S. interim standards. Of course, it is not my intention to endorse any specific system.

Chapter 1 is an introduction to wireless communication technologies and systems. After a brief summary of existing cellular, cordless phone, and wireless local area network (LAN) systems in the United States, Europe, and Japan, the current status of third-generation wireless communication systems is summarized.

Chapter 2 discusses propagation path characteristics specific to wireless mobile communications. This discussion includes statistics of propagation path characteristics, the time-variant nature of the propagation path, and the application of this knowledge to the development of demodulators and anti-fading techniques. In wireless communication systems, modulation/demodulation, synchronization, access, and duplex techniques are very important in achieving spectral efficient and high-quality wireless communication systems. However, we cannot directly apply wireless techniques for fixed links to wireless mobile communication systems because techniques used in wireless mobile communications are severely affected by propagation path characteristics. Therefore, chapter 3 through chapter 10 discuss how to apply these techniques to wireless mobile communication system design considering propagation path characteristics.

In chapter 3, we discuss basic modulation and demodulation schemes and bit error rate (BER) performances under additive white Gaussian noise (AWGN) and flat Rayleigh fading conditions. Although actual BER performance depends on the applied anti-fading and synchronization techniques, this chapter will discuss only theoretical performance, which will be used in the following chapters as reference BER performances.

When we apply anti-fading techniques to wireless communication systems, we have to consider whether the radio channel is treated as a flat fading channel or a frequency-selective fading channel. In chapter 4, we discuss anti-flat Rayleigh fading techniques. Although differential detection is a simple detection scheme for Gaussian-filtered minimum shift keying (GMSK) and quaternary phase shift keying (QPSK), the coherent detection scheme is preferable for improving receiver sensitivity of these modulation schemes as well as for applying quadrature amplitude modulation (QAM) to mobile communications using flat Rayleigh fading compensation techniques. The anti-flat Rayleigh fading techniques are categorized as the pilot signal-aided and nonpilot signal-aided techniques. Pilot signal-aided techniques are becoming more and more preferred for high-capacity systems. Therefore, chapter 4 details both pilot signal- and nonpilot signal-aided anti-flat Rayleigh fading techniques. It also discusses diversity combining techniques that are recognized as very powerful techniques to improve transmission quality under fading conditions.

Chapter 5 discusses anti-frequency-selective fading techniques. After a brief discussion of a frequency-selective fading channel model suitable for designing anti-frequency-selective fading techniques, we discuss the decision feedback equalizer (DFE), the maximum likelihood sequence estimation (MLSE), and adaptive array antenna techniques including their parameter optimization strategies. For wireless multimedia communication systems, although various anti-fading techniques discussed in chapters 4 and 5 are very helpful for improving receiver sensitivity, their performance is not sufficient to achieve high-quality nonvoice data transmission. To cope with this requirement, in chapter 6 we will discuss error control techniques that include both forward error correction (FEC) and automatic repeat request (ARQ) techniques.

In addition to the selection of anti-fading techniques, we have to carefully select access and duplex schemes. Chapter 7 discusses the basic features of access and duplex schemes.

We also have to take synchronization schemes into account for the wireless system design because synchronization circuits, such as frame synchronization, symbol synchronization, and so on, are severely affected by fading. In chapter 8, we will discuss how to design these circuits to be operated even under fading conditions.

Chapter 9 discusses how to verify the effect of key technologies as well as how to specify their parameters using computer simulation, laboratory experiments, and field trials. Then chapter 10 will cover the design of radio links, including coverage estimation and cell reuse planning. Chapters 11 through 14 discuss how air interfaces of standardized digital wireless communication systems are designed.

In chapters 15 through 17, we will discuss system design strategies for future wireless communication systems. Chapter 15 covers design strategies for TDMA-based systems, and chapter 16 is concerned with design strategies for CDMA-based systems. Chapter 17, the final chapter, discusses flexible radio interface design strategies for future wireless multimedia communication systems; the conventional radio link design strategies mentioned in chapters 15 and 16 are insufficient for future systems due to a lack of flexibility against dynamically changing traffic, quality of service (QoS) requirements, and channel conditions peculiar to global wireless communication systems. In chapter 17 we will discuss intelligent radio transmission technologies and the radio highway network as examples of flexible radio interface design techniques.

If you are using this text at the university student level, you may wish to concentrate on chapters 1–5 and 9–10 for a one- or two-semester course. You could include chapters 6–8 if time permits. For industrial engineers using this book, chapters 4–8, 15, and 16 will be the most beneficial. It is especially important for cellular system engineers to fully understand the system design procedures shown in chapters 15 and 16. This book is intended to provide a wide range of information about wireless communication technologies to various types of engineers. Chapters 1 through 10 provide the basics of wireless communication technologies. Therefore, graduate-level students at universities and freshmen in industry should first concentrate on chapters 1 through 10 in order to understand the problems originating from the propagation path and how to solve them when designing wireless communication systems. For those who are in charge of designing wireless communication subsystems in the industry, chapters 15 through 17 (in addition to chapters 1–10) will aid in understanding the basic concept of radio subsystem designs and the evaluation procedures for system capacity, its QoS, and so on.

Acknowledgments

First, I am deeply grateful to Dr. Kamilo Feher, editor of the Feher/Prentice Hall Digital and Wireless Communication Series, and professor of the University of California, Davis, for encouraging me to write this book. I also express my sincere appreciation to Professor Norihiko Morinaga, Department of Communications Engineering, Faculty of Engineering, Osaka University, for giving me the opportunity to study the emerging and exciting wireless multimedia communication field. This book also includes a lot of research activities from the Communications Research Laboratory (CRL), Ministry of Posts and Telecommunications, for which I worked during 1982–1993; Digital & Communication Laboratory of the University of California, Davis, where I was a visiting researcher during 1991–1992; and my current research group at Osaka University since 1993. I would like to give special thanks to Professor Mitsuo Yokoyama of Toyohashi University of Technology, who was my supervisor when I started my research at CRL, for his valuable guidance, encouragement, and support. Thanks also to Dr. Hideichi Sasaoka, Mr. Fumito Kubota, Mr. Eimatsu Moriyama, and Mr. Yukiyoshi Kamio, all of whom are at CRL; and Professor Takeo Ohgane of Hokkaido University (who was at CRL) for their cooperation in developing various key technologies for wireless communication systems. My research at CRL was supported by many research associates from national research institutes or universities. I would like to thank Mr. Makito Nakajima of National Police Agency; Mr. Norihito Kinoshita of Matsushita Communications Industrial Co. Ltd.; Mr. Terumi Sunaga and Mr. Takayuki Nagayasu of Mitsubishi Electric Co. Ltd. I would also like to express my appreciation to members of my research group at Osaka University. I owe special thanks to Dr. Sadayuki Abeta and Dr. Toyoki Ue for their help in reviewing and editing my manuscript. Ms. Karen Gettman, executive editor of Prentice Hall, helped me a lot during all phases of the preparation of my book. Finally, I wish to thank my wife, Masako, and all my family for their help and encouragement during the writing phases of this book.

Dr. Seiichi Sampei

Introduction

In this chapter, we will discuss the background of land mobile communication technologies. After a brief historical review of land mobile communication systems, we will discuss recent development of digital wireless communication systems including features, covered services, and restrictions of the operational environments. Of great interest at present is the development of third-generation wireless communication systems that will support global, personal, and multimedia communication services with a quality of services (QoS) equivalent to that of fixed networks, we will discuss one of the currently developed third-generation systems called Future Public Land Mobile Telecommunication Systems (FPLMTS) that is being standardized in Task Group 8/1 (TG-8/1) of the International Telecommunication Union Radio Section (ITU-R).

1.1 HISTORICAL REVIEW OF LAND MOBILE COMMUNICATION SYSTEMS

Land mobile communication has a very long history that is almost the same as that of radio communication because *radio communication* means not only to release a terminal from a wire but also to accept terminal mobility. Mobile communication was mainly applied to military and public safety services until the end of World War II, but after the war it began to be applied to public telephone services.

The first public mobile phone service was the Mobile Telephone System (MTS) introduced in the United States in 1946. In this system, operation was simplex and call placement was handled by a manual operation. This system was then followed by a full duplex and automatic switching system—the Improved Mobile Telephone System (IMTS)—introduced in 1969 using a 450-megahertz (MHz) band. Although IMTS was widely introduced in the United States as a standard mobile phone system, it was not able to cope with rapidly increasing demand because it was a large-zone system, and its assigned bandwidth was not sufficient. Therefore, a more advanced and high-capacity land mobile communication system called Advanced Mobile Phone Service (AMPS), was introduced in 1983. The most important feature of AMPS is that it employs a cellular concept to achieve high system capacity. By the end of 1995, the total number of subscribers of the AMPS system reached around 30 million.

In Europe, each region or country introduced its own cellular phone services—such as Nordic Mobile Telephone (NMT) in the Nordic region, Total Access Communication System (TACS) in the United Kingdom, RC-2000 in France, C-net in Germany, and I-450 in Italy—since early 1980s. In Japan, Nippon Telegraph and Telephone (NTT) started cellular phone service in 1979, and a narrowband high-capacity system was introduced in 1989.

Although each analog cellular system has a certain capacity, it is not sufficient to satisfy rapidly increasing demand for cellular services. Therefore, Europe, the United States, and Japan have independently developed digital cellular systems as the second-generation cellular systems. In Europe, Group Special Mobile (GSM) was organized under the Conference of European Posts and Telecommunications (CEPT); it standardized the GSM system in 1988. Today, GSM stands for Global System for Mobile communications. In the United States, Telecommunications Industry Association (TIA) established a standardization committee for digital cellular systems, and interim standards for a time division multiple access (TDMA)-based digital cellular system (IS-54) and code division multiple access (CDMA)-based digital cellular system (IS-95) were issued in 1989 and 1993, respectively. In Japan, Research and Development Center for Radio Systems (RCR), now called Association of Radio Industries and Businesses (ARIB), organized a standardization committee for digital cellular systems and issued a standard called Personal Digital Cellular (PDC) system in 1992. Although there are not so many digital cellular subscribers in the United States, subscribers for GSM and PDC are rapidly increasing. As a result, the number of cellular subscribers for both analog and digital systems had reached 20 million in Europe and 10 million in Japan as of March 1996.

Although the cellular phone is the most popular land mobile communication system, there are other application fields of land mobile communication technologies—the cordless phone and wireless local area network (LAN). Table 1.1 shows a comparison of features of these three systems, where the cellular phone and cordless phone are analog systems and the wireless LAN is a digital system. The features that are the most different in regard to the cellular phone, cordless phone, and wireless LAN are terminal mobility and coverage. Basically, mobility of the cordless phone and wireless LAN is very low. Therefore, these systems do not require any location registration or handover functions, which means their system cost is much lower than that for cellular systems. For this reason, the cordless phone has spread more rapidly and widely than cellular systems.

Table 1.1 Feature comparison between cellular phone, cordless phone, and wireless LAN systems.

Items	Cellular Phone	Cordless Phone	Wireless LAN
Coverage	1–10 kilometers (km)	10–100 meters (m)	<30 m
Transmitter power	high	low	low
Terminal mobility	high	low	almost stationery
Mobility management	location registration handover	nothing	nothing
Bandwidth	narrow	narrow	wide
Radio link control	centralized system	distributed system	distributed system

Another feature difference between these three systems is the signal bandwidth. In the case of cellular phone or cordless phone systems, the signal bandwidth is very narrow—approximately 20–30 kilohertz (kHz)—because these systems are used only for voice transmission. On the other hand, the bandwidth for the wireless LAN system is very wide—more than 10 MHz—because high bit rate transmission with its bit rate of more than 1 megabits per second (Mbit/s) is required to satisfy intersystem matching between wireless and wired LANs. Fortunately, channel condition for wireless LAN is not so severe in comparison with that for cellular systems owing to its small-zone radius. Moreover, most of the wireless LAN systems employ a license-free band called Industry, Scientific and Medical (ISM) band. Therefore, wireless LAN with its bit rate of 0.1–1.0 Mbit/s has also become very popular recently.

1.2 DEVELOPMENT OF DIGITAL WIRELESS COMMUNICATION SYSTEMS

At present, various wireless mobile communication systems are changing from analog systems to digital systems. Until now, digital wireless communication technologies have developed mainly because of the need for increased system capacity for voice transmission. However, this is now changing. During the last two years, demand for multimedia transmission systems rather than dedicated voice transmission systems is rapidly increasing, even in the wireless communication systems, because of the rapid growth of multimedia communications via the Internet as well as emerging personal digital assistance (PDA) terminals.

The other important requirements for digital wireless systems include the increasing size of coverage and the availability of higher terminal mobility. One good example is the cordless phone system. At present, the zone radius for an analog cordless phone is limited to around 100 m. However, we may have a very simple question—could we use our cordless phone terminal outside our home or, if possible, anywhere outside our home? When we want to achieve almost the same digital cordless system coverage as that for cellular phone systems, we have to employ almost the same network as we do for the cellular system, which is very expensive. On the other hand, when we just want to increase coverage by a few times, we can achieve it by improving receiver sensitivity using digital transmission technologies.

In the late 1980s, most people thought that it was satisfactory for digital cordless phone systems to cover only limited public spots outside the home, such as railroad and bus stations. But, in more recent years, much wider coverage is required because such a digital cordless phone system is considered not only a digital version of cordless phone systems but also a part of personal communication systems. Of course, emerging modulation/access technologies as well as micro cellular technologies have had big impacts on this trend. The Digital European Cordless Telephone (DECT) system [1-1] and personal handyphone system (PHS) [1-2] are examples of such systems.

Other than these public wireless systems, digital private mobile radio (PMR) systems have also been developed in the last few years. The Extended Specialized Mobile Radio (ESMR) service is an example of this new digital private radio service provided by Nextel and Motorola in the United States [1-3]. This system supports integrated services of voice dispatch, cellular phone, and data services with a bit rate of up to 64 kilobits per second (kbit/s) using quadrature amplitude modulation (QAM), TDMA, and cellular technologies. Therefore, ESMR is now one of the competitors for cellular systems. In Japan, almost the same technologies are applied to the large-zone private mobile radio systems called digital multichannel access (MCA) system, as well as to a part of public safety systems [1-4].

When we compare digital cellular, digital cordless, and PMR systems from the viewpoint of terminal mobility and supported maximum bit rate, we can find very specific features of each system. Figure 1.1 summarizes the supported range of mobility and maximum bit rate for each system. In the case of analog cellular systems, the maximum bit rate is limited to several kbit/s, say 2.4 kbit/s, although very high terminal mobility is accepted. When we employ digital cellular systems, a much higher bit rate, say up to 9.6 kbit/s, is possible using one physical channel, and 19.2 kbit/s will also be achievable once it is possible to assign two physical channels for each subscriber.

In the case of PHS, DECT, and personal communication systems (PCS), terminal mobility is restricted to pedestrian speed to simplify radio resource management and call control, such as handover. On the contrary, they support a higher user bit rate, say 32 kbit/s using one physical channel and 64 kbit/s using two physical channels. Because 64 kbit/s is the same user rate as that for a narrowband ISDN B-channel, these systems are expected to be wireless modems for various ISDN-supported instruments.

When we employ digital MCA or ESMR systems, we can also transmit 64-kbit/s data with high terminal mobility. In the case of wireless LAN systems, the supported user rate is 1–10 Mbit/s as previously discussed.

While analog systems could be defined as first-generation wireless communication systems, these digital systems would be the second-generation systems. Although first- and second-generation systems were developed in each country or region, more and more there is a necessity for next-generation systems that feature global services, regardless of the original place of subscription; a higher user bit rate; and a transmission quality close to that of a fixed network. Thus, a third-generation system that is

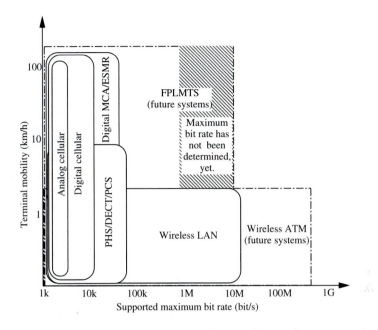

Fig. 1.1 Supported range of mobility and maximum bit rate for wireless communication systems.

expected to support global services with a high bit rate, high quality, and higher terminal mobility is now being developed. The expected range of the terminal mobility and the maximum bit rate of the third-generation systems are shown in Figure 1.1. Although its maximum supported bit rate has not yet been defined, it will be between 1 Mbit/s and 10 Mbit/s.

One of the third-generation systems is the Universal Mobile Telecommunication System (UMTS) developed by the CEPT [1-5]. At present, main tasks for this standardization process are conducted by the Special Mobile Group (SMG) Sub-Technical Committee (STC) of the European Telecommunications Standards Institute (ETSI), and very active committee meetings have been held several times a year. To promote this standardization process, CEPT also organized pan-European research projects called "R&D in Advanced Communications technologies in Europe" (RACE), followed by a new research project called "Advanced Communications Technologies and Services" (ACTS) [1-6].

Another third-generation system is the FPLMTS developed by the ITU-R; main tasks are conducted in the TG 8/1 of the ITU-R. At present, the TG 8/1 committee meeting is held twice a year and works toward creating recommendations for an FPLMTS standard. Although UMTS and FPLMTS have been developed by independent organizations, UMTS could be aligned with the worldwide standardization process of FPLMTS to provide global terminal roaming and user roaming between regions and between terminals.

One more future wireless communication system is the wireless asynchronous transfer mode (ATM) system that requires a bit rate of more than 100 Mbit/s. To support such high bit rate, however, we have to use a millimeter wave (>30 gigahertz [GHz]) band because it is almost impossible to prepare such wide bandwidth in the lower frequency band. At present, anti-frequency selective fading techniques as well as components for millimeter wave are extensively developed [1-7; 1.8; 1.9].

Although what is classified as UMTS or FPLMTS and when they will be specified and serviced is still ambiguous, we will discuss outlines of FPLMTS in the next section because the FPLMTS concept includes very important suggestions for future wireless communication systems.

1.3 OUTLINE OF FPLMTS

1.3.1 Historical Overview of FPLMTS

The ITU started standardization activity for FPLMTS in 1985. In the beginning, standardization was discussed in the Interm Working Party 8/13 (IWP 8/13) organized to investigate a concept of FPLMTS with particular regard to the overall objectives, suitable frequency bands, and the degree of compatibility or commonality. After the IWP 8/13 reported a system concept, design objectives, and spectrum requirements for FPLMTS in 1990 [1-10], the task was forwarded to the TG 8/1 which was established to formulate a set of recommendations for FPLMTS. After the first TG 8/1 meeting in May 1991, the 1992 World Administrative Radio Conference (WARC-92) assigned a 230-MHz radio spectrum (1885–2025 MHz and 2110–2200 MHz) as the worldwide common radio spectrum for FPLMTS [1-11]. This conference also decided that a portion of these frequencies are to be available for mobile satellite services after the year 2005 because not only geostationary earth orbit (GEO) satellites but also low earth orbit (LEO) satellites are considered to be very important components to support global services of FPLMTS in the ITU. After that, the spectrum for mobile satellite systems, decided by the 1995 World Radio communication Conference (WRC-95), are as follows:

☞ 1980–2010 MHz and 2170–2200 MHz bands are available for mobile satellite systems worldwide after January 1, 2000.

☞ 2010–2025 MHz and 2160–2170 MHz bands are available for mobile satellite systems after January 1, 2000, only in the United States and Canada, and after January 1, 2005, in other countries in Region 2.

After the spectrum allocation by the WARC-92 and WRC-95, ITU-R issued a series of recommendations shown in Table 1.2, which were drafted by TG 8/1.

1.3.2 System Concept of FPLMTS

The main objectives of FPLMTS shown in Recommendation 817 are summarized here.

Table 1.2 Recommendations for FPLMTS.

Document No.	Title	Year Issued
ITU-R M.687-1	Future public land mobile telecommunication systems (FPLMTS)	1990, rev. in 1992
ITU-R M.816	Framework for services supported on FPLMTS	1992
ITU-R M.817	FPLMTS network architecture	1992
ITU-R M.818-1	Satellite operation within FPLMTS	1992, rev. in 1994
ITU-R M.819-1	FPLMTS for developing countries	1992, rev. in 1994
ITU-R M.1034	Requirements for radio interface(s) and radio subsystem	1994
ITU-R M.1035	Framework for radio interface(s) and radio subsystem functionality for FPLMTS	1994
ITU-R M.1036	Spectrum considerations for implementation of FPLMTS in the 1885–2025 MHz and 2110–2200 MHz bands	1994
ITU-R M.1078	Security principles for FPLMTS	1994
ITU-R M.1079	Speech and voiceband data performance requirements for FPLMTS	1994
ITU-R M.1167	Framework for the satellite component of FPLMTS	1995
ITU-R M.1168	Framework for FPLMTS management	1995

General objective

☞ To provide various telecommunication services to users regardless of worldwide location or user's mobility, along with making efficient and economical use of the radio spectrum at an acceptable cost

☞ To provide, as far as practical, services with a quality of service comparable to fixed networks

☞ To accommodate a variety of mobile terminals ranging from a pocket-type terminal (personal station [PS]) to a terminal mounted in a vehicle (mobile station [MS])

☞ To provide for the continuing flexible extension of service provision, subject to the constraints of radio transmission, spectrum efficiency, and system economics

☞ To permit the use of the FPLMTS for the purpose of providing its services to fixed users, under conditions approved by the appropriate national or regional authority, either permanently or temporarily, either in rural or urban areas

☞ To adopt a phased approach for the definition of FPLMTS—Phase 1 includes services supported by user bit rates of up to 2 Mbit/s; Phase 2 includes services with
a much higher user rate (about 10 Mbit/s)

☞ To provide an open architecture and modular structure that will permit easy
introduction of technology advancements as well as different applications

Technical objectives

☞ To support integrated communication and signaling using signaling interface
standards in terms of the open system interconnection (OSI) reference model

☞ To provide service flexibility that permits the operation integration of services,
such as mobile phone, dispatch, paging, and data communication, or any combination thereof, as well as to provide an additional level of security

☞ To support terminals that allow the alternative use of terminal equipment in the
fixed ISDN network in various operational environments

When we summarize these objectives from the viewpoint of radio interface
design, the following are considered to be necessary requirements.

☞ To support global services by the combination of the terrestrial mobile wireless
network, satellite links, and fixed wireless or wired links

☞ To provide QoS equivalent to fixed networks

☞ To achieve high system capacity using high spectral efficient radio communication techniques

☞ To support variable bit rate transmission ranging from several kbit/s to several
Mbit/s with varying transmission quality by economically and effectively
combining various modulation/demodulation, fading compensation, and access
techniques

☞ To support services under various conditions of operational environments and
service requirements as well as to preserve radio interface flexibility for future
system upgrade

To satisfy these requirements, the following four radio interfaces are defined in ITU-R
Recommendation 687-1.

☞ **R1**. The radio interface between an MS and the base station (BS) that supports
terminals with high mobility

☞ **R2**. The radio interface between a PS and the cell station (CS) (BS for PS) that
supports terminals with low mobility

☞ **R3**. The radio interface between the satellite and the mobile earth station (MES)
or handheld-type personal earth station (PES) that supports wide area coverage
and a part of international roaming

☞ **R4**. The additional radio interface used for alerting (paging) in the call terminated at an FPLMTS terminal

One of very important features of FPLMTS is that terrestrial wireless systems, satellite networks, and fixed networks—it does not matter whether they are the existing infrastructure or not—are all integrated in the concept of FPLMTS. Although R1 to R4 interfaces are reasonably classified, a smaller number of radio interfaces is preferable from the viewpoint of terminal portability and terminal cost. Therefore, how to minimize the number of radio interfaces by maximizing commonality between these radio interfaces is now studied in the TG 8/1 [1-12; 1-13]. An example of FPLMTS, including radio interfaces R1–R4, is shown in Figure 1.2.

Another important issue for the development of radio interfaces is the selection of propagation path models. At present, the following four propagation path conditions are considered as the test environments to assess system performance [1-13].

☞ **Indoor office test environment.** Characterized by small cells and low transmit powers. The root mean square (rms) delay spread is relatively small, up to several hundred nanoseconds. Path loss is subject to scatter and attenuation by walls, floors, and storage. These obstacles also produce shadow fading. Its standard deviation is expected to be more than 10 decibels (dB).

☞ **Outdoor to indoor and pedestrian test environment.** Characterized by small cells and low transmit power. BS with relatively low-height antennas are located outdoors, and pedestrian users are located on the street and/or inside buildings and residences. Path loss is subject to the rule of R^4 in most cases, but the path loss follows the R^2 rule when there is a Fresnel zone clearance, and it is subject to the R^6 rule if obstacles between the BS and a terminal are densely distributed.

☞ **Vehicular test environment.** Characterized by relatively large cells and high transmit power. The rms delay spread ranges from several hundred nanoseconds

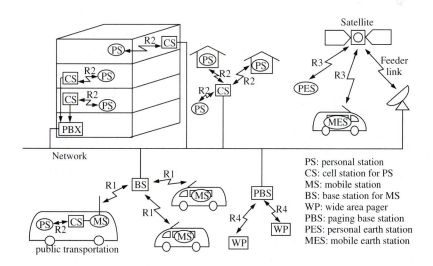

Fig. 1.2 An example of FPLMTS including radio interfaces R1–R4.

to more than 10 microseconds. In urban and suburban areas, average path loss is subject to the path loss rule of R^4, and log-normal fading with its standard deviation of 10 dB is very typical. In rural areas, average path loss may be subject to the path loss rule of R^2 when the path is in line-of-sight (LOS) conditions.

☞ **Satellite test environment**. Characterized by Nakagami-Rice fading because the received signal is composed of an LOS path and defused/reflected paths. The rms delay spread is very small, a few tens or hundreds of nanoseconds. Doppler shift in the carrier is a function of satellite velocity, whereas Doppler spread is dependent on the motion of the terminal.

At present, propagation path models for each environment are discussed in the TG 8/1 committee. They will be included in the series of FPLMTS recommendations [1–14]. Another important feature is that FPLMTS has to support various media, not only voice but also nonvoice data. For this purpose, we have to define evaluation criteria, such as

☞ BER for voice data
☞ BER for nonvoice data (it could be plural values depending on the kind of information)
☞ User rate
☞ Transmission processing delay time
☞ Handover requirement

With this background, TG 8/1 is now drafting a recommendation on guidelines for the evaluation of radio transmission technologies for FPLMTS. When this new draft recommendation is approved, TG 8/1 will call for a proposal of the radio transmission technology for FPLMTS, which will be followed by the radio interface evaluation and specification of radio interface for FPLMTS. However, the time schedule for this process is very ambiguous as of April 1996.

1.4 ORGANIZATION OF THIS BOOK

The main objective of this book is not to discuss which is better—TDMA or CDMA, GMSK or $\pi/4$-quaternary phase shift keying (QPSK), frequency division duplex (FDD), or time division duplex (TDD)—or simply to line up wireless communication technologies. The purpose is to understand what the key wireless communication technologies are, how these key technologies are now applied to second-generation wireless systems, and how to design third-generation global, wireless, personal, and multimedia systems like FPLMTS by efficiently and economically combining these key technologies. The book is organized as shown in Figure 1.3.

Contents of this book can be roughly divided into the following four areas:

☞ Propagation path characteristics (chapter 2)
☞ Basic technologies for wireless communication systems (chapters 3–10)

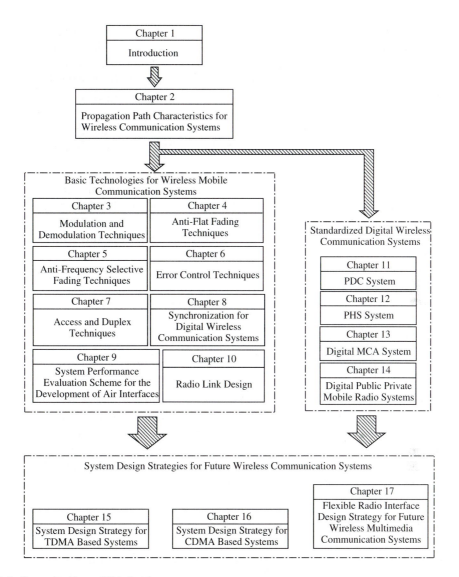

Fig. 1.3 Organization of this book.

☞ Standardized digital wireless communication systems (chapters 11–14)

☞ System design strategies for future wireless communication systems (chapters 15–17)

To design wireless communication systems, first of all we have to fully understand propagation path characteristics—statistics of propagation path characteristics, time variant nature of propagation paths, and how this knowledge is applied to the development of demodulators and anti-fading techniques. Therefore, chapter 2 discusses these issues.

In wireless communication systems, modulation/demodulation, synchronization, access, and duplex techniques are very important to achieving spectral efficient and high-quality wireless communication systems. However, we cannot directly apply wireless techniques used for the fixed links to the wireless mobile communication systems because they are severely affected by fading that is specific to wireless mobile communication channels. Therefore, chapters 3 through 8 discuss how to apply these techniques to the wireless mobile communication systems including anti-fading techniques. Moreover, chapter 9 discusses how to verify the effect of key technologies as well as how to specify their parameters using computer simulation, laboratory experiments, and field trials. Then chapter 10 discusses how to design radio links including coverage estimation and cell reuse planning.

Chapters 11 through 14 discuss how air interfaces of standardized digital wireless communication systems are designed. Although there are many standardized digital wireless communication systems in the world, this book intentionally discusses standardized systems in Japan, such as the PDC system and the PHS, for the following reasons:

☞ There are not many publications about Japanese standardized systems written in English although there are many books or documents about European and U.S. interim standards.

☞ We have to understand as many digital wireless standards as possible in order to know the advantages and disadvantages of these systems as well as to understand what kind of problems we have to overcome for the development of third-generation wireless communication systems.

Of course, this book is not intended to support any specific systems.

In chapters 15 through 17, we will discuss system design strategies for future wireless communication systems. In these chapters, we will discuss how to design TDMA-based and CDMA-based systems considering various requirements and constraint conditions specific to wireless systems. Moreover, we will discuss how to make more flexible radio interfaces to support future global, wireless, personal, and multimedia communication systems.

REFERENCES

1-1. DECT, "Digital European cordless system—common interface specifications," Code RES-3(89), DECT, 1989.

1-2. RCR, "Personal handy phone system," RCR STD-28, December 1993.

1-3. Davidson, A. and Marturano, L., "The impact of digital technologies on future land mobile spectrum requirements," 43rd IEEE Veh. Tech. Conf. (Secaucus, New Jersey), pp. 560–63, May 1993.

1-4. RCR, "Digital MCA system," RCR STD-32, November 1992.

1-5. Silva, J. S., Barani, B. and Fernandez, B. A., "European mobile communications on the move," *IEEE Communications Magazine*, pp. 60–69, February 1996.

1-6. Baier, P. W., Jung, P. and Klein, A., "Taking the challenge of multiple access for third-generation cellular mobile radio systems," *IEEE Communications Magazine*, pp. 82–89, February 1996.

1-7. Sato, K., Kojima, H., Manabe, T., Ihara, T., Kasashima, Y. and Yamaki, K., "Measurements of reflection characteristics and reflective indices of interior construction materials in millimeter-wave bands," 45th IEEE Veh. Tech. Conf. (Chicago, Illinois), pp. 449–53, July 1995.

1-8. Meinel, H. H., "Recent advances on millimeter wave PCN system development in Europe—an invited survey," 1995 IEEE MTT-S International Microwave Symposium (Orlando, Florida), pp. 401–4, May 1995.

1-9. Takimoto, Y., "Recent activities on millimeter wave indoor LAN system development in Japan," 1995 IEEE MTT-S International Microwave Symposium (Orlando, Florida), pp. 405–8, May 1995.

1-10. ITU-R Report 1153, "Future Public Land Mobile Telecommunication Systems," 1990.

1-11. Resolution No. 212 (WARC-92), "Implementation of Future Land Mobile Telecommunication Systems (FPLMTS)," 1992.

1-12. Recommendation ITU-R M.1035, "Framework for the radio interface(s) and radio subsystem functionality for future public land mobile telecommunication systems (FPLMTS)," 1994.

1-13. Recommendation ITU-R M.1034, "Requirements for the radio interface(s) for future public land mobile telecommunication systems (FPLMTS)," 1994.

1-14. ITU-R Document 8-1/TEMP/44-E, "Guidelines for evaluation of radio transmission technologies for FPLMTS," 1996.

Propagation Path Characteristics for Wireless Communication Systems

For wireless communication system design, it is very important to fully understand characteristics of the propagation path conditions because average path loss is extremely large due to low antenna height of the terminals in comparison with fixed radio links operated under LOS conditions, and channel condition is rapidly changing due to fast terminal mobility. Therefore, this chapter discusses the propagation path characteristics specific to wireless mobile communication systems.

2.1 OUTLINE OF THE PROPAGATION PATH CHARACTERISTICS FOR WIRELESS COMMUNICATION SYSTEMS

In land mobile communication systems, propagation path characteristics have a big impact on the system design issues. When a terminal is in an outdoor environment and its coverage is medium or large size (more than 1 km), the propagation path characteristics are considered to be under non-LOS (NLOS) conditions in most cases because a terminal is shadowed by the natural terrain and man-made constructions. Moreover, an NLOS condition is considered to be more severe than an LOS condition. Therefore, we usually assume only NLOS conditions when we estimate system capacity and link budget because very few areas are in an LOS condition.

On the other hand, in the case of micro/pico cell systems, the LOS condition becomes dominant. In the case of indoor systems, both the LOS and NLOS conditions coexist.

15

Propagation path characteristics can be divided into three main components in both outdoor and indoor cases:

☞ Path loss with respect to distance

☞ Shadowing

☞ Multipath fading

Figure 2.1 shows an example of received signal level variation. First of all, when we measure the received signal over a distance of a few hundred wavelengths, we can observe very deep and fast envelope fluctuation caused by the mutual interference between the received signal components incoming from all directions. Therefore, this variation is called *multipath fading*.

Even if we remove this fast variation by averaging the received signal level over a few hundred wavelengths, there still remains relatively slow signal level variation called *shadowing*, which is caused by the nonuniformity of the terrain features or man-made constructions. Because this probability density can be approximated by the log-normal distribution in most cases, this is called *log-normal fading*. This variation is also called short-term median-value variation or large-scale signal variation.

When we further average slowly varying signal variation, we can obtain the area average signal level. This area average is called *path loss* or long-term median-value variation and it varies with the distance between the BS and the terminal.

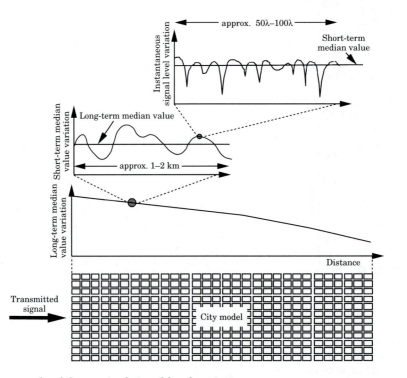

Fig. 2.1 An example of the received signal level variation.

Figure 2.2 shows an example of the received signal level variation obtained by field experiments. Figure 2.2(a) shows the received signal level variation averaged over a period of 1 second. Therefore, this variation includes both shadowing and path loss. When we further smooth this variation, we can obtain the variation as shown in Figure 2.2(b), which corresponds to the area average of the signal level variation determined by the path loss. Figure 2.2(c) shows the difference between Figure 2.2(a) and Figure 2.2(b), which corresponds to the variation due to shadowing. In this case, its standard deviation is 3.0 dB.

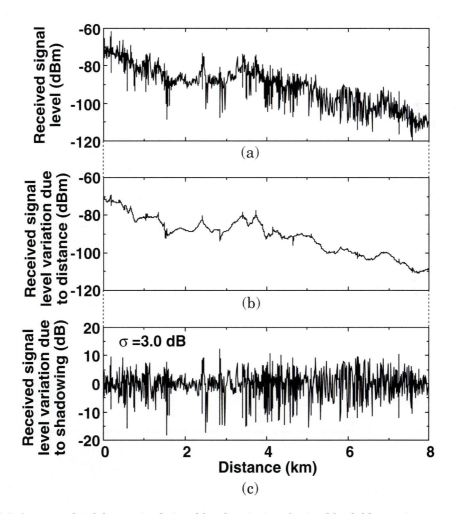

Fig. 2.2 An example of the received signal level variation obtained by field experiments.
(a) received signal level variation due to both shadowing and path loss.
(b) received signal level variation due to path loss.
(c) received signal level variation due to shadowing.

In the following sections, we will discuss these features and how to cope with these variations when we design a wireless communication system.

2.2 PATH LOSS

2.2.1 Outdoor Large-Zone Systems

When there are no obstacles around or between the BS and MS, the propagation path characteristics are subject to free space propagation. In this case, the path loss is given by

$$L_{pf} \text{ (dB)} = 32.44 + 20 \log_{10} f_c + 20 \log_{10} d \tag{2.1}$$

where
   f_c = carrier frequency (MHz)
   d = distance between BS and MS (km)

On the other hand, when there are many obstacles around or between the BS and MS, path loss is determined by many factors, such as irregular configuration of the natural terrain and irregularly arranged artificial structures.

Okumura et al. analyzed such complicated propagation path loss characteristics based on a large amount of empirical data around Tokyo, Japan [2-1]. First of all, they selected propagation path conditions and obtained the average path loss curves under flat urban areas as the standard propagation path conditions because most of the terminals are located in the urban areas. These curves are now called *Okumura curves*. Then, they obtained correction factors for the other propagation path conditions, such as

☞ Antenna height and frequency
☞ Suburban, quasi-open space, open space, or hilly terrain areas
☞ Diffraction loss due to mountains
☞ Sea or lake areas
☞ Road slope

Although Okumura curves are practical and effective when used to estimate coverage area for a system, it is not convenient to use them for the computational system designs including system parameter optimization. To solve this problem, Hata derived empirical formulas for the median path loss that are fit to Okumura curves [2-2]. The obtained formulas, called *Hata's equation*, are classified into three models—typical urban, typical suburban, and rural area models. The results are as follows.

Typical urban model

$$L_p \text{ (dB)} = 69.55 + 26.16 \log_{10} f_c + (44.9 - 6.55 \log_{10} h_b) \log_{10} d$$
$$- 13.82 \log_{10} h_b - a(h_m) \tag{2.2}$$

where

f_c = carrier frequency (MHz)

d = distance between base and mobile stations (km)

and $a(h_m)$ is the correction factor for MS antenna height given by

$$a(h_m) = \begin{cases} 8.29[\log_{10}(1.54h_m)]^2 - 1.1 & (f_c \leq 200 \text{ MHz}) \\ 3.2[\log_{10}(11.75h_m)]^2 - 4.97 & (f_c \geq 400 \text{ MHz}) \end{cases} \tag{2.3}$$

(for large cities)

$$a(h_m) = [1.1\log_{10}(f_c) - 0.7]h_m - [1.56\log_{10}(f_c) - 0.8] \tag{2.4}$$

(for small and medium-size cities)

Typical suburban model

$$L_{ps} = L_p - 2\{\log_{10}(f_c/28)\}^2 - 5.4 \tag{2.5}$$

where L_p is given by equation (2.2) and a(h_m) in equation (2.2) is given by equation (2.4).

Rural area model

$$L_{po} = L_p - 4.78(\log_{10}f_c)^2 + 18.33\log_{10}f_c - 40.94 \ [dB] \tag{2.6}$$

where L_p is given by equation (2.2) and a(h_m) for equation (2.2) is given by equation (2.4).

2.2.2 Indoor Systems

Propagation path characteristics for indoor communication systems are very unique compared to outdoor systems because there are so many obstacles that reflect, diffract, or shadow the transmitted radio waves, such as the wall, ceiling, floor, and various office furniture. First, let's discuss the relationships between radio operation environments, room configuration, and zone configurations. Table 2.1 summarizes these relationships, where definitions of the room and zone configurations are as follows, and images of zone configurations are shown in Figure 2.3.

Room configurations

1. Large room without partition; density of obstacles: low
2. Large room with soft partition; density of obstacles: low-middle

Table 2.1 Relationship between operational environments, room configurations, and zone configurations.

Radio Operating Environments		Room Configurations					Zone Configurations					
		1	2	3	4	5	1	2	3	4	5	6
Business area	Large factory		o	o					o	o	o	
	Large office	o	o	o	o	o			o	o	o	
	Department store		o	o						o	o	
	Station, airport		o			o			o	o	o	
	Event hall	o	o	o	o	o			o	o	o	
	Shopping mall		o		o	o			o	o	o	
	Supermarket			o				o	o	o	o	
	Small factory					o	o	o	o			o
	Small office				o	o	o	o				o
	Small shop				o	o	o	o				o
Residential area	House in urban or suburban area				o	o	o					o
	House in rural area				o	o						o
	Apartment house				o	o				o	o	o

3. Large room without partition; density of obstacles: heavy
4. Small room; density of obstacles: low
5. Small room; density of obstacles: high

Zone configuration

1. Extra-large-zone system; the BS is located outside the buildings, and it covers several buildings
2. Large-zone system; one BS is installed inside a building, and it covers the whole building
3. Middle-zone systems; one BS covers several rooms; usually, there are several BSs inside the building
4. Small-zone system; one BS covers one room
5. Microzone system; several BSs cover one room
6. Self-organized distributed system; the same system as the existing cordless telephone

When room size is large and population density is relatively high, middle to microzone systems are preferable from the viewpoint of system cost and traffic den-

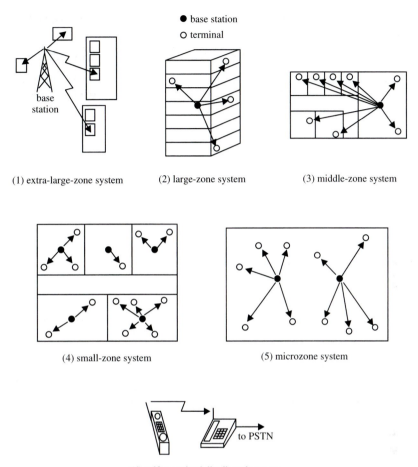

(1) extra-large-zone system   (2) large-zone system   (3) middle-zone system

(4) small-zone system   (5) microzone system

(6) self-organized distributed system

Fig. 2.3 System configurations for indoor wireless systems.

sity. On the other hand, when the room size is small, a large-zone or self-organized system is preferable.

Shopping malls are considered to be very special operational environments because various sizes of the rooms coexist in a very big building. A supermarket is also a very special operational environment because, even though distribution of the obstacles is heavier than a department store, its traffic density is considered to be almost the same or less compared to the department store. Therefore, zone configurations for these environments greatly depend on the scale of the shops.

In residential areas, extra-large-zone or self-organized distributed systems could be used, especially in the urban or suburban areas. In apartment houses, we may be able to apply large-zone systems. However, in rural areas, we can use only the wired system or the conventional cordless telephone.

As shown in Table 2.1, items 1–6 of the zone configurations are the most primitive models for indoor PCS. Therefore, we will discuss the propagation path characteristics of these models.

2.2.2.1 Extra-Large-Zone System   In extra-large-zone systems, the propagation path can be divided into the path outside a building and penetration into the building. As a result, path loss for extra-large-zone systems [2-3; 2-4; 2-5] can be expressed as

$$L_p(r) = L_r(r_0)\left(\frac{r}{r_0}\right)^{\alpha_1} L_B(r_0)\left(\frac{r}{r_0}\right)^{\alpha_2} A_F$$ (2.7)

where

$L_r(r_0)$ = path loss attenuation due to propagation at a distance $r = r_0$
$L_B(r_0)$ = attenuation due to building at $r = r_0$
α_1 = attenuation factor of the propagation path loss with respect to distance
α_2 = building attenuation factor
A_F = building penetration loss

$L_r(r_0)$ and $L_B(r_0)$ are determined by the frequency and the density of obstacles nearby [2-3]. The difference is that $L_r(r_0)$ increases with frequency, whereas $L_B(r_0)$ decreases. Therefore, the additional path loss at a higher frequency could be offset by lower building attenuation [2-3].

α_1 is determined by the distribution of buildings between the base station and each terminal. In the case of LOS conditions, α_1 takes a value of around 2.0, and, in the case of non-LOS conditions, α_1 takes a value in the range of $3 \leq \alpha_1 \leq 6$, the exact value being dependent on the obstacles around the building. On the other hand, α_2 depends less on the distance than α_1 and usually takes a value in the range of 0.5–1.5 [2-3].

A_F depends on the antenna height difference between the BS and each terminal as well as materials for windows. When the transmitter and receiver antennas are located at the same height, A_F becomes minimal. On the other hand, when the antenna height difference is increased, A_F is also increased [2-4].

2.2.2.2 Large-Zone Systems   Large-zone systems cover all the terminals in a building by a BS in the building. Therefore, it is an extreme case of a middle-zone system in which a BS covers several rooms in the building. The large-zone system may be effective for wireless private branch exchange (PBX) systems in a building with relatively low terminal density.

Path loss for this system is given by

$$L_p(r) = L_r(r_0)\left(\frac{r}{r_0}\right)^{\alpha_p}$$ (2.8)

where α_p is 2–3 when the transmitter and the terminal are located on the same floor, and $\alpha_p \geq 3$ when they are located on the different floors [2-6; 2-7; 2-8].

In the large-zone systems, location of the BS is very important. Generally speaking, the middle floor is the best position for the BS to maximize the coverage [2-6]. Of course, it depends on the room size and the arrangement of the rooms in the building.

2.2.2.3 Middle-Zone System   Middle-zone systems represent one of the most practical and widely applicable zone configurations for indoor systems as shown in Table 2.1. One of its path loss models is given by [2-8]

$$L_p(r) = \left(\frac{4\pi f_c r}{c} \right)^2 F(r)^{k_1} W(r)^{k_2} R(r) \tag{2.9}$$

where

c	= velocity of light
f_c	= carrier frequency
$F(r)$	= floor attenuation
$W(r)$	= wall attenuation
$R(r)$	= reflection loss
k_1	= number of floors transversed
k_2	= number of walls transversed

$F(r)$ is usually 20–40 dB and less dependent on r [2-9]. In the middle-zone system, it is practical and economical to restrict coverage to within the same floor. For this purpose, larger $F(r)$ is preferable. Although floor attenuation in dB linearly increases with the number of floors in most cases, it shows some nonlinearity with respect to the number of floors because of the power leakage through stairways or windows [2-7].

$W(r)$ is a very important factor in determining the coverage. Table 2.2 shows attenuation factors for various wall materials. When we want to cover a relatively large coverage area, a moderate value of $W(r)$ is preferable. On the other hand, a larger value is necessary when we want to restrict zone radius. Rappaport conducted extensive studies of this attenuation factor and reported the results [2-9].

$R(r)$ is a reflection loss. In the case of indoor communications, however, $R(r)$ is sometimes a very small value, especially when the transmitted signal is propagated along the corridor, because the radiated wave outside a corridor is relatively small.

Table 2.2 Attenuation of the materials for the wall.

Material	Attenuation
Wood (15 millimeters [mm])	2.5–3.5 dB [2-10]
Plasterboard	0.2–3.5 dB [2-10]
Concrete block	8.0–15 dB [2-11]
Glass wool heat insulation	≈38 dB [2-10]

Moreover, when we take diffraction effect into account for the estimation of the path loss, we can estimate the path loss more accurately. Lafortune and Lecours show an example of the path loss estimation based on this idea and showed that its estimation error is less than 3 dB [2-12]. Furthermore, when we include attenuation factors due to some obstacles such as storage racks and soft partitions in the rooms, we can further improve path loss estimation.

2.2.2.4 Small-Zone Systems
Small-zone systems are very effective for achieving low outage probability in a large building with high traffic density. Because one BS covers one room, high wall attenuation as well as high floor attenuation are required.

Path loss for this system greatly depends on the number of obstacles between the BS and the terminal. Iwama et al. investigated the path loss characteristics in small-zone systems [2-13]. In this paper, the propagation path attenuation is classified into LOS, NLOS1 (radio signal is shadowed by only one obstacle), and NLOS2 (radio signal is shadowed by two or more obstacles). The results show that the path loss decay factor (α_p) is around 2 in the LOS conditions, whereas it is $\alpha_p \approx 3$ in the case of NLOS1 and $\alpha_p \approx 4$ in the case of NLOS2.

As mentioned before, high attenuation for walls and floors is strongly required for small-zone systems. Moriyama et al. reported some interesting results on the propagation path characteristics in a shielded building [2-14]. The measured room is constructed of the materials shown in Table 2.3.

In the measured room, there was no office furniture or soft partitions. The results show that the path loss does not depend on the distance between the transmitter and receiver because the transmitted radio wave is completely restricted in the room owing to shielding.

In the case of small-zone systems, employing a higher frequency band, such as microwave and millimeter wave band, is also a good strategy to improve system capacity because the path loss is much larger than that for UHF waves [2-15; 2-16]. Moreover, a higher frequency band is also effective for high bit rate transmission because delay spread becomes very small owing to its large path loss [2-17].

2.2.2.5 Microzone Systems
Microzone systems, in which several BS are installed in a room, are effective for covering a large business office with high terminal density. Path loss for this system is almost the same as that for middle-zone or

Table 2.3 Materials for the shielded building used for experiments.

Items	Material
Floor and ceiling	deck plate
Walls	metal net
Windows	shielding glass coated by transparent shielding films
Doors	steel plate

small-zone systems except that smaller α is more probable in the case of microzone systems.

Figure 2.4 shows an example of microzone systems. The room is sectored by soft partitions, and there are many obstacles that could shadow terminals. Because such rooms include LOS and non-LOS conditions, we have to carefully select locations of the BSs to achieve very low outage probability.

The best criteria for the number and the locations of the BSs is to minimize the outage probability with a smaller number of BSs.

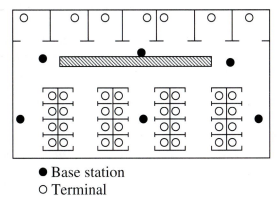

● Base station
○ Terminal

Fig. 2.4 An example of microzone systems.

2.3 SHADOWING

Because the path loss we have discussed represents the average propagation path characteristics, the actual local mean received signal level fluctuation is caused by many factors. We usually call this fluctuation *large-scale signal variation* or just *shadowing* [2-18].

Large-scale signal variation is also called *log-normal fading* because its probability density function (p.d.f.) can usually be expressed as

$$p(X) = \frac{1}{\sqrt{2\pi}\sigma_0} \exp\left[-\frac{(X - X_m)^2}{2\sigma_0{}^2}\right] \tag{2.10}$$

where
    X = $10\log_{10}(x)$ [dBm]
    X_m = area average signal level (dBm)
    x = true received signal level (milliwatts [mW])
    σ_0 = standard deviation in dB

When we express the p.d.f. of the received signal level in terms of mW, it is given by

$$p(x) = \frac{1}{\sqrt{2\pi}\sigma x} \exp\left\{ -\frac{1}{2\sigma^2} \left(\ln^2 \frac{x}{x_m} \right) \right\}$$ (2.11)

where
x_m = long-term average received signal level [mW]
σ = $\ln(10)\sigma_0/10$

Standard deviation σ_0 greatly depends on the environment of the service area. In the case of large-zone outdoor systems, it is 6–10 dB [2-1; 2-3; 2-19]. When we employ large-zone indoor systems, the standard deviation becomes very large, more than 10 dB in some cases because LOS and NLOS conditions coexist in many cases [2-6; 2-20]. When we can independently estimate the path loss for LOS and NLOS conditions, or we can independently install a BS in LOS and NLOS conditions, we can reduce standard deviation [2-13]. However, it is not practical from a system integration point of view because LOS and NLOS areas in each zone could be changed when office furniture is rearranged or office storage is added in the room. When we introduce small-zone or microzone configurations in indoor systems, we can reduce standard deviation because channel characteristics for small-zone or microzone configuration are more uniform than that for large-zone indoor systems.

2.4 FLAT FADING

In most of the wireless communication systems, the propagation path is considered to be in NLOS conditions because the antenna height of the terminal is very low (1–2 m in most cases). When a terminal moves in such an NLOS condition, it receives multiple signal components composed of the diffracted, reflected, and scattered waves due to many obstacles around the terminals. When a terminal is in an indoor environment, the obstacles are walls, ceiling, and floor. On the other hand, when a terminal is in an outdoor environment, obstacles are buildings, poles, and any other man-made obstacles as well as natural obstacles.

2.4.1 Rayleigh Fading Phenomenon

A very simple outdoor propagation path condition model is shown in Figure 2.5. When the i-th path is arriving at an angle of θ_i with respect to the motion of the mobile unit, its frequency is shifted by

$$f_i = f_d \cos \theta_i$$ (2.12)

$$f_d = \frac{v}{c} f_c$$ (2.13)

due to the Doppler effect, where f_d is the maximum Doppler frequency, v (m/s) is the vehicular speed, c is the speed of light, and f_c is the carrier frequency [2-21].

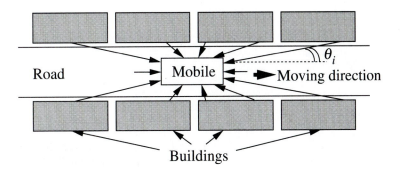

Fig. 2.5 Propagation path condition model.

Let us assume that the i-th path with its amplitude and phase of A_i and ϕ_i is coming from an angle of θ_i, there are no path length differences between any two paths, and the transmitted signal is expressed as

$$s_T(t) = \mathrm{Re}\big[m(t)\exp(j2\pi f_c t)\big] \tag{2.14}$$

The received signal is then given by

$$
\begin{aligned}
s_R(t) &= \mathrm{Re}\left[\sum_{i=-\infty}^{\infty} A_i m(t)\exp\{j2\pi(f_c + f_d\cos\theta_i)t + \phi_i\}\right] \\
&= \mathrm{Re}\big[c(t)m(t)\exp(j2\pi f_c t)\big]
\end{aligned}
\tag{2.15}
$$

where

$$c(t) = c_I(t) + j \cdot c_Q(t) \tag{2.16}$$

$$c_I(t) = \sum_{i=-\infty}^{\infty} A_i \cos(2\pi f_d \cos\theta_i t + \phi_i) \tag{2.17}$$

$$c_Q(t) = \sum_{i=-\infty}^{\infty} A_i \sin(2\pi f_d \cos\theta_i t + \phi_i) \tag{2.18}$$

and the average received signal power is given by

$$b_0 = \frac{1}{2}\sum_{i=-\infty}^{\infty} A_i^2 \tag{2.19}$$

Therefore, we can express the fading channel as shown in Figure 2.6. Because the number of incoming paths are large, we can treat $c_I(t)$ and $c_Q(t)$ as the Gaussian random variables from the central limit theorem. Their joint p.d.f. is expressed as

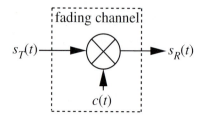

Fig. 2.6 Flat fading channel model.

$$p(c_I,\ c_Q) = \frac{1}{2\pi b_0^2}\exp\left(-\frac{c_I^2 + c_Q^2}{2b_0^2}\right) \tag{2.20}$$

When we assume that θ_i and ϕ_i are uniformly distributed from 0 to 2π, and each path has equal power ($A_i = A$), power spectrum of $c(t)$ is given by

$$S(f) = \frac{b_0}{\pi f_d\sqrt{1-\left(\dfrac{f}{f_d}\right)^2}} \tag{2.21}$$

2.4.2 Nakagami-Rice Fading Phenomenon

In the microcell systems, the received signal is composed of a strong direct wave and many reflected, scattered, or diffracted waves because LOS conditions are dominant in such systems. When the transmitted signal is given by equation (2.14), a direct wave with its amplitude of A_{dir} is arriving at the angle of θ_{di}, and the other components are expressed as the complex random Gaussian variables, the received signal can be expressed as

$$\begin{aligned}
s_R(t) &= \mathrm{Re}\Big[A_{dir}m(t)\exp\big\{j2\pi(f_c + f_d\cos\theta_{dir})t\big\}\Big] \\
&\quad + \mathrm{Re}\left[\sum_{i=-\infty}^{\infty}A_i m(t)\exp\big\{j2\pi(f_c + f_d\cos\theta_i)t + \phi_i\big\}\right] \\
&= \mathrm{Re}\Big[\big\{A_{dir}\exp(j2\pi f_d\cos\theta_d t) + c_{rand}(t)\big\}m(t)\exp(j2\pi f_c t)\Big] \\
&= \mathrm{Re}\big[c(t)m(t)\exp(j2\pi f_c t)\big]
\end{aligned} \tag{2.22}$$

where $c_{rand}(t)$ is the variation of the nondirect waves expressed as

$$c_{rand}(t) = c_{randI}(t) + j\cdot c_{randQ}(t) \tag{2.23}$$

$$c_{randI}(t) = \sum_{i=-\infty}^{\infty}A_i\cos(2\pi f_d\cos\theta_i t + \phi_i) \tag{2.24}$$

$$c_{randQ}(t) = \sum_{i=-\infty}^{\infty} A_i \sin(2\pi f_d \cos\theta_i t + \phi_i) \qquad (2.25)$$

and fading variation $c(t)$ is given by

$$c(t) = A_{dir} \exp\left(j2\pi f_d \cos\theta_d t\right) + c_{rand}(t) \qquad (2.26)$$

In the case of Nakagami-Rice fading, power spectrum density of $c(t)$ is given by

$$S(f) = b_{dir}\delta\left(f_d \cos\theta_d\right) + \frac{b_{rand}}{\pi f_d \sqrt{1 - \left(\dfrac{f}{f_d}\right)^2}} \qquad (2.27)$$

The first term in equation (2.27) represents a direct wave component, and the second one represents nondirect wave components. b_{dir} and b_{rand} represent the power of the direct wave and the total power of the nondirect waves, and the power ratio of them

$$K = \frac{b_{dir}}{b_{rand}} = \frac{A_{dir}^2}{2b_{rand}} \qquad (2.28)$$

is called Rician factor. Because nondirect waves are subject to a complex Gaussian random process, p.d.f. of $c(t)$ is given by

$$p(c_I, \ c_Q) = \frac{1}{2\pi b_0^2} \exp\left(\frac{\left(c_I - A_{dir}\right)^2 + c_Q^2}{2b_0^2}\right) \qquad (2.29)$$

A channel model for Nakagami-Rice fading is also expressed as shown in Figure 2.6, in which $A_{dir} = 0$ ($K = 0$) corresponds to a Rayleigh fading channel, and $b_{rand} = 0$ ($K = \infty$) corresponds to an additive white Gaussian noise (AWGN) channel.

2.4.3 Envelope and Phase Variation by Fading

2.4.3.1 Rayleigh Fading   When we observe a complex random process on the polar coordinate instead of the orthogonal coordinate, we can observe envelope and phase variations of the fading. Let us assume that

$$c(t) = c_I(t) + j \cdot c_Q(t) = r(t)\exp(j\theta(t)) \qquad (2.30)$$

In this case, $r(t)$ and $\theta(t)$ are calculated by

$$r(t) = \sqrt{c_I^2(t) + c_Q^2(t)} \qquad (2.31)$$

$$\theta(t) = \tan^{-1}\left(\frac{c_Q(t)}{c_I(t)}\right) \tag{2.32}$$

and joint p.d.f. of r and θ can be obtained as

$$p(r,\theta) = p(r)p(\theta)$$

$$= \frac{r}{2\pi b_0} \exp\left[-\frac{r^2}{2b_0}\right] \tag{2.33}$$

$$p(r) = \frac{r}{b_0} \exp\left(\frac{r^2}{2b_0}\right) \tag{2.34}$$

$$p(\theta) = \frac{1}{2\pi} \tag{2.35}$$

Equation (2.34) shows that p.d.f. of the envelope is Rayleigh distribution, and that of the phase is uniform distribution with its range of $(-\pi, \pi)$. Therefore this fading is called Rayleigh fading.

2.4.3.2 Nakagami-Rice Fading When the fading variation is subject to Nakagami-Rice fading, p.d.f. of its envelope and phase are given by

$$p(r,\theta) = \frac{r}{2\pi b_{rand}} \exp\left[-\frac{r^2 + A_{dir}^2 - 2A_{dir}r\cos\theta}{2b_{rand}}\right] \tag{2.36}$$

Because r and θ are not independent as shown in equation (2.36), we have to note that $p(r, \theta) \neq p(r)p(\theta)$. The p.d.f.s of r and θ are given by

$$p(r) = \frac{r}{b_{rand}} I_0\left[\frac{A_{dir}r}{b_{rand}}\right] \exp\left[-\frac{r^2 + A_{dir}^2}{2b_{rand}}\right]$$

$$= \frac{r}{b_{rand}} I_0\left[\frac{r\sqrt{2K}}{\sqrt{b_{rand}}}\right] \exp\left[-K - \frac{r^2}{2b_{rand}}\right] \tag{2.37}$$

$$p(\theta) = \frac{1}{2\pi} \exp\left[-\frac{A_{dir}^2}{2b_{rand}}\right]$$

$$+ \frac{A_{dir}\cos\theta}{2\sqrt{2\pi b_{rand}}}\left\{1 + \mathrm{erf}\left[\frac{A_{dir}\cos\theta}{\sqrt{2b_{rand}}}\right]\right\} \exp\left[-\frac{A_{dir}^2}{2b_{rand}}\right] \tag{2.38}$$

$$= \frac{1}{2\pi}\exp(-K)\left\{1 + \sqrt{\pi K}\cos\theta\left[1 + \mathrm{erf}(\sqrt{K}\cos\theta)\right]\exp(K\cos^2\theta)\right\}$$

Figure 2.7 shows (a) $p(r)$ and (b) $p(\theta)$ in Nakagami-Rice fading environments with a parameter of K. When there is no direct wave ($K = 0$), $p(r)$ and $p(\theta)$ are identical to that for Rayleigh fading, and they are independent with each other. On the other hand, when the direct wave is strong ($K \gg 1$), $I_0(x)$ and erf(x) can be approximated by

$$I_0(x) \approx \frac{\exp(x)}{\sqrt{2\pi x}} \tag{2.39}$$

$$\mathrm{erf}(x) \approx 1 - \frac{\exp(-x^2)}{x\sqrt{\pi}} \tag{2.40}$$

where $I_0(x)$ is the zero-order modified Bessel function of the first kind. As a result, $p(r)$ and $p(\theta)$ can be approximated by

$$p(r) = \frac{1}{\sqrt{2\pi b_{rand}}} \exp\left[-\frac{(r - A_{dir})^2}{2b_{rand}}\right] \tag{2.41}$$

$$p(\theta) = \sqrt{\frac{K}{\pi}} \cos\theta \exp\left[-K\sin^2\theta\right] \tag{2.42}$$

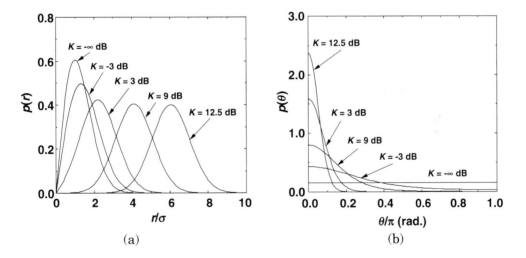

Fig. 2.7 p.d.f.s of the (a) envelope variation and (b) phase variation under Nakagami-Rice fading.

2.4.4 p. d. f. of SNR

Another important statistical value is the probability density function of the received signal-to-noise power ratio (SNR). When the instantaneous SNR and its average value are expressed as γ and γ_0, its p.d.f. for Rayleigh fading is given by

$$p(\gamma) = \frac{1}{\gamma_0} \exp\left(-\frac{\gamma}{\gamma_0}\right) \tag{2.43}$$

When the fading statistics are subject to Nakagami-Rice fading, p.d.f. of its SNR is given by

$$p(\gamma) = \frac{(1+K)}{\gamma_0} \exp\left[-K - \frac{\gamma}{\gamma_0}(1+K)\right] I_0\left[2\sqrt{\frac{\gamma}{\gamma_0}(K^2 + K)}\right] \tag{2.44}$$

These p.d.f.s will be used to obtain the average bit error rate (BER) in Rayleigh or Nakagami-Rice fading environments.

2.4.5 Second-Order Statistics of the Fading Variation

P.d.f.s of $r(t)$, $\theta(t)$, $c_I(t)$, and $c_Q(t)$ are the first-order statistics. These values are useful to evaluate average quality of the communication systems, such as the average BER. However, for the development of anti-fading techniques, we have to consider second-order statistics of fading, such as fading correlation and level crossing rate. Among them, fading correlation is one of the very important parameters.

2.4.5.1 Autocorrelation Function of $c(t)$ The autocorrelation function of fading $R(\tau)$ is given by the inverse Fourier transform of its power spectrum as

$$R(\tau) = R_I(\tau) + j \cdot R_Q(\tau)$$
$$= \int_{-f_d}^{f_d} S(f) e^{j2\pi f \tau} df \tag{2.45}$$

where

$$R_I(\tau) = \int_{-f_d}^{f_d} S(f) \cos(2\pi f \tau) df$$
$$= E\left[c_I(t)c_I(t+\tau)\right] = E\left[c_Q(t)c_Q(t+\tau)\right] \tag{2.46}$$

$$R_Q(\tau) = \int_{-f_d}^{f_d} S(f) \sin(2\pi f \tau) df$$
$$= E\left[c_I(t)c_Q(t+\tau)\right] = -E\left[c_Q(t)c_I(t+\tau)\right] \tag{2.47}$$

When $S(f)$ is given by equation (2.21), $R_I(\tau)$ and $R_Q(\tau)$ are obtained as

$$R_I(\tau) = b_0 J_0(2\pi f_d \tau) \tag{2.48}$$

$$R_Q(\tau) = 0 \tag{2.49}$$

where $J_0(\cdot)$ is Bessel function of the first kind of zero order.

When we consider fading correlation, we usually use the correlation coefficient given by

$$\rho_I(\tau) = R_I(\tau) / b_0 = J_0(2\pi f_d \tau) \tag{2.50}$$

This equation shows the correlation coefficient in terms of time. On the other hand, when a terminal is moving at a constant speed v (m/s), we can translate this correlation coefficient to one in terms of spatial distance as

$$
\begin{aligned}
\rho_I(d) &= J_0\left(2\pi \frac{v f_c}{c} \tau \right) \\
&= J_0\left(2\pi \frac{d}{\lambda} \right),
\end{aligned}
\tag{2.51}
$$

where

$$d = v\tau \tag{2.52}$$

These autocorrelation functions are used to evaluate tracking ability of anti-fading techniques. For example, for the application of the Rayleigh fading estimator, we can analyze its estimation error by using these autocorrelation functions.

2.4.5.2 Envelope Autocorrelation Function
For design of a diversity receiver, envelope autocorrelation would be very important because branch selection or branch signal combining are carried out based on the received signal level of each branch.

The envelope autocorrelation function is given by

$$
\begin{aligned}
L_e(\tau) &= E\big[\{r(t) - E(r(t))\}\{r(t + \tau) - E(r(t + \tau))\}\big] \\
&= E[r(t)r(t + \tau)] - [E(r(t))]^2 \\
&= \frac{\pi}{8b_0} g^2(\tau) \\
&= \frac{\pi}{8b_0} J_0^2(2\pi f_d \tau)
\end{aligned}
\tag{2.53}
$$

Therefore, its correlation coefficient is given by

$$\rho_r(\tau) = J_0^2(2\pi f_d \tau) \tag{2.54}$$

This correlation coefficient can also be converted to one in terms of spatial distance d as

$$\rho_r(d) = J_0^2(2\pi \frac{d}{\lambda}) \tag{2.55}$$

2.4.6 Fading Observation Strategy: Which Is Better for Fading Observation, the Orthogonal Coordinate or the Polar Coordinate?

In the preceding section, we discussed that the fading variation can be observed on both the orthogonal coordinate and polar coordinate. At this stage, we might wonder which is better for observation of and compensation for fading—the orthogonal coordinate or the polar coordinate.

Figure 2.8 shows the configuration of the fading monitor (a) on the orthogonal coordinate and (b) on the polar coordinate, and Figure 2.9 shows an example of the computer-simulated fading variation (a) measured on the orthogonal coordinate and (b) measured on the polar coordinate.

When we observe fading variation on the orthogonal coordinate, spectrums of $c_I(t)$ and $c_Q(t)$ are limited to $[-f_d, f_d]$. In most cases, f_d is up to 100–200 Hz. Therefore, the fading variation measured on the orthogonal coordinate is very smooth as shown in Figure 2.9(a). As a result, we can accurately estimate $c_I(t)$ and $c_Q(t)$ by suppressing noise using a low-pass filter (LPF) with its bandwidth of around f_d. Of course, envelope and phase components can easily be obtained by using the estimated $c_I(t)$ and $c_Q(t)$ as shown in Figure 2.8(b). On the other hand, when we observe fading variation

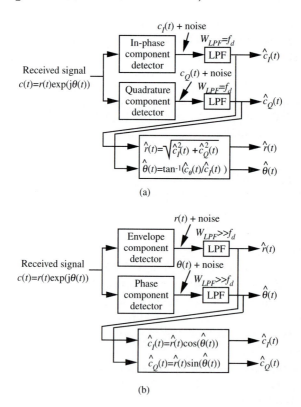

Fig. 2.8 Configuration of the fading monitor (a) on the orthogonal coordinate and (b) polar coordinate.

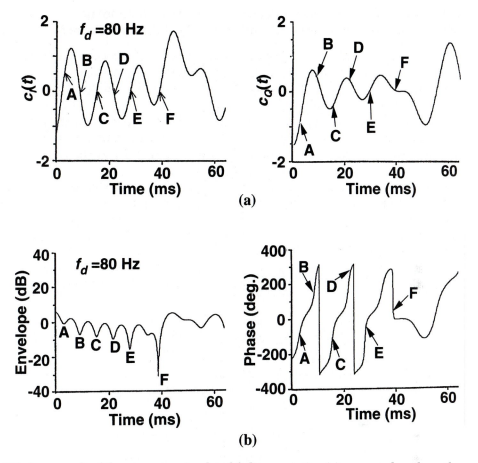

Fig. 2.9 An example of the computer-simulated fading variation (a) measured on the orthogonal coordinate and (b) measured on the polar coordinate.

on the polar coordinate, variations of $r(t)$ and $\theta(t)$ look faster than those of $c_I(t)$ and $c_Q(t)$ as shown in Figure 2.9(b). Why does this happen?

When we compare equations (2.51) and (2.55), we can find that the deviation of $\rho_r(\tau)$ is smaller than that of $\rho_I(\tau)$, which means that the power spectrum for the envelope component includes a higher frequency component than those for $c_I(t)$ and $c_Q(t)$ because power spectrum is given by the Fourier transform of correlation. Therefore, fading variation observed on the polar coordinate looks faster than that observed on the orthogonal coordinate.

These results mean that the bandwidth for the LPF in Figure 2.8(b) should be much wider than f_d. As a result, more noise is included in the estimated values of the $r(t)$ and $\theta(t)$ components than in the estimated values of the $c_I(t)$ and $c_Q(t)$ components. Therefore, a fading monitor on the orthogonal coordinate is considered to be better strategy than that on the polar coordinate.

Historically, it was more popular to have a fading monitor on the polar coordinate than on the orthogonal coordinate because gain control devices, such as the automatic gain controller (AGC) and attenuator, and phase control devices, such as the voltage-controlled oscillator (VCO) and phase shifter, were widely used in various radio systems. Actually, most of the fading compensation techniques developed until the mid-1980s were based on the analog signal processing techniques processed on the polar coordinate. However, since the mid-1980s, when the digital signal processor (DSP) was getting more popular and its application was extensively developed, fading compensation techniques on the orthogonal coordinate have been getting more and more popular. Actually, most of the recent fading compensation techniques, such as pilot signal-aided techniques, Rake diversity for CDMA systems, and the adaptive equalizer for TDMA systems, are all signal processing techniques on the complex coordinate. These techniques will be detailed in chapters 4 and 5.

2.4.7 Simulation Model for Rayleigh Fading

2.4.7.1 Multitone Method for Omnidirectional Antenna   When we conduct computer simulations or laboratory experiments, we usually use a fading simulator. When we assume that

☞ MSs use an omnidirectional antenna

☞ the initial phase of the signal is a uniformly distributed random variable

☞ the arrival angle is uniformly distributed

☞ the number of paths is N

complex fading variation expressed by the equivalent low-pass system is given by

$$c(t) = \frac{1}{\sqrt{N}} \sum_{n=1}^{N} e^{j 2\pi f_d \cos(2\pi n/N)t + j\phi_n}$$

(2.56)

In the case of an omnidirectional antenna, an incoming path from a direction of θ and one from a direction of $-\theta$ have the same Doppler frequency. Therefore, we can combine such overlapped terms in equation (2.56). As a result, $c(t)$ can be expressed as

$$c(t) = c_I(t) + j \cdot c_Q(t)$$

$$= \left[\sqrt{\frac{2}{N_1+1}} \sum_{n=1}^{N_1} \cos\left(\frac{\pi n}{N_1}\right) \cos\left\{ 2\pi f_d \cos\left(\frac{2\pi n}{N_1}\right) t \right\} + \frac{1}{\sqrt{N_1+1}} \cos(2\pi f_d t) \right]$$

$$+ j \sqrt{\frac{2}{N_1}} \sum_{n=1}^{N_1} \sin\left(\frac{\pi n}{N_1}\right) \cos\left\{ 2\pi f_d \cos\left(\frac{2\pi n}{N_1}\right) t \right\}$$

(2.57)

Where $N/2$ is an odd number, N_1 is given by

$$N_1 = \frac{1}{2}\left(\frac{N}{2} - 1\right) \tag{2.58}$$

In this case, the following relations are satisfied.

$$\mathrm{E}\left[c_I^2(t)\right] = \mathrm{E}\left[c_Q^2(t)\right] = \frac{1}{2} \tag{2.59}$$

$$\mathrm{E}\left[c_I(t)c_Q(t)\right] = 0 \tag{2.60}$$

Figure 2.10 shows a configuration of a fading simulator using this scheme [2-21].

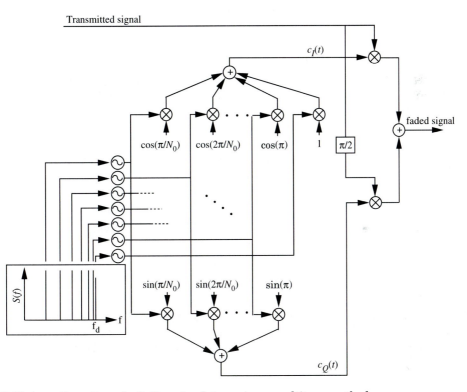

Fig. 2.10 A configuration of a fading simulator using a multitone method.

2.4.7.2 PN Method   As we have discussed before, $c(t)$ is subject to a band-limited complex random Gaussian process when the number of paths (N) is very large. Therefore, we can produce $c(t)$ by generating two independent pseudorandom binary sequences (PRBSs) at a high clock rate and filtering them by an LPF with its transfer function of

$$C(f) = \begin{cases} \left[1 - \left(\dfrac{f}{f_d} \right)^2 \right]^{-1/4} & ; \quad -f_d \le f \le f_d \\[2ex] 0; & \text{otherwise} \end{cases}$$

(2.61)

Figure 2.11 shows a configuration of a fading simulator using a pseudorandom noise (PN) technique [2-22]. $c_I(t)$ and $c_Q(t)$ are generated by the PN method. When we generate a p-stage PN sequence, its spectrum is given by

$$S_{PN}(f) = \frac{1}{N_p^2}\delta(f) + \frac{N_p+1}{N_p^2}\left[\frac{\sin(\pi f T_p)}{(\pi f T_p)}\right]^2 \sum_{\substack{k=-\infty \\ k\ne 0}}^{\infty} \delta\left(f - \frac{k}{N_p T_p}\right)$$

(2.62)

where $N_p = 2^p - 1$ is the PN code length and T_p is the PN bit duration. In this case, the following are required to generate a complex fading envelope variation.

☞ The fading variation's spectrum is flat at least in the signal bandwidth.
☞ Its probability density function is Gaussian distribution.

Equation (2.62) shows that, when T_p is reduced, its spectrum gets flatter in the assigned signal bandwidth. Therefore, shorter T_p is preferable from the viewpoint of spectrum flatness. On the other hand, to satisfy the second requirement, we have to increase the number of line spectra in the signal bandwidth according to the law of large numbers. It can be achieved by increasing N_p as shown in equation (2.62). Therefore, PN codes for the AWGN generation should satisfy the following:

☞ Operation clock rate for the PN code generator should be high; for example, more than 1 MHz
☞ Order of the PN code generator polynomial (p) should be large; for example, more than 20.

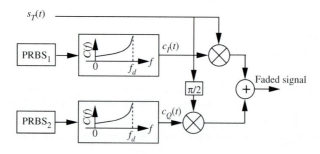

Fig. 2.11 Configuration of a fading simulator using a PN method.

2.4.7.3 Repeating Function of Fading Simulator   For the development of anti-fading techniques, we have to understand what is fading and, in which case, the error, especially burst error, occurs. One very important advantage of the PN method is that we can easily reproduce the same fading variation just by resetting the initial values of the shift registers for PRBSs. Figure 2.12 shows a configuration of a fading simulator with repeating function. The initial values of PRBS shift registers (PNI_{init} and PNQ_{init}) are loaded when a preset signal is triggered by the fading repeating counter. When the repeated duration is T_m, the same fading variation is repeated every T_m second.

2.4.8 Example of Fading Simulator and Its Performance

Kamio et al. shows an example of the implemented fading simulator with repeating function [2-22]. The implemented fading simulator has the following features.

☞ Four independent fading path simulators are prepared.

☞ All the paths can simulate Rayleigh fading, and two of them can also simulate Nakagami-Rice fading.

☞ Arbitrary correlation values between paths are adjustable.

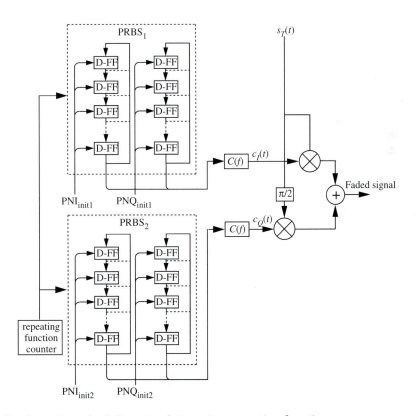

Fig. 2.12 Configuration of a fading simulator using repeating function.

☞ It has two delay lines to simulate two-ray Rayleigh fading.

☞ The same fading variation can be repeated every 0.5, 1, 2, or 4 seconds.

Figure 2.13 shows its configuration and Table 2.4 shows its specifications [2-23]. This system can be divided into the delayed wave generation unit, fading variation

Table 2.4 Specifications of the fading simulator.

Input/Output Interface	
Bandwidth	250 kHz (±0.5 dB)
Input/output	0 dBm/70 MHz/50Ω
Delayed Wave Generation Unit	
Number of paths	2
Delay time of the delayed wave	4 μs (±3%) using SAW device
Shadow Fading Generator	
Number of paths	4
Dynamic range	>40 dB
Variation speed	0–3 Hz
Standard deviation	0–8 dB
Average signal level setting	−30–0 dBm
Fading correlation setting range	0–1 (0.01 step)
Rapid Fading Generator	
Number of paths	4
Variation speed	0–100 Hz (1-Hz step)
p.d.f. of the envelope variation	Rayleigh distribution
Deviation from theory	±1 dB (in the range of −32 dB – +8 dB)
Doppler shift for Rician fading	−100 Hz–100 Hz (only for path #1 and #3)
Rician factor (K)	−20 dB–20 dB, $K = \infty$ and $-\infty$ are selectable
Fading correlation setting range	0–1 (0.01 step)
Repeating Function	
Repeated cycle	0.5, 1, 2, and 4 (no repetition is selectable)

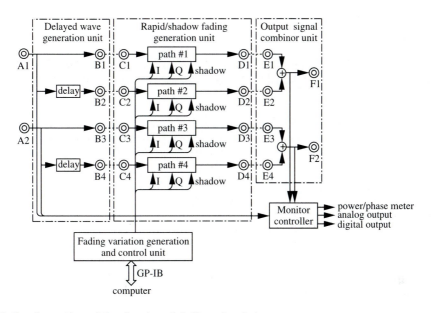

Fig. 2.13 Configuration of the developed fading simulator.

generation and control unit, rapid/shadow fading generation unit, output signal combiner unit, and display unit.

2.4.8.1 Delayed Wave Generation unit   In this unit, the modulated signal is delayed by surface acoustic wave (SAW) devices. Delay time of the delayed path is fixed at 4 μs, and only the two-ray model can be simulated.

2.4.8.2 Fading Variation Generation and Control Unit   In this unit, shadowing as well as the complex envelope variations of rapid fading are generated. Figure 2.14 shows the configuration of this unit, where DSPs (μPD77P20; NEC Corporation) are used for all the processing.

The Gaussian noise generators #1–#4 produce a complex envelope for Rayleigh fading using the aforementioned PN method, where f_d is controlled by changing the clock rate of the digital filter in each Gaussian noise generator. The generated signals are then sent to the rapid fading generators in which correlation between two arbitrary paths are calculated and the complex envelope for each path is generated. When Nakagami-Rice fading is selected to be generated, a frequency-shifted specular component is generated at the specular signal generator, and it is sent to the rapid fading generator. This function is valid only for path #1 and path #3.

The shadow fading generator produces the envelope components of shadow fading by generating a band-limited complex Gaussian random signal using PN method, and calculating the envelope value of the generated signal.

The central control processor (Intel 8085) controls all the functions listed below.

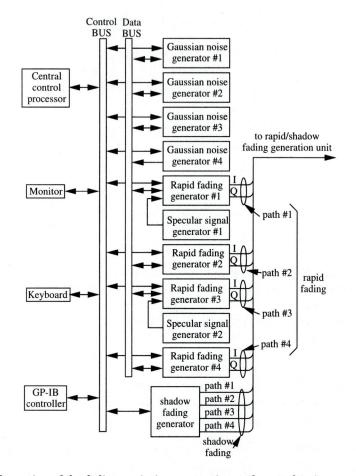

Fig. 2.14 Configuration of the fading variation generation and control unit.

☞ Repeating function

☞ Fading parameter control from an external computer via GP-IB or from keypad

☞ Display control for the fading parameters

To repeat a same fading variation, a timing signal with its period equal to the repeated period is generated at the central control processor, and it is fed to the interrupt port of the related DSPs to reload initial values for the related PN generators and restart all the PN generators at the same time.

2.4.8.3 Rapid/Shadow Fading Generation Unit   Figure 2.15 shows the configuration of this unit. Rapid fading is generated at the quadrature modulator, and shadow fading is generated by controlling the variable attenuator.

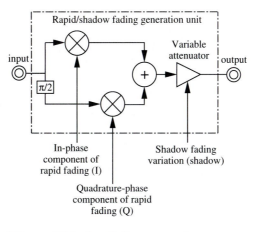

Fig. 2.15 Configuration of the rapid/shadow fading generator.

2.4.8.4 Output Signal Combiner Unit This unit is used to produce a two-ray Rayleigh fading variation. In this case, the direct and delayed signals produced at the delayed wave generation unit are connected to the rapid/shadow fading generators in the rapid/shadow fading generator unit, and their output signals are connected to the input ports of the output signal combiner unit to generate a two-ray Rayleigh fading signal.

2.4.8.5 Monitor Controller This unit generates envelope and phase variations of the generated fading in the analog and digital forms. These values are also shown on the power/phase meter.

2.4.8.6 Performance of the Developed Fading Simulator Figure 2.16 shows the (a) envelope variation, (b) phase variation, and (c) corresponding error bit position when we employ the developed fading simulator, where fading is repeated every 0.5 second, f_d = 40 Hz, and the used modem is differentially encoded binary phase shift keying with coherent detection (DEBPSK). As shown in Figure 2.16, the same envelope and phase variations are repeated every 0.5 second. At the same time, we can find that the error also repeatedly occurs at deeply faded points. These results show that the fading simulator with a repeating function is very helpful to analyze the relationship between the fading variation and the error sequence.

Figure 2.17 shows cumulative distributions of (a) the envelope component and (b) the phase component of Nakagami-Rice fading with a parameter of Rician factor K. As shown in this figure, the obtained results are very close to the theoretical curves.

2.4.9 Fading Variation for Directive Antenna

In the preceding discussion, we have assumed fading variation received by omni-directional antennas. On the other hand, when we employ directive antennas, characteristics of fading variation becomes quite different.

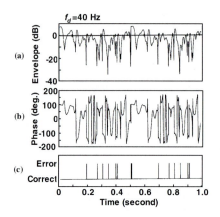

Fig. 2.16 An example of the fading variation and resultant bit error sequence of the developed fading simulator; (a) envelope, (b) phase, and (c) bit error, where fading is repeated every 0.5 second, f_d = 40 Hz, and the used modem is differentially encoded binary phase shift keying (DEBPSK) with coherent detection (from Ref. 2-22, © Institute of Electronics, Information and Communication Engineers, 1987).

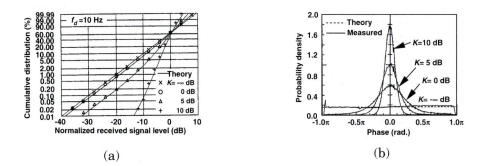

(a)   (b)

Fig. 2.17 Cumulative distribution of the (a) envelope component and (b) phase component of Nakagami-Rice fading with a parameter of Rician factor K (from Ref. 2-23, ©Communications Research Laboratory, 1991).

Let us assume that i-th path with its amplitude and phase of A_i and ϕ_i is coming from an angle of θ_i. When antenna gain for angle θ_i is given by $G(\theta_i)$, the received signal is given by

$$s_R(t) = \text{Re}\left[\sum_{i=-\infty}^{\infty} A_i G(\theta_i) m(t) \exp\{j2\pi(f_c + f_d \cos\theta_i)t + \phi_i\}\right]$$

$$= \text{Re}[c(t)m(t)\exp(j2\pi f_c t)]$$

(2.63)

where

$$c(t) = c_I(t) + j \cdot c_Q(t)$$

(2.64)

$$c_I(t) = \sum_{i=-\infty}^{\infty} A_i G(\theta_i) \cos(2\pi f_d \cos \theta_i t + \phi_i) \qquad (2.65)$$

$$c_Q(t) = \sum_{i=-\infty}^{\infty} A_i G(\theta_i) \sin(2\pi f_d \cos \theta_i t + \phi_i) \qquad (2.66)$$

When $G(\theta_i) = 1$ for all θ_i, equations (2.64)–(2.66) are identical to equations (2.16)–(2.18) which express fading variation for omnidirectional antenna cases.

Although directive antennas are usually employed in the BS for sector cell layout, they are also effective for mitigating fading variation in the MS.

Figure 2.18 shows (a) antenna directivity and (b) power spectral density of Rayleigh fading in the case of both omnidirectional antenna reception and directive antenna reception. When we employ a directive antenna, we can restrict direction of the received paths in a certain angle (θ_{dir}) as shown in Figure 2.18(a). In this case, Doppler shift of the received signal is restricted in the range $[f_d \cos\theta_{dir}, f_d]$ as shown in Figure 2.18(b). This means that the Doppler spread of the received signal using a directive antenna is smaller than that using an omnidirectional antenna although there exists a certain offset frequency (f_{off}). Therefore, when we can estimate f_{off} and cancel it out in the case of directive antenna reception, we can improve tracking ability of the fading variation estimator [2-24].

Figure 2.19 shows a configuration of the fading estimator in the case of directive antenna reception. Let us assume that the antenna directivity is given by Figure 2.18(a), channel variation is expressed as $c(t) = c_I(t) + jc_Q(t)$, power spectrum density function $S(f)$ and variation of $c_I(t)$ are given as shown in Figure 2.19, and fading variation can be estimated by pilot signal-aided techniques that will be discussed in chapter 4. First of all, f_{off} is estimated by monitoring zero-crossing rate and phase rotating

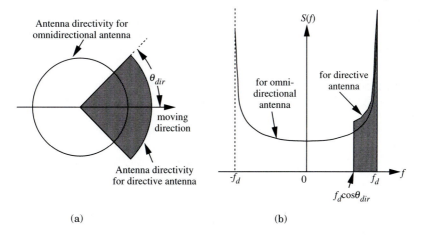

(a)   (b)

Fig. 2.18 (a) antenna directivity and (b) power spectral density of Rayleigh fading in the case of both omnidirectional antenna reception and directive antenna reception.

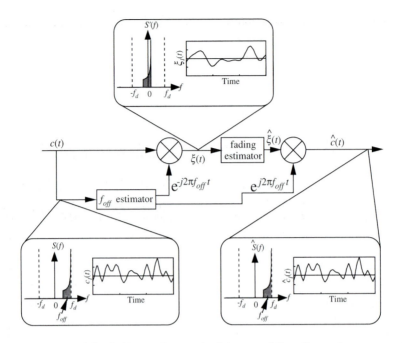

Fig. 2.19 Configuration of the fading estimator in the case of directive antenna reception.

direction using pilot signal-aided techniques. Then, the offset frequency is canceled out as

$$\xi(t) = c(t)e^{-j2\pi f_{off}t} \tag{2.67}$$

where $\xi(t)$ is the offset frequency-canceled fading variation. As a result, we can reduce fading variation speed as shown in Figure 2.19. After that, offset frequency-compensated fading variation is estimated by some means, where $\hat{\xi}(t)$ is the estimated value of $\xi(t)$, and actual fading variation is finally estimated as

$$\hat{c}(t) = \hat{\xi}(t)e^{j2\pi f_{off}t} \tag{2.68}$$

Because we can reduce channel variation speed using this scheme, we can improve accuracy of channel estimation even under fast terminal mobility conditions.

2.5 FREQUENCY SELECTIVE FADING

In the previous section, we discussed characteristics of flat fading in which arrival time difference between paths is negligibly small. However, when the bandwidth of the transmitted signal is very wide, we cannot neglect the arrival time difference, especially when the time difference becomes comparable to a symbol duration. In this case, when the transmitted signal is given by equation (2.14), the received signal can be expressed as

$$s_R(t) = \text{Re}\left[\sum_{k=0}^{\infty}\sum_{i=0}^{\infty} A_{ki} m(t-\tau_i)\exp\left\{j2\pi(f_c + f_d\cos\theta_{ki})t + \phi_{ki}\right\}\right]$$

$$= \text{Re}\left[\left\{\sum_{k=0}^{\infty} c_i(t)\delta(t-\tau_i)\otimes m(t)\right\}\exp(j2\pi f_c t)\right] \tag{2.69}$$

$$= \text{Re}\left[\left\{c(t;\tau)\otimes m(t)\right\}\exp(j2\pi f_c t)\right]$$

$$c_i(t) = \sum_{k=0}^{\infty} A_{ki}\exp\left\{j2\pi f_d\cos\theta_{ki}t + \phi_{ki}\right\} \tag{2.70}$$

$$c(t;\tau) = \sum_{i=0}^{\infty} c_i(t)\delta(\tau-\tau_i) \tag{2.71}$$

where A_{ki} and ϕ_{ki} are the amplitude and phase of i-th path coming from an angle of θ_{ki} with its delay time of τ_i; equation (2.70) shows that the power spectrum and the average power of $c_i(t)$ are given by

$$S_i(f) = \sum_{k=0}^{\infty} b_{0i}\delta(f - f_d\cos\theta_{ki}) \tag{2.72}$$

$$b_{0i} = \frac{1}{2}\sum_{k=0}^{\infty} A_{ki}^2 \tag{2.73}$$

and equation (2.71) means that the channel impulse response expressed as an equivalent low-pass system at a time $t = t_0$ is given by $c(t_0;\tau)$. The total received signal power is given by

$$b_0 = \sum_{i=0}^{\infty} b_{0i} = \sum_{i=0}^{\infty}\sum_{k=0}^{\infty} A_{ki}^2 \tag{2.74}$$

$c(t;\tau)$ is called an instantaneous delay profile,

$$\bar{c}(\tau) = \sum_{i=0}^{\infty} b_{0i}\delta(t-\tau_i) \tag{2.75}$$

is called the average power delay profile, and

$$S_{dD}(f) = \sum_{i=0}^{\infty}\sum_{k=0}^{\infty}\frac{A_{ki}^2}{2}\delta(f - f_d\cos\theta_{ki})\delta(\tau-\tau_i) \tag{2.76}$$

is called time delay/Doppler spectra function. When θ_{ki} and τ_i are continuously distributed, the time delay/Doppler spectra function becomes a continuous function.

Figure 2.20 shows examples of the time delay/Doppler spectra function obtained in the urban area of Tokyo, and Figure 2.21 shows a map of the measured area [2-25]. In this measurement, a complex delay profile is measured every 10.2 milliseconds (ms) using a spread spectrum technique, and the time delay/Doppler spectra function is obtained by Fourier-transforming 64 profiles using the fast Fourier transform (FFT). Therefore, each delay/Doppler spectra function is obtained every 652.8 ms, and these eight figures are continuously measured.

These figures show that the spectrum at around $\tau = 0$ includes various frequency components, whereas the delayed waves with longer delay time arrive from some specific angles. This result suggests that the p.d.f. of the amplitude for a short delayed

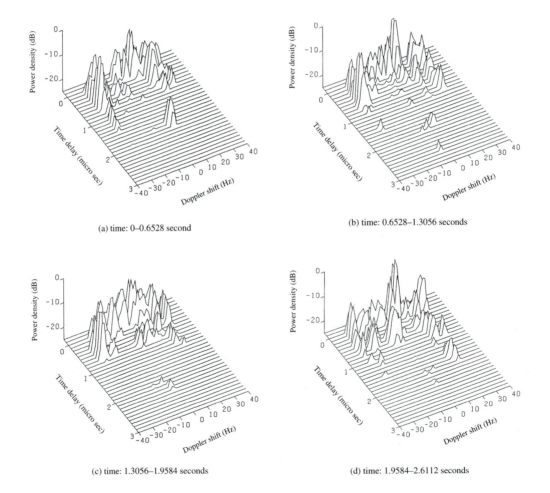

(a) time: 0–0.6528 second

(b) time: 0.6528–1.3056 seconds

(c) time: 1.3056–1.9584 seconds

(d) time: 1.9584–2.6112 seconds

Fig. 2.20 Examples of the time delay/Doppler spectra function obtained in the urban area of Tokyo.

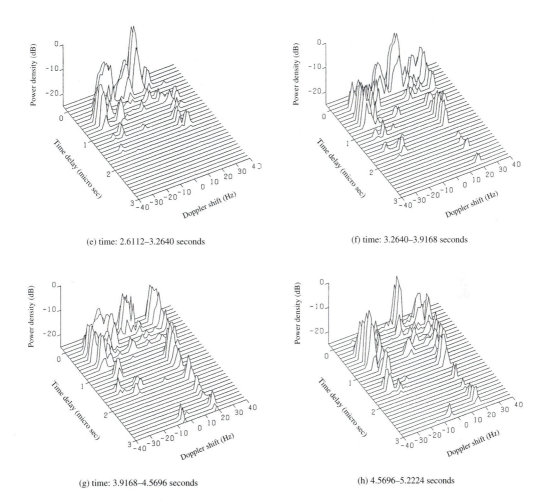

(e) time: 2.6112–3.2640 seconds

(f) time: 3.2640–3.9168 seconds

(g) time: 3.9168–4.5696 seconds

(h) 4.5696–5.2224 seconds

Fig. 2.20 Continued.

wave tends to be Rayleigh distribution, whereas that for a longer delayed wave tends to be Nakagami-Rician distribution. We can observe this tendency in many places. However, when we develop some anti-frequency selective fading techniques, we have to evaluate such techniques in the worst case. Therefore, we usually employ a time delay/Doppler spectra function given by

$$S_{dD}(f) = \sum_{i=0}^{\infty} \frac{b_{0i}}{\pi f_d \sqrt{1 - \left(\dfrac{f}{f_d}\right)^2}} \delta(\tau - \tau_i) \tag{2.77}$$

for the evaluation of various anti-frequency selective fading techniques (discussed in chapter 5). This power spectrum density function corresponds to the condition that

Buildings with 3 or more stories

Buildings with 2 or less stories

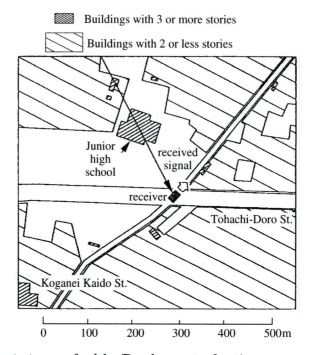

Fig. 2.21 Map of the test course for delay/Doppler spectra function measurement (from Ref. [2-25]) (©Institute of Electronics, Information and Communication Engineers, 1989).

the arrival angle for each delayed component is uniformly distributed. In this case, direct and delayed waves are subject to Rayleigh fading because each of them is composed of many paths coming from various directions.

From a practical point of view, there are many propagation path models based on the extensive measurement of the delay profile. We will discuss some examples.

2.5.1 Two-Ray Rayleigh Model

One of the most popular frequency-selective fading models is the two-ray Rayleigh model, which is also called the double-spike Rayleigh model [2-26]. This model's instantaneous delay profile and delay time delay/Doppler spectra function are given by

$$c(t; \tau) = c_0(t)\delta(\tau) + c_1(t)\delta(\tau - \tau_1) \tag{2.78}$$

$$S_{dD}(f) = \frac{b_{00}}{\pi f_d \sqrt{1 - \left(\dfrac{f}{f_d}\right)^2}} \delta(\tau) + \frac{b_{01}}{\pi f_d \sqrt{1 - \left(\dfrac{f}{f_d}\right)^2}} \delta(\tau - \tau_1) \tag{2.79}$$

where $c_0(t)$ and $c_1(t)$ are subject to Gaussian distribution. This model is valid when a symbol duration is so long that the expected maximum delay time of the delayed wave is, at most, one symbol duration. For example, this model is used for the evaluation of U.S. IS-54 and Japanese PDC systems.

One of the most important advantages of the two-ray Rayleigh model is that it is simple and effective for analyzing basic operation of some anti-frequency-selective fading techniques. For example, we sometimes use this model to investigate the path diversity effect of the adaptive equalizer. Furthermore, no anti-frequency fading technique is effective unless it normally operates at least under two-ray Rayleigh channel condition. Therefore, when we want to develop an anti-frequency-selective fading technique, we usually evaluate the basic operation under the two-ray Rayleigh model first, and then we apply a more practical channel model, such as the multi-ray Rayleigh model.

2.5.2 Multi-Ray Rayleigh Model

The multi-ray Rayleigh model is an extention of the two-ray Rayleigh model. Because channel impulse response can be modeled by the discrete impulse response in the case of digital modulated signal transmission as we will discuss in chapter 5, any time the variant impulse response can be modeled as

$$c(t; \tau) = \sum_{i=0}^{N-1} c_i(t) \delta(\tau - iT_s) \qquad (2.80)$$

and its time delay/Doppler spectra function is given by

$$S_{dD}(f) = \sum_{i=0}^{N-1} \frac{b_{0i}}{\pi f_d \sqrt{1 - \left(\dfrac{f}{f_d}\right)^2}} \delta(\tau - iT_s) \qquad (2.81)$$

where the average signal power of each component, b_{0i} ($0 \le i \le N - 1$), is arbitrarily determined case by case.

2.6 RELATIONSHIP BETWEEN PROPAGATION STUDY, RADIO SUBSYSTEM DESIGN, RADIO LINK DESIGN, AND NETWORK DESIGN

In this chapter, we have discussed propagation path characteristics specific to wireless mobile communication systems. However, we also have to understand how to apply the knowledge of propagation path characteristics to the radio interface design. In other words, we have to understand the relationship between the study of propagation path characteristics and wireless communication system design.

Figure 2.22 shows this relationship. In this figure, system design is further divided into the following fields:

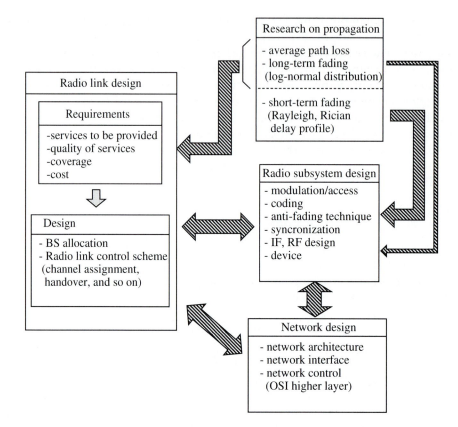

Fig. 2.22 Relation between study on propagation and wireless communication system design.

☞ Radio link design

☞ Radio subsystem design

☞ Network design

In the radio link design process, we have to determine locations of the BSs; the radio link control scheme, such as the channel assignment scheme, the handover scheme, and so on; considering the required services, quality of services, coverage of the system, and system cost. For this purpose, we have to preliminarily determine some appropriate path loss models and statistics of long-term fading using a database of the topological map and man-made constructions as well as that of the propagation path characteristics obtained by extensive field measurements. How to design a radio link will be discussed in chapter 10.

On the other hand, we also need knowledge of short-term fading to design radio subsystems because short-term fading directly affects the operation of various parts in the receivers, such as the demodulator, frame, symbol timing synchronization circuits, and so on. In other words, the main objective of radio subsystem design is to determine how to combine cost-effective anti-fading techniques in the receivers to achieve high receiver sensitivity. From this viewpoint, radio frequency or intermediate frequency (RF/IF) component as well as the DSP technologies also have very important roles to play in radio subsystem design. How to design modulation/access schemes including anti-fading techniques will be discussed in chapters 4–8.

Because receiver sensitivity is one of the important radio link design parameters, results of the radio subsystem design are also fed back to the radio link design strategies. Moreover, based on the results of radio link and radio subsystem design, the network for the wireless system is designed. Of course, requirements from network design are also fed back to the radio link design and to radio subsystem design.

REFERENCES

2-1. Okumura, T., Ohmori, E. and Fukuda, K., "Field strength and its variability in VHF and UHF land mobile service," Review of Electrical Communication Laboratory, Vol. 16, No. 9–10, pp. 825–73, September–October 1968.

2-2. Hata, M., "Empirical formula for propagation loss in land mobile radio services," IEEE Trans. Veh. Technol., Vol. VT-29, No. 3, pp. 317–25, August 1980.

2-3. Cox, D. C., Murray, R. R. and Norris, A. W., "Measurements of 800-MHz radio transmission into buildings with metallic walls," B.S.T., Vol. 62, No. 9, pp. 2695–717, November 1983.

2-4. Toledo, A. F. and Turkmani, A. M. D., "Propagation into and within buildings at 900, 1800 and 2300 MHz," 42d IEEE Veh. Tech. Conf. (Denver, Colorado), pp. 633–36, May 1992.

2-5. Cox, D. C., "Universal Digital Portable Radio Communications," Proc. IEEE, Vol. 75, No. 4, pp.436–77, April 1987.

2-6. Turkmani A. M. D. and Toledo, A. F. "Radio transmission at 1800 MHz into and within, multi-story buildings," IEE Proc.-I, Vol. 138, No. 6, pp. 577–84, December 1991.

2-7. Seidel, S. Y. and Rappaport, T. S., "900 MHz path loss measurements and prediction techniques for in-building communication system design," IEEE ICC'91 (Denver, Colorado), pp. 613–18, June 1991.

2-8. Lotse, F., Berg, J. E. and Bownds, R., "Indoor propagation measurements at 900 MHz," 42d IEEE Veh. Tech. Conf. (Denver, Colorado), pp. 629–32, May 1992.

2-9. Rappaport, T. S., Wireless communications, Prentice Hall, Upper Saddle River, New Jersey, 1996.

2-10. Okumura, Y. and Shinji, M., eds., "Basic land mobile communications," IEICE, October 1986.

2-11. Rappaport, T. S., "Indoor radio communications for factories of the future," IEEE Communications Magazine, Vol. 27, No. 5, pp. 15-24, May 1989.

2-12. Lafortune, J. F. and Lecours, M., "Measurement and modeling of propagation losses in a building at 900 MHz," IEEE Trans. Veh. Technol., Vol. 39, No. 2, pp. 101–8, May 1990.

2-13. Iwama, T., Saruwatari, T., Wakao, M., Ogawa, K. and Ariizumi, Y., "Experimental results of 1.2 GHz band premises data transmission system using GMSK modulation," IEEE GLOBE-COM'87 (Tokyo, Japan), pp. 1921–25, November 1987.

2-14. Moriyama, E., Mizuno, M., Nagata, Y., Furuya, Y., Kamiya, I. and Hattori, S., "2.6 GHz multipath characteristics measurement in a shielded building," 42d IEEE Veh. Tech. Conf. (Denver, Colorado), pp. 621–24, May 1992.

2-15. Kalivas, C. A., El-Tanany, M. and Mahmoud, S. A., "Millimeter-wave channel measurements for indoor wireless communications," 42d IEEE Veh. Tech. Conf. (Denver, Colorado), pp. 609–12, May 1992.

2-16. Janssen, G. J. M. and Prasad, R., "Propagation measurements in an indoor radio environment at 2.4 GHz, 4.75 GHz and 11.5 GHz," 42d IEEE Veh. Tech. Conf., pp. 617–20, May 1992.

2-17. Davies, R., Bensebti, M., Beach, M. A. and McGeehan, J. P., "Wireless propagation measurements in indoor multipath environments at 1.7 GHz and 60 GHz for small cell systems," IEEE ICC'91 (Denver, Colorado), pp. 589–93, June 1991.

2-18. Lee, W. C. Y., *Mobile communications design fundamentals* (2d Edition), John Wiley & Sons, New York, 1993.

2-19. Cox, D. C., Murray, R. and Norris, A., "800 MHz attenuation measured in and around suburban houses," AT&T Bell Laboratories Technical Journal, Vol. 63, No. 6, pp. 921–54, July–August, 1984.

2-20. Walker, E. H., "Penetration of radio signals into buildings in the cellular radio environment," B. S. T. J., Vol. 62, No. 9, pp. 2719–34, November 1983.

2-21. Jakes, W. C., ed., *Microwave mobile communications*, John Wiley, New York, 1974.

2-22. Kamio, Y., Sampei, S., Sasaoka, H. and Yokoyama, M., "A new type fading simulator with DSP," Trans. IEICE, Vol. E70, No. 4, pp. 379–82, April 1987.

2-23. Kamio, Y., Sampei, S., Sasaoka, H. and Yokoyama, M., "Experimental equipment—Fading simulator," *Reviw of the Communications Research Laboratory*, Vol. 37, No. 1, pp. 187–92, February 1991.

2-24. Suzuki, T., Sampei, S. and Morinaga, N., "Directive antenna diversity reception scheme for an adaptive modulation system in high mobility land mobile communications," IEICE Trans. Commun., Vol. E79-B, No. 3, pp. 335–41, March 1996.

2-25. Ohgane, T., Sampei, S., Kamio, Y., Sasaoka, H. and Mizuno, M., "UHF urban characteristics in wideband mobile radio communications," Trans. IEICE (B-II), Vol. J72-B-II, No. 2, pp.63–71, February 1989.

2-26. Ikegami, F. and Yoshida, S., "Analysis of multipath propagation structure in urban mobile radio environments," IEEE Trans. Antennas Propag., Vol. AP-28, No. 10, pp. 1119–26, July 1980.

Modulation and Demodulation Techniques

In wireless personal communication systems, selection of the modulation scheme is very important to achieving high capacity and supporting high-grade services. However, when we want to employ a modulation scheme to wireless personal communication systems, we have to solve so many problems, such as how to achieve a large-capacity system using a limited spectrum, how to offer a good quality of services (QoS), how to make small terminals, and so on.

To clearly answer these questions and to develop really convenient wireless systems, system engineers have to fully understand what the features of each modulation scheme are and what kind of anti-fading techniques are effective for the modulation scheme. Anti-fading techniques will be discussed in chapters 4–6. Therefore, this chapter covers the basic performance of each modulation scheme including BER performance under AWGN and Rayleigh fading conditions. But first, here's a brief overview of the development of modulation techniques for wireless personal communication systems.

3.1 Outline of Modulation and Demodulation Techniques

To design convenient wireless personal communication systems, it is very important to select modulation and demodulation techniques based on the following three requirements:

☞ **High spectral efficiency.** In wireless personal communication systems, most of the subscribers are expected to be in high subscriber density areas, such as on the downtown streets, in business offices, or on public transportation. Therefore, we have to maximize spectral efficiency in terms of the number of voice channels per unit bandwidth per unit area (ch/Hz/m^2) for voice transmission or spectral efficiency in terms of bit/s/Hz/m^2 for data transmission.

☞ **High power efficiency.** Terminals for wireless personal communication systems should be small, say 100–200 grams (g) under strict limitation of battery size. To satisfy this requirement, power efficiency for the transmitter amplifier should be high.

☞ **High fading immunity.** Terminals for wireless personal communication systems are located in multipath fading environments. Therefore, they should be operated under time-varying multipath (both flat Rayleigh and frequency-selective fading) conditions.

Figure 3.1 shows classification of the modulation schemes based on these three requirements. During late 1970s and early 1980s, constant envelope digital modulation schemes using a C-class amplifier were studied aiming at achieving high power-efficient terminals with high robustness to signal level variation due to fading. As a result, Gaussian-filtered minimum shift keying (GMSK) [3-1; 3-2], phase-locked loop

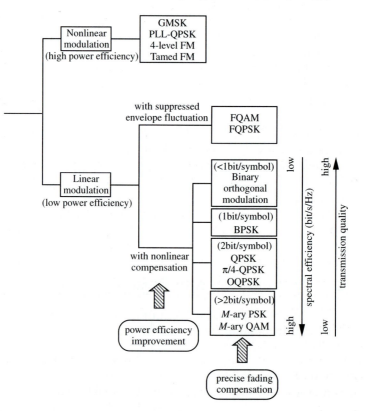

Fig. 3.1 Classification of the modulation schemes.

quaternary phase shift keying (PLL-QPSK) [3-3], 4-level frequency modulation (FM) [3-4], and Tamed FM [3-5] were developed. Among them, GMSK was selected as the modulation scheme for the European digital cellular system called the GSM system [3-6] and the DECT system [3-7].

In the mid-1980s, when the capacity limitation of analog cellular systems was getting more severe, developments of linear modulation having 2 bit/s/Hz transmission ability were started [3-8]. For the linear modulation to be applied to land mobile communication systems, however, we have to achieve high spectral efficiency and high power efficiency at the same time. For this purpose, there are two strategies.

☞ To suppress envelope fluctuation by introducing cross correlation between the adjacent symbols
☞ To apply a nonlinear distortion compensator to the transmitter amplifier

Feher's QPSK (FQPSK) and FQAM [3-9] are categorized in the former strategy. Because the phase of these modulations are smoothly changed with keeping its envelope almost constant, their spectra are very compact even if they are amplified by a nonlinear amplifier.

As for the latter strategy, there are various types of nonlinear distortion compensators, such as a predistortion type [3-8; 3-11], Cartician loop [3-12], feed-forward type [3-13], and linear amplification with nonlinear components (LINC) techniques [3-14]. Owing to these nonlinear compensation techniques, $\pi/4$-QPSK technologies were extensively developed in late 1980s to early 1990s. As a result, $\pi/4$-QPSK was applied to the Japanese [3-15] and North American [3-16] digital cellular standards.

Since the late 1980s, there has been development of modulation schemes with much higher spectral efficiency (>2 bit/s/Hz), such as 16QAM [3-17; 3-18; 3-19]. Because information is included in both amplitude and phase components of such modulated signals, very precise fading compensation techniques are required for demodulation. To solve this problem, pilot tone-aided techniques, such as transparent-tone in band (TTIB) [3-17], and pilot symbol-aided techniques [3-18] were developed. Details of these fading compensation techniques will be introduced in chapter 4. Owing to these developments, 16QAM or multi-carrier 16QAM (M16QAM) have been applied to the Japanese multichannel access (MCA) system [3-20], the Japanese public PMR systems [3-21], and the ESMR system in the U.S. [3-22].

To employ M-ary PSK and M-ary QAM, we can achieve higher spectral efficiency in terms of bit/s/Hz than BPSK and QPSK. On the other hand, transmission quality (receiver sensitivity) gets worse with an increasing modulation level. Therefore, we have to select a modulation scheme considering the trade-off between transmission quality and spectral efficiency in terms of bit/s/Hz.

When transmission quality is a very important factor, we can achieve a high-quality transmission by sacrificing spectral efficiency. Binary orthogonal modulation is such a modulation scheme [3-23].

From these historical reviews of the modulation techniques, we can see that we now have a wide variety of modulation schemes for wireless personal communication systems. Furthermore, for the design of a TDMA system, we can apply different types

of modulation schemes to each part of a burst signal according to the required reliability. For example, 16QAM is applied to the traffic channel, QPSK is applied to the associated control channel, BPSK is applied to the synchronization word, and binary
orthogonal modulation with its transmission ability of less than 1 bit/s/Hz is applied
to very important signals. Therefore, it is imperative that we fully understand the features and performance of each modulation scheme.

   With this as background, this chapter discusses the basic performance of each
modulation scheme including BER performance under AWGN and Rayleigh fading
conditions.

3.2 BPSK

3.2.1 BPSK with Coherent Detection

   BPSK is the most primitive modulation scheme for wireless communication systems. Figure 3.2 shows a configuration of the BPSK modem. Let's assume that the
source bit is transmitted every T_b second (bit rate is $1/T_b$ bit/s) and i-th source bit is
expressed as a_i. In the BPSK modulator, the carrier phase is modulated according to
the source bit as $\theta_D(t) = 0$ when the source bit is 1, and $\theta_D(t) = \pi$ when the source bit is
0. Therefore, the BPSK-modulated signal can be expressed as

$$
\begin{aligned}
s_T'(t) &= A\cos(2\pi f_c t + \theta_D(t)) \\
&= A\cos(\theta_D(t))\cos(2\pi f_c t) - A\sin(\theta_D(t))\sin(2\pi f_c t) \\
&= u_I(t)\cos(2\pi f_c t) - u_Q(t)\sin(2\pi f_c t)
\end{aligned}
\tag{3.1}
$$

where A is the amplitude of the modulated signal, f_c is carrier frequency, and

$$
u_I(t) = A\cos(\theta_D(t))
\tag{3.2}
$$

$$
u_Q(t) = A\sin(\theta_D(t))
\tag{3.3}
$$

Because $u_Q(t)$ is always 0 in the case of BPSK, $s'_T(t)$ can be simplified as

$$
s_T'(t) = u_I(t)\cos(2\pi f_c t)
\tag{3.4}
$$

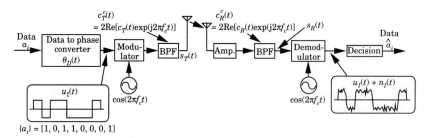

Fig. 3.2 Configuration of the BPSK modem.

In the BPSK modulator, $u_I(t) = 1$ corresponds to $\theta_D(t) = 0$ and $u_I(t) = -1$ corresponds to $\theta_D(t) = \pi$, and $\theta_D(t)$ is changed every T_b second.

Now, let us assume that the waveform of $u_I(t)$ is given by

$$u_I(t) = A \sum_{n=-\infty}^{\infty} b_n g(t - nT_b) \tag{3.5}$$

where

$$b_n = \cos(\theta_D(nT_b)) \tag{3.6}$$

$$g(t) = \begin{cases} 1; & -\dfrac{T_b}{2} \leq t \leq \dfrac{T_b}{2} \\ 0; & \text{otherwise} \end{cases} \tag{3.7}$$

Figure 3.3 shows waveforms of $\{a_n\}$, $\{b_n\}$, and $u_I(t)$ in the case of $\{a_n \mid 0 \leq n \leq 7\} = [1, 0, 1, 1, 0, 0, 0, 1]$ as an example. The modulated signal is then band-limited by a band-pass filter (BPF) with its impulse response of

$$c_T^c(t) = 2\,\mathrm{Re}\!\left[c_T(t)\exp(j2\pi f_c t)\right] \tag{3.8}$$

to suppress the out-of-band radiation, where $c_T(t)$ is the impulse response of the BPF expressed as an equivalent low-pass system. As a result, the transmitted signal is given by

$$s_T(t) = [u_I(t) \otimes c_T(t)]\cos(2\pi f_c t) \tag{3.9}$$

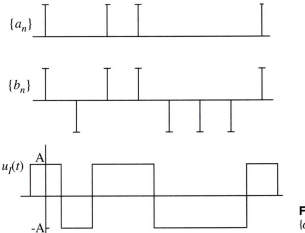

Fig. 3.3 Waveform of $u_I(t)$ in the case of $\{a_n \mid 0 \leq n \leq 7\} = [1, 0, 1, 1, 0, 0, 0, 1]$.

At the receiver, the received signal is amplified and band-limited by the BPF with its impulse response of

$$c_R^c(t) = 2\,\mathrm{Re}\!\left[c_R(t)\exp(j2\pi f_c t)\right] \tag{3.10}$$

to suppress the out-of-band noise, where $c_R(t)$ is the impulse response of BPF expressed as an equivalent low-pass system. As a result, the received signal, after band-limitation by the BPF, is given by

$$
\begin{aligned}
s_R(t) &= s_T(t) + n(t) \\
&= [u_I(t) \otimes c_T(t) \otimes c_R(t)]\cos(2\pi f_c t) \\
&\quad + n_I(t)\cos(2\pi f_c t) - n_Q(t)\sin(2\pi f_c t) \\
&= \left[u_I(t) \otimes c_T(t) \otimes c_R(t) + n_I(t)\right]\cos(2\pi f_c t) - n_Q(t)\sin(2\pi f_c t)
\end{aligned}
\tag{3.11}
$$

where

$$n(t) = n_I(t)\cos(2\pi f_c t) - n_Q(t)\sin(2\pi f_c t) \tag{3.12}$$

and $n_I(t)$ and $n_Q(t)$ are in-phase and quadrature components of noise expressed as an equivalent low-pass system. SNR of $s_R(t)$ is given by

$$SNR = \gamma = \frac{A^2}{2\sigma^2} \tag{3.13}$$

where

$$\sigma^2 = \mathrm{E}[n_I^2(t)] = \mathrm{E}[n_Q^2(t)] = \mathrm{E}[n^2(t)] \tag{3.14}$$

When we can perfectly regenerate the carrier component, the coherently demodulated baseband signal is then given by

$$\hat{u}_I(t) = u_I(t) \otimes c_T(t) \otimes c_R(t) + n_I(t) \tag{3.15}$$

Now, let's disregard the effect of the transmitter and receiver filters, say, $c_T(t) = c_R(t) = \delta(t)$. In this case, we can express the demodulated baseband signal as

$$\hat{u}_I(t) = u_I(t) + n_I(t) \tag{3.16}$$

The waveform of $\hat{u}_i(t)$ is shown in Figure 3.4. This waveform is sampled every T_b second (at $t = nT_b$), and its polarity is detected to regenerate the source data sequence as

$$
\hat{a}_i = \begin{cases} 1; & \hat{u}_I(nT_b) \ge 0 \\ 0; & \hat{u}_I(nT_b) < 0 \end{cases}
\tag{3.17}
$$

The BER of this signal is given by

$$
\begin{aligned}
\text{BER} &= p(\hat{u}_I(nT_b) < 0 \,|\, \theta_D(nT_b) = 0)p(\theta_D(nT_b) = 0) \\
&\quad + p(\hat{u}_I(nT_b) \geq 0 \,|\, \theta_D(nT_b) = \pi)p(\theta_D(nT_b) = \pi)
\end{aligned}
\tag{3.18}
$$

where

$$
p(\hat{u}_I(nT_b) \,|\, \theta_D(nT_b) = 0) = \frac{1}{\sqrt{2\pi}\sigma} \exp\!\left(-\frac{(\hat{u}_I(nT_b) - A)^2}{2\sigma^2}\right)
\tag{3.19}
$$

$$
p(\hat{u}_I(nT_b) \,|\, \theta_D(nT_b) = \pi) = \frac{1}{\sqrt{2\pi}\sigma} \exp\!\left(-\frac{(\hat{u}_I(nT_b) + A)^2}{2\sigma^2}\right)
\tag{3.20}
$$

$$
p(\theta_D(nT_b) = 0) = 0.5
\tag{3.21}
$$

$$
p(\theta_D(nT_b) = \pi) = 0.5
\tag{3.22}
$$

Therefore, the BER for BPSK with coherent detection (P_{CBPSK}) is given by

$$
\begin{aligned}
P_{CBPSK}(\gamma) &= \frac{1}{2}\int_{-\infty}^{0} \frac{1}{\sqrt{2\pi}\sigma} \exp\!\left(-\frac{(x - A)^2}{2\sigma^2}\right) dx \\
&\quad + \frac{1}{2}\int_{0}^{\infty} \frac{1}{\sqrt{2\pi}\sigma} \exp\!\left(-\frac{(x + A)^2}{2\sigma^2}\right) dx \\
&= \frac{1}{2}\operatorname{erfc}\!\left(\frac{A}{\sqrt{2}\sigma}\right) = \frac{1}{2}\operatorname{erfc}\!\left(\sqrt{\gamma}\right)
\end{aligned}
\tag{3.23}
$$

3.2.2 Optimum Receiver Filter

In the foregoing discussion, we have disregarded the effect of the transmitter and receiver BPFs. In practice, however, these filters are essential because noise power σ^2 is given by $N_0 B$ where N_0 is the noise power spectrum density and B is the receiver filter bandwidth.

Now, let us discuss the optimization of the receiver filter and focus on a single pulse waveform centered on $t = 0$. This waveform can be expressed as $u_{I0}(t) = b_0 g(t)$.

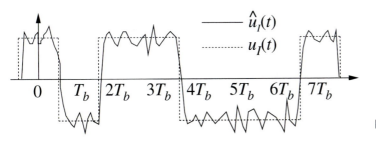

Fig. 3.4 Waveform of $\hat{u}_i(t)$.

When $u_{I0}(t)$ is filtered by an LPF with its impulse response of $c_R(t)$, the output waveform is given by

$$\hat{u}_a(t) = u_{I0}(t) \otimes c_R(t) = \int_{-\infty}^{\infty} u_{I0}(\tau)c_R(t-\tau)d\tau$$

$$= b_0 \int_{-\infty}^{\infty} C_R(f)G(f)\exp(j2\pi ft)df \tag{3.24}$$

where $C_R(f)$ and $G(f)$ are Fourier transforms of $c_R(t)$ and $g(t)$. The signal power of a sampled value at $t = 0$ (s) is then given by

$$S = |u_a(0)|^2$$

$$= \left| b_0^2 \int_{-\infty}^{\infty} C_R(f)G(f)df \right|^2 \tag{3.25}$$

$$= \left| \int_{-\infty}^{\infty} C_R(f)G(f)df \right|^2$$

On the other hand, the noise power is given by

$$N = N_0 \int_{-\infty}^{\infty} |C_R(f)|^2 df \tag{3.26}$$

As a result, the SNR for the sampled signal at $t = 0$ is given by

$$\gamma = \frac{S}{N} = \frac{\left| \int_{-\infty}^{\infty} C_R(f)G(f)df \right|^2}{N_0 \int_{-\infty}^{\infty} |C_R(f)|^2 df} \tag{3.27}$$

When we employ Schwarz inequality given by

$$\left| \int_{-\infty}^{\infty} C_R(f)G(f)df \right|^2 \le \left[\int_{-\infty}^{\infty} |C_R(f)|^2 df \right]\left[\int_{-\infty}^{\infty} |G(f)|^2 df \right] \tag{3.28}$$

we can obtain the upper bound of γ as

$$\gamma \le \frac{\left[\int_{-\infty}^{\infty} |C_R(f)|^2 df \right]\left[\int_{-\infty}^{\infty} |G(f)|^2 df \right]}{N_0 \int_{-\infty}^{\infty} |C_R(f)|^2 df}$$

$$= \frac{1}{N_0} \int_{-\infty}^{\infty} |G(f)|^2 df = \frac{E_b}{N_0} \tag{3.29}$$

where

$$E_b = \int_{-\infty}^{\infty} |G(f)|^2 df \tag{3.30}$$

is energy per bit, and equality holds when

$$C_R(f) = kG^*(f) \tag{3.31}$$

Without loss of generality, we can assume $k = 1$. In this case, by applying the inverse Fourier transform to $C_R(f)$, we can obtain the optimum impulse response of the receiver filter as

$$
\begin{aligned}
c_R(t) &= \int_{-\infty}^{\infty} G^*(f) \exp(j2\pi ft) df \\
&= \left\{ \int_{-\infty}^{\infty} G(f) \exp(-j2\pi ft) df \right\}^* = g^*(-t)
\end{aligned}
\tag{3.32}
$$

When $g(t)$ is a real, even function, $c_R(t)$ is given by

$$c_R(t) = g(t) \tag{3.33}$$

With these results, we can conclude that the optimum $c_R(t)$ is equal to the waveform of transmitted signal pulse $g(t)$, at which γ is equal to E_b/N_0. In the case of rectangular pulse transmission systems, a filtering process using the optimum $c_R(t)$ is equivalent to integrating the received baseband signal from $t = (i - 1/2)T_b$ to $t = (i + 1/2)T_b$ to obtain i-th sample. Therefore, such a filter is called an *integrate and dump* (I&D) *filter*.

3.2.3 Optimization of the Transmitter and Receiver Filters

3.2.3.1 Nyquist Filter
In practical radio systems, the presence of transmitter BPF is essential to save spectrum as much as possible. However, such band-limitation could degrade the transmission performance due to intersymbol interference (ISI). Therefore, we have to reduce the signal bandwidth as much as possible without producing any ISI.

The received baseband signal after demodulation is expressed as

$$\bar{u}_I(t) = \sum_{n=-\infty}^{\infty} b_n g_1(t - nT_b) + n_I(t) \tag{3.34}$$

where

$$g_1(t) = g(t) \otimes c_T(t) \otimes c_R(t) \tag{3.35}$$

To prevent ISI, $g_1(t)$ should satisfy the following conditions:

$$g_1(t) = \begin{cases} 1; & t = 0 \\ 0; & t = nT_b \ (n \neq 0) \end{cases} \tag{3.36}$$

The Nyquist filter is a filter that satisfies such conditions. Its frequency response is given by

$$G_1(f) = \begin{cases} 1; & 0 \le |f| \le \dfrac{(1-\alpha)}{2T_b} \\[2ex] \cos^2\left[\dfrac{T_b}{4\alpha}\left\{2\pi|f| - \dfrac{\pi(1-\alpha)}{T_b}\right\}\right]; & \dfrac{(1-\alpha)}{2T_b} \le |f| \le \dfrac{(1+\alpha)}{2T_b} \\[2ex] 0; & |f| \le \dfrac{(1+\alpha)}{2T_b} \end{cases} \qquad (3.37)$$

where α is a roll-off factor that determines the channel bandwidth. Figure 3.5 shows (a) frequency response $G_1(f)$ and (b) impulse response of the roll-off filter $g_1(t)$ with a parameter of α. In the case of $\alpha = 0$, frequency response is identical to the ideal LPF, which achieves the minimum bandwidth with ISI-free conditions. In this case, however, its impulse response lasts for a long time. On the other hand, when $\alpha = 1$, the bandwidth is twice as wide as that for the ideal LPF, although its impulse response quickly decays with time.

Because the maximum SNR is obtained by

$$C_R(f) = G(f)C_T(f) = \sqrt{G_1(f)} \qquad (3.38)$$

and $G(f)$ for the nonreturn to zero (NRZ) pulse [3-24] is given by

$$G(f) = \frac{\sin(\pi f T_b)}{\pi f T_b} \qquad (3.39)$$

the optimum $C_T(f)$ and $C_R(f)$ are given by

$$C_T(f) = G^{-1}(f)\sqrt{G_1(f)} = \frac{\pi f T_b}{\sin(\pi f T_b)}\sqrt{G_1(f)} \qquad (3.40)$$

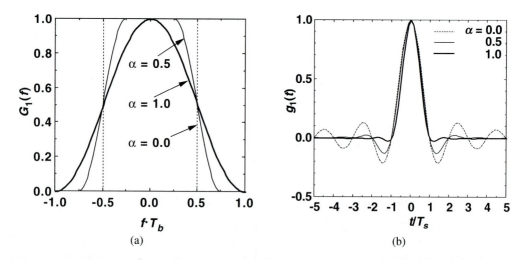

Fig. 3.5 Characteristics of roll-off filter: (a) frequency response and (b) impulse response.

$$C_R(f) = \sqrt{G_1(f)} \tag{3.41}$$

where $G^{-1}(f)$ is an equalizer that converts the waveform from NRZ pulse to impulse. Therefore, $G^{-1}(f)$ can be equivalently implemented by a sampler that samples the NRZ sequence every T_b second. In other words, when we define

$$u_I(t) = \sum_{n=-\infty}^{\infty} b_n \delta(t - nT_b) \tag{3.42}$$

instead of the waveform defined by equation (3.5), the optimum $C_T(f)$ and $C_R(f)$ are given by

$$C_T(f) = C_R(f) = \sqrt{G_1(f)}$$

$$= \begin{cases} 1; & 0 \le f \le \dfrac{(1-\alpha)}{2T_b} \\[2ex] \cos\left[\dfrac{T_b}{4\alpha}\left\{2\pi f - \dfrac{\pi(1-\alpha)}{T_b}\right\}\right]; & \dfrac{(1-\alpha)}{2T_b} \le f \le \dfrac{(1+\alpha)}{2T_b} \\[2ex] 0; & \dfrac{(1+\alpha)}{2T_b} \le f \end{cases} \tag{3.43}$$

This filter is called a root roll-off filter or root Nyquist filter, and its impulse response is given by

$$c_T(t) = c_R(t)$$

$$= \frac{1}{\pi t} \frac{1}{1-\left(\dfrac{4\alpha t}{T_b}\right)^2} \sin\left\{2\pi(1-\alpha)\frac{1}{T_b}\right\}$$

$$+ \frac{1}{\pi} \frac{\dfrac{4\alpha}{T_b}}{1-\left(\dfrac{4\alpha t}{T_b}\right)^2} \cos\left\{2\pi(1+\alpha)\frac{t}{T_b}\right\} \tag{3.44}$$

In this case, the BER for BPSK with coherent detection under the AWGN condition is the same as that given by equation (3.23).

3.2.4 Differentially Encoded BPSK with Coherent Detection

In practical BPSK systems, a reference carrier for coherent detection is regenerated at the receiver using a carrier regeneration circuit, such as squaring loop and Costas loop. When we apply these methods, however, π [rad] of phase ambiguity arises in the case of BPSK (which will be discussed in chapter 4).

Differential encoding is a method of mitigating such a phase ambiguity effect. At the differential encoder, the source bit sequence is encoded by the following rule:

$$d_n = d_{n-1} \oplus a_n \tag{3.45}$$

and the carrier phase is modulated by the following rule:

$$\theta_D(nT_s) = \begin{cases} 0; & d_n = 1 \\ \pi; & d_n = 0 \end{cases} \tag{3.46}$$

From equations (3.45) and (3.46), we can find that the carrier phase stays at the same value when $a_n = 0$ and π [rad] phase shift occurs when $a_n = 1$. In other words, the data are included in the phase transition instead of the phase position.

At the receiver, after the received signal is coherently detected, the regenerated sequence $\{\hat{d}_n\}$ is decoded by the following rule:

$$\hat{a}_n = \hat{d}_{n-1} \oplus \hat{d}_n \tag{3.47}$$

When phase of the regenerated carrier has π [rad] offset for some reason, $\hat{d}_{n-1}$ and $\hat{d}_n$ are inverted as

$$\hat{d}_{n-1} = d_{n-1} \oplus 1 \tag{3.48a}$$

$$\hat{d}_n = d_n \oplus 1 \tag{3.48b}$$

In this case, although the demodulated phase is π [rad] shifted from the original positions, differential detection can cancel out this phase shift as follows:

$$\begin{aligned} \hat{a}_n &= \hat{d}_{n-1} \oplus \hat{d}_n \\ &= (d_{n-1} \oplus 1) \oplus (d_n \oplus 1) \\ &= a_n \end{aligned} \tag{3.49}$$

As a result, we can find that, although $\{\hat{d}_n\}$ includes phase ambiguity, the data sequence can be correctly detected when the source data are differentially encoded. Especially, when we transmit signals over Rayleigh fading channels, the phase of a carrier regenerator is sometimes shifted by π [rad] due to fast phase variation by the fading. Therefore, differential encoding is necessary for land mobile communication systems if the modem does not employ any pilot signal-aided fading compensation scheme (this will be discussed in chapter 4).

Although differential encoding is effective for mitigating carrier phase ambiguity, it degrades the BER performance by the error propagation effect due to differential encoding. As shown in equation (3.47), when a bit in the sequence of $\{\hat{d}_n\}$ is errored, it will affect both $\hat{a}_n$ and $\hat{a}_{n+1}$. In other words, 1 bit of detection error before differential decoding results in a 2-bit error after differential decoding. However, when consecutive 2 bits in the sequence of $\{\hat{d}_n\}$, say $\hat{d}_{n-1}$ and $\hat{d}_n$, are errored, $\hat{a}_n$ can be correctly detected. Therefore, the correct detection probability for differentially encoded BPSK is given by

$$P_c(\gamma) = \left(1 - P_{CBPSK}(\gamma)\right)^2 + P_{CBPSK}^2(\gamma) \tag{3.50}$$

As a result, the BER for differentially encoded BPSK with coherent detection (P_{DEBPSK}) is given by

$$
\begin{aligned}
P_{DEBPSK}(\gamma) &= 1 - P_c(\gamma) \\
&= 2P_{CBPSK}(\gamma)\left(1 - P_{CBPSK}(\gamma)\right) \\
&\cong \mathrm{erfc}\left(\sqrt{\gamma}\right)
\end{aligned}
\tag{3.51}
$$

3.2.5 BER Performance of BPSK with Differential Detection

Another method of mitigating carrier phase ambiguity is differential detection. When differential encoding is applied to BPSK modulation, the data are included in the phase transition. Therefore, when we can detect the amount of phase transition between consecutive 2 bits, we can detect the transmitted data sequence.

Let's disregard distortion due to transmitter and receiver filters to simplify this discussion. The received signal is given by

$$
\begin{aligned}
s_R(t) &= \left[u_I(t) + n_I(t)\right]\cos(2\pi f_c t) - n_Q(t)\sin(2\pi f_c t) \\
&= X_1 \cos(2\pi f_c t) - Y_1 \sin(2\pi f_c t) \\
&= R(t)\cos\left(2\pi f_c t + \theta(t)\right)
\end{aligned}
\tag{3.52}
$$

and its 1-bit delayed signal is given by

$$
\begin{aligned}
s_R(t - T_b) &= \left[u_I(t - T_b) + n_I(t - T_b)\right]\cos\left(2\pi f_c(t - T_b)\right) \\
&\quad - n_Q(t - T_b)\sin\left(2\pi f_c(t - T_b)\right) \\
&= \left[u_I(t - T_b) + n_I(t - T_b)\right]\cos(2\pi f_c t) \\
&\quad - n_Q(t - T_b)\sin(2\pi f_c t) \\
&= X_2 \cos(2\pi f_c t) - Y_2 \sin(2\pi f_c t) \\
&= R(t - T_b)\cos\left(2\pi f_c t + \theta(t - T_b)\right)
\end{aligned}
\tag{3.53}
$$

where

$$X_1 = u_I(t) + n_I(t) = R(t)\cos\left(\theta(t)\right) \tag{3.54}$$

$$Y_1 = n_Q(t) = R(t)\sin\left(\theta(t)\right) \tag{3.55}$$

$$X_2 = u_I(t - T_b) + n_I(t - T_b) = R(t - T_b)\cos\left(\theta(t - T_b)\right) \tag{3.56}$$

$$Y_2 = n_Q(t - T_b) = R(t - T_b)\sin(\theta(t - T_b)) \tag{3.57}$$

and $f_c T_b = 1$ is assumed.

At the differential detector, $s_R(t)$ and $s_R(t - T_b)$ are multiplied and low-pass filtered to obtain the baseband signal given by

$$
\begin{aligned}
v(t) &= \text{LPF}\big[s_R(t)s_R(t - T_b)\big] \\
&= R(t)R(t - T_b)\cos(\theta(t) - \theta(t - T_b)) \\
&= X_1 X_2 + Y_1 Y_2
\end{aligned}
\tag{3.58}
$$

As shown in equations (3.45) and (3.46), signal phase is constant when the transmitted bit is logical 0 and it is changed by π [rad] when the transmitted bit is logical 1. Therefore, the transmitted data can be detected by the following rule:

$$
a_n = \begin{cases} 1; & v(nT_b) < 0 \\ 0; & v(nT_b) \geq 0 \end{cases}
\tag{3.59}
$$

As a result, the BER can be obtained as

$$
\begin{aligned}
P_b &= \Pr(X_1 X_2 + Y_1 Y_2 < 0 \,|\, a_n = 0)\Pr(a_n = 0) \\
&\quad + \Pr(X_1 X_2 + Y_1 Y_2 > 0 \,|\, a_n = 1)\Pr(a_n = 1) \\
&= \Pr(X_1 X_2 + Y_1 Y_2 < 0 \,|\, a_n = 0)
\end{aligned}
\tag{3.60}
$$

$X_1 X_2 + Y_1 Y_2$ can also be expressed as

$$
\begin{aligned}
&X_1 X_2 + Y_1 Y_2 \\
&= \frac{1}{4}\Big[\big\{(X_1 + X_2)^2 + (Y_1 + Y_2)^2\big\} - \big\{(X_1 - X_2)^2 + (Y_1 - Y_2)^2\big\}\Big] \\
&= |U_1 + jV_1|^2 - |U_2 + jV_2|^2 \\
&= r_1^2 - r_2^2
\end{aligned}
\tag{3.61}
$$

where

$$U_1 = \frac{1}{2}(X_1 + X_2) \tag{3.62}$$

$$U_2 = \frac{1}{2}(X_1 - X_2) \tag{3.63}$$

$$V_1 = \frac{1}{2}(Y_1 + Y_2) \tag{3.64}$$

$$V_2 = \frac{1}{2}(Y_1 - Y_2) \tag{3.65}$$

$$r_1 = |U_1 + jV_1| \tag{3.66}$$

$$r_2 = |U_2 + jV_2| \tag{3.67}$$

U_1 is a random variable with its average of A and its variance of $\sigma^2/2$, whereas U_2, V_1, and V_2 are 0-mean Gaussian random variables with a variance of $\sigma^2/2$, where σ^2 is the noise power. Using these relations, the p. d. f. for r_1 and r_2 are given by

$$p(r_1) = \frac{2r_1}{\sigma^2} I_0\left(\frac{2Ar_1}{\sigma^2}\right) \exp\left(-\frac{r_1^2 + A^2}{\sigma^2}\right) \tag{3.68}$$

$$p(r_2) = \frac{2r_2}{\sigma^2} \exp\left(-\frac{r_2^2}{\sigma^2}\right) \tag{3.69}$$

Therefore, the BER of BPSK with differential detection (P_{DBPSK}) is given by

$$\begin{aligned}
P_{DBPSK}(\gamma) &= \Pr(r_1 < r_2) \\
&= \int_0^\infty p(r_1)\left[\int_{r_1}^\infty p(r_2)dr_2\right]dr_1 \\
&= \int_0^\infty \frac{2r_1}{\sigma^2} I_0\left(\frac{2Ar_1}{\sigma^2}\right) \exp\left(-\frac{r_1^2 + A^2}{\sigma^2}\right) \exp\left(-\frac{r_1^2}{\sigma^2}\right) dr_1 \\
&= \frac{1}{2}\exp(-\gamma)\int_0^\infty x I_0\left(\sqrt{2\gamma}x\right) \exp\left(-\frac{x^2 + 2\gamma}{2}\right) dx \\
&= \frac{1}{2}\exp(-\gamma)
\end{aligned} \tag{3.70}$$

where

$$\gamma = \frac{A^2}{2\sigma^2} \tag{3.71}$$

$$x = \frac{2r_1}{\sigma} \tag{3.72}$$

When a root Nyquist filter is employed for the transmitter and receiver filters, γ is identical to E_b/N_0.

3.2.6 BER Performance in Rayleigh Fading Channel

When a modulated signal is transmitted over a Rayleigh fading channel, the received signal after the receiver BPF is given by

$$s_R(t) = \text{Re}\big[c(t)z(t)\exp\big(j2\pi f_c t\big) + n(t)\exp\big(j2\pi f_c t\big)\big]$$

$$= \text{Re}\big[r(t)z(t)\exp\big(j2\pi f_c t + j\theta(t)\big) \tag{3.73}$$

$$+ n'(t)\exp\big(j2\pi f_c t + j\theta(t)\big)\big]$$

$$n'(t) = n(t)\exp\big(-j\theta(t)\big) \tag{3.74}$$

where

$c(t)$ = fading variation subject to complex Gaussian random process

$r(t)$ = envelope component of $c(t)$

$\theta(t)$ = phase component of $c(t)$

$z(t)$ = transmitted baseband signal band-limited by the transmitter and receiver filters expressed as

$$z(t) = u_I(t) \otimes c_T(t) \otimes c_R(t) \tag{3.75}$$

When we can perfectly regenerate the carrier component expressed as

$$s_0(t) = \text{Re}\big[\exp\big(j2\pi f_c t + j\theta(t)\big)\big] \tag{3.76}$$

with no phase ambiguity, we can obtain the received baseband signal expressed as

$$v(t) = r(t)z(t) + \text{Re}\big[n'(t)\big] \tag{3.77}$$

Under Rayleigh fading channel conditions, the envelope of the received signal is randomly changed with its p.d.f. of

$$p(r) = \frac{2r}{r_0^2}\exp\left(-\frac{r^2}{r_0^2}\right) \tag{3.78}$$

where r_0 is the average value of r.

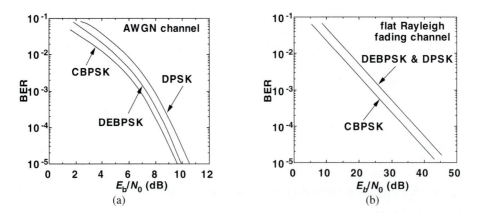

Fig. 3.6 BER performances of CBPSK, DEBPSK, and DBPSK (a) under AWGN conditions and (b) under Rayleigh fading conditions.

At the sampling timing, amplitude of $z(t)$ is A, and variance of $\text{Re}[n'(t)]$ is σ^2, which gives the average SNR of $\gamma_0 = A^2/2\sigma^2$. When $p(r)$ is converted to the p.d.f. of the received SNR, it is given by

$$p(\gamma) = \frac{1}{\gamma_0} \exp\left(-\frac{\gamma}{\gamma_0}\right) \tag{3.79}$$

Therefore, the average BER for CBPSK transmitted over a Rayleigh fading channel is given by

$$
\begin{aligned}
P_{CBPSK}^{Ray}(\gamma_0) &= \int_0^\infty P_{CBPSK}(\gamma)p(\gamma)dr \\
&= \int_0^\infty \frac{1}{2}\text{erfc}\left(\sqrt{\gamma}\right)\frac{1}{\gamma_0}\exp\left(-\frac{\gamma}{\gamma_0}\right)d\gamma \\
&= \frac{1}{2}\left[1 - \frac{1}{\sqrt{1+1/\gamma_0}}\right]
\end{aligned}
\tag{3.80}
$$

In the same manner, BER performance of differentially encoded BPSK with coherent detection (P_{DBPSK}^{Ray}) and that with differential detection (P_{DBPSK}^{Ray}) are given by

$$
\begin{aligned}
P_{DEBPSK}^{Ray}(\gamma) &= \int_0^\infty \text{erfc}\left(\sqrt{\gamma}\right)\frac{1}{\gamma_0}\exp\left(-\frac{\gamma}{\gamma_0}\right)d\gamma \\
&= \left[1 - \frac{1}{\sqrt{1+1/\gamma_0}}\right]
\end{aligned}
\tag{3.81}
$$

$$
\begin{aligned}
P_{DBPSK}^{Ray}(\gamma) &= \int_0^\infty \frac{1}{2}\exp(-\gamma)\frac{1}{\gamma_0}\exp\left(\frac{\gamma}{\gamma_0}\right)d\gamma \\
&= \frac{1}{2}\left[\frac{1}{1+\gamma_0}\right]
\end{aligned}
\tag{3.82}
$$

When a root Nyquist filter is employed as the transmitter and receiver filters, γ_0 is equivalent to the average E_b/N_0. Figure 3.6 shows theoretical BER performances of CBPSK, DEBPSK, and DPSK (a) under AWGN conditions and (b) under Rayleigh fading conditions.

3.3 GMSK

3.3.1 Modulation Process

GMSK is minimum shift keying (MSK) (frequency modulation with its modulation index of 0.5) with a premodulation Gaussian filter [3-1]. Figure 3.7 shows a configuration of the GMSK modulator. When the transmitted data sequence is $\{a_n\}$, a waveform of the transmitted bit sequence can be expressed as

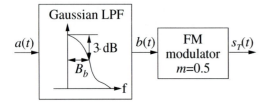

Fig. 3.7 Configuration of the GMSK modulator.

$$a(t) = \sum_{k=-\infty}^{\infty} b_k u(t - kT_b) \tag{3.83}$$

$$b_k = \begin{cases} 1; & a_k = 1 \\ -1; & a_k = 0 \end{cases} \tag{3.84}$$

$$u(t) = \begin{cases} 1; & 0 \le t < T_b \\ 0; & otherwise \end{cases} \tag{3.85}$$

When $a(t)$ is filtered by a Gaussian LPF with its 3-dB bandwidth of B_b, its output is given by

$$b(t) = a(t) \otimes c_b(t) \tag{3.86}$$

$$c_b(t) = \sqrt{\frac{2\pi}{\ln 2}} B_b \exp\left(-\frac{2\pi^2}{\ln 2} B_b^2 t^2 \right) \tag{3.87}$$

$b(t)$ is then fed to the FM modulator with its modulation index of 0.5 to obtain the transmitted signal expressed as

$$s_T(t) = A \cdot \cos\left(2\pi f_c t + \phi(t) + \theta_0\right) \tag{3.88}$$

$$\phi(t) = \frac{\pi}{2T_b} \int_{-\infty}^{t} b(\tau)d\tau \tag{3.89}$$

Figure 3.8 shows a signal state diagram of MSK. When $a_n = 1$, signal phase is shifted with an angle frequency of $0.5\pi/T_b$, and it is rotated with an angle frequency of $-0.5\pi/T_b$ when $a_n = 0$.

As for the demodulation scheme, we can apply three types of demodulation schemes for GMSK: coherent detection, differential detection, and frequency discriminator detection. These will be explained in the following sections.

3.3.2 GMSK with Coherent Detection

Figure 3.9 shows a receiver configuration of GMSK with coherent detection [3-1]. The received signal is first filtered by a BPF to pick up the desired signal as well as to

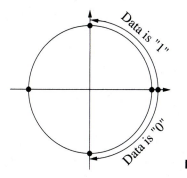

Fig. 3.8 Signal state diagram of MSK.

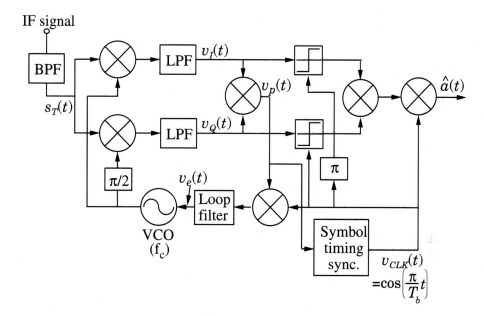

Fig. 3.9 Receiver configuration of GMSK with coherent detection.

suppress adjacent channel interference (ACI) and noise. When the output signal of the VCO is expressed as

$$s_{VCO}(t) = \cos(2\pi f_c t + \theta_1) \tag{3.90}$$

the LPF outputs of I- and Q-arms are given by

$$v_I(t) = \frac{1}{2} A \cos(\phi(t) + \varDelta\theta) \tag{3.91}$$

$$v_Q(t) = \frac{1}{2} A \sin(\phi(t) + \varDelta\theta) \tag{3.92}$$

where

$$\Delta\theta = \theta_0 - \theta_1 \tag{3.93}$$

When we multiply $v_I(t)$ and $v_Q(t)$, we can obtain

$$v_p(t) = \frac{1}{8}A^2 \sin(2\phi(t) + 2\Delta\theta) \tag{3.94}$$

Using this signal, the symbol timing clock

$$v_{CLK}(t) = \cos\left(\frac{\pi}{T_s}t\right) \tag{3.95}$$

is first regenerated by the symbol timing synchronization circuits (symbol timing sync).

Suppose that the bandwidth of the premodulation Gaussian filter is infinite (that corresponds to MSK), and the initial phase is 0 [rad] for analysis convenience. In this case, $\phi(t)$ can be expressed as

$$\phi(t) = \begin{cases} a_n\pi\dfrac{t-nT_b}{2T_b} + \begin{pmatrix} 0 \\ \pi \end{pmatrix} + \Delta\theta \; ; & n: \text{ even} \\[4mm] a_n\pi\dfrac{t-nT_b}{2T_b} + \begin{pmatrix} \pi/2 \\ -\pi/2 \end{pmatrix} + \Delta\theta \; ; & n: \text{ odd} \end{cases} \tag{3.96}$$

When we substitute equation (3.96) into equation (3.94), multiply $v_p(t)$ and $v_{CLK}(t)$, and remove the high-frequency component by a loop filter, we can obtain

$$\begin{aligned} v_e(t) &= \left[\frac{A^2}{16}\left\{\sin\left(\frac{(a_n+1)\pi}{T_b}t + 2\Delta\theta\right) + \sin\left(\frac{(a_n-1)\pi}{T_b}t + 2\Delta\theta\right)\right\}\right]_{LPF} \\ &= \frac{A^2}{16}\sin(2\Delta\theta) \end{aligned} \tag{3.97}$$

When $v_e(t)$ is fed to the VCO, the VCO signal phase is controlled to keep the average $v_e(t)$ at 0. As a result, $v_I(t)$ and $v_Q(t)$ are obtained as

$$v_I(t) = \cos(\phi(t)) \tag{3.98}$$

$$v_Q(t) = \sin(\phi(t)) \tag{3.99}$$

When n is an odd number, if the transmitted data is logical 1, $\phi(t)$ stays in the first or third quadrants of the phase plane, and if the transmitted data is logical 0, it stays in the second or fourth quadrants. On the other hand, when n is an even number and the transmitted data is logical 1, $\phi(t)$ stays in the second or fourth quadrants, and it stays in the first or third quadrants when the transmitted data is logical 0. As a result, the transmitted data can be estimated as

$$\hat{a}(t) = \text{sgn}(v_I(t)) \cdot \text{sgn}(v_Q(t)) \cdot v_{CLK}(t) \tag{3.100}$$

In the case of MSK, the BER performance is the same as that of DEBPSK [3-1]. On the other hand, when the baseband signal before an FM modulator is band-limited by a Gaussian LPF, the BER performance is approximated by

$$P_{GMSK}(\gamma) \cong \text{erfc}\left(\sqrt{\beta\gamma}\right) \tag{3.101}$$

where β is a factor determined by the degradation due to the premodulation filter.

Table 3.1 summarizes the bandwidth of the premodulation filter $(B_b T_b)$ and β. This table also shows the relationship between $B_b T_b$ and the 99.99% bandwidth of the GMSK-modulated signal. As we can see from this table, a narrower premodulation filter can reduce the bandwidth of the modulated signal by sacrificing the BER performance. Figure 3.10 shows the BER performances of GMSK $(B_b T_b = 0.25)$ with coherent detection under AWGN and Rayleigh fading conditions.

3.3.3 GMSK with Differential Detection

One of the serious problems for GMSK with coherent detection, as with any other modulation scheme, is the irreducible error due to fast phase variation by Rayleigh fading. Differential detection is one of the demodulation schemes to mitigate the

Table 3.1 Relationship between $B_b T_b$, 99.99% bandwidth of the transmitted signal, and β.

$B_b T$	99.99% Bandwidth	β
0.20	1.22	0.76
0.25	1.37	0.84
0.30	1.41	0.89
0.40	1.80	0.94
0.50	2.08	0.97

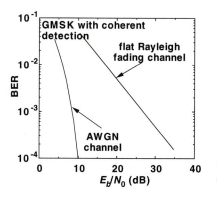

Fig. 3.10 BER performances of GMSK $(B_b T_b = 0.25)$ with coherent detection.

irreducible errors. There are several types of differential detection schemes for GMSK, such as 1-bit differential detection and 2-bit differential detection.

3.3.3.1 1-Bit Differential Detection   In the (G)MSK modulation process, the signal phase is shifted by $\pi/2$ [rad] when the source bit is logical 1, and it is shifted by $-\pi/2$ [rad] when the source bit is logical 0. Therefore, when the received signal is multiplied by the 1-bit delayed received signal, and the polarity of the obtained baseband signal

$$v(t) = \sin\big(\phi(t) - \phi(t - T_b)\big) \tag{3.102}$$

is detected, we can regenerate the transmitted data sequence with almost the same process as that for DPSK. However, when the bandwidth of the premodulation filter becomes narrower, the BER performance of 1-bit differential detection becomes much worse due to severe ISI. For example, the required E_b/N_0 for BER = 10^{-3} are 10.8 dB, 12.7 dB, and 22.3 dB for $B_bT_b = \infty$, 0.5, and 0.25, respectively, under AWGN conditions [3-24].

To solve this problem, 2-bit differential detection was developed [3-24]. In the 2-bit differential detection circuit, the received signal is multiplied by the 2-bit delayed signal. As a result, we can obtain the baseband signal given by

$$v(t) = \cos\big(\phi(t) - \phi(t - 2T_b)\big) \tag{3.103}$$

We can find from equation (3.89) that

$$\phi((n + 1)T_b) - \phi((n - 1)T_b) = \begin{cases} \pi; \ b_n = b_{n+1} = 1 \\ 0; \ b_n = 1 \ \text{and} \ b_{n+1} = -1 \\ \quad \text{or} \ b_n = -1 \ \text{and} \ b_{n+1} = 1 \\ -\pi; \ b_n = b_{n+1} = -1 \end{cases} \tag{3.104}$$

These results mean that

$$v(nT_b) = \cos\big(\phi((n + 1)T_b) - \phi((n + 1)T_b)\big) = \begin{cases} 1; \ b_n b_{n+1} = -1 \\ -1; \ b_n b_{n+1} = 1 \end{cases} \tag{3.105}$$

Therefore, if the source bit sequence $\{a_n\}$ is differentially encoded as

$$a'_n = a'_{n-1} \oplus a_n \tag{3.106}$$

and $\{b_n\}$ is assigned as

$$b_n = \begin{cases} 1; \ a'_n = 1 \\ -1; \ a'_n = 0 \end{cases} \tag{3.107}$$

we can regenerate the transmitted bit sequence as

$$\hat{a}_n = \begin{cases} 1; & v(nT_b) \geq 0 \\ 0; & v(nT_b) < 0 \end{cases} \tag{3.108}$$

The most important advantage of 2-bit differential detection is that the BER performance is much better than that of 1-bit differential detection. For example, the required E_b/N_0 for BER = 10^{-3} are 9.8 dB, 11.2 dB, and 14.8 dB for $B_bT_b = \infty$, 0.5, and 0.25, respectively, under AWGN conditions [3-25].

Figure 3.11 shows BER performance of GMSK ($B_bT_b = 0.25$) with 2-bit differential detection under AWGN and flat Rayleigh fading conditions.

3.3.4 GMSK with Frequency Discriminator Detection

As we discussed in this chapter, frequency discriminator detection is also applicable to GMSK. However, when a very narrow premodulation filter is employed, the BER performance for discriminator detection is very poor. To overcome this problem, discriminator detection with a decision feedback equalizer (DFE) was proposed [3-26]. Figure 3.12 shows the obtained BER performance under AWGN and flat Rayleigh fading conditions.

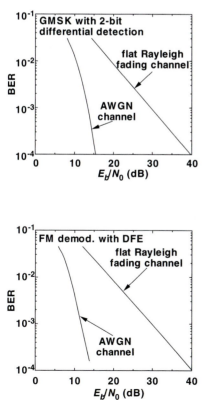

Fig. 3.11 BER performance of GMSK ($B_bT = 0.25$) with 2-bit differential detection.

Fig. 3.12 BER performance of GMSK ($B_bT_b = 0.25$) with frequency discriminator and 2-bit DFE.

3.4 QPSK

3.4.1 Absolute Phase Encoding and Differential Phase Encoding

QPSK is a modulation scheme that transmits 2-bit information using four states of phase. As in the case of BPSK, both the absolute phase encoding and differential encoding are applicable to QPSK.

In the case of absolute phase encoding, Gray encoding is one of the most preferable mapping schemes. The relationship between 2-bit information and the assigned phase is shown in Table 3.2, and the corresponding signal state diagram is shown in Figure 3.13(a).

One of the most distinct feature of the Gray encoding is that one symbol error results in only a 1-bit error because the Hamming distance between the adjacent phases is only 1 as shown in Figure 3.13(a).

On the other hand, when the carrier reference for coherent detection includes phase ambiguity, we have to employ differential encoding. Fig. 3.13 (b) shows a signal state diagram of the differentially encoded QPSK (DEQPSK). In this case, the phase for n-th symbol ϕ_n is given by

$$\phi_n = \phi_{n-1} + d\phi_n \tag{3.109}$$

Table 3.2 Relationship between 2-bit information and the assigned phase for Gray encoding.

2-Bit Information	ϕ
00	$-3\pi/4$
01	$3\pi/4$
10	$-\pi/4$
11	$\pi/4$

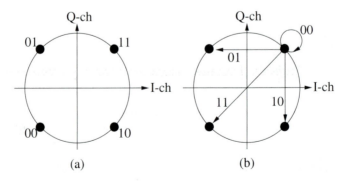

(a)   (b)

Fig. 3.13 Signal state diagram of QPSK: (a) absolute phase encoding and (b) differential encoding.

where the amount of phase shift ($d\phi$) corresponding to 2-bit information is shown in Table 3.3.

Table 3.3 Relationship between 2-bit information and phase transition for differential encoding.

2-Bit Information	$d\phi$
00	0
01	$+\pi/2$
10	$-\pi/2$
11	π

3.4.2 QPSK Modulation Process

Figure 3.14 shows a configuration of the QPSK modulator. At the serial to parallel (S/P) converter, the serial data sequence is converted to a 2-bit parallel data sequence. At this stage, a symbol duration of the S/P converter output (T_s) is twice as long as a bit duration (T_b). At the baseband signal generator, a baseband signal is generated, where Gray encoding or differential encoding is employed. As a result, the baseband signal expressed as

$$b(t) = \sum_{i=-\infty}^{\infty} b_i \delta(t - iT_s) \tag{3.110}$$

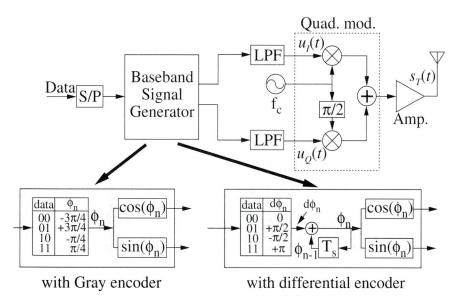

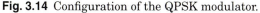

Fig. 3.14 Configuration of the QPSK modulator.

$$b_n = b_{In} + j \cdot b_{Qn}$$
$$= \cos(\phi_n) + j \cdot \sin(\phi_n) \tag{3.111}$$

is generated at the baseband signal generator (BSG), where $\delta(t)$ is Dirac's delta function, and b_n is the transmitted symbol at $t = nT_s$ that takes a value of $\exp(j\pi/4)$, $\exp(j3\pi/4)$, $\exp(-j\pi/4)$, or $\exp(j3\pi/4)$. This signal is then filtered by the root Nyquist LPF with its impulse response of

$$c_T(t) = \frac{1}{\pi t} \frac{1}{1 - \left(\dfrac{4\alpha t}{T_s}\right)^2} \sin\left\{2\pi(1-\alpha)\frac{t}{T_s}\right\}$$

$$+ \frac{1}{\pi} \frac{\dfrac{4\alpha}{T_s}}{1 - \left(\dfrac{4\alpha t}{T_s}\right)^2} \cos\left\{2\pi(1+\alpha)\frac{t}{T_s}\right\} \tag{3.112}$$

The band-limited baseband signal can be expressed as

$$u(t) = u_I(t) + j \cdot u_Q(t)$$
$$= b(t) \otimes c_T(t) \tag{3.113}$$

Figure 3.15 shows waveforms of the input data sequence, output of the S/P converter, output of the baseband signal generator, and output of the LPF for (a) Gray-encoded QPSK and (b) differentially encoded QPSK, where the BSG includes a sampler to convert an NRZ data sequence to an impulse data sequence. For differential encoding, the initial phase is assumed to be $\pi/4$.

At the quadrature modulator, a carrier with its frequency of f_c is modulated by $u(t)$ and transmitted, where the transmitted signal can be expressed as

$$s_T(t) = \mathrm{Re}[u(t)\exp(2\pi f_c t)]$$
$$= u_I(t)\cos(2\pi f_c t) - u_Q(t)\sin(2\pi f_c t) \tag{3.114}$$

3.4.3 Absolute Phase-Encoded QPSK with Coherent Detection

When we can regenerate the carrier component with no phase ambiguity—for example, a pilot signal is embedded in the transmitted signal—we can apply coherent detection with absolute phase decoding. Figure 3.16 shows its configuration. The received signal after the BPF is given by

$$s_R(t) = \mathrm{Re}[c(t)z(t)\exp(j2\pi f_c t) + n(t)\exp(j2\pi f_c t)]$$
$$= \mathrm{Re}[r(t)z(t)\exp(j2\pi f_c t + j\theta(t))$$
$$+ n'(t)\exp(j2\pi f_c t + j\theta(t))] \tag{3.115}$$

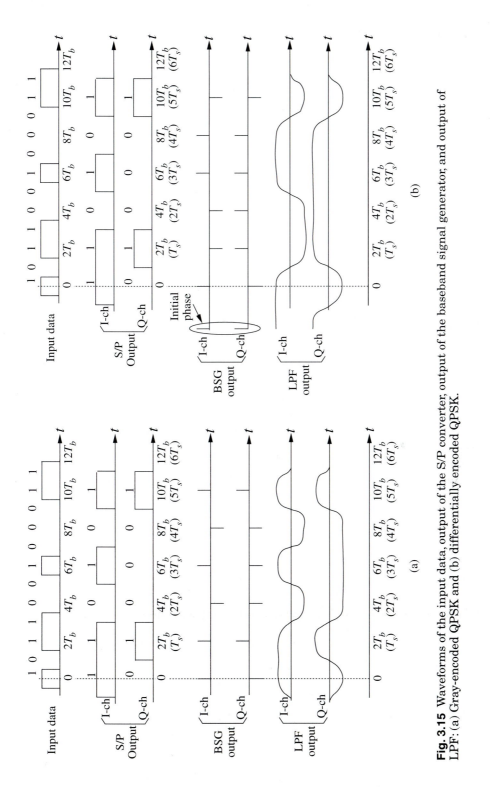

Fig. 3.15 Waveforms of the input data, output of the S/P converter, output of the baseband signal generator, and output of LPF: (a) Gray-encoded QPSK and (b) differentially encoded QPSK.

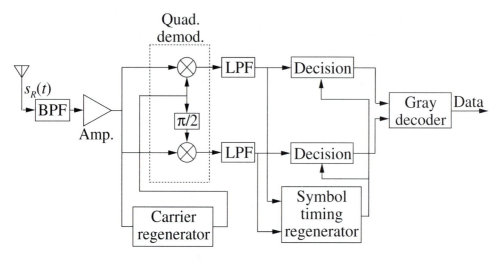

Fig. 3.16 Configuration of the QPSK receiver with absolute phase coherent detection.

$$n'(t) = n(t)\exp(-j\theta(t)) \tag{3.116}$$

When we can perfectly regenerate the carrier component expressed as

$$s_o(t) = \exp(j2\pi f_c t + j\theta(t)) \tag{3.117}$$

at the receiver, we can obtain the received baseband signal expressed as

$$\begin{aligned} v(t) &= v_I(t) + j \cdot v_Q(t) \\ &= r(t)z(t) + n'(t) \end{aligned} \tag{3.118}$$

by the quadrature demodulation using $s_o(t)$.

After the symbol timing is regenerated by using $v(t)$, the transmitted symbol sequence $\{\hat{b}_n\}$ is obtained by threshold detectors as

$$\mathrm{Re}[\hat{b}_n] = \hat{b}_{In} = \mathrm{sgn}(v_I(nT_s)) \tag{3.119}$$

$$\mathrm{Im}[\hat{b}_n] = \hat{b}_{Qn} = \mathrm{sgn}(v_Q(nT_s)) \tag{3.120}$$

where

$$\mathrm{sgn}(x) = \begin{cases} 1; & x \geq 0 \\ -1; & x < 0 \end{cases} \tag{3.121}$$

Using the detected data $(\hat{b}_{In}, \hat{b}_{Qn})$, the transmitted phase information $\hat{\phi}_n$ is estimated and the transmitted data $(\hat{a}_{2n-1}, \hat{a}_{2n})$ is regenerated.

When $r(t)$ is constant, say $r(t) = 1$, it is equivalent to the AWGN conditions. In this case, $v_I(t)$ and $v_Q(t)$ are Gaussian random variables with the joint probability density function of

$$p(v_I, \ v_Q) = \frac{1}{2\pi\sigma^2} \exp\left(\frac{\left(v_I - A/\sqrt{2}\right)^2 + \left(v_Q - A/\sqrt{2}\right)^2}{2\sigma^2} \right) \tag{3.122}$$

The probability that the received symbol is correctly detected is therefore given by

$$P_c = \Pr\left(v_I(t) > 0, \ v_Q(t) > 0 \middle| \phi(t) = \frac{\pi}{4} \right)$$
$$= \left[1 - \frac{1}{2}\operatorname{erfc}\left(\frac{A}{2\sigma}\right) \right]^2 = \left[1 - \frac{1}{2}\operatorname{erfc}\left(\sqrt{\frac{\gamma_s}{2}}\right) \right]^2 \tag{3.123}$$

where γ_s is energy per symbol to the noise spectral density. In the case of QPSK, a symbol includes 2 bits. Therefore, when energy per bit to the noise spectral density is expressed as γ, $\gamma_s = 2\gamma$ is satisfied. Furthermore, one symbol error results in only a 1-bit error when Gray encoding is applied. Consequently, BER for Gray-encoded QPSK with coherent detection (CQPSK) is given by

$$P_{CQPSK}(\gamma) = \frac{1}{2}\left[1 - P_c\right]$$
$$= \frac{1}{2}\operatorname{erfc}\left(\sqrt{\gamma}\right) - \frac{1}{8}\operatorname{erfc}^2\left(\sqrt{\gamma}\right) \tag{3.124}$$

When the propagation path is subject to Rayleigh fading, the BER is given by

$$P_{CQPSK}^{Ray}(\gamma_0) = \int_0^\infty P_{CQPSK}(\gamma) \frac{1}{\gamma_0} \exp\left(\frac{\gamma}{\gamma_0}\right) d\gamma$$
$$\cong \frac{1}{2}\left[1 - \frac{1}{\sqrt{1 + 1/\gamma_0}} \right] \tag{3.125}$$

3.4.4 Coherent Detection with Differential Decoding

When we cannot remove phase ambiguity, we have to apply differential *encoding* at the modulator and differential *decoding* at the demodulator. The receiver configuration for coherent detection with differential decoding is the same as that with the Gray-encoding system except that the differential decoder is employed instead of the Gray decoder. Figure 3.17 shows a configuration of the differential decoder. Let's assume that the detected symbol at $t = (n-1)T_s$ and $t = nT_s$ can be expressed as

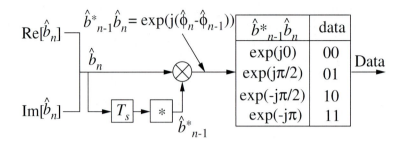

Fig. 3.17 Configuration of the differential decoder.

$$\hat{b}_{n-1} = \exp(j\hat{\phi}_{n-1}) \tag{3.126}$$

$$\hat{b}_n = \exp(j\hat{\phi}_n) \tag{3.127}$$

Differential decoding is an inverse operation of differential encoding. At the differential decoder, the amount of phase transition between $t = (n - 1)T_s$ and nT_s is obtained by multiplying the complex conjugate of $\hat{b}_{n-1}$ times $\hat{b}_n$ as

$$\hat{b}^*_{n-1}\hat{b}_n = \exp(j(\hat{\phi}_n - \hat{\phi}_{n-1})) \tag{3.128}$$

The transmitted 2-bit data are decoded using the table shown in Figure 3.17.

When differential encoding is applied instead of Gray encoding, the BER is twice as large as that for Gray-encoded QPSK. Therefore, the BER of differentially encoded QPSK with coherent detection (DEQPSK) under the AWGN conditions is given by

$$P_{DEQPSK}(\gamma) = 2P_{CQPSK}(\gamma)$$
$$= \mathrm{erfc}\left(\sqrt{\gamma}\right) - \frac{1}{4}\mathrm{erfc}^2\left(\sqrt{\gamma}\right) \tag{3.129}$$

and the BER of differentially encoded QPSK with coherent detection under Rayleigh fading conditions is given by

$$P_{DEQPSK}^{Ray}(\gamma_0) = \int_0^\infty P_{DEQPSK}(\gamma)\frac{1}{\gamma_0}\exp\left(\frac{\gamma}{\gamma_0}\right)d\gamma$$
$$\cong \left[1 - \frac{1}{\sqrt{1 + 1/\gamma_0}}\right] \tag{3.130}$$

3.4.5 QPSK with Differential Detection

Figure 3.18 shows a receiver configuration of QPSK with differential detection (DQPSK). Let the received signal be given by

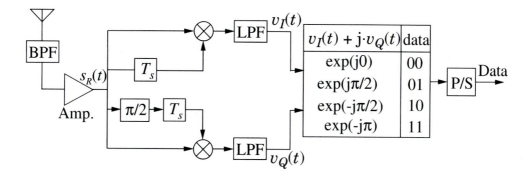

Fig. 3.18 Configuration of the QPSK receiver with a differential detector.

$$s_R(t) = \big(z_I(t) + n_I(t)\big)\cos(2\pi f_c t)$$
$$- \big(z_Q(t) + n_Q(t)\big)\sin(2\pi f_c t) \qquad (3.131)$$

At the upper arm of the differential detector, one symbol period-delayed received signal, expressed as

$$s_R(t - T_s) = \big(z_I(t - T_s) + n_I(t - T_s)\big)\cos(2\pi f_c t)$$
$$- \big(z_Q(t - T_s) + n_Q(t - T_s)\big)\sin(2\pi f_c t) \qquad (3.132)$$

is multiplied to the received signal, and it is filtered by an LPF to obtain the following signal, where $f_c T_s = 1$ is assumed.

$$v_I(t) = \big(z_I(t) + n_I(t)\big)\big(z_I(t - T_s) + n_I(t - T_s)\big)$$
$$+ \big(z_Q(t) + n_Q(t)\big)\big(z_Q(t - T_s) + n_Q(t - T_s)\big)$$
$$= \big(z_I(t)z_I(t - T_s) + z_Q(t)z_Q(t - T_s)\big) \qquad (3.133)$$
$$+ \big(z_I(t)n_I(t - T_s) + z_I(t - T_s)n_I(t) + n_I(t)n_I(t - T_s)\big)$$
$$+ \big(z_Q(t)n_Q(t - T_s) + z_Q(t - T_s)n_Q(t) + n_Q(t)n_Q(t - T_s)\big)$$

In this equation, the first term corresponds to $\exp(\phi(t) - \phi(t - T_s))$ whereas the second and third terms represent noise and distortion. Because $n_I(t)n_I(t - T_s)$ and $n_Q(t)n_Q(t - T_s)$ are small relative to the other distortion terms, we can disregard them. Therefore, $v_I(t)$ can be approximated by

$$v_I(t) = \cos\big(\phi(t) - \phi(t - T_s)\big)$$
$$+ \big(z_I(t)n_I(t - T_s) + z_I(t - T_s)n_I(t)\big) \qquad (3.134)$$
$$+ \big(z_Q(t)n_Q(t - T_s) + z_Q(t - T_s)n_Q(t)\big)$$

In the same manner, one symbol period-delayed and $-\pi/2$-shifted received signal expressed as

$$\text{Re}\left[s_R(t-T_s)\exp\left(-j\frac{\pi}{2}\right)\right]$$
$$=\left(z_I(t-T_s)+n_c(t-T_s)\right)\sin(2\pi f_c t) \tag{3.135}$$
$$+\left(z_Q(t-T_s)+n_Q(t-T_s)\right)\cos(2\pi f_c t)$$

is multiplied to the received signal at the lower arm of the differential detector, and it is low-pass filtered to obtain the following signal:

$$v_Q(t)=\left(z_I(t)+n_I(t)\right)\left(z_Q(t-T_s)+n_Q(t-T_s)\right)$$
$$-\left(z_Q(t)+n_Q(t)\right)\left(z_I(t-T_s)+n_I(t-T_s)\right)$$
$$=\sin\left(\phi(t)-\phi(t-T_s)\right) \tag{3.136}$$
$$+\left(z_I(t)n_Q(t-T_s)+z_Q(t-T_s)n_I(t)\right)$$
$$-\left(z_Q(t)n_I(t-T_s)+z_I(t-T_s)n_Q(t)\right)$$

What we have to be aware of is that the noise components in $v_I(t)$ and in $v_Q(t)$ are not independent from each other, which makes derivation of the BER very complicated. Therefore, only the result is given here [3-27]. That is

$$P_{DQPSK}(\gamma)=Q(a,b)-\frac{1}{2}I_0(ab)\exp\left\{-\frac{1}{2}(a^2+b^2)\right\} \tag{3.137}$$

$$a=\sqrt{2\gamma_b(1-\frac{1}{\sqrt{2}})} \tag{3.138}$$

$$b=\sqrt{2\gamma_b(1+\frac{1}{\sqrt{2}})} \tag{3.139}$$

where $Q(a,b)$ is Marcum's Q-function defined as

$$Q(a,b)=\int_b^\infty tI_0(at)\exp\left\{-\frac{1}{2}(a^2+t^2)\right\}dt \tag{3.140}$$

When γ_b is high, the BER under AWGN conditions is approximated by

$$P_{DQPSK}(\gamma)=\frac{1}{2}\text{erfc}\left(2\sin\frac{\pi}{8}\sqrt{\gamma}\right) \tag{3.141}$$

Under Rayleigh fading conditions, the BER is given by

$$P_{DQPSK}^{Ray}(\gamma_0) = \frac{1}{2}\left[1 - \frac{2\gamma_0}{\sqrt{2(1+2\gamma_0)^2 - 4\gamma_0^2}}\right] \tag{3.142}$$

Figure 3.19 shows BER performances of CQPSK, DEQPSK, and DQPSK (a) under AWGN conditions and (b) under flat Rayleigh fading conditions. As we can see from this figure, the BER performance of CQPSK is much better than that of the DEQPSK and DQPSK under both conditions. Furthermore, when we compare the performances of DEQPSK and DQPSK, although the BER performance of DEQPSK is better than that of DQPSK under AWGN conditions, it is the same as that of DQPSK under flat Rayleigh fading conditions. Therefore, DQPSK has been popularly employed in land mobile communication systems. In other words, when we want to employ CQPSK rather than DEQPSK, we have to employ some pilot signal-aided scheme to bring out advantages of the coherent detection scheme.

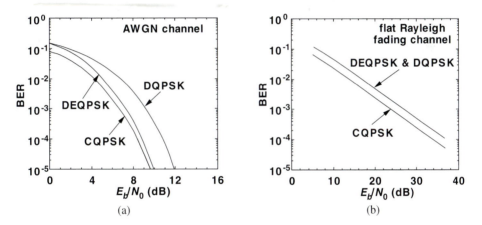

Fig. 3.19 BER performances of CQPSK, DEQPSK, and DQPSK (a) under AWGN conditions and (b) under Rayleigh fading conditions.

3.5 π/4-QPSK

3.5.1 Absolute Phase-Encoded π/4-QPSK

π/4-shift QP-SK is categorized as a modified QPSK. As for the encoding, both absolute phase encoding and differential phase encoding are available for π/4-QPSK.

Figure 3.20 shows a Gray encoding scheme for π/4-QPSK. Its difference from that for QPSK is that in-phase and quadrature-phase (I-Q) axes of 0 and π/2 and those of −π/4 and π/4 are alternately changed every T_s second [3-28]. Table 3.4 shows the relationship between the source bit (a_{2n-1}, a_{2n}) and the output phase, and Figure 3.21 shows a signal state diagram for absolute phase-encoded π/4-QPSK. When $t = 2mT_s$ (m = 0, 1, 2, ...), four phases represented by a black circle in Figure 3.21 (a) are used to

Table 3.4 Relationship between the source bit and output phase for the absolute phase-encoded π/4-QPSK.

(a_{2n-1}, a_{2n})	$t = 2mT_s$	$t = (2m + 1)T_s$
00	$-3\pi/4$	π
01	$3\pi/4$	$\pi/2$
10	$-\pi/4$	$-\pi/2$
11	$\pi/4$	0

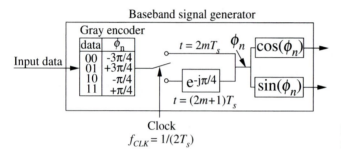

Fig. 3.20 Gray encoding scheme for π/4-QPSK.

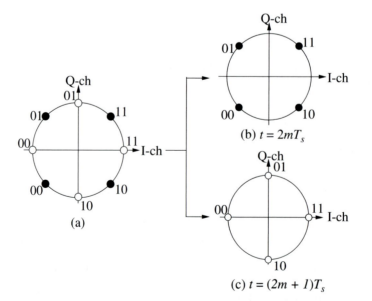

(a)

(b) $t = 2mT_s$

(c) $t = (2m + 1)T_s$

Fig. 3.21 Signal state diagram for absolute phase-encoded π/4-QPSK.

transmit 2-bit information. Therefore, in this case, the signal state diagram for π/4-QPSK is the same as that for QPSK as shown in Figure 3.21 (b). On the other hand, when t = $(2m + 1)T_s$, I-Q axes are rotated by −π/4 [rad]. Therefore, four phases represented by a white circle in Figure 3.21 (a) are used. As a result, its signal state diagram is given as shown in Figure 3.21 (c).

Because the envelope of the π/4-QPSK signal never goes through 0 amplitude, π/4-QPSK mitigates spectral spreading caused by the nonlinearity of the transmitter amplifier, which results in relatively high transmitter power efficiency. The configuration of the π/4-QPSK transmitter is almost the same as that for QPSK shown in Figure 3.14 except for the encoder.

Figure 3.22 shows a receiver configuration of the absolute phase-encoded π/4-QPSK. After the carrier regeneration, the received signal is coherently demodulated at the complex demodulator. In this receiver, we have to regenerate symbol timing and frame timing to take I-Q axes rotation synchronization. Using the regenerated I-Q axes rotation synchronization, the received baseband signal phase is shifted by 0 for $t = 2mT_s$ and π/4 for $t = (2m + 1)T_s$ at the phase shifter. Because we can convert the I-Q axes of −π/4 and π/4 to the axes of 0 and π/2 with this process, we can apply the same detection rule as we use for QPSK to regenerate the transmitted data sequence.

Because the difference between π/4-QPSK and QPSK is whether I-Q axes are alternately rotated or not, and the effect of this I-Q axes rotation can be perfectly canceled out provided that the frame synchronization is perfectly taken at the receiver, the BER performance of absolute phase-encoded π/4-QPSK is the same as that for the absolute phase-encoded QPSK.

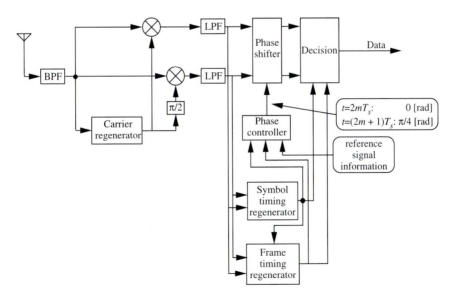

Fig. 3.22 Receiver configuration of the absolute phase-encoded π/4-QPSK with coherent detection.

3.5.2 Differential Phase-Encoded π/4-QPSK

The signal state diagram of differentially encoded π/4-QPSK is shown in Figure 3.23. At the modulation stage, phase for n-th symbol is calculated as

$$\phi_n = \phi_{n-1} + d\phi_n \tag{3.143}$$

where $d\phi_n$ is given by Table 3.5. Using this phase, the baseband signal expressed as

$$b(t) = b_I(t) + jb_Q(t)$$
$$= \sum_{i=-\infty}^{\infty} \cos(\phi_i)\delta(t - iT_s) + j \sum_{i=-\infty}^{\infty} \sin(\phi_i)\delta(t - iT_s) \tag{3.144}$$

After this signal is filtered by the LPF (root Nyquist filter), the carrier is modulated by this signal and transmitted.

Table 3.5 Relationship between 2-bit information and phase transition for π/4-QPSK.

(a_{2n-1}, a_{2n})	$d\phi_n$
00	$+\pi/4$
01	$+3\pi/4$
10	$-\pi/4$
11	$-3\pi/4$

Figure 3.24 shows the receiver configuration of differential phase-encoded π/4-QPSK [3.29]. At first, the phase of the received signal is rotated by $-\pi/4$ [rad] at every symbol timing. With this process, the relationship between 2-bit information and phase transition after $-\pi/4$-rotation is given as shown in Table 3.6. Therefore, we can apply the same differential detector as that for DEQPSK after the $-\pi/4$ rotation stage. In this case, the BER is the same as that for DEQPSK because $-\pi/4$ rotation does not cause any BER performance degradation at all.

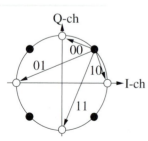

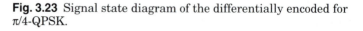

Fig. 3.23 Signal state diagram of the differentially encoded for π/4-QPSK.

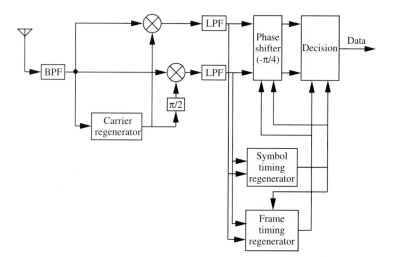

Fig. 3.24 Receiver configuration of differentially encoded $\pi/4$-QPSK with coherent detection.

Table 3.6 Relationship between 2-bit information and phase transition after $-\pi/4$ rotation.

(a_{2n-1}, a_{2n})	$d\phi_n$
00	0
01	$+\pi/2$
10	$-\pi/2$
11	$-\pi$

3.5.3 Differential Detection

When the source data sequence is differentially encoded, we can also employ differential detection. For the same reason as that of the differentially encoded $\pi/4$-QPSK with coherent detection, the BER of the $\pi/4$-QPSK with differential detection ($\pi/4$-DQPSK) is the same as that of DQPSK under both AWGN and Rayleigh fading conditions [3-30; 3-31].

3.6 *M*-ARY PSK

3.6.1 Modulator for *M*-ary PSK

In the case of *M*-ary PSK, a symbol can transmit $\log_2 M$-bit of information, and each symbol takes one of the phases given by

$$\phi = \frac{2\pi}{M}(m-1), \quad m = 1,\ 2,\ \ldots,\ M \tag{3.145}$$

Therefore, the baseband signal is given by

$$
\begin{aligned}
b(t) &= b_I(t) + jb_Q(t) \\
&= \sum_{i=-\infty}^{\infty} \cos(\phi_i)\delta(t - iT_s) + j\sum_{i=-\infty}^{\infty} \sin(\phi_i)\delta(t - iT_s)
\end{aligned}
\tag{3.146}
$$

This signal is then band-limited by a root Nyquist filter with its impulse response of $c_T(t)$. Using this signal, the carrier is modulated and transmitted.

Both the absolute phase encoding and differential encoding are available for M-ary PSK. Table 3.7 shows the relationship between the input data and phase position for Gray-encoded 8PSK. Table 3.8 shows the same relationship for 16PSK.

When differential encoding is applied to cope with phase ambiguity of the regenerated carrier components, the phase transition between any two consecutive symbols ($d\phi_n$) is generated, and its accumulated value given by

$$\phi_n = \phi_{n-1} + d\phi_n \tag{3.147}$$

is generated at the phase generator, where the relationship between the source bits and $d\phi_n$ is given by Table 3.9 for 8PSK and Table 3.10 for 16PSK. After this signal is generated, the same process as that for Gray-coded M-ary PSK is carried out.

3.6.2 Coherent Detection for *M*-ary PSK (MCPSK)

The symbol error rate (SER) for Gray-encoded M-ary PSK with coherent detection is given by

Table 3.7 Relationship between 3 bits of source data and the output phase for Gray-coded 8PSK.

$(a_{3n-2},\ a_{3n-1},\ a_{3n})$	ϕ_n
000	$\pi/8$
001	$3\pi/8$
011	$5\pi/8$
010	$7\pi/8$
110	$-7\pi/8$
111	$-5\pi/8$
101	$-3\pi/8$
100	$-\pi/8$

Table 3.8 Relationship between 4 bits of source data and the output phase for Gray-coded 16PSK.

$(a_{4n-3}, a_{4n-2}, a_{4n-1}, a_{4n})$	ϕ_n
0000	$\pi/16$
0001	$3\pi/16$
0011	$5\pi/16$
0010	$7\pi/16$
0110	$9\pi/16$
0111	$11\pi/16$
0101	$13\pi/16$
0100	$15\pi/16$
1100	$-15\pi/16$
1101	$-13\pi/16$
1111	$-11\pi/16$
1110	$-9\pi/16$
1010	$-7\pi/16$
1011	$-5\pi/16$
1001	$-3\pi/16$
1000	$-\pi/16$

Table 3.9 Relationship between 3 bits of source data and $d\phi_n$ for differentially encoded 8PSK.

$(a_{3n-2}, a_{3n-1}, a_{3n})$	$d\phi_n$
000	0
001	$\pi/4$
011	$\pi/2$
010	$3\pi/4$
110	π
111	$-3\pi/4$
101	$-\pi/2$
100	$-\pi/4$

Table 3.10 Relationship between 4 bits of source data and $d\phi_n$ for differentially encoded 16PSK.

$(a_{4n-3}, a_{4n-2}, a_{4n-1}, a_{4n})$	$d\phi_n$
0000	0
0001	$\pi/8$
0011	$\pi/4$
0010	$3\pi/8$
0110	$\pi/2$
0111	$5\pi/8$
0101	$3\pi/4$
0100	$7\pi/8$
1100	π
1101	$-7\pi/8$
1111	$-3\pi/4$
1110	$-5\pi/8$
1010	$-\pi/2$
1011	$-3\pi/8$
1001	$-\pi/4$
1000	$-\pi/8$

$$
\begin{aligned}
SER_{MCPSK} &= \Pr\left[|\theta| > \frac{\pi}{M}\right] \\
&= \Pr\left[z_Q > z_I \tan\frac{\pi}{M}\right] + \Pr\left[z_Q < -z_I \tan\frac{\pi}{M}\right] \\
&\quad - \Pr\left[z_I \tan\frac{\pi}{M} \le z_Q \le -z_I \tan\frac{\pi}{M}\right].
\end{aligned}
\tag{3.148}
$$

In this equation, we can disregard the third term because it is much smaller than the first and second terms. Moreover, the first and second terms are the same. Therefore, the SER_{MCPSK} can be approximated by

$$
SER_{MCPSK}(\gamma_s) = \text{erfc}\left(\sqrt{\gamma_s}\,\sin\frac{\pi}{M}\right)
\tag{3.149}
$$

where γ_s is the energy per symbol to the noise spectral density. When Gray coding is employed, one symbol error results in only a 1-bit error in most cases. Therefore, the

BER is $1/\log_2 M$ times the SER. Moreover, $\gamma\,(E_b/N_0)$ is also $1/\log_2 M$ times γ_s. Therefore the BER versus γ for Gray-coded *M*-ary PSK (P_{MCPSK}) is given by

$$P_{MCPSK}(\gamma) = \frac{1}{\log_2 M}\,\mathrm{erfc}\!\left(\sqrt{\gamma\log_2 M}\,\sin\frac{\pi}{M}\right) \tag{3.150}$$

When differential encoding is applied to *M*-ary PSK with coherent detection, the BER is twice as large as that for Gray-encoded *M*-ary PSK. Therefore, its BER (P_{DEMPSK}) is given by

$$P_{DEMPSK}(\gamma) = \frac{2}{\log_2 M}\,\mathrm{erfc}\!\left(\sqrt{\gamma\log_2 M}\,\sin\frac{\pi}{M}\right) \tag{3.151}$$

where γ is the E_b/N_0.

When the Gray-encoded *M*-ary PSK signal is transmitted over the Rayleigh fading channel and it is detected by absolute phase coherent detection, the BER is given by

$$P_{MCPSK}^{Ray}(\gamma_0) = \frac{1}{\log_2 M}\left[1 - \frac{1}{\sqrt{1 + 1\big/\left(\log_2 M\cdot\sin^2\dfrac{\pi}{M}\cdot\gamma_0\right)}}\right] \tag{3.152}$$

and the BER for differentially encoded *M*-ary PSK with coherent detection under Rayleigh fading conditions is given by

$$P_{DEMPSK}^{Ray}(\gamma_0) = \frac{2}{\log_2 M}\left[1 - \frac{1}{\sqrt{1 + 1\big/\left(\log_2 M\cdot\sin^2\dfrac{\pi}{M}\cdot\gamma_0\right)}}\right] \tag{3.153}$$

where γ_0 is the average E_b/N_0.

Figure 3.25 shows BER performances of Gray-encoded *M*-ary PSK with absolute phase coherent detection (a) under AWGN conditions and (b) under flat Rayleigh fading conditions, whereas Figure 3.26 shows BER performances of differentially encoded *M*-ary PSK with coherent detection (a) under AWGN conditions and (b) under flat Rayleigh fading conditions.

More general BER performance analysis of *M*-ary coherent PSK in the *m*-distributed fading channel is shown in Ref. [3-32].

3.6.3 Differential Detection for *M*-ary PSK (MDPSK)

Differential detection is a very effective demodulation scheme, especially for *M*-ary PSK, because its receiver configuration is very simple.

Let the demodulated baseband signal be expressed as

$$v(t) = z(t) + n(t)$$
$$= R(t)\exp\big(\phi(t) + d\theta(t)\big) \tag{3.154}$$

where $\phi(t)$ is the modulated phase variation and $R(t)$ and $d\theta(t)$ are random amplitude and phase variations due to noise. When $\phi(t)$ and $d\theta(t)$ at $t = nT_s$ are expressed as ϕ_n and $d\theta_n$, the phase difference after differential detection is given by

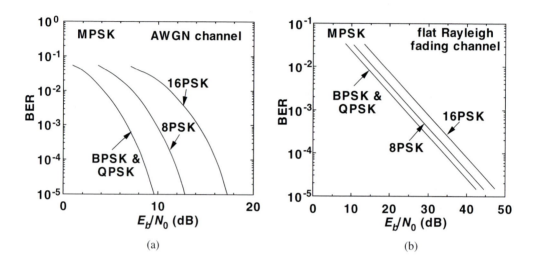

Fig. 3.25 BER performance of Gray-encoded M-ary PSK with absolute phase coherent detection (a) under AWGN conditions and (b) under flat Rayleigh fading conditions.

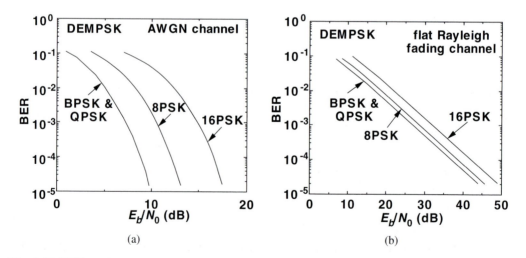

Fig. 3.26 BER performance of differentially encoded M-ary PSK with coherent detection (a) under AWGN conditions and (b) under flat Rayleigh fading conditions.

$$d\hat{\phi}_n = \theta_n - \theta_{n-1}$$
$$= (\phi_n - \phi_{n-1}) + (d\theta_n - d\theta_{n-1}) \qquad (3.155)$$
$$= d\phi_n + (d\theta_n - d\theta_{n-1})$$

where the first term is the phase difference between consecutive two transmitted symbols and the second term is the phase distortion. $d\theta_n$ and $d\theta_{n-1}$ are subject to Nakagami-Rice phase distribution with its p.d.f. of

$$p(d\theta|\gamma_s)$$
$$= \frac{1}{2\pi} e^{\gamma_s} \left[1 + \sqrt{\pi\gamma_s} \cos(d\theta) \left\{ 1 + \operatorname{erf}\left(\sqrt{\gamma_s}\cos(d\theta)\right) \right\} e^{\gamma_s \cos^2(d\theta)} \right] \qquad (3.156)$$

where γ_s is E_s/N_0. Therefore, the approximated p.d.f. for $\varphi_n = (d\theta_n - d\theta_{n-1})$ is given by [3-33]

$$p(\varphi_n \mid \gamma_s) \cong \frac{\gamma_s}{2} e^{-\gamma_s} \left[\cos\varphi_n I_0(\gamma_s \cos\varphi_n) + I_1(\gamma_s \cos\varphi_n) \right]$$
$$\cong \frac{\sqrt{\gamma_s}}{\sqrt{2\pi}} \cos\varphi_n e^{-\gamma_s \sin^2 \frac{\varphi_n}{2}} \qquad (3.157)$$

Using this equation, we can obtain the SER with respect to γ_s as

$$SER_{MDPSK}(\gamma_s) = 1 - 2\int_0^{\pi/M} p(\varphi_n \mid \gamma_s) d\varphi_n$$
$$= \operatorname{erfc}\left(\sqrt{2\gamma_s} \sin\frac{\pi}{2M} \right) \qquad (3.158)$$

Therefore, the BER versus E_b/N_0 (γ) for *M*-ary PSK with differential detection (P_{MDPSK}) is given by

$$P_{MDPSK}(\gamma) = \frac{1}{\log_2 M} \operatorname{erfc}\left(\sqrt{2\log_2 M \cdot \gamma} \sin\frac{\pi}{2M} \right) \qquad (3.159)$$

When *M*-ary PSK signal with differential detection is transmitted over the Rayleigh fading channel, the BER is given by

$$P_{MDPSK}^{Ray}(\gamma_0) = \frac{1}{\log_2 M} \left[1 - \frac{1}{\sqrt{1 + 1/\left(2\log_2 M \cdot \sin^2 \frac{\pi}{2M} \cdot \gamma_0\right)}} \right] \qquad (3.160)$$

where γ_0 is the average E_b/N_0.

Figure 3.27 shows BER performance of differentially encoded *M*-ary PSK with differential detection (a) under AWGN conditions and (b) under flat Rayleigh fading conditions.

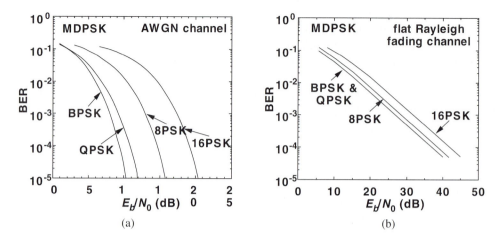

(a)   (b)

Fig. 3.27 BER performance of differentially encoded M-ary PSK with differential detection (a) under AWGN conditions and (b) under flat Rayleigh fading conditions.

3.7 M-ARY QAM

In the M-ary QAM, a symbol is generated according to $\log_2 M$-bit of the source data. Because coherent detection is essential for square-QAM, carrier regeneration techniques using pilot signal-assisted schemes (which will be discussed in chapter 4) are necessary for M-ary QAM. On the other hand, when we can employ a coherent detection scheme assisted by a pilot signal, we can apply Gray encoding with absolute phase coherent detection, which gives better BER performance than differential encoding with coherent detection or differential detection. Therefore, we will discuss only the Gray-encoded M-ary QAM with coherent detection.

Figure 3.28 shows signal state diagrams of (a) 16QAM, (b) 64QAM, and (c) 256QAM. In M-ary QAM, a serial data sequence is converted to $\log_2 M$-bit of parallel

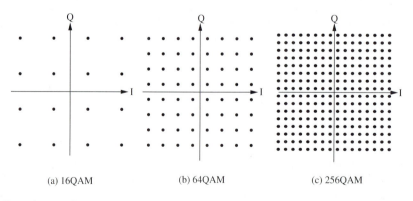

(a) 16QAM   (b) 64QAM   (c) 256QAM

Fig. 3.28 Signal state diagrams of (a) 16QAM, (b) 64QAM, and (c) 256QAM.

data, and $\log_2 M$-bit data are divided into two groups; for example, (a_{4n-3}, a_{4n-2}) and (a_{4n-1}, a_{4n}) in the case of 16QAM. Then the former group is assigned to the in-phase channel and the latter to the quadrature channel. Tables 3.11 through 3.13 show this mapping rule, where Gray encoding is employed, and signal Euclidean distance is defined as 2 in this case.

In the case of *M*-ary QAM, each symbol has a different SER. Figure 3.29 shows the first quadrant of a 16QAM signal state diagram. Let's define that the SER for symbols a to d are P_a, P_b, P_c, and P_d, and noise power is σ^2. SER for each symbol is given by

$$P_a = 1 - \int_0^{2A} \frac{1}{\sqrt{2\pi}\sigma} e^{-\frac{(z_I - A)^2}{2\sigma^2}} dz_I \int_0^{2A} \frac{1}{\sqrt{2\pi}\sigma} e^{-\frac{(z_Q - A)^2}{2\sigma^2}} dz_Q$$
$$= 2\,\text{erfc}\left(\frac{A}{\sigma\sqrt{2}}\right) - \text{erfc}^2\left(\frac{A}{\sigma\sqrt{2}}\right)$$

(3.161)

Table 3.11 An example of Gray mapping rule for 16QAM.

In-Phase Channel		Quadrature Channel	
(a_{4n-3}, a_{4n-2})	Amplitude	(a_{4n-1}, a_{4n})	Amplitude
00	−3	00	−3
01	−1	01	−1
11	1	11	1
10	3	10	3

Table 3.12 An example of Gray mapping rule for 64QAM.

In–Phase Channel		Quadrature Channel	
$(a_{6n-5}, a_{6n-4}, a_{6n-3},)$	Amplitude	$(a_{6n-2}, a_{6n-1}, a_{6n})$	Amplitude
000	−7	000	−7
001	−5	001	−5
011	−3	011	−3
010	−1	010	−1
110	1	110	1
111	3	111	3
101	5	101	5
100	7	100	7

Table 3.13 An example of Gray mapping rule for 256QAM.

In–Phase Channel		Quadrature Channel	
$(a_{8n-7}, a_{8n-6}, a_{8n-5}, a_{8n-4})$	Amplitude	$(a_{8n-3}, a_{8n-2}, a_{8n-1}, a_{8n})$	Amplitude
0000	−15	0000	−15
0001	−13	0001	−13
0011	−11	0011	−11
0010	−9	0010	−9
0110	−7	0110	−7
0111	−5	0111	−5
0101	−3	0101	−3
0100	−1	0100	−1
1100	1	1100	1
1101	3	1101	3
1111	5	1111	5
1110	7	1110	7
1010	9	1010	9
1011	11	1011	11
1001	13	1001	13
1000	15	1000	15

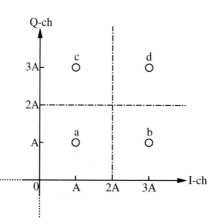

Fig. 3.29 First quadrant of 16QAM signal state diagram.

$$P_b = 1 - \int_{2A}^{\infty} \frac{1}{\sqrt{2\pi}\sigma} e^{-\frac{(z_I - 3A)^2}{2\sigma^2}} dz_I \int_0^{2A} \frac{1}{\sqrt{2\pi}\sigma} e^{-\frac{(z_Q - A)^2}{2\sigma^2}} dz_Q$$

$$= 1 - \left[1 - \frac{1}{2}\mathrm{erfc}\left(\frac{A}{\sigma\sqrt{2}}\right) \right]\left[1 - \mathrm{erfc}\left(\frac{A}{\sigma\sqrt{2}}\right) \right] \qquad (3.162)$$

$$= \frac{3}{2}\mathrm{erfc}\left(\frac{A}{\sigma\sqrt{2}}\right) - \frac{1}{2}\mathrm{erfc}^2\left(\frac{A}{\sigma\sqrt{2}}\right)$$

$$P_c = 1 - \int_0^{2A} \frac{1}{\sqrt{2\pi}\sigma} e^{-\frac{(z_I - A)^2}{2\sigma^2}} dz_I \int_{2A}^{\infty} \frac{1}{\sqrt{2\pi}\sigma} e^{-\frac{(z_Q - 3A)^2}{2\sigma^2}} dz_Q$$

$$= \frac{3}{2}\mathrm{erfc}\left(\frac{A}{\sigma\sqrt{2}}\right) - \frac{1}{2}\mathrm{erfc}^2\left(\frac{A}{\sigma\sqrt{2}}\right) \qquad (3.163)$$

$$P_d = 1 - \int_{2A}^{\infty} \frac{1}{\sqrt{2\pi}\sigma} e^{-\frac{(z_I - 3A)^2}{2\sigma^2}} dz_I \int_{2A}^{\infty} \frac{1}{\sqrt{2\pi}\sigma} e^{-\frac{(z_Q - 3A)^2}{2\sigma^2}} dz_Q$$

$$= \mathrm{erfc}\left(\frac{A}{\sigma\sqrt{2}}\right) - \frac{1}{4}\mathrm{erfc}^2\left(\frac{A}{\sigma\sqrt{2}}\right) \qquad (3.164)$$

When selection probability for each symbol is the same, the average signal power is given by

$$S = \frac{1}{4}\frac{A^2 + A^2}{2} + \frac{1}{2}\frac{A^2 + 9A^2}{2} + \frac{1}{4}\frac{9A^2 + 9A^2}{2} = 5A^2 \qquad (3.165)$$

Therefore, E_b/N_0 is given by

$$\gamma = \frac{E_b}{N_0} = \frac{1}{4}\frac{E_s}{N_0} = \frac{S}{4\sigma^2} = \frac{5}{2}\frac{A^2}{2\sigma^2} \qquad (3.166)$$

Furthermore, the BER for Gray-encoded 16QAM is 25% of the SER. As a result, the BER for Gray-encoded 16QAM (P_{16QAM}) is given by

$$P_{16QAM}(\gamma) = \frac{1}{4}\frac{P_a + P_b + P_c + P_d}{4}$$

$$= \frac{3}{8}\mathrm{erfc}\left(\sqrt{\frac{2}{5}\gamma}\right) - \frac{9}{64}\mathrm{erfc}^2\left(\sqrt{\frac{2}{5}\gamma}\right) \qquad (3.167)$$

The BER for 64QAM and 256QAM can be obtained in the same manner.

$$P_{64QAM}(\gamma) = \frac{7}{24}\mathrm{erfc}\left(\sqrt{\frac{1}{7}\gamma}\right) - \frac{49}{384}\mathrm{erfc}^2\left(\sqrt{\frac{1}{7}\gamma}\right) \qquad (3.168)$$

$$P_{256QAM}(\gamma) = \frac{15}{64}\operatorname{erfc}\left(\sqrt{\frac{4}{85}\gamma}\right) - \frac{225}{2048}\operatorname{erfc}^2\left(\sqrt{\frac{4}{85}\gamma}\right) \tag{3.169}$$

When the M-ary QAM signal is transmitted over a Rayleigh fading channel, the BER for each modulation is given by

$$P_{16QAM}^{Ray}(\gamma_0) = \frac{3}{8}\left[1 - \frac{1}{\sqrt{1+5/(2\gamma_0)}}\right] \tag{3.170}$$

$$P_{64QAM}^{Ray}(\gamma_0) = \frac{7}{24}\left[1 - \frac{1}{\sqrt{1+7/\gamma_0}}\right] \tag{3.171}$$

$$P_{256QAM}^{Ray}(\gamma_0) = \frac{15}{64}\left[1 - \frac{1}{\sqrt{1+85/(4\gamma_0)}}\right] \tag{3.172}$$

where γ_0 is the average E_b/N_0.

Figure 3.30 shows the BER performance of Gray-encoded QPSK, 16QAM, 64QAM, and 256QAM (a) under AWGN conditions and (b) under flat Rayleigh fading conditions.

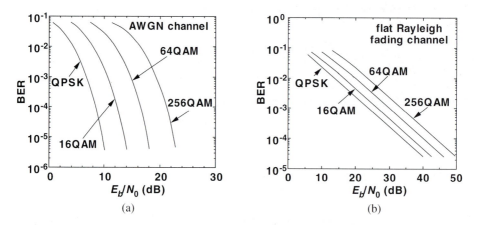

Fig. 3.30 BER performances of QPSK, 16QAM, 64QAM, and 256QAM (a) under AWGN conditions and (b) under flat Rayleigh fading conditions.

3.8 BINARY ORTHOGONAL MODULATION

The binary orthogonal modulator transmits k-bit of information using 2^k-bit of orthogonal code, where two arbitrary codewords are orthogonal with each other. 2^k-bit of orthogonal code $\{s_1, s_2, ..., s_m\}$ $(m = 2^k)$ can be generated by using $2^k \times 2^k$. The Hadamard matrix follows.

$$H_k = \begin{bmatrix} H_{k-1} & H_{k-1} \\ H_{k-1} & \overline{H}_{k-1} \end{bmatrix} = \begin{bmatrix} s_1 \\ s_2 \\ \cdot \\ \cdot \\ \cdot \\ s_{2^k} \end{bmatrix} \tag{3.173}$$

where

$$H_1 = \begin{bmatrix} 0 & 0 \\ 0 & 1 \end{bmatrix} \tag{3.174}$$

For example, 4-bit orthogonal code ($k = 2$) can be obtained as

$$H_2 = \begin{bmatrix} 0 & 0 & 0 & 0 \\ 0 & 1 & 0 & 1 \\ 0 & 0 & 1 & 1 \\ 0 & 1 & 1 & 0 \end{bmatrix} = \begin{bmatrix} s_1 \\ s_2 \\ s_3 \\ s_4 \end{bmatrix} \tag{3.175}$$

where s_1, s_2, s_3, and s_4 correspond to the source information bits of 00, 01, 10, and 11, respectively.

Figure 3.31 shows a configuration of the binary orthogonal modulator. The information bit sequence is converted to k-bit of parallel data, and one of the 2^k-bit codewords is generated according to this k-bit data followed by the BPSK modulator.

Now, the baseband signal of the j-th BPSK-modulated orthogonal codeword is expressed as

$$s_j(t) = \sum_{l=1}^{2^k} \sqrt{P} b_l^j u(t - lT_c) \tag{3.176}$$

$$u(t) = \begin{cases} 1; & -\dfrac{T_c}{2} \le t \le \dfrac{T_c}{2} \\ 0; & \text{otherwise} \end{cases} \tag{3.177}$$

where b_l^j means the amplitude of l-th bit in the j-th codeword, P is the signal power, and T_c is a bit duration of the orthogonal code. Because a codeword consists of 2^k-bit, a symbol duration of the codeword T_s is given by $T_s = 2^k T_c$. Figure 3.32 shows waveforms of the source information bit sequence and the codeword sequence, where $k = 2$.

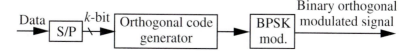

Fig. 3.31 Configuration of the binary orthogonal modulator.

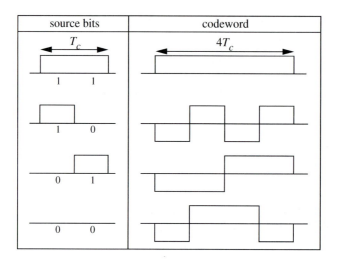

Fig. 3.32 Waveforms of the source information bit sequence and the codeword sequence, where $k = 2$.

Figure 3.33 shows the receiver configuration of the binary orthogonal modulation. After the received signal is down-converted to the baseband, the transmitted codeword is detected by the maximum likelihood (ML) detection as follows.

The j-th codeword signal is transmitted over an AWGN channel, and the received baseband signal can be expressed as

$$r_j(t) = s_j(t) + n(t) \tag{3.178}$$

At the ML detector, a correlation is made between the received signal and $s_i(t)$ ($1 \leq i \leq 2^k$), and $s_i(t)$ that gives the maximum correlation value is selected as the transmitted codeword signal.

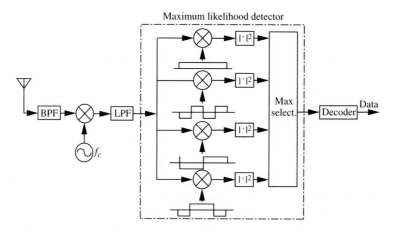

Fig. 3.33 Receiver configuration of the binary orthogonal modulation.

The correlation between $r_j(t)$ and $s_i(t)$ is given by

$$
\begin{aligned}
q_i &= \int_0^{T_s} r_j(t)s_i(t)dt \\
&= \int_0^{T_s} s_j(t)s_i(t)dt + \int_0^{T_s} n(t)s_i(t)dt
\end{aligned}
\tag{3.179}
$$

The average value of q_i is given by

$$
\overline{q}_i = \begin{cases} PT_s; & i = j \\ 0 & ; \ i \neq j \end{cases}
\tag{3.180}
$$

and its variance is given by $\sigma^2 = N_0 PT_s/2$, where N_0 is the noise power spectral density.

Now, let's assume that s_1 is transmitted. The probability that the received signal is correctly decoded is given by

$$
\begin{aligned}
P_c &= \left(q_2 < q_1, \ q_3 < q_1, \ \dots, \ q_{2^k} < q_1 \right) \\
&= \int_{-\infty}^{\infty} p(q_1)dq_1 \int_{-\infty}^{q_1} p(q_2)dq_2 \int_{-\infty}^{q_1} p(q_3)dq_3 \cdots \int_{-\infty}^{q_1} p(q_{2^k})dq_{2^k} \\
&= \int_{-\infty}^{\infty} \frac{1}{\sqrt{2\pi}\sigma} e^{-\frac{(q_1-E_s)^2}{2\sigma^2}} \left[\int_{-\infty}^{q_1} \frac{1}{\sqrt{2\pi}\sigma} e^{-\frac{q_2^2}{2\sigma^2}} dq_2 \right]^{2^k-1} dq_1 \\
&= \int_{-\infty}^{\infty} \frac{1}{\sqrt{2\pi}} e^{-\frac{x^2}{2}} \left[\frac{1}{2}\mathrm{erfc}\left(-\frac{x+\sqrt{2\gamma_s}}{\sqrt{2}} \right) \right]^{2^k-1} dx
\end{aligned}
\tag{3.181}
$$

where $E_s = PT_s$ corresponds to energy per code symbol, $\gamma_s = E_s/N_0$, and

$$
x = \frac{q_1 - E_s}{\sigma}
\tag{3.182}
$$

Therefore, the SER of binary orthogonal modulation is given by

$$
\begin{aligned}
SER_{OR}(\gamma_b) &= 1 - P_c \\
&= 1 - \int_{-\infty}^{\infty} \frac{1}{\sqrt{2\pi}} e^{-\frac{x^2}{2}} \left[\frac{1}{2}\mathrm{erfc}\left(-\frac{x+\sqrt{2k\gamma_b}}{\sqrt{2}} \right) \right]^{2^k-1} dx
\end{aligned}
\tag{3.183}
$$

where $\gamma_b = \gamma_s/k$ means energy per information bit to the noise spectral density.

When one symbol error occurred, all symbol errors are equiprobable. Furthermore, there are $_kC_n$ ways in which n-bit out of k may be in error. Therefore, we can obtain the conditional probability that n-bit out of k is in error after one symbol is in error as

$$P(n) = \frac{{}_kC_n}{\sum\limits_{n=1}^{k}{}_kC_n} \tag{3.184}$$

Using this equation, the average number of bit errors in a symbol error is given by

$$P(\text{bit / symbol}) = \sum_{n=1}^{k} nP(n) = \frac{1}{k}\frac{\sum\limits_{n=1}^{k} n\left({}_kC_n\right)}{\sum\limits_{n=1}^{k}{}_kC_n} = \frac{2^{k-1}}{2^k - 1} \tag{3.185}$$

Using equations (3.183) and (3.185), we can obtain the BER performance of the binary orthogonal modulation as

$$P_{OR}(\gamma_b)$$

$$= \frac{2^{k-1}}{2^k - 1}\left[1 - \int_{-\infty}^{\infty}\frac{1}{\sqrt{2\pi}}e^{-\frac{x^2}{2}}\left[\frac{1}{2}\text{erfc}\left(-\frac{x + \sqrt{2k\gamma_b}}{\sqrt{2}}\right)\right]^{2^k - 1}dx\right] \tag{3.186}$$

Figure 3.34 shows BER performance of the binary orthogonal modulation scheme under AWGN conditions. As we can see from this figure, the BER performance is improved by increasing k. When $k = 5$ is employed, the BER performance improvement over its performance when $k = 1$ is about 5 dB in the range of $10^{-5} \le$ BER $\le 10^{-3}$. On the other hand, the improvement is small when BER $> 10^{-2}$. Most of all, when BER $> 10^{-1}$, the BER performance for $k = 5$ is worse than that for $k = 1$. Therefore, binary orthogonal modulation is effective for smaller BER.

Figure 3.35 shows the BER performance under flat Rayleigh fading conditions. Because BER performance improvement at small E_b/N_0 is very small even if larger k

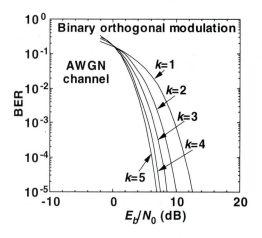

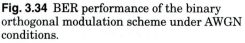

Fig. 3.34 BER performance of the binary orthogonal modulation scheme under AWGN conditions.

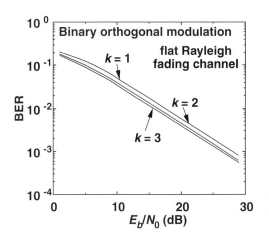

Fig. 3.35 BER performance of the binary orthogonal modulation scheme under flat Rayleigh fading conditions.

is employed, the BER performance improvement is not so large under flat Rayleigh fading conditions.

3.9 OTHER MODULATION SCHEMES

In this chapter, we have discussed modulation and demodulation schemes in terms of linear modulation schemes because

☞ Many standardized systems employ linear modulation schemes.

☞ Application of linear modulation is quite a new trend in the field of wireless communication systems.

☞ Modulation with a higher modulation level has a potential to achieve a high bit rate and a high spectrally efficient system, although it depends on the anti-fading compensation techniques, nonlinear compensation techniques, and the service demand.

Therefore, in the following chapters, we will concentrate on how to design wireless communication systems, along with anti-fading techniques, error control techniques, synchronization techniques, and so on. Fortunately, Feher [3-9] discusses the application of constant (quasi-constant) envelope modulation techniques, such as the FQPSK family, and he also explains their applications to the wireless communication systems in detail. Please refer to this book if you are interested in these technical fields.

REFERENCES

3-1. Murota, K. and Hirade, K., "GMSK modulation for digital mobile radio telephony," IEEE Trans. Commun., Vol. COM-29, No. 7, pp. 1044–50, July 1980.

3-2. Kinoshita, K., Hata, M. and Nagabuchi, H., "Evaluation of 16kbit/s digital voice transmission," IEEE Trans. Commun., Vol. VT-33, No. 4, pp. 321–26, November 1984.

3-3. Honma, K., Murata E. and Riko, Y., "On a method of constant envelope modulation for digital mobile communications," IEEE ICC'80, p. 24.1.1–24.1.5, June 1980.

3-4. Akaiwa, Y., Takase, I., Kojima, S., Ikoma, M. and Saegsa, N., "Performance of baseband bandlimited multi-level FM with discriminator detection for digital mobile telephony," Trans. IECE Japan, Vol. E64, pp. 463–69, July 1981.

3-5. Jager, F. D. and Dekker, C. B., "Tamed frequency modulation, a novel method to achieve spectrum economy in digital transmission," IEEE Trans. Commun., Vol. COM-26, No. 5, pp. 534–42, May 1978.

3-6. G.S.M., "Physical layer on the radio-path: G.S.M. system," G.S.M. Recommendation, Vol. G, G.S.M. Standard Committee, July 1988.

3-7. DECT, "Digital European cordless system—common interface specifications," Code RES-3(89), DECT, 1989.

3-8. Akaiwa, Y. and Nagata, Y., "Highly efficient digital mobile communications with a linear modulation method," *IEEE Journal of Selected Areas in Communications*, Vol. 5, No. 5, pp. 890–95, June 1987.

3-9. Feher, K., *Wireless digital communications, modulation and spread spectrum applications*, Prentice Hall, Upper Saddle River, New Jersey, 1995.

3-10. Stapleton, S. P. and Costescu, F. C., "An adaptive predistorter for a power amplifier based on adjacent channel emission," IEEE Trans. Veh. Technol., Vol. VT-41, No. 1, pp. 49–56, February 1992.

3-11. Cavers, J. K., "Amplifier linearization using a digital predistorter with fast adaptation and low memory requirements," IEEE Trans. Veh. Technol., Vol. 39, No. 4, pp. 374–82, November 1990.

3-12. Briffa, M. A. and Faulkner, M., "Dynamically biased cartesian feedback linearization," 43rd IEEE Veh. Tech. Conf. (Secaucus, New Jersey), pp. 672–75, May 1993.

3-13. Cox, D. C., "Linear amplification with nonlinear components," IEEE Trans. Commun., Vol. COM-22, pp. 1942–45, December 1974.

3-14. Narahashi, S. and Nojima, T., "Extremely low-distortion multi-carrier amplifier, self-adjusting feed-forward (SAFF) amplifier," IEEE ICC'91 (Denver, Colorado), pp. 1485–90, June 1991.

3-15. RCR, "Digital cellular telecommunication systems," RCR STD-27, April 1991.

3-16. EIA, "Dual-mode subscriber equipment compatibility specification," EIA specification IS-54, EIA project number 2215, Washington, D.C., May 1990.

3-17. Bateman, A., "Feedforward transparent tone-in-band: Its implementation and applications," IEEE Trans. Veh. Technol., Vol. 39, No. 3, pp. 235–43, August 1990.

3-18 Sampei, S. and Sunaga, T., "Rayleigh fading compensation for QAM in land mobile radio communications," IEEE Trans. Veh. Technol., Vol. 42, No. 2, pp. 137–47, May 1993.

3-19. Hanzo, L., Steele, R. and Fortune, P. M. "A subband coding, BCH coding and 16-QAM system for mobile radio speech applications," IEEE Trans. Veh. Technol., Vol. 39, No. 4, pp. 327–39, November 1990.

3-20. RCR, "Digital MCA system," RCR STD-32, November 1992.

3-21. RCR, "Digital Public Private Mobile Radio Systems," RCR STD-29, December 1993.

3-22. Davidson, A. and Marturano, L., "Impact of digital techniques on future LM spectrum requirements," *IEEE Vehicular Technology Society News*, pp. 14–30, May 1993.

3-23. Lindsey, W. C. and Simon, M. K., *Telecommunication system engineering,* Prentice Hall, Englewood Cliffs, New Jersey, 1973.

3-24. Feher, K., *Digital communications*, Prentice Hall, Englewood Cliffs, New Jersey, 1981.

3-25. Ogose, S. and Murota, K., "Differentially encoded GMSK with 2-bit differential detection," Rev. Electric. Commun. Lab., Vol. 32, No. 6, pp. 1295–1304, December 1983.

3-26. Ohno, K. and Adachi, F., "Performance analysis of GMSK frequency detection with decision feedback equalization in digital land mobile radio," Proc. IEE, part-F, Vol. 135, pp. 199–207, June 1988.

3-27. Proakis, J. G., *Digital communications* (2d edition), McGraw-Hill, New York, 1989.

3-28. Liu, C. L., and Feher, K., "π/4-QPSK modems for satellite sound/data broadcast system," IEEE Trans. Broadcast. Technol., Vol. 37, No. 1, pp. 1-8, January 1991.

3.29. Matsumoto, Y., Kubota, S. and Kato, S., "A new burst coherent demodulator for microcellular TDMA/TDD systems," IEICE Trans. Commun., Vol. E77-B, No. 7, pp. 927–33, July 1994.

3-30. Adachi, F. and Ohno, K., "BER performance of QDPSK with post detection diversity reception in mobile radio channels," IEEE Trans. Veh. Technol., Vol. 40, No. 1, pp. 237–49, February 1991.

3-31. Adachi, F. and Ohno, K., "Post detection MRC diversity for π/4-QDPSK mobile radio," Electron Letter, Vol. 27, No. 18, pp. 1642–43, August 1991.

3-32. Miyagaki, Y., Morinaga, N. and Namekawa, T., "Error probability characteristics for CPSK signal through m-distributed fading channel," IEEE Trans. Commun., Vol. COM-26, No. 1, pp. 88–100, January 1978.

3-33. Miyagaki, Y., "Error rate analysis in MDPSK and MCPSK with a pilot on satellite-aircraft multipath channel," Trans. IEICE (B-II), Vol. J72-B-II, No. 12, pp. 640–47, December 1989.

Anti-Flat Fading Techniques

$\mathbf{I}$n narrowband land mobile communication systems, no terminal works normally without fading compensation techniques. For the application of frequency or phase modulation techniques in which no information is included in the signal amplitude, we have to compensate for phase distortion due to fading. On the other hand, for the application of amplitude modulation or quadrature amplitude modulation techniques, we have to compensate for both amplitude and phase distortion due to fading.

Until the mid-1980s, very few studies had been done on fading compensation techniques because differential detection or FM discriminator schemes were considered to be the most appropriate demodulation scheme for land mobile communication systems because of the following:

☞ Their circuit configurations are very simple.

☞ Receivers using differential detection or frequency discriminator detection give better BER performance than those using the coherent demodulators.

However, since the mid-1980s, studies on coherent detection have been started because its optimal performance is superior to that of differential detection or frequency discriminator detection if we have some precise fading compensation techniques. For this purpose, a lot of studies have been done on the improvement of coherent demodulators; i.e., fading compensation techniques for coherent demodulators.

Fading compensation techniques for coherent demodulators are roughly divided into two groups.

☞ Nonpilot signal-aided techniques

☞ Pilot signal-aided techniques

We will discuss these fading compensation techniques in this chapter, including problems of conventional coherent demodulators.

Although these techniques are indispensable in establishing radio links in flat Rayleigh fading environments, improvement of receiver sensitivity is another important issue for wireless communication systems. Among various anti-flat fading techniques, diversity combining is well known as a very effective means of improving receiver sensitivity. Therefore, we will also discuss diversity combining techniques in this chapter.

4.1 PROBLEMS OF CONVENTIONAL COHERENT RECEIVERS

Figure 4.1 shows a typical receiver configuration of a coherent demodulator. To simplify our discussion, we will look at its operation in the case of M-ary PSK demodulators. In this case, the received signal is given by

$$s_R(t) = \mathrm{Re}[c(t)\exp(j\theta_D(t))\exp(j2\pi f_c t)]$$
$$= \mathrm{Re}[r(t)\exp(j2\pi f_c t + j\theta(t) + j\theta_D(t))] \tag{4.1}$$

$$\theta_D(t) = \pm(2k-1)\pi/M, \ (k = 1, \ 2, \ \cdots, M/2) \tag{4.2}$$

where

    $c(t)$ = complex fading variation

    $r(t)$ = envelope component of $c(t)$

    $\theta(t)$ = phase component of $c(t)$

    $\theta_D(t)$ = phase variation by the phase modulation

The received signal is first filtered by a BPF to pick up the spectrum around the desired signal. Since information resides in only the phase component of the transmitted signal, we usually remove amplitude variation using an automatic gain controller (AGC) or a hard limiter. Furthermore, the frequency of the received signal is con-

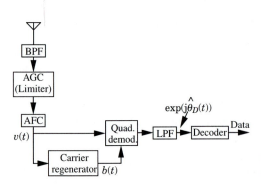

Fig. 4.1 Receiver configuration of the coherent demodulator.

trolled at a proper frequency by the automatic frequency controller (AFC). As a result, the signal after the AFC is given by

$$v(t) = \text{Re}[\exp(j2\pi f_c t + j\theta(t) + j\theta_D(t))] \tag{4.3}$$

where noise is disregarded in equation (4.3) to simplify explanation.

At the carrier regenerator, a reference carrier expressed as

$$b(t) = \exp(j2\pi f_c t + j\hat{\theta}(t)) \tag{4.4}$$

is generated, where $\hat{\theta}(t)$ is the estimated value of $\theta(t)$. At the quadrature demodulator, coherent detection is carried out using $b(t)$ to obtain the complex baseband given by

$$\begin{aligned} z(t) &= \exp[j\theta_D(t) + j(\theta(t) - \hat{\theta}(t))] \\ &= \exp[j\hat{\theta}_D(t)] \end{aligned} \tag{4.5}$$

where $\hat{\theta}_D(t)$ is the estimated value of $\theta_D(t)$. $z(t)$ is then quantized to regenerate the transmitted data sequence. As shown in equation (4.5), the performance of coherent detection depends on the estimation accuracy of $\hat{\theta}_D(t)$.

Figure 4.2 shows a carrier regeneration circuit for M-ary PSK called PLL with a times M ($\times M$) multiplier [4-1]. There are various types of carrier regeneration circuits for M-ary PSK, such as the Costas loop. However, their operation is equivalent to the PLL with a $\times M$ multiplier [4-1]. Therefore, we will discuss only the basic operation of the PLL with a $\times M$ multiplier.

Because $\theta_D(t)$ is given by equation (4.2), the following equation is satisfied.

$$\exp^M[j\theta_D(t)] = -1 \tag{4.6}$$

This equation means that, when M-th harmonic of the $\times M$ multiplier output is picked up and its phase is shifted by π [rad], we can obtain

$$b_1(t) = \exp(j2\pi M f_c t + jM\theta(t)) + \text{distortion} \tag{4.7}$$

where distortion is caused by components other than the M-th harmonics.

Let's assume that the output of the VCO is given by

$$b_2(t) = \exp(j2\pi M f_c t + jM\hat{\theta}(t)) \tag{4.8}$$

At the phase detector (PD), phase difference between $b_1(t)$ and $b_2(t)$ given by

$$\Delta\theta(t) = M(\theta(t) - \hat{\theta}(t)) \tag{4.9}$$

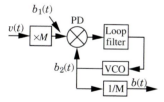

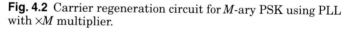

Fig. 4.2 Carrier regeneration circuit for M-ary PSK using PLL with $\times M$ multiplier.

is detected. Because $\Delta\theta(t)$ includes actual phase difference and distortion, a loop filter is inserted after the PD to suppress this distortion. The VCO controls its phase to keep the filtered phase error at 0. As a result, the carrier component of the received signal is regenerated by using an M-divider as

$$b(t) = \exp(j2\pi f_c t + j\hat{\theta}(t)) \tag{4.10}$$

The most important parameter of the PLL is the noise bandwidth of the loop given by

$$B_L = \int_0^\infty |H(j2\pi f)|^2 df \tag{4.11}$$

where $H(j2\pi f)$ is the closed loop transfer function of PLL.

Figure 4.3 shows an example of the envelope and phase variation due to Rayleigh fading, where $f_d = 40$ Hz. As we can see from Figure 4.3, the phase variation due to fading includes very high frequency components. Therefore, wider B_L is required to keep up with the fast phase variation.

Figure 4.4 shows (a) regenerated phase and (b) phase estimation error of the PLL when a continuous wave (CW) transmitted over the Rayleigh fading channel is fed to the PLL, where $B_L = 1$ kHz, $f_d = 40$ Hz, and the fading variation is the same as the one shown in Figure 4.3 [4-2]. In Figure 4.4 (b), we can find that a very large phase estimation error sometimes occurs. At these points, the PLL is in the out-of-locked condition because the PLL cannot perfectly keep up with fast phase variation due to the phase error averaging function of the PLL. This synchronization loss causes so-called irreducible errors in the case of digital transmission. When we increase B_L, we can reduce the probability of irreducible errors. However, the wider B_L increases distortion due to noise in the regenerated carrier. Therefore, there exists the optimum B_L. When the symbol transmission rate is around 20 ksymbol/s, the optimum value is approximately 1 kHz.

Figure 4.5 shows BER performance under AWGN and flat Rayleigh fading condition ($f_d = 80$ Hz) of 16 symbols/s QPSK with coherent detection using a Costas loop and BER performance with differential detection, where B_L for the coherent detection cir-

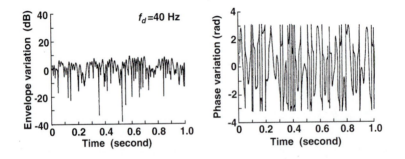

Fig. 4.3 An example of amplitude and phase variation due to Rayleigh fading.

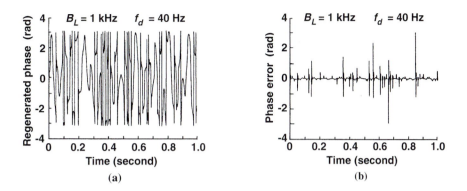

Fig. 4.4 Operation of the phaselocked loop: (a) regenerated phase (b) phase estimation error

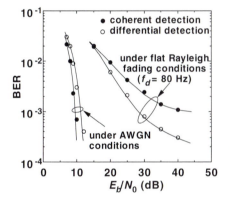

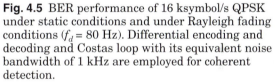

Fig. 4.5 BER performance of 16 ksymbol/s QPSK under static conditions and under Rayleigh fading conditions (f_d = 80 Hz). Differential encoding and decoding and Costas loop with its equivalent noise bandwidth of 1 kHz are employed for coherent detection.

cuit is 1 kHz. Although the BER performance of coherent detection shows better performance than that of differential detection under AWGN conditions, it is worse than that of differential detection under flat Rayleigh fading conditions. Therefore, the differential detection was considered to be more appropriate than coherent detection for the land mobile communication systems.

4.2 NONPILOT SIGNAL-AIDED FADING COMPENSATION TECHNIQUES

4.2.1 Modified PLL

The main cause of coherent detection degradation is that the PLL sometimes loses synchronization when phase error in the PLL is very large. The modified PLL [4-2; 4-3] has an instantaneous phase error monitoring function to detect a large phase error that could cause out-of-locked conditions. When it detects a very large instantaneous phase error, the VCO output phase is compulsorily shifted to reduce the phase error.

Figure 4.6 shows a configuration of the modified PLL for QPSK in which an adaptive carrier tracking (ACT) circuit is added to the conventional Costas loop. The circuit enclosed by the dashed lines is the same as the conventional Costas loop for QPSK. Because basic operation of this circuit is almost the same as the operation of the conventional Costas loop, the modified PLL also has phase ambiguity. Therefore, differential encoding and differential decoding are essential for this circuit.

Let's assume that the symbol timing of the n-th symbol is $t_n = nT_s$, and

$$t_n^- = t_n - dt \tag{4.12}$$

$$t_n^+ = t_n + dt \tag{4.13}$$

where the transmitted symbol of the QPSK signal is selected from $\exp(j\pi/4)$, $-\exp(j\pi/4)$, $\exp(j3\pi/4)$, and $-\exp(j3\pi/4)$. When the phase estimation error is $\Delta\theta(t_n^-)$ at $t = t_n^-$, the received baseband signal at $t = t_n^-$ is given by

$$\hat{z}(t_n^-) = z(t_n)\exp(j\Delta\theta(t_n^-)) \tag{4.14}$$

First, the ACT controller detects in which quadrant $\hat{z}(t_n^-)$ exists. When it is in the q-th quadrant ($1 \le q \le 4$), the ACT controller immediately sends a control signal to the phase shifter to satisfy that the phase of $\hat{z}(t_n^+)$ becomes $(q-1)\pi/2 + \pi/4$. The relationship between $\hat{z}(t_n^-)$, $\hat{z}(t_n^+)$, $\Delta\theta(t_n^-)$ and the the amount of shifted phase at the phase shifter $\Delta\phi_n$ are given as follows:

$$\begin{aligned}\hat{z}(t_n^+) &= \hat{z}(t_n^-)\exp(-j\Delta\phi_n) \\ &= z(t_n)\exp(j\Delta\theta(t_n^-) - j\Delta\phi_n) \\ &= \exp(j\frac{(q-1)}{2}\pi + \frac{\pi}{4}) \end{aligned} \tag{4.15}$$

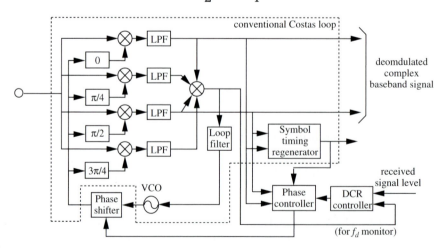

Fig. 4.6 Configuration of the adaptive carrier tracking (ACT) circuit (from Ref. 4-3, © Institute of Electrical and Electronic Engineers, 1989).

The reference carrier phase is controlled by both VCO and ACT, and these two controls could make the loop operation unstable. To avoid this instability, ACT is performed only when $\Delta\theta(t_n^-) \geq \phi_p$, ($\phi_p$, is called the phase margin) is satisfied.

Figure 4.7 shows the BER performance of QPSK using a Costas loop with ACT and the BER performance of QPSK using a conventional Costas loop under AWGN and flat Rayleigh fading conditions, where $f_d T_s = 2.5 \times 10^{-3}$. As we can see in this figure, the ACT circuit suppresses irreducible errors due to fading by one order of magnitude at high E_s/N_0. However, the BER performance of the Costas loop with ACT is worse than the BER performance of the conventional Costas loop at lower E_s/N_0 under the flat Rayleigh fading condition as well as at any E_s/N_0 under AWGN conditions. To overcome this problem, the carrier recovery circuit shown in Figure 4.6 also has the dual-mode carrier recovery (DCR) controller that selects the conventional Costas loop mode or ACT mode. Because this decision depends on f_d and the received signal level, the DCR monitors these signals and makes a decision for the mode selection.

Figure 4.8 shows the BER performance of the QPSK with coherent detection using the DCR circuit, where $f_d T_s = 2.5 \times 10^{-3}$. We can see from this figure that the

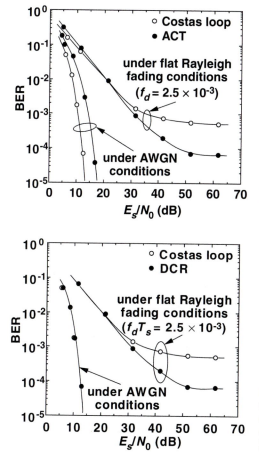

Fig. 4.7 BER performance of differentially encoded QPSK with coherent detection using the ACT circuit (from Ref. 4-3, © Institute of Electrical and Electronic Engineers, 1989).

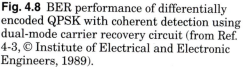

Fig. 4.8 BER performance of differentially encoded QPSK with coherent detection using dual-mode carrier recovery circuit (from Ref. 4-3, © Institute of Electrical and Electronic Engineers, 1989).

BER performance of the DCR is better than that of the conventional Costas loop at any E_s/N_0 under both AWGN and flat Rayleigh fading conditions.

Although we have discussed ACT and DCR circuits for QPSK, these techniques can also be applicable to any other modulation schemes, such as GMSK, Tamed FM, and $\pi/4$-QPSK.

4.2.2 Carrier Regeneration Based on Linear Mean Square (LMS) Estimation

As we have discussed in chapter 2, the flat Rayleigh fading phenomenon is interpreted such that a time-varying complex gain is multiplied to the transmitted signal. Therefore, if we can find an inverse of the multiplied complex gain, we can compensate for fading just by multiplying it by the received signal.

Figure 4.9 shows the receiver configuration with a fading compensator based on LMS estimation [4-4]. The received signal is picked up by the BPF, its envelope variation is suppressed by the AGC, its frequency drift is compensated for by using an AFC, and then the received signal is down-converted to the baseband using a local oscillator. Because this local oscillator is not perfectly synchronized to the phase variation due to fading, the received baseband signal is given by

$$
\begin{aligned}
u(t) &= u_I(t) + j \cdot u_Q(t) \\
&= c(t)z(t)
\end{aligned}
\tag{4.16}
$$

When we can find a complex gain $h(t)$ that satisfies

$$
h(t) = c^{-1}(t) = r^{-1}(t)\exp(-j\theta(t))
\tag{4.17}
$$

we can compensate for both envelope and phase distortion caused by fading. Let's discuss the compensation means principle.

Figure 4.10 shows the configuration of a fading compensation circuit based on the LMS estimation. Let's assume that the sampling timing of the received baseband signal is $t = nT_s$, where T_s is a symbol duration. Moreover, let's assume that c_n, z_n, u_n, and h_n mean $c(nT_s)$, $z(nT_s)$, $u(nT_s)$, and $h(nT_s)$, respectively. Using these notations, the baseband signal after fading compensation y_n is given by

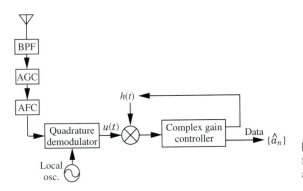

Fig. 4.9 Receiver configuration with a fading compensator based on LMS algorithm.

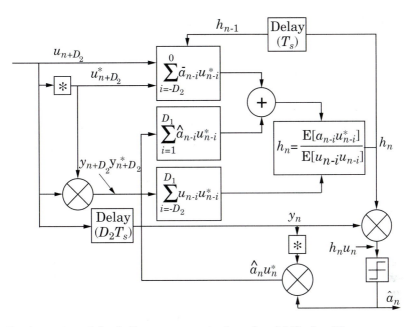

Fig. 4.10 Configuration of the fading compensator based on LMS algorithm.

$$y_n = y_{In} + j \cdot y_{Qn}$$
$$= h_n u_n \tag{4.18}$$

Using this signal, the transmitted symbol is detected by

$$\hat{a}_n = \text{sgn}(y_{In}) + j \cdot \text{sgn}(y_{Qn}) \tag{4.19}$$

where

$$\text{sgn}(x) = \begin{cases} 1 & (x \geq 0) \\ -1 & (x < 0) \end{cases} \tag{4.20}$$

and its estimation error is given by

$$e_n = \hat{a}_n - y_n \tag{4.21}$$

Based on the LMS estimation, the optimum h_n is a value that minimizes

$$E_n = E\left[|e_n|^2\right]$$
$$= E\left[|\hat{a}_n - y_n|^2\right] \tag{4.22}$$
$$= E\left[|\hat{a}_n - h_n u_n|^2\right]$$

This solution is given by Wiener-Hopf's equation [4-5] as

$$h_n = \frac{E\left[\hat{a}_n u_n^*\right]}{E\left[u_n u_n^*\right]}$$ (4.23)

where $E[\cdot]$ means an ensemble average of $[\cdot]$, and a^* means a complex conjugate of a.

Because the fading variation is very fast, the averaging period should be sufficiently short. Let's define the averaging period for h_n as $t = (n - D_1)T_s$ to $(n + D_2)T_s$. In this case, the number of samples to be averaged is given by $N = D_1 + D_2 + 1$. The relationship between D_1, D_2, and N is shown in Table 4.1. To estimate the optimum h_n using data for $(n - D_1)T_s \leq t \leq (n + D_2)T_s$, the circuit shown in Figure 4.10 includes a delay line with a delay time of $D_2 T_s$.

Table 4.1 Relationship between D_1, D_2, and N.

N	1	2	3	4	5
D_1	0	1	1	2	2
D_2	0	0	1	1	2

In equation (4.23), we can easily obtain its denominator as

$$E[u_n u_n^*] = \sum_{i=-D_2}^{D_1} u_{n-i} u_{n-i}^*$$ (4.24)

For the numerator in equation (4.23), on the other hand, although we have already obtained a_{n-i} ($i > 0$) at $t = nT_s$, we have not yet detected the symbol of a_{n-i} ($i \leq 0$) at $t = nT_s$. Therefore, we will suboptimally detect them as follows.

First, h_{n-1} and u_{n-i} ($i \leq 0$) are multiplied as

$$\begin{aligned}\bar{y}_{n-i} &= \bar{y}_{In-i} + j \cdot \bar{y}_{Qn-i} \\ &= h_{n-1} u_{n-i} \qquad (-D_2 \leq i \leq 0)\end{aligned}$$ (4.25)

and the result is quantized as

$$\begin{aligned}\bar{a}_{n-i} &= \bar{a}_{In-i} + j \cdot \bar{a}_{Qn-i} \\ &= \mathrm{sgn}(\bar{y}_{In-i}) + j \cdot \mathrm{sgn}(y_{Qn-i}) \quad (-D_2 \leq i \leq 0)\end{aligned}$$ (4.26)

to obtain the suboptimally estimated value of a_{n-i} ($-D_2 \leq i \leq 0$).

The numerator in equation (4.23) can then be obtained as

$$E[a_n u_n^*] = \sum_{i=-D_2}^{0} \bar{a}_{n-i} u_{n-i}^* + \sum_{i=1}^{D_1} \hat{a}_{n-i} u_{n-i}^*$$ (4.27)

For the circuit to be applied, differential encoding and decoding are essential because we cannot remove phase ambiguity. In this circuit, N has a big impact on the performance. Figure 4.11 shows the BER performance of this circuit with a parameter of N (a) under AWGN conditions and (b) under flat Rayleigh fading conditions ($f_d T_s = 5.0 \times 10^{-3}$). Under AWGN conditions, larger N gives better performance. On the other hand, under flat Rayleigh fading conditions, the BER performance at higher E_b/N_0 (at 40 dB in Figure 4.11) degrades with increasing N. To achieve satisfactory BER performance in both conditions, $N = 3$ is considered to be optimum.

Figure 4.12 shows BER performance of QPSK with differential detection, conventional Costas loop, and the LMS-type ($N = 3$) fading compensation schemes (a) under AWGN conditions and (b) under flat Rayleigh fading conditions ($f_d T_s = 5.0 \times 10^{-3}$). Under AWGN conditions, BER performance of the LMS-type fading compensation scheme is almost the same as that of the conventional Costas loop, and it is better than that of differential detection. On the other hand, under flat Rayleigh fading conditions, BER performance of the LMS-type fading compensation scheme is almost the same as that of differential detection, and it is 3–5 dB better than that of the conventional Costas loop. Therefore, the LMS type is considered to be superior to the differential detection under both static and flat Rayleigh fading conditions.

4.2.3 Summary of the Nonpilot Signal-Aided Fading Compensation Techniques

The modified PLL and LMS estimation techniques introduced here are very effective for achieving coherent detection in wireless communication systems. However, they have a common drawback—they need differential encoding and decoding that doubles the BER. Although the degradation due to differential encoding and decoding is very small under AWGN conditions (less than 0.5 dB), it becomes very

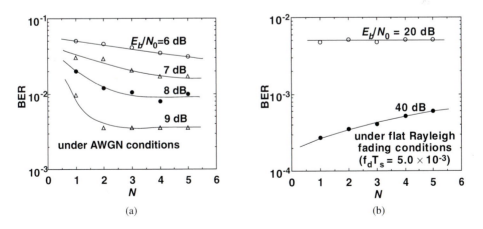

Fig. 4.11 BER performance of differentially encoded QPSK with coherent detection using the LMS-type fading compensation (a) under static conditions and (b) under flat Rayleigh fading conditions ($f_d T_s = 5.0 \times 10^{-3}$).

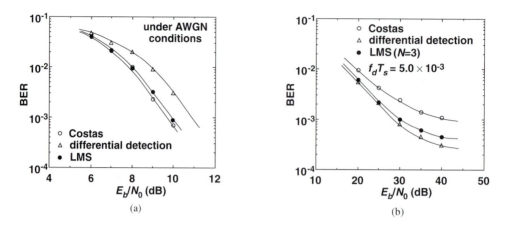

Fig. 4.12 BER performance comparison between differential detection, coherent detection with Costas loop, and coherent detection with LMS-type fading compensation scheme, where 32 ksymbol/s differentially encoded QPSK is employed (a) under static conditions and (b) under Rayleigh fading conditions ($f_d = 80$ Hz).

large (3 dB) under flat Rayleigh fading conditions. Therefore, absolute phase mapping is preferable to improve receiver sensitivity of the terminals under flat Rayleigh fading conditions. Pilot symbol-aided calibration techniques will solve this problem.

4.3 PILOT SIGNAL-AIDED CALIBRATION TECHNIQUES

When we want to apply absolute phase encoding schemes, we have to estimate carrier frequency as well as its phase variation due to fading with no ambiguity. Moreover, we also have to estimate amplitude variation if we want to employ QAM as a modulation scheme.

To accurately estimate fading variation, pilot signal-aided calibration techniques are widely used in wireless communication systems. There are three types of the pilot signal-aided calibration techniques.

☞ Pilot tone-aided techniques in which one or more tone (CW) signal(s) and the information signal are multiplexed in the frequency domain (frequency division multiplexing [FDM] type)

☞ Pilot symbol-aided techniques in which a known pilot symbol sequence and the information symbol sequence are multiplexed in the time domain (time division multiplexing [TDM] type)

☞ Pilot code-aided technique in which a spread spectrum signal using a spreading code orthogonal to that for the information (traffic) channel(s) and the traffic channel are multiplexed (code division multiplexing [CDM] type)

Figure 4.13 shows classification of the pilot signal-aided calibration techniques. We will discuss each technique in detail.

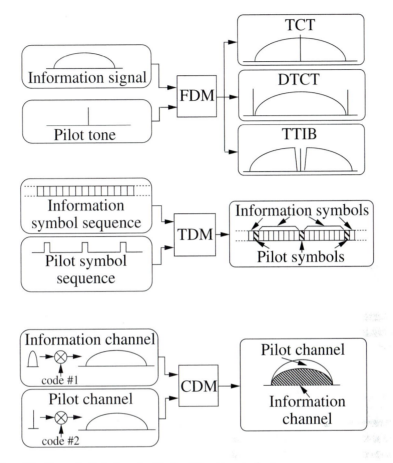

Fig. 4.13 Classification of pilot signal-aided calibration techniques.

4.3.1 Pilot Tone-Aided Techniques (FDM Type)

When we transmit a carrier component simultaneously with the modulated signal over a flat Rayleigh fading channel, we can exactly regenerate a reference signal of the received signal with no phase ambiguity. In this case, however, it is desirable to make them orthogonal with each other because we have to discriminate these two components at the receiver.

The simplest way to achieve orthogonality between a modulated signal and its pilot tone is proposed by Yokoyama [4-6], in which a reference carrier is transmitted in the quadrature channel of the BPSK-modulated channel (BPSK+CW scheme). In this scheme, the transmitted signal is expressed as

$$s_T(t) = s_S(t) + s_P(t) \tag{4.28}$$

where $s_S(t)$ is the BPSK-modulated signal expressed as

$$s_S(t) = A_S \cos(2\pi f_c t + \phi_D(t)) \tag{4.29}$$

$$\phi_D(t) = 0, \ or \ \pi \tag{4.30}$$

and $s_P(t)$ is its pilot tone signal expressed as

$$s_P(t) = -A_P \sin(2\pi f_c t) \tag{4.31}$$

Figure 4.14 shows a configuration of the transmitter and receiver [4-6]. Because $s_S(t)$ and $s_P(t)$ experience the same fading variation when the propagation path characteristic is flat fading, the received signal after BPF (its bandwidth is W_{IF}) is given by

$$
\begin{aligned}
s_R(t) = r(t)A_S a(t)\cos(2\pi f_c t + \theta(t)) - r(t)A_P \sin(2\pi f_c t + \theta(t)) \\
+ n_c(t)\cos(2\pi f_c t) - n_s(t)\sin(2\pi f_c t)
\end{aligned} \tag{4.32}
$$

where the first term corresponds to the BPSK-modulated signal, the second term corresponds to the pilot signal, and third and fourth terms correspond to the noise components. $a(t)$ represents the baseband signal of the BPSK signal including distortion due to transmitter and receiver filters.

At the receiver, the pilot signal is picked up by a narrowband BPF (BPF$_N$), with its bandwidth of W_N followed by a $\pi/2$-phase shifter. At this stage, W_N has to be wider than $2f_d$ to pick up the fading variation exactly. As a result, the obtained pilot tone component is given by

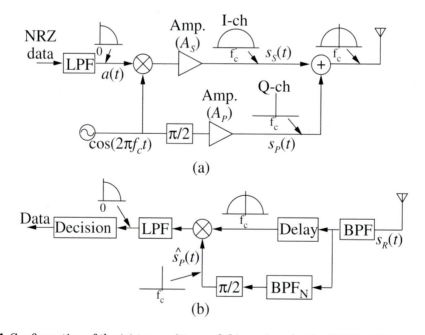

Fig. 4.14 Configuration of the (a) transmitter and (b) receiver for the BPSK+CW system.

$$\hat{s}_P(t) = r(t)A_P \cos(2\pi f_c t + \theta(t)) + \text{distortion} \tag{4.33}$$

in which distortion is caused by noise as well as by the BPSK-modulated component. $s_R(t)$ is then coherently demodulated using this signal to regenerate the transmitted data sequence. At this stage, $s_R(t)$ has to be delayed to compensate for the delay time produced by the BPF_N.

In the BPSK+CW scheme, optimization of the ratio of the pilot signal power (P_P) to the total power (P_T) is a key parameter in determining its performance. Ref. [4-7] shows that the theoretical BER performance of the BPSK+CW scheme under flat Rayleigh fading conditions is given by

$$P_{non}(r_0) = \frac{1}{2}\left[1 - \frac{1}{1 + \dfrac{a + k(1 - 2\sqrt{a})}{k(1-k)r_0}}\right] \tag{4.34}$$

where

$$k = \frac{P_P}{P_T} = \frac{A_p^2}{\left(A_S^2 + A_P^2\right)} \tag{4.35}$$

$$a = \frac{W_N}{W_{IF}} \ll 1 \tag{4.36}$$

Equation (4.34) shows that the BER performance depends on a and k. Figure 4.15 shows BER versus $k = P_P/P_T$ with a parameter of a, where $W_{IF}T_s = 1.0$ is assumed. This figure shows that

☞ the optimum k (k_{opt}) becomes small with decreasing a because the SNR of the filtered pilot tone is improved with narrower W_N.

☞ BER is degraded with decreasing k at $k < k_{opt}$ because the SNR of the filtered pilot tone is degraded.

☞ BER is degraded with increasing k at $k > k_{opt}$ because power redundancy is increased.

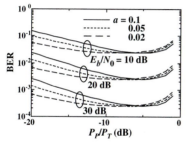

Fig. 4.15 BER versus k performance with a parameter of a, where $W_{IF}T_s = 1.0$.

For these reasons, Hou et al. [4-7] found that the optimum k was around −10 dB.

Although the BPSK+CW scheme is very simple to achieve pilot tone-assisted coherent demodulation, we cannot apply it to the quadrature modulation scheme, such as QPSK. A more complex scheme applicable to the quadrature modulation scheme is the tone calibration technique (TCT) proposed by Davarian [4-8]. In the TCT scheme, a spectrum notch is created at the center of the spectrum by using the balanced encoding scheme such as the Manchester code to keep orthogonality between the modulated signal and the pilot tone.

Figure 4.16 shows the configuration of the (a) transmitter and (b) receiver using TCT. The transmitted data sequence is split into in-phase and quadrature streams. A balanced encoding is applied to intentionally create a null at the center of the spectrum followed by an LPF. Then, quadrature modulation and a pilot tone insertion are performed. At the receiver, after the transmitted spectrum is picked up by a BPF with its bandwidth of W_{IF}, the pilot tone is picked up by a BPF with its bandwidth of W_N and hard-limited to make its amplitude constant, where W_N is much narrower than W_{IF}. After this process, the received signal is coherently detected using the filtered pilot tone, and the transmitted data sequence is regenerated. At this stage, although the in-phase channel of the demodulated signal includes direct current (DC) offset due to the pilot tone, it can be removed by using matched filter techniques or some simple DC offset techniques.

Optimization of the ratio of the pilot power to the total power (P_P/P_T) was theoretically investigated in Ref. [4-8]. The results show that, when $W_N/W_{IF} = 0.05$, the optimum P_P/P_T is around 0.25, which degrades the BER performance by less than 2.0

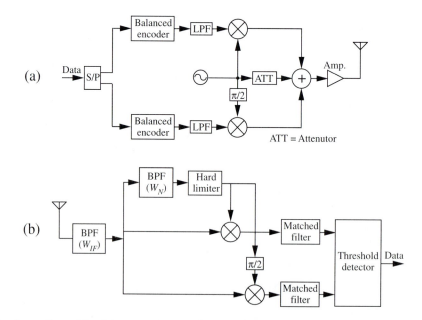

Fig. 4.16 Configuration of the (a) transmitter and (b) receiver using TCT.

dB, and the degradation does not depend on the E_b/N_0. When $W_N/W_{IF} = 0.02$, the optimum P_P/P_T is around 0.15, and its degradation is only 1.0 dB.

Although Manchester coding can easily create a spectral notch, it doubles its spectrum. Another spectral-nulling technique is binary block coding (BBC) [4-9; 4-10]. A BBC coder divides the data sequence into $(N - 1)$-discrete signaling blocks and insert one redundant status symbol (+1). First, it calculates a digital sum (DS) for each block and compares the result with the accumulated running digital sum (RDS) up to the previous block. When they have the same polarity, all N symbols in the block are inverted before transmission. Otherwise, the block is transmitted as it is. Because this process limits the magnitude of the RDS to be less than N, it can create a spectrum null at 0 Hz.

When we select larger N, the spectral null gets narrower and its overhead becomes smaller. Ref. [4-9] shows that, when we assume the maximum Doppler frequency of around 100 Hz and the symbol rate of around 20 ksymbol/s, $N = 11$ can create sufficient spectrum null to insert a pilot tone, and a pilot tone with $P_P/P_T = -10$ dB can achieve coherent detection with its degradation of less than 1 dB from the theoretical one.

Another pilot tone-aided technique is the dual-tone calibration technique (DTCT) proposed by Simon [4-11], in which a tone is located on each side of the signal spectrum. Figure 4.17 shows a spectrum of the transmitted signal using the DTCT. Pilot tones are placed at the frequency of $f_c + f_p$ and $f_c - f_p$.

Figure 4.18 shows a receiver configuration for DTCT. BPF_{N1} picks up a pilot tone with a frequency of $f_c + f_p$, and BPF_{N2} picks up another pilot tone with a frequency of $f_c - f_p$. When these two signals are multiplexed, hard limited and frequency divided by two, we can obtain a reference signal including the phase variation due to fading.

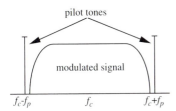

Fig. 4.17 Spectrum of the transmitted signal using DTCT.

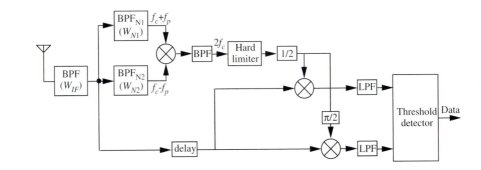

Fig. 4.18 Configuration of the transmitter and receiver using DTCT.

Ref. [4-11] shows that, although the optimum P_P/P_T for DTCT is almost the same as that of the TCT, it is more sensitive to the E_b/N_0.

The advantages of this method are that

☞ The required bandwidth is narrower than that for the TCT.

☞ It is easy to make orthogonality between pilot tones and signal spectrum.

However, this scheme has drawbacks, such as

☞ The pilot tones are more sensitive to the frequency shift and ACIs because they are located on the edge of the assigned band.

☞ Characteristic imbalances between BPF_{N1} and BPF_{N2} may cause some distortion.

☞ Frequency division causes phase ambiguity of the regenerated carrier. Therefore differential encoding and decoding are necessary, which doubles the BER in comparison with the performance with absolute phase encoding and decoding.

The transparent tone-in band (TTIB) scheme proposed by Bateman et al. [4-12; 4-13; 4-14] is another pilot tone scheme that combines advantages of the TCT and DTCT.

☞ A pilot tone is located at the center of the spectrum, which makes the pilot tone less sensitive to the frequency shift and ACI.

☞ A pilot tone and signal spectrum is completely orthogonal with each other owing to the signal spectrum splitting technique.

Figure 4.19 shows a configuration of the transmitter using TTIB. The baseband signal $A(f)$ is first split into two components, $B(f)$ and $C(f)$, using a pair of quadrature mirror filters (QMFs), $H_{1-}(f)$ and $H_{1+}(f)$, as

$$B(f) = H_{1-}(f)A(f) \tag{4.37}$$

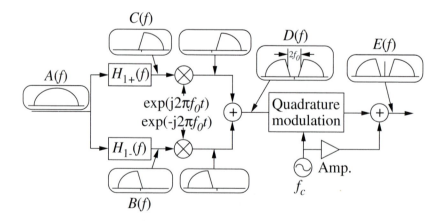

Fig. 4.19 Configuration of the transmitter using the TTIB technique.

$$C(f) = H_{1+}(f)A(f) \tag{4.38}$$

where $H_{1-}(f)$ and $H_{1+}(f)$ satisfy the following:

$$H_{1-}(f) = H_{1+}(-f) \tag{4.39}$$

$$\left|H_{1-}(f)\right|^2 + \left|H_{1+}(f)\right|^2 = 1 \tag{4.40}$$

To create a spectral null at the center of $A(f)$, $B(f)$ and $C(f)$ are frequency shifted by $-f_0$ and f_0, respectively. The frequency shift value, f_0, is set to accommodate the expected Doppler spread. The TTIB-processed baseband signal is thus given by

$$\begin{aligned} D(f) &= B(f + f_0) + C(f - f_0) \\ &= A(f + f_0)H_{1-}(f + f_0) + A(f - f_0)H_{1+}(f - f_0) \end{aligned} \tag{4.41}$$

Then, quadrature modulation and pilot tone insertion are performed and transmitted. As a result, the spectrum of the transmitted signal is given by

$$\begin{aligned} E(f) &= \frac{1}{2}\left[D(f - f_c) + D(f + f_c)\right] \\ &\quad + \frac{1}{2}\left[A_p\delta(f - f_c) + A_p\delta(f + f_c)\right] \end{aligned} \tag{4.42}$$

Figure 4.20 shows a configuration of the TTIB receiver. The received signal is split into two paths. One path uses a band elimination filter (BEF) to remove the pilot signal, and the second path picks up the pilot tone using a BPF. Using this pilot tone, the envelope variation due to fading is estimated and compensated for. At the same time, the reference carrier for quadrature demodulation is produced by hard limiting this signal. As a result, the TTIB baseband signal can be obtained. The complementary frequency translation process is then carried out using frequencies $f_0 + f_{off}$ and $-(f_0 + f_{off})$, where f_{off} represents an arbitrary frequency error between frequency translators of the transmitter and receiver. The lower subband signal after frequency translation is given by

$$F(f) = D(f - (f_0 + f_{off})) \tag{4.43}$$

and the upper subband signal after frequency translation is given by

$$G(f) = D(f + (f_0 + f_{off})) \tag{4.44}$$

These signals are then filtered to pick up each subband signal as

$$\begin{aligned} K(f) &= F(f)H_{1-}(f) \\ &\cong A(f - f_{off})\left|H_{1-}(f)\right|^2 \end{aligned} \tag{4.45}$$

$$\begin{aligned} L(f) &= G(f)H_{1+}(f) \\ &\cong A(f + f_{off})\left|H_{1+}(f)\right|^2 \end{aligned} \tag{4.46}$$

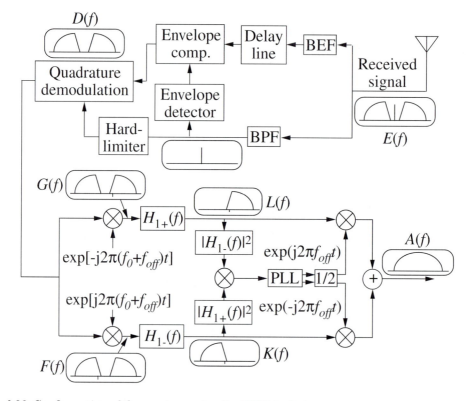

Fig. 4.20 Configuration of the receiver using the TTIB technique.

where an approximation of $H_{1-}(f + f_{off}) = H_{1-}(f)$ and $H_{1+}(f + f_{off}) = H_{1+}(f)$ are used in equations (4.45) and (4.46). Because $K(f)$ and $L(f)$ include offset frequency f_{off}, we have to estimate and compensate for f_{off}. Estimation of f_{off} is carried out this way.

First, the lower subband is band-limited by a filter with its transfer function of $|H_{1+}(f)|^2$, and the upper subband is band-limited by a filter with its transfer function of $|H_{1-}(f)|^2$. Because the difference product of these two waveforms includes components having the frequencies of $+2f_{off}$ and $-2f_{off}$, and because they are not affected by fading, we can pick up these components using PLL. The desired components, $+f_{off}$ and $-f_{off}$, can be obtained by the frequency division. In this process, however, f_{off} and $-f_{off}$ have phase ambiguity caused by the frequency division. In the case of DTCT, we have to apply differential encoding and decoding because phase ambiguity is dynamically changed due to fading. On the other hand, phase ambiguity in the case of TTIB is caused only by spectral splitting, which means that phase ambiguity is stational. Therefore, we can remove this ambiguity by inserting a small amount of redundant reference symbols into the transmitted data sequence. After f_{off} in $K(f)$ and $L(f)$ is compensated for, these two signals are added to reproduce the transmitted signal. As for the P_P/P_T ratio, around -10 dB is reported to be optimum [4-13].

The features of the TTIB include

☞ A spectral null with an arbitrary bandwidth can easily be created.
☞ Distortion due to the spectral null creation is very small.
☞ Phase as well as amplitude distortion due to fading can be accurately compensated for.
☞ It is applicable to any modulation scheme.
☞ Absolute phase-encoded coherent detection is possible by inserting a small amount of redundant reference symbols.

Owing to these features, the TTIB is widely accepted as one of the most promising pilot tone-aided fading compensation techniques, especially for spectral efficient modulation schemes, such as 16QAM [4-14; 4-15].

Figure 4.21 shows laboratory experimental results of the BER performance of 16QAM with TTIB under flat Rayleigh fading conditions [4-15]. The results show that the measured BER performance agrees very well with the theoretical performance.

4.3.2 Pilot Symbol-Aided Techniques (TDM Type)

One very important characteristic of fading is that its variation is very smooth owing to its spectrum limitation in the range of $f_c - f_d \le f \le f_c + f_d$. Pilot symbol-aided schemes [4-16; 4-17; 4-18; 4-19] actively use this characteristic. Although the pilot symbol-aided scheme can be applied to any modulation scheme, we will discuss its principles in the case of 16QAM.

Figure 4.22 shows a constellation variation of 16QAM signal under flat Rayleigh fading conditions, where

$z(t)$ = transmitted complex baseband signal expressed as

$$z(t) = z_I(t) + j \cdot z_Q(t) \tag{4.47}$$

$c(t)$ = complex fading variation expressed as

$$c(t) = c_I(t) + j \cdot c_Q(t) \tag{4.48}$$

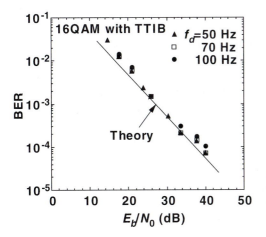

Fig. 4.21 BER performance of 16QAM using the TTIB technique in Rayleigh fading environments (from Ref. 4-15, © Institute of Electrical and Electronic Engineers, 1991).

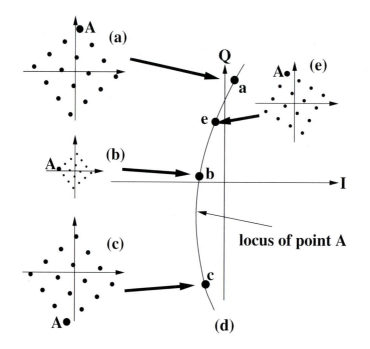

Fig. 4.22 Constellation variation due to fading and the concept of pilot symbol-aided fading estimation technique.

$u(t)$ = received complex baseband signal expressed as

$$u(t) = u_I(t) + j \cdot u_Q(t) \tag{4.49}$$

When $z(t)$ is transmitted over the fading channel, $u(t)$ is given by

$$u(t) = c(t)z(t) + n(t) \tag{4.50}$$

where $n(t)$ is the additive white Gaussian noise expressed as an equivalent low-pass system. This equation shows that the fading variation changes both the size and angle of the constellation of the received signal. Moreover, this equation suggests that, if we regularly transmit a known pilot symbol a_f—for example, a symbol having the maximum amplitude in the first quadrant (point A in Figure 4.22)—we can know exactly the signal state diagram at every pilot symbol timing as shown in Figure 4.22(a)–(c). Moreover, we know that the fading variation is subject to the band-limited Gaussian random process, which means the variation is very smooth. Therefore, we can exactly estimate the locus of point A as shown in Figure 4.22(d). As a result, the signal state diagram for nonpilot symbol timings can be estimated as shown in Figure 4.22(e). Practically, we can estimate the fading variation this way.

At the pilot symbol timing $t = t_1$, we can measure $c(t_1)$ at the receiver as

$$\hat{c}(t_1) = u(t_1)/a_F + n(t_1)/a_F \tag{4.51}$$

Equation (4.51) means that $\hat{c}(t_1)$ is the sampled data of $c(t_1)$ and it is corrupted by noise $(n(t)/a_F)$. Now, let's assume that an information symbol sequence with its symbol rate of R_s (ksymbol/s) and a pilot symbol sequence with its symbol rate of R_p (ksymbol/s) are multiplexed in the time domain, and the relationship between their symbol rates is given by

$$R_s = (N-1)R_p \qquad (4.52)$$

Figure 4.23 shows this multiplexing process. A pilot symbol is regularly inserted every $(N-1)$ symbols. It means that the symbol rate after multiplexing becomes $NR_s/(N-1)$ ksymbol/s. Because a pilot symbol is inserted every $T_F = 1/R_p$ second, we can measure the fading variation every T_F second at the receiver. Figure 4.24 shows an example of the fading variation estimation procedure using this pilot symbol-aided technique. When in-phase and quadrature-phase components of fading are changing as shown in Figure 4.24(a), the sampled data sequence

$$\bar{c}(t) = \sum_{k=-\infty}^{\infty} c(kT_F)\delta(t-kT_F) \qquad (4.53)$$

can be obtained as shown in Figure 4.24(b). Since the fading variation is subject to a band-limited Gaussian random process as we have discussed before, we can estimate $c(t)$ by interpolating $\bar{c}(t)$ as shown in Figure 4.24(c).

For the pilot symbol-aided fading estimation techniques, accuracy of the fading estimation depends on the selection of interpolation techniques. Theoretically, the optimum interpolation scheme is the Nyquist interpolation. Using this interpolation, the fading variation at $t = kT_F + mT_s = (k + m/N)T_F$ is estimated as

$$\hat{c}((k+m/N)T_F) = \sum_{k=-\infty}^{\infty} \frac{\sin((k+m/N)\pi)}{(k+m/N)\pi}\bar{c}(kT_F) \qquad (4.54)$$

In this case, the optimum pilot symbol period is given by $T_F = 1/(2f_{dmax})$, where f_{dmax} is the expected maximum value of f_d. Because this process corresponds to transmit $\bar{c}(t)$

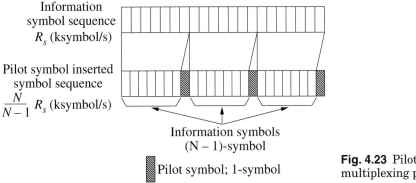

Fig. 4.23 Pilot symbol multiplexing process.

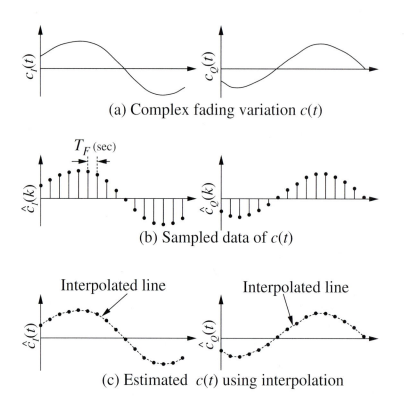

(a) Complex fading variation $c(t)$

(b) Sampled data of $c(t)$

(c) Estimated $c(t)$ using interpolation

Fig. 4.24 Fading variation estimation procedure using the pilot symbol-aided technique: (a) fading variation, (b) measured sequence of the fading variation, and (c) estimated fading variation using interpolation.

through an ideal LPF with its bandwidth of f_{dmax}, it can regenerate exactly the fading variation $c(t)$.

Although Nyquist interpolation gives very accurate interpolation performance with the smallest redundancy, it requires a very long processing delay due to its long impulse response. When we accept more redundancy, we can apply a more simple interpolation scheme. Gaussian interpolation is one of the simpler interpolation schemes. The estimated fading variation at $t = (k + m/N)T_F$ using second-order Gaussian interpolation is given by

$$\hat{c}((k+m/N)T_F) = \sum_{i=-1}^{1} Q_k(m)\bar{c}((k+i)T_F) \tag{4.55}$$

where

$$Q_{-1}(m) = \frac{1}{2}\left\{\left(\frac{m}{N}\right)^2 - \left(\frac{m}{N}\right)\right\} \tag{4.56a}$$

$$Q_0(m) = 1 - \left(\frac{m}{N}\right)^2 \qquad (4.56b)$$

$$Q_1(m) = \frac{1}{2}\left\{\left(\frac{m}{N}\right)^2 + \left(\frac{m}{N}\right)\right\} \qquad (4.56c)$$

When we apply first-order interpolation, equations (4.57a)–(4.57c) are used instead of equations (4.56a)–(4.56c).

$$Q_{-1}(m) = 0 \qquad (4.57a)$$

$$Q_0(m) = 1 - \left(\frac{m}{N}\right) \qquad (4.57b)$$

$$Q_1(m) = \frac{m}{N} \qquad (4.57c)$$

When we apply zeroth-order interpolation, equations (4.58a)–(4.58c) are used instead of equations (4.56a)–(4.56c).

$$Q_{-1}(m) = 0 \qquad (4.58a)$$

$$Q_0(m) = 1 \qquad (4.58b)$$

$$Q_1(m) = 0 \qquad (4.58c)$$

Table 4.2 shows BER performance of 16QAM with the parameters of E_b/N_0 and the interpolation order, where $f_d \cdot T_F = 8 \times 10^{-2}$, $f_d \cdot T_s = 5 \times 10^{-3}$, and $f_{off} = 0$ Hz. In the case of $E_b/N_0 = 20$ and 30 dB, the performances of first- and second-order interpolations are superior to that of the zeroth-order interpolation, and the performances of the first- and second-order interpolations are almost the same.

In the case of $E_b/N_0 = 40$ dB, on the other hand, the performance of the second-order interpolation is better than that of the first-order interpolation. Thus, second-order interpolation is considered to be sufficient. In the following discussion in this section, second-order interpolation will be used.

In the case of second-order Gaussian interpolation, it is equivalent to pass $\bar{c}(t)$ through an LPF with its impulse response of

Table 4.2 Relationship between BER and interpolation order for 16QAM [4-18].

Order	$E_b/N_0 = 20$ dB	$E_b/N_0 = 30$ dB	$E_b/N_0 = 40$ dB
0	5.3×10^{-2}	4.9×10^{-2}	4.8×10^{-2}
1	5.0×10^{-3}	5.2×10^{-4}	1.5×10^{-4}
2	5.2×10^{-3}	5.4×10^{-4}	4.9×10^{-5}

$$
g(t) = \begin{cases}
\dfrac{1}{2}\left[\left(\dfrac{t}{T_F}\right)^2 + 3\left(\dfrac{t}{T_F}\right) + 2\right], & for -T_F \le t \le 0 \\[2ex]
1 - \left(\dfrac{t}{T_F}\right)^2, & for\ \ 0 \le t \le T_F \\[2ex]
\dfrac{1}{2}\left[\left(\dfrac{t}{T_F}\right)^2 - 3\left(\dfrac{t}{T_F}\right) + 2\right], & for T_F \le t \le 2T_F
\end{cases}
\tag{4.59}
$$

Ref. [4-18] shows that the maximum value of f_d to be compensated for is $0.1/T_F$ in the case of second-order Gaussian interpolation, and it is $0.5/T_F$ in the case of ideal Nyquist interpolation. This means that the second-order Gaussian interpolation requires five times more redundancy than the ideal Nyquist interpolation although its processing delay is small. Of course, accuracy of interpolation becomes higher with increasing interpolation order, although it requires longer processing delay time. To reduce such redundancy without increasing processing delay so much, another interpolation function, such as Lagrange and ideal LPF with raised cosine window, can be used [4-20]. The results show that Lagrange interpolation can cope with 50% higher maximum Doppler frequency than second-order Gaussian interpolation with the same redundancy. Another approach toward less redundant interpolation techniques is proposed by Cavers. The proposed method employs a maximum likelihood function to interpolate the measured fading variation. The result shows very good estimation accuracy with small redundancy [4-19].

Figure 4.25 shows computer-simulated BER performances of QPSK, 16QAM, 64QAM, and 256QAM modems using pilot symbol-aided fading compensation in flat Rayleigh fading environments, where Gray coding with absolute phase coherent detection is employed for all the modulation scheme, second-order interpolation is applied, $f_d T_s = 2.5 \times 10^{-3}$, and $f_d T_F = 4 \times 10^{-2}$. Although the theoretical lines are not written in this figure to prevent complication of the drawing, degradation for all modulation schemes are about 2 dB. The causes of this degradation are as follows [4-18]:

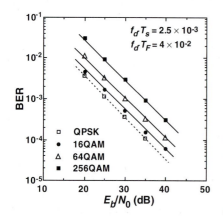

Fig. 4.25 Computer-simulated result of the BER performance of Gray-coded QPSK, 16QAM, 64QAM, and 256QAM using the pilot symbol-aided fading compensation technique with a second-order Gaussian interpolation, where $f_d T_s = 2.5 \times 10^{-2}$ and $f_d T_F = 4 \times 10^{-2}$.

☞ Power loss due to insertion of the pilot symbols

☞ Degradation due to inaccuracy of the interpolation

☞ Degradation by noise included in pilot symbols

Ref. [4-18] shows that the total degradation is theoretically obtained as approximately 2 dB. Therefore, Figure 4.25 is considered to show the theoretical limit of the BER performances of pilot symbol-aided QPSK, 16QAM, 64QAM, and 256QAM.

This technique is also examined by the laboratory and field experiments in the case of 16QAM/FDMA and 16QAM/TDMA systems and is confirmed to be very effective as a flat Rayleigh fading compensation technique in land mobile communications [4-18; 4-21; 4-22].

4.3.3 Pilot Code-Aided Techniques (CDM Type)

The concept of pilot code-aided techniques is practically applied to the direct sequence/code division multiple access (DS/CDMA) systems, in which a pilot channel is multiplexed with the traffic channel(s) using a spreading code orthogonal with those of the traffic channels. For the application of a pilot code-aided scheme in the downlink, the BS multiplexes all the traffic channels and a pilot channel by multiplying orthogonal codes like the Walsh code with each traffic channel and a pilot channel. In this case, each traffic channel can commonly use the pilot channel as a propagation path sounder because all the traffic channels and the pilot channel are transmitted from the same BS [4-23].

On the other hand, in the uplink, each traffic channel has to prepare its own pilot channel because propagation path characteristics of each traffic channel are independent of the others [4-24]. Therefore, the basic concept of the CDM-type pilot signal insertion schemes for downlink and uplink are quite different.

Moreover, the most important advantage of the pilot code-aided technique is that we can obtain the delay profile of the channel for the received signal; thereby the delayed paths are coherently combined using Rake diversity techniques [4-25; 4-26]. However, this section only explains the basic idea of pilot code-aided techniques under flat fading conditions, and its application to frequency-selective fading will be discussed in chapter 5.

4.3.3.1 Pilot Channel Multiplexing for Downlink   Figure 4.26 shows a transmitter configuration of the pilot channel-multiplexed CDMA for the downlink. Let's assume that the baseband signal is given by

$$a(t) = \sum_{k=-\infty}^{\infty} a_k \delta(t - nT_k) \tag{4.60}$$

$$a_k = a_{Ik} + j \cdot a_{Qk} \tag{4.61}$$

where a_{Ik} and a_{Qk} are in-phase and quadrature-phase components of the k-th symbol of the traffic channel, T_s is a symbol duration, and $\delta(t)$ is Dirac's delta function. On the other hand, the baseband signal of the pilot channel is given by

$$b(t) = 1 + j \cdot 0 \tag{4.62}$$

To keep orthogonality between these signals, each of the traffic and pilot channels is multiplied by a different Walsh code. Let's assume that the Walsh code for the traffic channel $w^T(t)$ and for the pilot channel $w^P(t)$ are given by

$$w^T(t) = \sum_{n=-\infty}^{\infty} w^T_{(n \bmod N)} \delta(t - nT_c) \tag{4.63}$$

$$w^P(t) = \sum_{n=-\infty}^{\infty} w^P_{(n \bmod N)} \delta(t - nT_c) \tag{4.64}$$

where N is the number of symbols in the Walsh code, and T_c is a symbol duration of the Walsh code. At this stage, orthogonality between the traffic and pilot channels is satisfied. When there are many traffic channels, we have to prepare different Walsh codes for each of the traffic channels to keep orthogonality between traffic channels. Let's assume an all 1 pattern as the Walsh code for the pilot channel to simplify discussion.

Then, a pseudorandom noise (PN) sequence expressed as

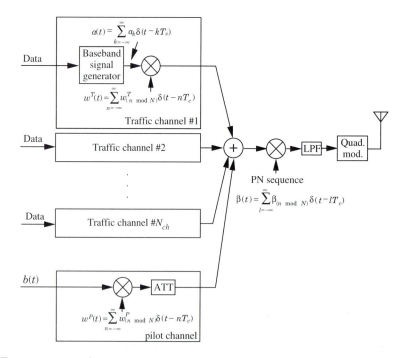

Fig. 4.26 Transmitter configuration of the pilot channel-multiplexed CDMA for the downlink.

$$\beta(t) = \sum_{l=-\infty}^{\infty} \beta_{(l \mod N_p)} \delta(t - lT_c)$$

(4.65)

$$= \sum_{l=-\infty}^{\infty} \beta_{I(l \mod N_p)} \delta(t - lT_c) + j \cdot \sum_{l=-\infty}^{\infty} \beta_{Q(l \mod N_p)} \delta(t - lT_c)$$

$$\beta_{(l \mod N_p)} = \beta_{I(l \mod N_p)} + j \cdot \beta_{Q(l \mod N_p)}$$

(4.66)

is multiplied by the Walsh code-multiplied baseband signals of both the traffic and pilot channels, where N_p is the number of PN symbols in one period. Let's assume that its chip duration is the same as that of the Walsh code (T_c). These signals are then band-limited by the LPF and are fed to the quadrature modulator. The modulated signals of the traffic channel and pilot channel after spreading are given by

$$s^T(t) = \sum_{k=-\infty}^{\infty} \sum_{l=0}^{N-1} a_k w_l^T \beta_{(kN+l \mod N_p)} c_T(t - kT_s - lT_c) \exp(j2\pi f_c t)$$

(4.67)

$$s^P(t) = \sum_{k=-\infty}^{\infty} \sum_{l=0}^{N-1} \beta_{(kN+l \mod N_p)} c_T(t - kT_s - lT_c) \exp(j2\pi f_c t)$$

(4.68)

$$= \sum_{n=-\infty}^{\infty} \beta_{(n \mod N_p)} c_T(t - nT_c) \delta(t - nT_c) \exp(j2\pi f_c t)$$

where $n = kN + l$, $c_T(t)$ is the impulse response of the transmitter LPF, and f_c is carrier frequency.

When traffic and pilot signals are transmitted via a multipath fading channel with its impulse response of $c_{ch}(t)$, the received signal of the traffic channel $q^T(t)$ and that of the pilot channel $q^P(t)$ are given by

$$q^T(t) = \sum_{k=-\infty}^{\infty} \sum_{l=0}^{N-1} a_k w_l^T \beta_{(kN+l \mod N_p)} c_1(t - kT_s - lT_c) \exp(j2\pi f_c t)$$

(4.69)

$$q^P(t) = \sum_{k=-\infty}^{\infty} \sum_{l=0}^{N-1} \beta_{(kN+l \mod N_p)} c_1(t - kT_s - lT_c) \exp(j2\pi f_c t)$$

(4.70)

where

$$c_1(t) = c_T(t) \otimes c_{ch}(t)$$

(4.71)

and the effect of noise is eliminated in equations (4.69) and (4.70) to simplify discussion. These signals are then band-limited by a BPF with its impulse response expressed as an equivalent low-pass system of $c_R(t)$, and down-converted to the baseband using a quadrature demodulator. The obtained baseband signals of the traffic and pilot channels are then given by

$$u^T(t) = \sum_{k=-\infty}^{\infty} \sum_{l=0}^{N-1} a_k w_l^T \beta_{(kN+l \mod N_p)} c(t - kT_s - lT_c) \tag{4.72}$$

$$u^P(t) = \sum_{k=-\infty}^{\infty} \sum_{l=0}^{N-1} \beta_{(kN+l \mod N_p)} c(t - kT_s - lT_c)$$

$$= \sum_{n=-\infty}^{\infty} \beta_{(n \mod N_p)} c(t - nT_c) \tag{4.73}$$

where

$$c(t) = c_T(t) \otimes c_{ch}(t) \otimes c_R(t) \tag{4.74}$$

First, let's focus on the pilot channel. As will be mentioned in chapter 5, when a digital modulated signal is transmitted via the multipath channel with its impulse response of $c(t)$ and $c(t)$ is time invariant, $c(t)$ can be identified as a discrete channel model given by

$$c(t) = \sum_{i=-\infty}^{\infty} c_i \delta(t - iT_c) \tag{4.75}$$

When equation (4.75) is substituted into equation (4.73), we can express $u^P(t)$ as

$$u^P(t) = \sum_{n=-\infty}^{\infty} \beta_{(n \mod N_p)} \sum_{i=-\infty}^{\infty} c_i \delta(t - (n+i)T_c)$$

$$= \sum_{i=-\infty}^{\infty} c_i \sum_{n=-\infty}^{\infty} \beta_{(n \mod N_p)} \delta(t - (n+i)T_c) \tag{4.76}$$

The despreading process is then carried out by multiplying the complex conjugate of a perfectly chip-timing-synchronized PN sequence. When we multiply $u^P(t)$ and the complex conjugate of the PN delayed by mT_c, we can obtain the following:

$$u^P(t)\beta^*(t - mT_c) = \sum_{i=-\infty}^{\infty} c_i \sum_{n=-\infty}^{\infty} \beta_{(n \mod N_p)} \delta(t - (n+i)T_c)$$

$$\times \sum_{j=-\infty}^{\infty} \beta^*_{(j \mod N_p)} \delta(t - (j+m)T_c)$$

$$= \sum_{i=-\infty}^{\infty} c_i \sum_{n=-\infty}^{\infty} \sum_{j=-\infty}^{\infty} \beta_{(n \mod N_p)} \beta^*_{(j \mod N_p)}$$

$$\times \delta(t - (n+i)T_c) \delta(t - (j+m)T_c) \tag{4.77}$$

$$= \sum_{i=-\infty}^{\infty} c_i \sum_{n=-\infty}^{\infty} \beta_{(n \mod N_p)} \beta^*_{(n+i-m \mod N_p)}$$

$$\times \delta(t - (n+i)T_c)$$

Because $\beta(t)$ is the PN sequence, the following relationship is satisfied.

$$\int_{t-N_pT_c/2}^{t+N_pT_c/2} \sum_{n=-\infty}^{\infty} \beta_{(n \text{ mod } N_p)} \beta^*_{(n+i-m \text{ mod } N_p)} \delta(t-(n+i)T_c)$$

$$= \begin{cases} -1 & (i \neq m) \\ N_p & (i-m) \end{cases} \tag{4.78}$$

Thus, when $u^P(t)\beta^*(t-mT_c)$ is integrated over $[t-N_pT_c/2, t+N_pT_c/2]$, for each m, we can estimate the channel impulse response as follows:

$$\hat{c}(t) = \sum_{i=-\infty}^{\infty} \hat{c}_m \delta(t-mT_c) \tag{4.79}$$

$$\hat{c}_m = \frac{1}{N_p} \int_{t-N_pT_c/2}^{t+N_pT_c/2} u^P(t)\beta^*(t-mT_c) \tag{4.80}$$

When the propagation path characteristics are flat Raleigh fading with its impulse response of

$$c(t) = c_0\delta(t) = r_0 \exp(j\theta_0)\delta(t) \tag{4.81}$$

we can achieve PN synchronization by scanning m from 0 to $N_p - 1$ and find an m that gives the maximum value of $|\hat{c}_m|^2$.

After that, the complex conjugate of the delay-time-compensated PN sequence and $w^T(t)$ are multiplied by $u^T(t)$ as

$$u^T(t)\beta^*(t)w^T(t)$$

$$= \sum_{k=-\infty}^{\infty} \sum_{l=0}^{N-1} a_k |w_l^T|^2 \left|\beta_{(kN+l \text{ mod } N_p)}\right|^2 c(t-kT_s-lT_c)$$

$$= N \sum_{k=-\infty}^{\infty} a_k c(t-kT_s-lT_c) \tag{4.82}$$

$$= N \sum_{k=-\infty}^{\infty} a_k c_0 \delta(t-kT_s)$$

Then we can obtain the transmitted symbol sequence by multiplying $1/N$, $c_0^*/|c_0|^2$ and $u^T(t)\beta^*(t)w^T(t)$ as

$$\hat{a}(t) = \sum_{k=-\infty}^{\infty} \hat{a}_k \delta(t-kT_s)$$

$$= \frac{c_0^*}{N|c_0|^2} u^T(t)\beta^*(t)w^T(t) \tag{4.83}$$

In this process, we have assumed that $c(t)$ is time invariant. Even if $c(t)$ is time variant, we can similarly estimate time variant $c(t)$ if we select $N_p T_c$ that is so short that we can assume $c(t)$ to be constant during the period. When the variation of $c(t)$ is much faster, we can estimate $c(t)$ by interpolating its estimated sequence using higher-order interpolation as in the case of TDM-type channel estimation.

In the previous discussion, the number of symbols in a PN frame and integrated period in equation (4.80) is assumed to be the same. However, some practical systems, such as IS-95 [4-27], apply long code (having a very long PN sequence period). In such cases, we have to integrate only a part of a PN sequence.

4.3.3.2 Pilot Channel Multiplexing for Uplink
In the downlink, we can assign very high pilot channel power. For example, when $p\%$ of the total transmitting power is assigned to the pilot channel and N_{ch} of traffic channels are transmitted from the BS, the power ratio of the pilot channel to each traffic channel is given by $pN_{ch}/(100 - p)$. For example, when $p = 20\%$ and $N_{ch} = 100$, the power of the pilot channel is 14 dB higher than that of each traffic channel. Therefore, the CDM-type pilot channel multiplexing method can achieve very accurate fading compensation without degrading spectral efficiency and power efficiency so much in the downlink.

On the other hand, in the uplink, because the received signal at the BS experiences different propagation path characteristics, each channel has to prepare its own pilot channel associated with the traffic channel. As a result, it is often perceived that the noncoherent detection techniques, such as differential phase shift keying (DPSK) with post-detection integrator (DPSK/PDI) [4-28; 4-29] and M-ary CDMA [4-27; 4-30], rather than pilot code-aided fading compensation are preferable in the uplink.

The suppressed pilot channel method is a solution to cope with this problem [4-24]. The suppressed pilot channel method prepares a pilot channel for each of the traffic channels, and its power is suppressed to a certain level. Basic configuration of the transmitter for a suppressed pilot channel system is almost the same as that for the downlink except that the pilot channel power is suppressed.

Figure 4.27 shows a receiver configuration of the suppressed pilot channel system. After the desired signal is picked up by the BPF, the signal is down-converted to the baseband by the quadrature demodulator, band-limited by the root Nyquist LPF, analog-to-digital (A/D) converted, and stored in the memory. First, this signal is fed to the matched filter to obtain a delay profile of the received signal. At this stage, however, the SNR of this delay profile is very low because the power of the pilot channel is suppressed. Fortunately, the fading variation is not so fast—up to 100 Hz in most cases. Therefore, we can improve the SNR of the delay profile by coherently accumulating delay profiles at the complex delay profile estimator. This concept is shown in Figure 4.28. For example, when the chip rate is 4.096 Mchips/s and the period of spreading code is 256, we can obtain a delay profile every 62 µs. On the other hand, 1 ms of the delay profile period is sufficient to cope with fading variation of up to $f_d = 100$ Hz, as we discussed earlier in this chapter. Therefore, we can accumulate about 16 delay profiles to estimate more reliable delay profiles.

Now, let's assume that the received signal power at the BS is perfectly controlled to be at a certain level; there is no intercell interference; the number of the traffic channel processed at the BS is N_{ch}; process gain for the traffic channel is G_p; and the

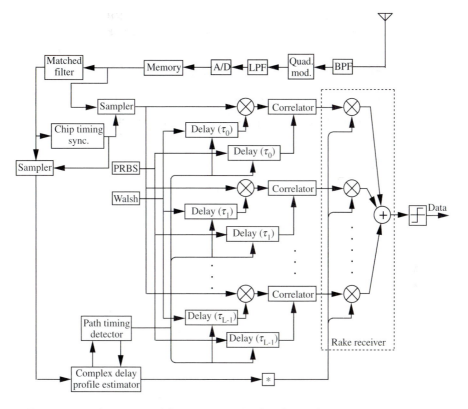

Fig. 4.27 Receiver configuration of the suppressed pilot channel system.

power ratio of the pilot channel to the pilot plus traffic channels ($P_P/(P_T + P_P)$) is β. The SNR of the pilot channel is then given by

$$SNR_p = \frac{G_p \beta N_{prof}}{(N_{ch} - 1)} = \beta N_{prof} \times SNR_T \tag{4.84}$$

where SNR_T is the SNR of the traffic channel. Therefore, if we can select N_{prof} to satisfy $\beta N_{prof} > 1$, we can obtain coherent amplitude and phase reference for the pilot code-aided fading compensation scheme with higher SNR than that of the traffic channel.

Figure 4.29 shows a computer-simulated result of the BER versus β performance with a parameter of N_{prof} under AWGN conditions [4-24]. The result shows that (N_{prof}, β) = (32, −10 dB) or (16, −8 dB) gives the best BER performance.

4.4 DIVERSITY COMBINING

As we have discussed, propagation path condition is determined by the path loss, shadowing, and multipath fading. Among them, multipath fading is caused by a terminal running through standing waves of the multipath. Therefore, when a mobile terminal antenna moves to a different route, the terminal experiences different fading variations.

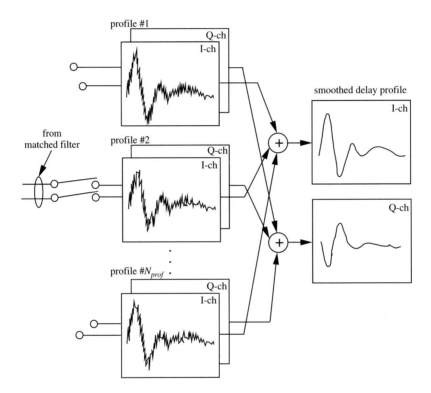

Fig. 4.28 Concept of the complex delay profile estimation using coherent accumulation of the delay profiles.

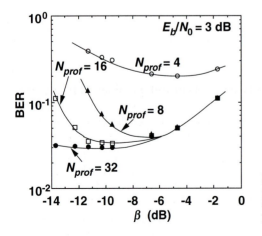

Fig. 4.29 BER versus β performance with a parameter of N_{prof} under AWGN conditions (from Ref. 4-24, © Institute of Electrical and Electronic Engineers, 1994).

Diversity techniques actively use the nature of propagation path characteristics to improve receiver sensitivity [4-31; 4-32]. For example, when we employ two antennas and locate them on the terminal with a certain antenna separation, these two antennas experience different fading variations when the terminal moves. Figure 4.30 shows an example of the received signal level variations of these two antennas. As shown in this figure, although the received signal level of antenna #1 is faded, that for antenna #2 is not faded in most cases. Therefore, when we select one of the antennas having the higher received signal level (which corresponds to space diversity with selective combining), we can reduce the probability for deep fading as shown by the heavy line in Figure 4.30. These diversity schemes are called *microscopic diversity* because they are intended to suppress rapid fading variation caused by the microscopic configuration of the construction around the terminal. Although they are effective in suppressing multipath fading, they are less helpful in compensating for large-scale signal variation (shadowing) because more than 10m of antenna spacing is necessary to achieve a low correlation of shadowing. To compensate for both multipath fading and shadowing, *macroscopic diversity*, which uses two or more cell sites (BSs) is applied to obtain plural uncorrelated shadow fading variations. Because it uses several cell sites, it is also called a *site diversity*.

In the following sections, we will discuss diversity schemes followed by the configurations and the performance of both microscopic diversity and macroscopic diversity.

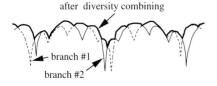

Fig. 4.30 An example of the received signal level variation at each diversity branch, and the basic concept of diversity reception.

4.4.1 Diversity Schemes

Among many diversity schemes, space diversity, polarization diversity, frequency diversity, time diversity, path diversity, and directional diversity are popular. Table 4.3 shows their features.

4.4.1.1 Space Diversity
Space diversity can obtain independent fading variations by using spatially separated antennas. Because its receiver configuration is relatively simple, because any number of the diversity branches are selectable, and since it does not require extra spectrum nor power, space diversity is the most popular diversity scheme. For terminals, antenna spacing of longer than $\lambda/2$ is sufficient to achieve very low fading correlation between branches, whereas 50λ to 100λ of antenna spacing is necessary at the BS [4.32]. Macroscopic diversity is an extension of space diversity because the only difference is whether or not the diversity antennas are in the same cell site.

Table 4.3 Features of each diversity scheme.

Diversity Scheme	Merits	Demerits
Space Diversity	Easy to design. Any number of diversity branches are (*L*) selectable. No extra power nor bandwidth is necessary. Applicable to macroscopic diversity.	Hardware size could be large (depends on device technologies). Large antenna spacing is necessary for microscopic diversity at the BS.
Polarization Diversity	No space is necessary. No extra bandwidth is necessary.	Only two branch diversity schemes are possible. 3 dB more power is necessary.
Frequency Diversity	Any number of diversity branches (*L*) are selectable.	*L* times more power and spectrum are necessary.
Time Diversity	No space is necessary. Any number of diversity branches (*L*) are selectable. Hardware is very simple.	*L* times more spectrum are necessary. Large buffer memory is necessary when f_d is small.
Directional (Angle) Diversity	Doppler spread is reducible.	Diversity gain depends on the obstacles around the terminal. Applicable only to the terminal.
Path Diversity	No space is necessary. No extra power nor bandwidth is necessary.	Diversity gain depends on the delay profile.

4.4.1.2 Polarization Diversity   When a signal is transmitted by two polarized antennas and received by two polarized antennas, we can obtain two uncorrelated fading variations because the vertical and horizontal polarized components experience different fading variation due to different reflection coefficients of the building walls. There are two distinct features of this scheme: we can install two polarization antennas at the same place, and it does not require any extra spectrum. In other words, we do not have to be careful about the antenna separation as in the case of space diversity. However, this scheme can achieve only two branch diversity schemes. One more drawback is that we have to transmit 3 dB more power because we have to feed signals to both polarization antennas at the transmitter.

4.4.1.3 Frequency Diversity   When a narrowband signal is transmitted over a frequency-selective fading channel, we can obtain independent fading variations if their frequency separation is larger than the coherence bandwidth. Although this scheme can easily obtain any number of diversity branches (*L*), it degrades system capacity because a channel occupies *L* times more bandwidth to achieve *L*-branch frequency diversity. Moreover, it requires *L* times more power. Therefore, this scheme is not applied much to land mobile communications in which spectrum and power savings are the most important issues. However, fading variation independence between sufficiently separated frequency components is a very important effect for land mobile communication technologies. This is called the *frequency diversity effect*. For example, multicarrier transmission [4-33; 4-34] and frequency hopping [4-35] techniques utilize this effect.

4.4.1.4 Time Diversity As discussed in chapter 2, the fading correlation coefficient $\rho(\tau)$ is low when $f_d\tau > 0.5$. Therefore, when an identical message is transmitted over different time slots with a time slot interval of more that $0.5/f_d$, we can obtain diversity branch signals. Although this scheme requires L times more spectrum, it has a very attractive advantage—its hardware is very simple because all of the process is carried out at the baseband. Therefore, time diversity is effective for the CDMA systems in which bandwidth expansion of the source signal is not a problem [4-36; 4-37]. However, it is less effective when the terminal speed is very slow because a very long time slot interval is necessary to obtain sufficient diversity gain. Moreover, when the terminal is standing still, we cannot obtain diversity gain at all.

4.4.1.5 Directional Diversity Because received signals at the terminal consist of reflection, diffraction, or scattered signals around the terminal, they come from different incident angles. When we can resolve the received signal by using directive antennas, we can obtain independently faded signals because all the paths coming from different angles are mutually independent. This scheme is, however, applicable only to the terminal because the received signal from a terminal comes from only limited directions at the BS. When we employ a directive antenna we can reduce Doppler spread for each branch as discussed in chapter 2.

4.4.1.6 Path Diversity Path diversity is a diversity combining scheme that resolves direct and delayed components and coherently combines them. Therefore, this scheme is called *implicit diversity* because diversity branches are created after the signal reception. The adaptive equalizer and Rake diversity are classified as path diversity schemes; we will discuss them in chapter 5.

The most distinct feature of this method is that no extra antenna, power, or spectrum are necessary. To design such a diversity scheme, however, we have to pay attention to the propagation path conditions because path diversity is less effective when the channel is under flat Rayleigh fading conditions.

4.4.2 Diversity Combining Schemes

4.4.2.1 Selective Combining In selective combining, the diversity branch having the strongest received signal level is selected. When the instantaneously received signal level is subject to Rayleigh fading, and the p.d.f. of the received signal level in each branch is assumed to be the same, the p.d.f. of the received signal level for k-th branch (γ_k) and its cumulative distribution are given by

$$p(\gamma_k) = \frac{1}{\gamma_0}\exp\left(-\frac{\gamma_k}{\gamma_0}\right) \tag{4.85}$$

$$P(\gamma_k \le x) = 1 - \exp\left(-\frac{x}{\gamma_0}\right). \tag{4.86}$$

When L-branch selective combining is employed, its cumulative distribution is the probability that the signal levels of all the branches go below a certain level x. Therefore, it is given by

$$P_{sel}(\gamma \le x) = P(\gamma_1 \le x) \cdot P(\gamma_2 \le x) \cdots P(\gamma_L \le x) = \prod_{k=1}^{L}\left[1 - e^{-x/\gamma_0}\right] \tag{4.87}$$

and corresponding p.d.f. is given by

$$p_{sel}(\gamma) = \frac{L}{\gamma_0} e^{-\gamma/\gamma_0}\left[1 - e^{-\gamma/\gamma_0}\right]^{L-1} \tag{4.88}$$

As discussed in chapter 3, BER for GMSK, MCPSK, MDPSK, and M-QAM under AWGN conditions can be expressed in the form of $a \cdot \mathrm{erfc}(\sqrt{b\gamma})$ or $a \cdot e^{-b\gamma}$. When BER in the static condition is given by $a \cdot \mathrm{erfc}(\sqrt{b\gamma})$, the average BER for L-branch selection diversity under Rayleigh fading conditions is given by

$$
\begin{aligned}
&P^{MSEL}(\gamma_0) \\
&= \int_0^\infty a \cdot \mathrm{erfc}\left(\sqrt{b\gamma}\right)\frac{L}{\gamma_0}e^{-\frac{\gamma}{\gamma_0}}\left[1 - e^{-\frac{\gamma}{\gamma_0}}\right]^{L-1} d\gamma \\
&= \sum_{m=0}^{L-1} aL(-1)^m \binom{L-1}{m}\frac{1}{m+1}\int_0^\infty \mathrm{erfc}\left(\sqrt{b\gamma}\right)\frac{(m+1)}{\gamma_0}e^{-\frac{\gamma}{\gamma_0}(m+1)} d\gamma \\
&= aL\sum_{m=0}^{L-1}(-1)^m \binom{L-1}{m}\frac{1}{m+1}\left[1 - \frac{1}{\sqrt{1+(m+1)/(b\gamma_0)}}\right]
\end{aligned}
\tag{4.89}
$$

where γ_0 is the average E_b/N_0.

When BER under AWGN conditions is given by $a \cdot e^{-b\gamma}$, the BER for L-branch selection diversity under Rayleigh fading conditions is given by

$$
\begin{aligned}
&P^{MSEL}(\gamma_0) \\
&= \int_0^\infty a \cdot e^{-b\gamma}\frac{L}{\gamma_0}e^{-\frac{\gamma}{\gamma_0}}\left[1 - e^{-\frac{\gamma}{\gamma_0}}\right]^{L-1} d\gamma \\
&= \sum_{m=0}^{L-1} aL(-1)^m \binom{L-1}{m}\frac{1}{m+1}\int_0^\infty e^{-b\gamma}\frac{(m+1)}{\gamma_0}e^{-\frac{\gamma}{\gamma_0}(m+1)} d\gamma \\
&= aL\sum_{m=0}^{L-1}(-1)^m \binom{L-1}{m}\frac{1}{m+1}\left[\frac{1}{1+b\gamma_0/(m+1)}\right]
\end{aligned}
\tag{4.90}
$$

4.4.2.2 Maximal Ratio Combining In the selection combining scheme, only one of the diversity branch signals are used for demodulation, which means that the rest of the branch signals are discarded at the diversity combiner. Thus, if all the branch signals are coherently combined with an appropriate weighting coefficient for each branch signal, performance improvement could be expected. Let the weight of k-th branch be g_k. Because the co-phased signal of each branch can be expressed as

$$y_k(t) = r_k(t)z(t) \tag{4.91}$$

where $z(t)$ is the transmitted baseband signal band-limited by the transmitter and receiver filters, the combined output can be expressed as

$$v(t) = \sum_{k=1}^{L} g_k(t)r_k(t)z(t) = z(t)\sum_{k=1}^{L} g_k(t)r_k(t) \tag{4.92}$$

where $g_k(t)$ is the weighting factor for k-th branch. Therefore, the signal power after combining is given by

$$S = E\left[z^2(t)\right]\left|\sum_{k=1}^{L} g_k(t)r_k(t)\right|^2 \tag{4.93}$$

From Schwarz's inequality, the following is satisfied [4-32].

$$\left|\sum_{k=1}^{L} g_k(t)r_k(t)\right|^2 \leq \left|\sum_{k=1}^{L} g_k^2(t)\right|\left\|\sum_{k=1}^{L} r_k^2(t)\right| \tag{4.94}$$

where equality holds when $g_k(t) = r_k(t)$. This means that, when each branch signal is weighted by a factor proportional to the received envelope level and combined, the SNR at the output of the diversity combiner can be maximized. Such a combining scheme is called *maximal ratio combining*. The instantaneous SNR after maximal ratio combining is given by

$$\gamma = \sum_{k=1}^{L} \gamma_k \tag{4.95}$$

Therefore, the p.d.f. after combining is given by

$$p(\gamma) = \frac{1}{(L-1)!}\frac{\gamma^{L-1}}{\gamma_0^L}e^{-\gamma/\gamma_0} \tag{4.96}$$

When the BER under static conditions is given by $a \cdot \text{erfc}(\sqrt{b\gamma})$, the BER for L-branch maximal ratio combining diversity under the Rayleigh fading conditions is given by

$$P^{MMAX}(\gamma_0) = a\left[1 - \frac{1}{\sqrt{1+1/(b\gamma_0)}}\sum_{k=0}^{L-1}\frac{(2k-1)!!}{(2k)!!}\frac{1}{(b\gamma_0+1)^k}\right] \tag{4.97}$$

When the BER in the static condition is given by $a \cdot e^{-b\gamma}$, BER for L-branch maximal ratio combining diversity under the Rayleigh fading condition is given by

$$P^{MMAX}(\gamma_0) = a\left[\frac{(L-1)!}{(1+b\gamma_0)^L}\right] \tag{4.98}$$

Table 4.4 summarizes theoretical BER performances for various types of modulation/demodulation schemes with L-branch diversity reception in the case of selective combining and maximal ratio combining.

Figures 4.31, 4.32, and 4.33 show theoretical BER performance of BPSK, QPSK, and M-ary QAM with a parameter of the number of diversity branches under flat Rayleigh fading conditions, where selective combining is employed for each case. In any case, absolute phase coherent detection gives the best performance. On the other hand, the performance difference between differentially encoded coherent detection and differential detection increases with M. Therefore, coherent detection is considered to be more desirable when we employ a large number of diversity branches.

Figure 4.34 shows BER performance difference between selective combining and maximal ratio combining for CQPSK with a parameter of the number of diversity branch. When the number of diversity branches is two, the performance of maximal

Table 4.4 BER performance with diversity reception.

Modulation/Demodulation	a	b	Equation Select. Comb.	Equation Max. Ratio
BPSK with coherent detection	1/2	1	(4.89)	(4.97)
DEBPSK with coherent detection	1	1	(4.89)	(4.97)
DEBPSK with differential detection	1/2	1	(4.90)	(4.98)
CQPSK with coherent detection	1/2	1	(4.89)	(4.97)
DEQPSK with coherent detection	1	1	(4.89)	(4.97)
DEQPSK with differential detection	1/2	$4\sin^2(\pi/8)$	(4.89)	(4.97)
M-ary PSK with coherent detection	$1/\log_2 M$	$\log_2 M \sin^2(\pi/M)$	(4.89)	(4.97)
M-ary DEPSK with coherent detection	$2/\log_2 M$	$\log_2 M \sin^2(\pi/M)$	(4.89)	(4.97)
16QAM with coherent detection	3/8	2/5	(4.89)	(4.97)
64QAM with coherent detection	7/24	1/7	(4.89)	(4.97)
256QAM with coherent detection	15/64	4/85	(4.89)	(4.97)

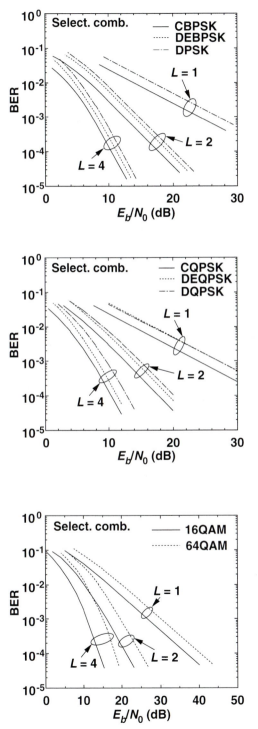

Fig. 4.31 Theoretical BER performance of BPSK selective combining diversity receiver with a parameter of the number of diversity branches.

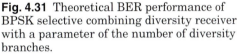

Fig. 4.32 Theoretical BER performance of QPSK selective combining diversity receiver with a parameter of the number of diversity branches.

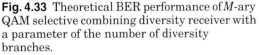

Fig. 4.33 Theoretical BER performance of M-ary QAM selective combining diversity receiver with a parameter of the number of diversity branches.

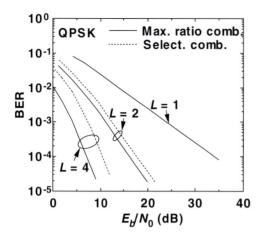

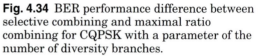

Fig. 4.34 BER performance difference between selective combining and maximal ratio combining for CQPSK with a parameter of the number of diversity branches.

ratio combining is superior to that of selective combining by only 1.5 dB. However, when 4-branch diversity is employed, the difference becomes about 3 dB. Therefore, maximal ratio combining is preferable when we employ diversity branches of more than two—for example, in the case of uplink reception.

4.4.2.3 Equal Gain Combining   Equal gain combining is a combining scheme that coherently combines all the branches with the same weighting factor. In other words, this process just co-phases all the branches and adds them up without weighting. Therefore, this configuration is simpler than that for maximal ratio combining. However, instead of such circuit simplification, SNR, at the output of the diversity combiner, is degraded by 0.5 dB (for the 2-branch case) to 1 dB (for the 10-branch case) compared to that for maximal ratio combining.

Although study on the implementation of equal gain combining has been done in the past, such study is not being done much at present because

☞ Selection diversity is the most cost-effective diversity combining scheme.

☞ Owing to the recent development of the pilot signal-aided scheme as well as DSP technologies, we can implement maximum combining schemes with a simple hardware configuration [4-38], which is especially effective for the BS to implement 3-branch diversity or more.

4.4.3 Configuration for Maximal Ratio Combining Diversity Receiver Using a Pilot Symbol-Aided Calibration Technique

In space diversity receivers, maximal ratio combining has hardly ever been used in land mobile communication systems, although it gives the maximum SNR of the combined signal, because the co-phasing and weighting stages were considered to be very complicated for hardware implementation. However, emerging DSP technologies make it possible to achieve complicated signal processing using only one or a few DSP chips. As a result, maximal ratio combining becomes feasible using a pilot signal-aided channel sounding technique with simple hardware.

Figure 4.35 shows a configuration of the diversity receiver using DSP techniques. The received signal at k-th branch is given by

$$s_{Rk}(t) = \text{Re}\big[c_k(t)z(t)\exp(j2\pi f_c t)\big] \tag{4.99}$$

where $c_k(t)$ represents a complex fading variation for k-th branch, $z(t)$ is the transmitted baseband signal, and f_c is the carrier frequency.

In each branch, spectrum around the desired signal is picked up by the BPF and amplified to a certain level by the AGC. For the AGC amplifier to be applied, we have to maintain the relative power levels between any branch signals; i.e., each branch signal should have the same gain. Therefore, the AGC amplifier has to control the gain to keep summation of all the received signal levels constant [4-38].

Figure 4.36 shows configurations of the AGC amplifier that satisfy this requirement. The configuration in Figure 4.36(a) was originally proposed by Suzuki [4-39]. In this figure, all the branch signals are frequency division multiplexed, and the multiplexed signal is then fed to the AGC amplifier [4-22; 4-38]. In this case, the multiplexed signal can be expressed as

$$s_{RL}(t) = \sum_{i=1}^{L} s_{Ri}(t)\exp(j2\pi f_{AGC}(i-1)t) \tag{4.100}$$

where f_{AGC} is a channel spacing for multiplexing, and the number of branches is L. Because the power of $s_{RL}(t)$ is equal to the sum of all the branch signals, this AGC satisfies the above-mentioned requirement.

Figure 4.36(b) shows another type of AGC amplifier [4-21; 4-40]. In this case, the AGC in each branch is controlled to keep the sum of all the received signal power constant. These two AGC configurations are both evaluated by laboratory and field experiments and their feasibility was confirmed under actual propagation path conditions.

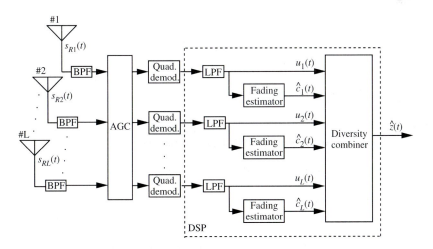

Fig. 4.35 A configuration of the diversity receiver using DSP techniques.

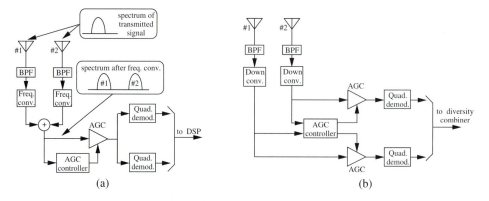

Fig. 4.36 Configurations of the AGC amplifier for the maximal ratio combining space diversity: (a) frequency-shift-type AGC and (b) common AGC control signal feeding type.

After signals are amplified, they are quadrature demodulated and low-pass filtered to obtain the received baseband signal. The baseband signals of all the branches are then fed to the diversity combiner, and the transmitted data sequence is regenerated using the output of the diversity combiner.

Figure 4.37 shows a configuration of the diversity stage. At the fading estimator, the multiplexed pilot signal is extracted and a fading variation is estimated using the extracted pilot signals. As a pilot signal-aided channel sounding scheme, FDM, TDM, or CDM types are applicable.

Let the estimated complex fading variation and received complex baseband signal for k-th branch be $\hat{c}_k(t)$ and $u_k(t)$. At the diversity combiner, maximal ratio combining is carried out using the following equation [4-38].

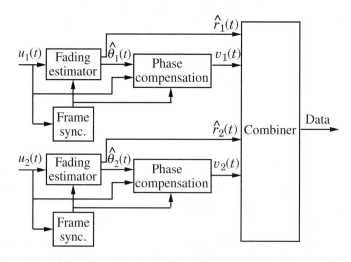

Fig. 4.37 Configuration of the diversity stage (from Ref. 4-38, © Institute of Electrical and Electronic Engineers, 1993).

$$\hat{z}(t) = \sum_{k=1}^{M} \hat{c}_k^*(t) u_k(t) \Big/ \sum_{k=1}^{M} \left| \hat{c}_k(t) \right|^2 \qquad (4.101)$$

In the case of such DSP-based diversity combining techniques, dynamic range of the DSP is limited by the resolution of the A/D converter. Unfortunately, it could degrade the BER performance because dynamic range of the fading is very wide. As a result, the optimization of the AGC time constant (τ_{AGC}) and the resolution of the A/D converter are very important factors.

Figure 4.38 shows the BER performance versus τ_{AGC} with parameters of E_b/N_0 and f_d, where resolution of the A/D converter is 12-bit resolution, and the modulation scheme is 16QAM. As shown in this figure, $\tau_{AGC} = 10$ ms gives the optimum value. The cause of degradation at $\tau_{AGC} > 10$ ms is due to insufficient dynamic range in the diversity stage, and that at $\tau_{AGC} < 10$ ms is due to the fact that the AGC also compensates for the envelope variation due to modulation.

Figure 4.39 shows the laboratory experimental results of the BER performance of 16QAM with maximal ratio combining space diversity. The results show that the

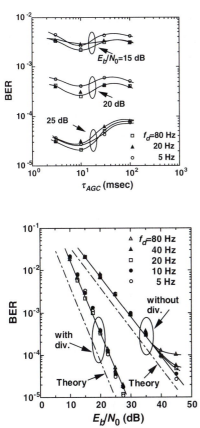

Fig. 4.38 BER performance versus τ_{AGC} with parameters of the E_b/N_0 and f_d, where the resolution of the A-D converter is 12-bit, and the modulation scheme is 16QAM (from Ref. 4-38, © Institute of Electrical and Electronic Engineers, 1993).

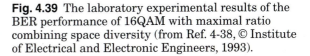

Fig. 4.39 The laboratory experimental results of the BER performance of 16QAM with maximal ratio combining space diversity (from Ref. 4-38, © Institute of Electrical and Electronic Engineers, 1993).

obtained BER performance is degraded by only 3 dB, 2 dB of which is due to the pilot symbol insertion and 1 dB of which is due to the hardware imperfection.

4.4.4 Transmitter Diversity

Although space diversity is very effective for improving BER performance, its most important drawback is that it requires plural receivers, which increases weight and size of terminals. One solution for this problem is transmission diversity. In this scheme, the BS receives a signal by using conventional diversity reception. In the downlink, on the other hand, the BS selects one of the transmitter antennas that could reach a terminal with better propagation path conditions. Such a diversity scheme is called *transmitter diversity*. Using this transmitter diversity, the terminal can obtain diversity gain using a conventional nondiversity receiver.

To achieve such transmitter diversity, however, the diversity gain depends on how to accurately estimate the downlink propagation path conditions. The time division duplex (TDD) is a very useful duplex scheme for this purpose.

Figure 4.40 shows an example of the frame format for a four-channel TDMA/TDD system, where the first four slots are used for reception and the latter four slots are used for transmission. Because the same frequency is employed in the uplink and downlink, when the time difference between the transmission and reception time slots is sufficiently short, correlation between the transmission and reception time slots is expected to be very high, which is called *reciprocity principle* [4-32]. Therefore, we can estimate the channel condition for the downlink by using the measured channel condition of the uplink [4-41].

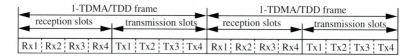

Fig. 4.40 An example of the frame format for a four-channel TDMA/TDD system.

4.4.5 Macroscopic Diversity

Although space diversity is a very effective technique for compensating for rapid fading, it is helpless to compensate for log-normal fading or path loss due to distance. To compensate for them, macroscopic diversity techniques that obtain independent diversity paths by using plural BSs are very effective. Because this diversity employs several cell sites, it is also called *site diversity*.

In cellular systems, *handover* is a kind of macroscopic diversity. When a terminal is located near the fringe area and it is moving toward another cell, the terminal transmits a handover request signal to the BS with some other necessary information, such as the received signal level from the adjacent cells. When the BS receives such a request, the message is forwarded to the mobile telephone switching office (MTSO), and the MTSO will make a decision whether it is possible or not, as well as to which cell the terminal should be switched. If it is possible, the MTSO sends the handoff messages to both the present BS and new BS.

The handover can be classified into two types—the hard handover and soft handover [4-42]. In the case of hard handover, when both of the BSs receive handover permission from the MTSO, the present BS releases the radio link and the new BS establishes the radio link at the same time. Therefore, the hard handover is considered to be a switching macroscopic diversity with a very long time constant.

On the other hand, in the case of soft handover, the MTSO establishes a new radio link while keeping the current radio link. Therefore, the soft handover is considered to be a selective combining macroscopic diversity with a short time constant.

REFERENCES

4-1. Gardner, F. M., *Phaselock techniques*, John Wiley & Sons, New York, 1979.

4-2. Sampei, S., "Phase-lock loop for fading channel," Technical report of IECE of Japan, CS84-38, June 1984.

4-3. Saito, S. and Suzuki, H., "Fast carrier-tracking coherent detection with dual-mode carrier recovery circuit for land mobile radio transmission," *IEEE J. of Select. Area in Commun.*, Vol. 7, No. 1, pp.130–39, January 1989.

4-4. Sampei, S., "QPSK coherent detection method for land mobile radio communications using linear mean square estimation," Trans, IEICE (B-II), Vol. J72-B-II, No. 4, pp. 125–32, April 1989.

4-5. Godard, D., "Channel equalization using a Kalman filter for fast data transmission," *IBM J. Res. Develop.*, Vol. 18, No. 3, pp. 267–73, May 1974.

4-6. Yokoyama, M., "BPSK with sounder to combat Rayleigh fading in mobile radio communications," IEEE Trans. Veh. Tech., Vol. VT-34, No. 1, pp. 35–40, February 1985.

4-7. Hou, X. Y., Wu, K. R., Morinaga, T. and Namekawa, T., "Bit error rate performance of fading BPSK system with sounder in the presence of cochannel interference," Trans. IEICE (E), Vol. E70, No. 1, pp. 42–48, January 1987.

4-8. Davarian, F., "Mobile digital communication via tone calibration," IEEE Trans. Veh. Tech., Vol. VT-36, No. 2, pp. 55–62, May 1987.

4-9. Leung, P. S. K. and Feher, K., "Transparent tone in band (TTIB) aided GMSK/MSK modem system," 39th IEEE Veh. Tech. Conf. (San Francisco, California), pp. 249–55, May 1989.

4-10. Cavers, J. K., Marchetto, R. F. and Carlson, S. D., "A new spectral notch generator for pilot tone systems," 40th IEEE Veh. Tech. Conf. (Orlando, Florida), pp. 547–51, May 1990.

4-11. Simon, M. K., "Dual-pilot tone calibration technique," IEEE Trans. Veh. Tech., Vol. VT-35, No. 2, pp. 63–70, May 1986.

4-12. Bateman, A., Lightfoot, G., Lymer, A. and McGeehan, J. P., "Speech and data communications over 942 MHz TAB and TTIB single sideband mobile radio systems incorporating feedforward signal regeneration," IEEE Trans. Veh. Technol., Vol. VT-34, No. 1, pp. 13–21, February 1985.

4-13. Martin, P. M., Bateman, A., McGeehan, J. P. and Marvill, J. D., "The implementation of a 16QAM mobile data system using TTIB based fading correction techniques," 38th IEEE Veh. Tech. Conf. (Philadelphia, Pennsylvania), pp. 71–76, June 1988.

4-14. Bateman, A., "Feedforward transparent tone-in-band: Its implementations and applications," IEEE Trans. Veh. Technol. Vol. 39, No. 3, pp. 235–43, August 1990.

4-15. Martin, P. M. and Bateman, A., "Practical results for a generic modem using linear mobile radio channels," 41st IEEE Veh. Tech. Conf. (St. Louis, Missouri), pp. 386–90, May 1991.

4-16. Moher, M. L. and Lodge, J. H., "TCMP—A modulation and coding strategy for Rician fading channels," *IEEE J. Select. Areas Commun.*, Vol. 7, No. 9, pp. 1347–55, December 1989.

4-17. Aghamohammadi, A., Mayer, H. and Ascheid, G., "A new method for phase synchronization and automatic gain control of linearly modulated signals in frequency-flat fading channels," Trans. IEEE Commun., Vol. COM-39, No. 1, pp. 25–29, January 1991.

4-18. Sampei, S. and Sunaga, T., "Rayleigh fading compensation for QAM in land mobile radio communications," Trans. IEEE Veh. Technol., Vol. 42, No. 2, pp. 137–47, May 1993.

4-19. Cavers, J. K. , "An analysis of pilot symbol assisted modulation for Rayleigh fading channels," Trans. IEEE Veh. Tech., Vol. VT-40, No. 4, pp. 686–93, November 1991.

4-20. Subasinghe-Dias, D., "New techniques for the improvement of capacity of digital mobile communication systems," Doctoral thesis, Dept. of Electrical and Computer Engineering, University of California, Davis, May 1992.

4-21. Sampei, S., Kamio, Y. and Sasaoka, H., "Field experiments on pilot symbol aided 16QAM for land mobile communications," Electron. Letters, Vol. 28, No. 23, pp. 2198–99, November 1992.

4-22. Kinoshita, N. et al., "Field experiments on 16QAM/TDMA and trellis coded 16QAM/TDMA systems for digital land mobile radio communications," IEICE Trans. Commun., Vol. E77-B, No. 7, pp. 911–20, July 1994.

4-23. Salmasi, A. and Gilhousen, K. S., "On the system design aspects of code division multiple access (CDMA) applied to digital cellular and personal communication networks," 41st IEEE Veh. Tech. Conf. (St. Louis, Missouri), pp. 57-62, May 1991.

4-24. Abeta, S., Sampei, S. and Morinaga, N., "DS/CDMA coherent detection system with a suppressed pilot channel," GLOBECOM'94 (San Francisco, California), pp. 1622–26, November 1994.

4-25. Turin, G. L., "Introduction to spread spectrum antimultipath techniques and their application to urban digital radio," Proc. IEEE, Vol. 66, No. 11, pp. 1468–96, November 1978.

4-26. Turin, G. L., "The effects of multipath and fading on the performance of direct-sequence CDMA systems," IEEE Trans. Veh. Technol., Vol. VT-33, No. 3, pp. 213–19, August 1984.

4-27. TIA/EIA, "Mobile station-base station compatibility standard for dual mode wideband spread spectrum cellular system," TIA/EIA IS-95, 1993.

4-28. Higashi, A. and Matsumoto, T., "Comparison of diversity combining schemes for PDI reception in DS/CDMA mobile radio," SITA'91, TH-E-II-3, December 1991.

4-29. Sanada, Y., Kajiwara, A. and Nakagawa, M., "Adaptive RAKE diversity for mobile communications," IEICE Trans. Commun., Vol. E76-B, No. 8, pp. 1002–7, August 1993.

4-30. Sasaki, S., Zhu, J. and Marubayashi, G., "Performance of the parallel combinatory spread spectrum multiple access communication system with the error control technique," IEEE ISSSTA'92 (Yokohama, Japan), pp. 159–62, November 1992.

4-31. Jakes, W. C., ed., *Microwave mobile communications,* John Wiley & Sons, New York, 1974.

4-32. Lee, W. C. Y., *Mobile communications design fundamentals* (2d edition), John Wiley & Sons, New York, 1993.

4-33. Kamio, Y., "Performance of trellis coded modulation using multi-frequency channels in land mobile communications," 40th IEEE Veh. Tech. Conf. (Orlando, Florida), pp. 455–58, May 1990.

4-34. Hara, S., Fukui, K., Okada, M. and Morinaga, N., "Multicarrier modulation techniques for broadband indoor wireless communications," PIMRC'93 (Yokohama, Japan), pp. 132–36, September 1993.

4-35. Kamio, Y. and Sampei, S., "Performance of a trellis-coded 16QAM/TDMA system for land mobile communications," Trans. IEEE Veh. Technol., Vol. 43, No. 3, pp. 528–36, August 1994.

4-36. Kubota, S., Kato, S. and Feher, K., "A time diversity CDMA scheme employing orthogonal modulation for time variant channels," 43rd IEEE Veh. Tech. Conf. (Secaucus, New Jersey), pp. 444–47, May 1993.

4-37. Meyer, M., "Improvement of DS-CDMA mobile communications systems by symbol splitting," 45th IEEE Veh. Tech. Conf. (Chicago, Illinois), pp. 689–93, July 1995.

4-38. Sunaga, T. and Sampei, S., "Performance of multi-level QAM with maximal ratio combining space diversity for land mobile communications," IEEE Trans. Veh. Technol., Vol. 42, No. 3, pp. 294–301, August 1993.

4-39. Suzuki, H., "A configuration of diversity combining by using an RLS equalizer for mobile radio transmission," presented at the 1989 Autumn Nat. Conv. Rec. of IEICE B-516, September 1989.

4-40. Sampei, S., Kamio, Y. and Sasaoka, H., "Field experiments on pilot symbol aided 16 QAM for land mobile communication systems," Review of the Communications Research Laboratory, Vol. 39, No. 2, pp. 83–93, June 1993.

4-41. Ue, T., Sampei, S. and Morinaga, N., "Symbol rate and modulation level controlled adaptive modulation/TDMA/TDD for personal communication systems," 45th IEEE Veh. Tech. Conf. (Chicago, Ilinois), pp. 306–10, July 1995.

4-42. Viterbi, A. J., CDMA, Principles of spread spectrum communication, Addison-Wesley Publishing Company (Reading, Massachusetts), 1995.

C H A P T E R   5

Anti-Frequency-Selective Fading Techniques

This chapter will discuss how to cope with frequency-selective fading when we design high-bit-rate wireless communication systems. For this purpose, we will first discuss a frequency-selective fading channel model including the effect of modulation and demodulation schemes because whether the channel has frequency selectivity or not depends not only on the channel characteristics themselves but also on the modem parameters. For example, when 90% value of the channel delay spread is 1 μs, we can regard this channel as a flat Rayleigh fading channel if the transmission symbol rate is 1 ksymbol/s. On the other hand, when the transmission symbol rate is 1 Msymbol/s, this channel should be treated as a frequency-selective fading channel.

Therefore, this chapter will discuss the adaptive equalizer, adaptive array antenna, and path diversity combining techniques as the anti-frequency-selective fading techniques for high-bit-rate transmissions.

5.1 FREQUENCY-SELECTIVE FADING CHANNEL MODEL

Figure 5.1 shows a frequency-selective fading channel model, including modulator and demodulator. Let the transmitted signal $s_T(t)$ be

$$s_T(t) = \mathrm{Re}[z_T(t)\exp(j2\pi f_c t)] \tag{5.1}$$

$$z_T(t) = c_T(t) \otimes a(t) \tag{5.2}$$

161

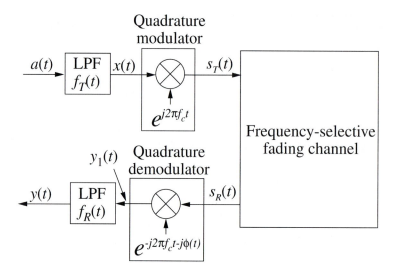

Fig. 5.1 Frequency-selective fading channel model, including modulator and demodulator.

$$a(t) = \sum_{i=-\infty}^{\infty} a_i \delta(t - iT_s) \qquad (5.3)$$

where

$z(t)$ = transmitted baseband signal band-limited by the transmitter filter
$a(t)$ = transmitted symbol sequence
$c_T(t)$ = impulse response of the transmitter filter

When an impulse response of a frequency-selective fading channel is given by

$$c_c(t) = 2\,\mathrm{Re}\big[c_B(t)\exp(j2\pi f_c t)\big] \qquad (5.4)$$

the received signal is given by

$$s_R(t) = s_T(t) \otimes c_c(t)$$

$$= \sum_{i=-\infty}^{\infty} a_i c_1(t - iT_s)\exp(j2\pi f_c t) \qquad (5.5)$$

where

$$c_1(t) = c_T(t) \otimes c_B(t) \qquad (5.6)$$

At the receiver, $s_R(t)$ is multiplied by a complex conjugate of a local oscillator given by

$$b(t) = \exp[\,j2\pi(f_c + f_{off})t + j\phi(t)] \qquad (5.7)$$

to down-convert the received signal to the baseband, where f_{off} is the offset frequency between the carrier frequency and the local oscillator frequency. The obtained signal is expressed as

$$y_1(t) = s_R(t)b^*(t)$$
$$= \sum_{i=-\infty}^{\infty} a_i c_1(t - iT_s)\exp(-j2\pi f_{off}t - j\phi(t))$$

(5.8)

This signal is then filtered by an LPF to suppress the out-of-band noise and ACI. The obtained received baseband signal is given by

$$y(t) = y_1(t) \otimes c_R(t)$$
$$= \sum_{i=-\infty}^{\infty} a_i c(t - iT_s)$$

(5.9)

$$c(t) = c_T(t) \otimes c_R(t) \otimes c_g(t)$$

(5.10)

$$c_g(t) = c_B(t)\exp(-j2\pi f_{off}t - j\phi(t))$$

(5.11)

where $c_R(t)$ is an impulse response of the received LPF.

Equation (5.11) shows that what we can measure as the channel characteristics is not the actual channel impulse response but the convolution of the impulse response of the transmitter filter, receiver filter, and channel impulse response including frequency offset. The adaptive equalizer and other types of anti-frequency-selective fading techniques compensate for the effect of such convoluted channel characteristics ($c(t)$). Therefore, we will treat $c(t)$ as the channel impulse response in the following discussion.

When we assume the optimal sampling timing as $t = nT_s$, $y(nT_s)$ is given by

$$y(nT_s) = \sum_{i=-\infty}^{\infty} a_i c((n-i)T_s)$$
$$= c(0)a_n + \sum_{\substack{i=-\infty \\ i\neq0}}^{\infty} a_{n-i}c(iT_s)$$

(5.12)

The first term is the desired component and the second term is the ISI. Furthermore, this equation shows that the impulse response of a frequency-selective fading channel for digital transmission systems can be identified as a discrete channel model. Therefore, we can apply a discrete channel model with a tap spacing of T_s for the performance evaluation of the adaptive equalizer.

In the discrete channel model, when $c(iT_s) = 0$ for $i \neq 0$ and $c(0)$ is the time variant with its p.d.f. of complex Gaussian distribution, the received signal is exactly under ISI-free condition. In this case, we can treat the propagation path characteristics as a flat Rayleigh fading. When $c(i)$ for $i = 0$ and 1 are the time variants with their p.d.f. of complex Gaussian distributions, and when $c(i)$ for the others are negligibly small, we have to consider only the first (direct) path and the second (delayed) path. This model is called the two-ray Rayleigh model. When we can identify the channel impulse response as the two-ray Rayleigh model, the BER performance depends only on the delay spread. When the maximum delay time is $\tau_{max} = mT_s$ $(m > 1)$, we have to use the $(m + 1)$-tap discrete channel model.

5.2 DELAY SPREAD IMMUNITY FOR DIVERSITY RECEIVERS

Diversity reception is well known as one of the simplest anti-frequency-selective fading techniques when delay spread is relatively small, say $0.1T_s$ or less, and maximum delay time of the delayed waves in less than T_s. Figure 5.2 shows irreducible BER versus normalized delay spread (τ_{rms}/T_s) performance under two-ray Rayleigh channel for Gray-encoded QPSK and 16QAM with coherent detection, where the roll-off factor of the root Nyquist filters in the transmitter and receiver (α) is 0.5. When diversity reception is not applied, tolerance of the normalized delay spread for irreducible BER of 10^{-2} is 0.085 in the case of QPSK and 0.06 in the case of 16QAM. On the other hand, when a diversity reception is employed, BER $< 10^{-2}$ is achieved even if the normalized delay spread is about 0.1. Therefore, we usually consider that we can regard a channel condition as a flat fading if the normalized delay spread is less than 0.1. On the other hand, when the normalized delay spread is expected to be larger than 0.1, we have to apply some anti-frequency-selective fading techniques.

5.3 GENERAL RECEIVER CONFIGURATION WITH THE ADAPTIVE EQUALIZER

Figure 5.3 shows a configuration of the receiver with an adaptive equalizer [5-1]. After the received signal is band-limited by a BPF to suppress the ACI as well as noise, the

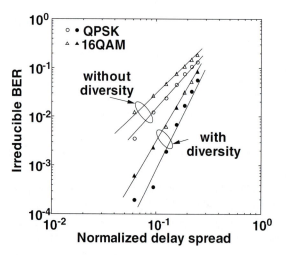

Fig. 5.2 Irreducible BER versus normalized delay spread (τ_{rms}/T_s) performance in the two-ray Rayleigh channel for QPSK and 16QAM with coherent detection, where the roll-off factor of the root Nyquist filters in the transmitter and receiver (α) is 0.5 (a) without diversity and (b) with diversity.

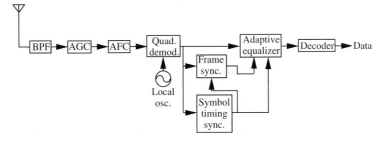

Fig. 5.3 Configuration of a receiver with an adaptive equalizer.

received signal level is controlled to a proper level by the AGC, and after the frequency offset between the locally oscillated frequency and the carrier frequency is reduced by the AFC, the received signal is quasi-coherently detected at the quadrature demodulator to obtain the received complex baseband signal. Using the obtained baseband signal, synchronization for symbol timing and frame timing is taken using the obtained baseband signal. Then distortion due to frequency-selective fading is compensated for at the adaptive equalizer, and the transmitted data sequence is regenerated by decoding the equalizer output.

Among various types of the adaptive equalizers, the decision feedback equalizer (DFE) and the maximum likelihood sequence estimation (MLSE) using the Viterbi algorithm (sometimes called the Viterbi equalizer) are the most popular equalizing schemes. In the following, we will discuss anti-frequency-selective fading strategies using DFE and MLSE.

5.4 DECISION FEEDBACK EQUALIZER

5.4.1 Configuration of DFE

Figure 5.4 shows a configuration of the DFE. The DFE consists of the following four stages:

1. **Equalizing filter.** Consists of a feed-forward (FF) filter with T_p-tap spacing ($T_p = T_s/p$) and a feedback (FB) filter with T_s-tap spacing. When $p = 1$, it is called a T_s-*spaced equalizer*, and an equalizer with $p \geq 2$ is called a *fractional-spaced equalizer*. Because the fractional-spaced equalizer can withstand timing jitter, we usually employ $p = 2$.

2. **Decision stage.** The equalizer output is decided at this stage to regenerate transmitted symbol sequence.

3. **Equalizing error estimator.** Error between the reference signal and the equalizer output (equalizing error) is estimated at this stage.

4. **Tap gain controller.** Tap gains in the equalizing filter are controlled to minimize the variance of the equalizing error.

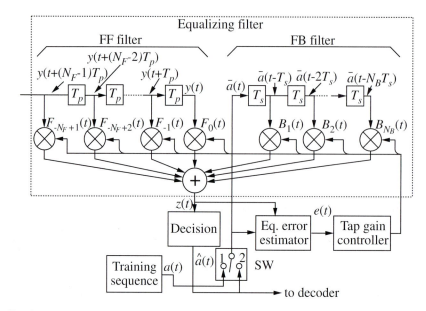

Fig. 5.4 Configuration of the DFE.

Let's assume that the input signal vector is

$$\boldsymbol{y}(t) = [y(t + (N_F - 1)T_p, \; y(t + (N_F - 2)T_p, \; \cdots, \; y(t),$$
$$\bar{a}(t - T_s), \; \bar{a}(t - 2T_s), \; \cdots, \; \bar{a}(t - N_B T_s)]^T \tag{5.13}$$

and the tap gain vector is

$$\boldsymbol{h}(t) = [F_{-N_F + 1}(t), \; F_{-N_F + 2}(t), \; \cdots, \; F_0(t), \; B_1(t), \; \cdots, \; B_{N_B}(t)]^T \tag{5.14}$$

where N_F is the number of feed-forward taps and N_B is the number of feedback taps. The equalizer output is then given by

$$z(t) = \sum_{i=-N_F}^{0} F_i(t) y(t - iT_p) + \sum_{k=1}^{N_B} B_k(t) \bar{a}(t - kT_s) \tag{5.15}$$
$$= \boldsymbol{h}^T(t) \boldsymbol{y}(t).$$

At the decision stage, the transmitted symbol is detected using threshold detectors. When the modulation scheme is QPSK, the transmitted symbol is detected as

$$\hat{a}_k = \hat{a}_{Ik} + j \cdot \hat{a}_{Qk}$$
$$= \text{sgn}[\text{Re}(z(kT_s))] + j \cdot \text{sgn}[\text{Im}(z(kT_s)] \tag{5.16}$$

When the modulation scheme is 16QAM, the decision is carried out as

$$\hat{a}_{Ik} = \begin{cases} 3; & 2 \le \mathrm{Re}[z(kT_s)] \\ 1; & 0 \le \mathrm{Re}[z(kT_s)] < 2 \\ -1; & -2 \le \mathrm{Re}[z(kT_s)] < 0 \\ -3; & \mathrm{Re}[z(kT_s)] < -2 \end{cases} \tag{5.17a}$$

$$\hat{a}_{Qk} = \begin{cases} 3; & 2 \le \mathrm{Im}[z(kT_s)] \\ 1; & 0 \le \mathrm{Im}[z(kT_s)] < 2 \\ -1; & -2 \le \mathrm{Im}[z(kT_s)] < 0 \\ -3; & \mathrm{Im}[z(kT_s)] < -2 \end{cases} \tag{5.17b}$$

In the same manner, we can detect the transmitted symbol for M-ary PSK or M-ary QAM modulations.

At the equalizing error estimator, the error is calculated between the ideal output and $z(t)$ given by

$$\begin{aligned} e(t) &= e_I(t) + j \cdot e_Q(t) \\ &= \overline{a}(t) - z(t) \end{aligned} \tag{5.18}$$

At the tap gain controller, $\boldsymbol{h}(t)$ is updated every T_s second to minimize variance of the estimation error. When the modulation scheme is QPSK or M-ary QAM, in which sampling timings of I-ch and Q-ch are the same, the tap gains are updated to minimize the following equation [5-2].

$$E_1 = \mathrm{E}[e_I^2(t) + e_Q^2(t)] \tag{5.19}$$

On the other hand, when the modulation scheme is Offset QPSK (OQPSK), MSK, or GMSK, in which the sampling timings of I-ch and Q-ch are staggered, Q-ch is distorted a lot by the ISI when I-ch is on the sampling timing (maximum eye-opening timing), and vise versa. Therefore, when I-ch is on the sampling timing

$$E_2 = \mathrm{E}[e_I^2(t)] \tag{5.20}$$

is used for tap gain updating, and

$$E_3 = \mathrm{E}[e_Q^2(t)] \tag{5.21}$$

is used for tap gain updating when Q-ch is on the sampling timing [5-2].

To achieve fast acquisition and tracking ability, recursive least squares (RLS) algorithm [5-3; 5-4] with a forgetting factor (λ) [5-5] is applied for the DFE in land mobile communication systems. In the tap gain updating algorithm, the adaptation is divided into two stages—training stage and tracking stage. During the training stage, a known symbol sequence (training sequence) is used as reference symbols as well as feedback symbols to rapidly converge tap gains to the optimum values. On the other hand, the detected symbol sequence is used during the tracking stage to keep up with the variation of the propagation path characteristics.

There are several types of burst structures according to the location of the training sequence. Figure 5.5 shows burst formats of (a) preamble type and (b) midamble type. The preamble-type format is suitable when we want to sequentially demodulate and equalize the received signal. In this case, equalizing for the training stage is carried out from the beginning of the preamble. During this training stage, the switch in Figure 5.4 is connected to 1. After the last data in the preamble are equalized, the switch is turned to 2, and the equalizing process is changed to the tracking stage.

When the receiver includes burst memory for the block demodulation [5-6], the midamble type gives better BER performance than the preamble type because correlation between the estimated propagation path characteristics at the training sequence and that for the other information symbols is higher than that in the case of the preamble-type burst format. When we apply the midamble type, the equalizing has two directions, forward equalizing for the latter part of data in the burst and backward equalizing for the former part as shown in Figure 5.5(b).

5.4.2 Basic Operation of the DFE

In land mobile communication systems, the DFE should normally operate under both minimum phase conditions in which the power of the direct path (D) is larger than that of the delayed path (U); (D > U, as shown in Figure 5.6[a]), and nonminimum phase conditions (D < U as shown in Figure 5.6[b]), because these two conditions frequently alternate with vehicular movement.

Figure 5.7 shows basic operations of a DFE. First, let's discuss the DFE operation under minimum phase conditions. Assume noise-free conditions to simplify discussion, and the channel impulse response as

$$\boldsymbol{c}(t) = c_0\delta(t) + c_1\delta(t - T_s) \tag{5.22}$$

The received signal coming through this channel is given by

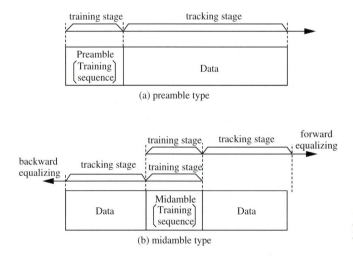

(a) preamble type

(b) midamble type

Fig. 5.5 Burst format: (a) preamble type ; (b) midamble type.

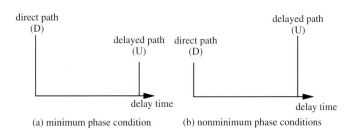

Fig. 5.6 Multipath fading channel conditions: (a) minimum phase conditions; (b) nonminimum phase conditions.

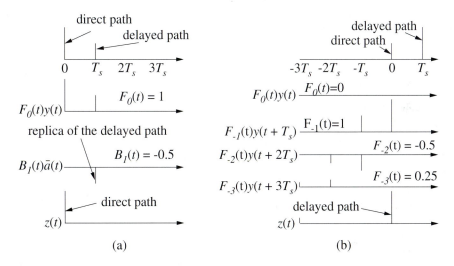

Fig. 5.7 Basic operations of a DFE (a) under minimum phase conditions and (b) under nonminimum phase conditions.

$$y(t) = c_0 a(t) + c_1 a(t - T_s)$$
$$= a(t) + \frac{1}{2} a(t - T_s) \tag{5.23}$$

Figure 5.7(a) shows how to combine tap data in the minimum phase condition. In this case, the delayed wave is canceled out by using FB taps. The first FB tap (with a tap gain of $B_1(t)$) regenerates the replica of one symbol-delayed signal ($B_1(t)\bar{a}(t - T_s)$) using the feedback symbol sequence $\{\bar{a}(t)\}$. Then this signal is combined with $y(t)$ as

$$z(t) = F_0(t) y(t) + B_1(t) \bar{a}(t - T_s)$$
$$= y(t) - \frac{1}{2} \bar{a}(t - T_s) \tag{5.24}$$
$$= a(t) + \frac{1}{2} \left\{ a(t - T_s) - \bar{a}(t - T_s) \right\}$$

where $F_0(t) = 1$ and $B_1(t) = -1/2$. When the feedback symbol sequence includes no error, the equalizer output sequence is the same as the transmitted symbol sequence. This equation also shows that, when the maximum delay time of the delayed save is $\tau_{max} = mT_s$, m-tap of FB taps are necessary.

When the propagation path is treated as the nonminimum phase condition with its impulse response of

$$\mathbf{c}(t) = c_0\delta(t) + c_1\delta(t - T_s)$$
$$= \frac{1}{2}\delta(t) + \delta(t - T_s) \tag{5.25}$$

the DFE operates to suppress the first path using FF taps. Figure 5.7(b) shows its operation. The second FF tap with its tap gain of $F_1(t)$ includes

$$y(t + T_s) = \frac{1}{2}a(t + T_s) + a(t) \tag{5.26}$$

and the third FF tap with its tap gain of $F_2(t)$ includes

$$y(t + 2T_s) = \frac{1}{2}a(t + 2T_s) + a(t + T_s) \tag{5.27}$$

When they are combined with their tap gains of $F_{-1}(t) = 1.0$ and $F_{-2}(t) = -1/2$, the combined signal is given by

$$z'(t) = y(t + T_s) - \frac{1}{2}y(t + 2T_s)$$
$$= a(t) + \left(\frac{1}{2}\right)^2 a(t + 2T_s) \tag{5.28}$$

Although equation (5.27) still includes an ISI component at $t = -2T_s$, it is smaller than the ISI before combining. When we use more FF taps, we can further suppress the ISI. For example, when we combine the fourth tap data with its tap gain of $F_{-3}(t) = 1/4$, the combined output becomes

$$z''(t) = y(t + T_s) - \frac{1}{2}y(t + 2T_s) + \frac{1}{4}y(t + 3T_s)$$
$$= a(t) + \left(\frac{1}{2}\right)^3 a(t + 3T_s) \tag{5.29}$$

We can express basic DFE operation in more general forms as follows [5-7]. Let's assume that the modulation scheme is QPSK, there is no decision error, the propagation path is a two-ray Rayleigh channel with its delay time of mT_s expressed as

$$c(t) = c_0\delta(t) + c_m\delta(t - mT_s) \tag{5.30}$$

and the DFE is a T_s-spaced DFE ($p = 1$). The received baseband signal transmitted over this channel is given by

$$y(t) = c_0 a(t) + c_m a(t - mT_s) + n(t) \tag{5.31}$$

When equation (5.31) is substituted into equation (5.15), we can obtain the equalizer output as

$$
\begin{aligned}
z(t) = {} & \sum_{i=0}^{N_F-1-m} \{c_0 F_{-i}(t) + c_m F_{-(i+m)}(t)\} a(t + iT_s) \\
& + \sum_{i=N_F-m+1}^{N_F-1} c_0 F_{-i}(t) a(t + iT_s) \\
& + \sum_{i=1}^{m} \{c_m F_{-(m-i)}(t) + B_i(t)\} a(t - iT_s) \\
& + \sum_{i=m+1}^{N_B} B_i a(t - iT_s) + \sum_{i=0}^{N_F-1} F_{-i}(t) n(t + iT_s)
\end{aligned}
\tag{5.32}
$$

In this case, the output SNR of the DFE is given by

$$
\begin{aligned}
\gamma = {} & \left\{ \left| c_0 F_0(t) + c_m F_{-m} \right|^2 \sigma_s^2 \right\} \Bigg/ \Bigg\{ \sum_{i=1}^{N_F-1-m} \left| c_0 F_{-i}(t) + c_m F_{-(i+m)}(t) \right|^2 \sigma_s^2 \\
& + \sum_{i=N_F-m+1}^{N_F-1} \left| c_0 F_{-i}(t) \right|^2 \sigma_s^2 + \sum_{i=1}^{m} \left| c_m F_{-(m-i)}(t) + B_i(t) \right|^2 \sigma_s^2 \\
& + \sum_{i=m+1}^{N_B} \left| B_i(t) \right|^2 \sigma_s^2 + \sum_{i=0}^{N_F-1} \left| F_i(t) \right|^2 \sigma_n^2 \Bigg\}
\end{aligned}
\tag{5.33}
$$

where $\sigma_s^2 = \mathrm{E}[\,|a(t)|^2\,]$, and $\sigma_n^2 = \mathrm{E}[\,|n(t)|^2\,]$.

Let the SNR of the first path and that of the second path be

$$\gamma_a = \frac{|c_0|^2 \sigma_s^2}{\sigma_n^2} \tag{5.34}$$

$$\gamma_b = \frac{|c_m|^2 \sigma_s^2}{\sigma_n^2} \tag{5.35}$$

and $L = \mathrm{Int}(N_F/m)$, where $\mathrm{Int}(a)$ is the integer part of a. The tap gains are controlled to maximize γ. Therefore, the optimum tap gains are obtained by partial differentiation of equation (5.33) by each tap gain as follows:

$$F_{-i}(t) = \begin{cases} \dfrac{A}{c_0} & (i = 0) \\[2ex] \dfrac{\gamma_b}{\gamma_a(1 + p_L \gamma_a)} \dfrac{c_0}{c_m} F_0(t) & (i = m) \\[2ex] -\dfrac{\gamma_b}{1 + p_{L-i+1}\gamma_b} \dfrac{c_0}{c_m} F_{m-i}(t) \\ \qquad (i = jm, \; j = 2, 3, \cdots, L) \\ 0 \qquad\qquad\qquad\qquad (i \neq jm) \end{cases}$$ (5.36a)

$$B_i(t) = \begin{cases} -c_m F_0(t) & (i = m) \\ 0 & (i \neq m) \end{cases}$$ (5.36b)

where p_n $(n = 1, 2, ..., L)$ is determined by the following series $\{p_n\}$

$$p_1 = 1$$ (5.37a)

$$p_{n+1} = \frac{1 + p_n \gamma_a}{1 + p_n \gamma_a + \gamma_b}$$ (5.37b)

$\{p_n\}$ is a positive and monotonically decreasing series, and it converges to

$$\lim_{n \to \infty} p_n = \frac{\gamma_a - \gamma_b - 1 + \sqrt{(\gamma_a - \gamma_b - 1)^2 + 4\gamma_a}}{2\gamma_a}$$ (5.38)

When these tap gains are used, the output SNR is given by

$$\gamma = \begin{cases} \gamma_a & (N_F < m) \\[1.5ex] \gamma_a + \dfrac{\gamma_b}{1 + p_L \gamma_a} & (N_F \geq m) \end{cases}$$ (5.39)

Equation (5.39) shows that the diversity gain can be obtained when $N_F \geq m$. However, the output SNR is slightly smaller than that of the ideal path diversity ($\gamma_a + \gamma_b$) due to the denominator in the second term of (5.39). We will discuss more on this degradation.

Figure 5.8 shows γ versus γ_b/γ_a with a parameter of $L = \text{Int}[N_F/m]$ for (a) $\gamma_a + \gamma_b = 10$ dB, (b) $\gamma_a + \gamma_b = 15$ dB, and (c) $\gamma_a + \gamma_b = 20$ dB. For example, when $\gamma_a + \gamma_b = 10$ dB and $L = 1$, the degradation of γ from $\gamma_a + \gamma_b$ is maximum at $\gamma_b/\gamma_a = 8$ dB. When L increases, this degradation becomes smaller, and γ_b/γ_a that gives the maximum degradation approaches 0 dB. This phenomenon can be observed for any $\gamma_a + \gamma_b$ values. Even though L is sufficiently large, there still exists 1–2 dB degradation at $\gamma_b/\gamma_a = 0$ dB, and this degradation becomes larger with increasing $\gamma_a + \gamma_b$.

Based on the foregoing discussions, we can conclude the following about the number of taps in the DFE:

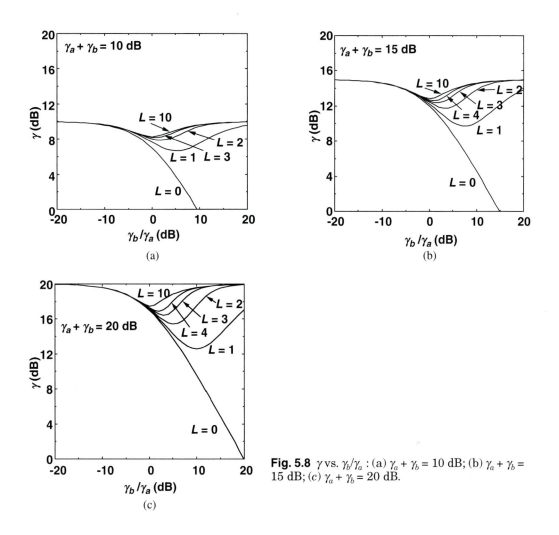

Fig. 5.8 γ vs. γ_b/γ_a : (a) $\gamma_a + \gamma_b = 10$ dB; (b) $\gamma_a + \gamma_b = 15$ dB; (c) $\gamma_a + \gamma_b = 20$ dB.

☞ Equation (5.36b) shows that the optimum value of N_B is m when the maximum delay time of the delayed wave is m-symbol duration.

☞ Equations (5.36a) and (5.39) show that at least $N_F = (m + 1)$ is necessary to obtain a diversity effect in the case of T_s-spaced DFE. In the case of fractional-spaced DFE, $N_F = p(m + 1)$ will be necessary. Equation (5.39) also shows that more than N_F taps are necessary to further improve SNR of the equalizer output.

5.4.3 RLS Algorithm

In land mobile communication systems, propagation path characteristics change very fast due to high terminal mobility. As a result, we need very fast initial acquisition and tracking ability for the tap gain updating. Therefore, we'll use the RLS algorithm (Kalman algorithm).

Let the $z(t)$, $e(t)$, $\bar{a}(t)$, $\boldsymbol{y}(t)$, and $\boldsymbol{h}(t)$ at $t = nT_s$ are represented by z_n, e_n, $\bar{a}_n$, $\boldsymbol{y}_n$, and $\boldsymbol{h}_n$, respectively. The DFE output and the estimation error are expressed as

$$z_n = \boldsymbol{h}_{n-1}^T \boldsymbol{y}_n \tag{5.40}$$

$$e_n = \bar{a}_n - z_n \tag{5.41}$$

The RLS algorithm recursively computes the optimum $\boldsymbol{h}_n$ that minimizes

$$E_n = \sum_{i=0}^{n-1} \lambda^i \left| e_{n-i} \right|^2 \tag{5.42}$$

as follows:

$$\boldsymbol{h}_0 = \left[F_{-N_F+1}(0),\ F_{-N_F+2}(0),\ \cdots,\ F_0(0),\ B_1(0),\ B_2(0),\ \cdots,\ B_{N_B}(0) \right]^T \tag{5.43}$$
$$= \left[0,\ 0,\ \cdots,\ 1,\ 0,\ 0,\ \cdots 0 \right]^T$$

$$\boldsymbol{P}_0 = \delta^{-1} \boldsymbol{I} \quad (\boldsymbol{I};\ \text{unit matrix}) \tag{5.44}$$

$$\boldsymbol{k}_n = \boldsymbol{P}_{n-1} \boldsymbol{y}_n (\boldsymbol{y}_n^{*T} \boldsymbol{P}_{n-1} \boldsymbol{y}_n + \lambda v)^{-1} \tag{5.45}$$

$$\boldsymbol{h}_n = \boldsymbol{h}_{n-1} + e_n \boldsymbol{k}_n \tag{5.46}$$

$$\boldsymbol{P}_n = (\boldsymbol{P}_{n-1} - \boldsymbol{k}_n \boldsymbol{y}_n^{*T} \boldsymbol{P}_{n-1}) \lambda^{-1} \tag{5.47}$$

where $\boldsymbol{y}_n^{*T}$ means a transpose conjugate of $\boldsymbol{y}_n$; $\boldsymbol{k}_n$ is the Kalman gain; $\boldsymbol{P}_n$ is the covariance matrix of $\boldsymbol{h}_n$; v is the variance of e_n; and λ is the forgetting factor ($0 < \lambda \leq 1$). With this algorithm, the obtained $\boldsymbol{h}_n$ satisfies the following equation [5-3]:

$$\boldsymbol{h}_n = \boldsymbol{R}_n^{-1} \boldsymbol{b}_n \tag{5.48}$$

$$\boldsymbol{R}_n = \sum_{i=1}^{n} \lambda^{n-i} \boldsymbol{y}_i \boldsymbol{y}_i^* + \delta \lambda^n \boldsymbol{I} \tag{5.49}$$

$$\boldsymbol{b}_n = \sum_{i=0}^{n} \lambda^{n-i} a_i^* \boldsymbol{y}_i \tag{5.50}$$

On the other hand, the optimum tap gain based on the linear least square estimation (Wiener-Hopf's equation) is given by [5-3]

$$\boldsymbol{h}_n = \boldsymbol{R}_n^{-1} \boldsymbol{b}_n$$
$$= \frac{E[a_n^* \boldsymbol{y}_n]}{E[\boldsymbol{y}_n \boldsymbol{y}_n^{*T}]} \tag{5.51}$$

where $\boldsymbol{R}_n = E[\boldsymbol{y}_n\boldsymbol{y}_n^{*T}]$ and $b_n = E[a_n^*\boldsymbol{y}_n]$. From comparison between equations (5.48) and (5.51), we can see that the RLS algorithm gives a solution of Wiener-Hopf's equation in which the ensemble average is converted to a time average using an exponentially weighted filter with its impulse response of

$$f_\lambda(t) = \sum_{i=0}^{\infty} \lambda^i \delta(t - iT_s) \tag{5.52}$$

to improve tracking ability of the algorithm. When smaller λ is employed, tracking ability of the DFE is improved because the averaging period becomes shorter. However, such tracking ability improvement sacrifices noise immunity because the number of data to be averaged becomes small. On the other hand, when larger λ is employed, although noise immunity is improved, its tracking ability is degraded.

5.4.3.1 Optimization of λ There are three main causes for performance degradation.

☞ Minimum mean square error (e_{mmse}) caused by noise
☞ Estimation error (e_{est}) caused by the misadjustment of $\boldsymbol{c}_n$ from its optimum value $\boldsymbol{c}_{opt}$
☞ Lag error (e_{lag}) caused by insufficient tracking ability of the DFE

Ref. [5-8] theoretically analyzes the variance of these errors by simplifying the channel variation using the Markov process. The results show that

$$\sigma_{mmse}^2 = \exp\left[\frac{1}{W} \int_{-W/2}^{W/2} \ln\left(\frac{N(f)}{S(f) + N(f)} \right) df \right] \tag{5.53}$$

$$\sigma_{est}^2 = \frac{1-\lambda}{1+\lambda}\left(1 + \frac{1-\lambda}{1+\lambda} \right) N_{total}\sigma_{mmse}^2 \tag{5.54}$$

$$\sigma_{lag}^2 = \frac{2(1-\rho)(1-\sigma_{mmse}^2)}{(1-\lambda^2)} \tag{5.55}$$

$$\rho = 1 - \frac{3(\pi f_d T_s)^2}{4(1-\lambda)} \tag{5.56}$$

$$N_{total} = N_F + N_B \tag{5.57}$$

where
   $S(f)$ = spectrum of the signal component
   $N(f)$ = spectrum of noise component
   W = bandwidth of the signal component

As shown in equations (5.54) and (5.55), λ has a big impact on performance. When λ is very close to 1, although σ_{est}^2 approaches 0, σ_{lag}^2 increases because larger λ means to average the optimum tap gain vector with longer time constant. On the other hand, smaller λ can reduce σ_{lag}^2, although it increases σ_{est}^2. Therefore, there exists an optimum λ, which is given by

$$\lambda_{opt} = 1 - \xi \sinh\left[\frac{\sinh^{-1}(4/\xi)}{3}\right] \tag{5.58}$$

$$\xi = \sqrt{\frac{3(\pi f_d T_s)^2 \Gamma}{N_{total}}} \tag{5.59}$$

where Γ is the energy per symbol to the noise spectral density (E_s/N_0). Equations (5.58) and (5.59) show that λ_{opt} decreases with increasing f_d or Γ, and it increases with N_{total}. Figure 5.9 shows the relationship between λ_{opt} and $f_d T_s$ with a parameter of Γ at $N_{total} = 11$.

Figure 5.10 shows the simulated results of BER versus λ performance of 128-ksymbol/s-QPSK with DFE (a) with a parameter of E_b/N_0 where $f_d T_s = 7.8125 \times 10^{-4}$, and (b) with a parameter of f_d, where $E_b/N_0 = 20$ dB. In this simulation, $N_{total} = 11$ and a propagation path model of a two-ray Rayleigh model ($D/U = 0$ dB and $\tau = T_s$) are employed. λ_{opt} that minimizes the BER in Figure 5.10 is almost the same as the values shown in Figure 5.9.

Figure 5.11 shows examples of BER performance of the same system for (a) $\lambda = 1.0$, (b) $\lambda = 0.94$, and (c) $\lambda = 0.88$. When $\lambda = 1$ and $f_d T_s = 7.813 \times 10^{-7}$ are employed, the performance is degraded by about 2 dB from the theoretical performance of the maximal ratio combining path diversity. This degradation is almost the same as the maximum degradation shown in Figure 5.8. This degradation increases with λ. When $\lambda = 0.94$ is used, the degradation is about 4–5 dB, and it becomes 5–6 dB in the case of $\lambda = 0.88$. On the other hand, the irreducible BER at high E_b/N_0 is lowered by reducing λ. When $f_d T_s = 1.172 \times 10^{-3}$, the irreducible BER is about 6.0×10^{-2} for $\lambda = 1.0$, 6.0×10^{-3} for $\lambda = 0.94$, and 2.0×10^{-3} for $\lambda = 0.88$, respectively.

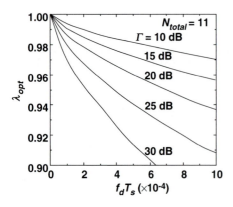

Fig. 5.9 Relationship between λ_{opt} and $f_d T_s$ with a parameter of Γ at $N_{total} = 11$.

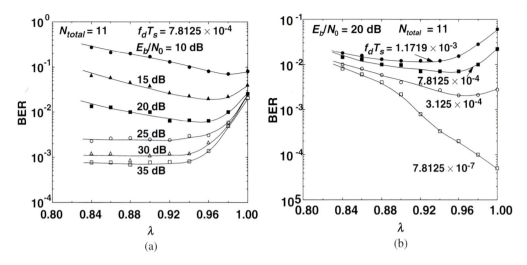

Fig. 5.10 Simulated results of BER versus λ performances of 128-ksymbol/s QPSK with DFE (a) with a parameter of E_s/N_0 where $f_d T_s = 7.8125 \times 10^{-4}$ and (b) with a parameter of f_d, where E_b/N_0 = 20 dB; in this simulation, $N_{total} = 11$ and a propagation path model of a two-ray Rayleigh model (D/U = 0 dB and $\tau = T_s$) are employed (from Ref. 5-7, © Institute of Electronics, Information and Communication Engineers, 1989).

5.4.3.2 Error Propagation Effect   During the training stage, the tap gain vector is exactly converged to a certain value because feedback symbols to the FB taps include no error. However, the feedback symbol during the tracking process may include error. When such an error occurs frequently, those errors lead the tap gains in the wrong direction, and it sometimes diverges the RLS algorithm. This effect is called the *error propagation effect*.

Figure 5.12 shows the performance of BER versus information symbol length in each burst, where E_b/N_0 = 20 dB, two-ray Rayleigh model (D/U = 0 dB, $\tau = T_s$) is assumed, and a frame format shown in Figure 5.6(a) is employed. The results show that BER increases with information symbol length due to the error propagation effect, and it is more remarkable at larger f_d. Therefore, we have to carefully select the information symbol length in each burst.

5.4.3.3 Symbol Rate Dependence of DFE   As shown in equation (5.55), σ_{lag}^2 decreases with the symbol rate (R_s) because fading variation during a symbol duration becomes slower. On the other hand, the amount of computation increases with R_s because more DFE taps are required. When the maximum delay time of the delayed wave is τ_{max}, the DFE has to compensate for $m = [\tau_{max}/T_s]$-symbol-delayed wave. Consequently, m-tap of the FB-tap and $2(m + 1)$-tap of the FF-tap are necessary when the tap spacing of the FF-tap is $T_s/2$.

At present, most DSPs can simultaneously operate multiplication, addition/subtraction, and data transfer between registers. Therefore, we can roughly evaluate the number of computations by the number of multiplications. In the RLS algorithm, the number of multiplication per symbol is given by

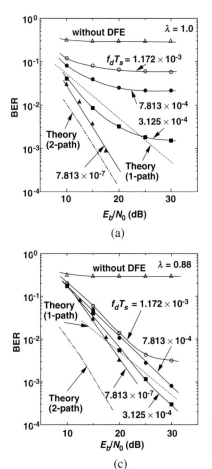

(a)

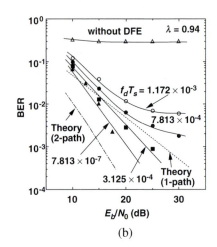

(b)

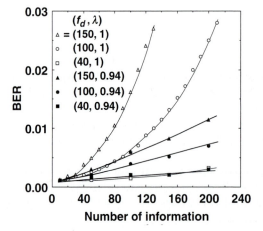

(c)

Fig. 5.11 An example of BER performance of QPSK with DFE system: (a) $\lambda = 1.0$, (b) $\lambda = 0.94$, and (c) $\lambda = 0.88$ (from Ref. 5-7, © Institute of Electronics, Information and Communication Engineers, 1989).

Fig. 5.12 BER versus information symbol length in each burst with a frame format shown in Figure 5.6(a), where $E_b/N_0 = 20$ dB, and two-ray Rayleigh model ($D/U = 0$ dB, $\tau = T_s$) is assumed (from Ref. 5-7, © Institute of Electronics, Information and Communication Enginners, 1989).

$$C_{RLS}(N_{total}) = 7N_{total}^2 + 12N_{total} \qquad (5.60)$$

Figure 5.13 shows the number of multiplications versus R_s for τ_{max} = 10, 25, and 50 µs in the case of the RLS algorithm. This figure shows that the number of multiplications increases with R_s and τ_{max} because more DFE taps are required. For example, when we compare the number of multiplications for R_s = 150 ksymbol/s and 500 ksymbol/s at τ_{max} = 10 µs, 6 times more taps are needed. With these results, we can conclude the following:

☞ When R_s is slow, faster tracking ability is necessary.

☞ When R_s is fast, simplification of the equalizing algorithm is necessary.

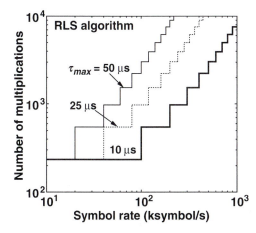

Fig. 5.13 Number of multiplications per symbol versus R_s for τ_{max} = 10, 25, and 50 µs in the case of RLS algorithm, where FF filter of the DFE is a $T_s/2$-spaced filter.

5.4.4 Strategy for the Reduction of Computations

There are mainly four strategies for the reduction of computation. The first is the reduction of redundancy in the RLS algorithm, such as the fast Kalman algorithm [5-5]. Although this algorithm is effective when the number of taps in the DFE is very large, the number of computations is not that reduced when the number of taps is small. For this reason, there are not too many studies in existence on the application of the fast Kalman algorithm to the land mobile communication systems.

Another method is the tap gain interpolation technique [5-9; 5-10]. When the symbol rate is very fast and a burst length is very short, the fading variation during each burst is considered to be very smooth. Therefore, when we regularly transmit a training sequence and obtain the optimum tap gain vector at each training sequences, we can estimate the fading variation between the training sequence by interpolating the obtained tap gain vectors.

The third way to reduce computations is to reduce the number of taps in the DFE. The bidirectional DFE (BDDFE) is an effective method to reduce the number of taps in the DFE [5-11; 5-12; 5-13; 5-14; 5-15]. The BDDFE consists of forward equalizing, in which the equalizing is carried out in the order of reception, and backward equalizing, in which the equalizing is carried out in the reverse order of reception. In this case, some nonminimum phase conditions in the forward equalizing is converted to the minimum phase condition in the backward equalizing. Consequently, we can reduce the number of FF taps that are used for equalization in the nonminimum phase condition as shown in Figure 5.5.

The fourth computation reduction method is the application of a high spectral efficient modulation scheme such as 16QAM [5-15; 5-16; 5-17]. In the DFE algorithm, the number of computations does not depend on the bit rate but on the symbol rate. When we increase the modulation level, we can reduce the symbol rate. For example, when we apply 16QAM instead of QPSK, we can reduce the symbol rate by half. Therefore, we can reduce the number of computations by increasing the modulation level.

We will discuss these techniques next.

5.4.5 Fast Kalman Algorithm

The tap data vector ($\boldsymbol{y}_n$) and tap vector ($\boldsymbol{h}_n$) for the fast Kalman algorithm [5-5] at $t = nT_s$ are given:

$$\boldsymbol{y}_n = [y(nT_s + (N_F - 1)T_p), \; y(nT_s + (N_F - 2)T_p), \; \cdots, \; y(nT_s)$$
$$\overline{a}((n-1)T_s), \; \overline{a}((n-2)T_s), \; \cdots, \; \overline{a}((n-N_B)T_s)]^T \tag{5.61}$$

$$\boldsymbol{h}_n = [F_{-N_F+1}(nT_s), \; F_{-N_F+2}(nT_s), \; \cdots, \; F_0(nT_s)$$
$$B_1(nT_s), \; B_2(nT_s), \; \cdots, \; B_{N_B}(nT_s)]^T \tag{5.62}$$

Let $\boldsymbol{\xi}_n$ be a q- (= p + 1) dimensional new input data vector at $t = (k + 1)T_s$ expressed as

$$\boldsymbol{\xi}_k = \begin{bmatrix} y(kT_s + (N_F + p - 1)T_p) \\ y(kT_s + (N_F + p - 2)T_p) \\ \vdots \\ y(kT_s + N_F T_p) \\ \cdots\cdots\cdots\cdots\cdots\cdots\cdots\cdots \\ \overline{a}_k \end{bmatrix} \tag{5.63}$$

and $\boldsymbol{\rho}_k$ is a q-dimensional shifted-out data vector at $t = (k + 1)T_s$ expressed as

$$\boldsymbol{\rho}_k = \begin{bmatrix} y(kT_s + (p - 1)T_p) \\ y(kT_s + (p - 2)T_p) \\ \vdots \\ y(kT_s) \\ \cdots\cdots\cdots\cdots\cdots\cdots\cdots\cdots \\ \overline{a}_{k-N_B} \end{bmatrix} \tag{5.64}$$

Moreover, we will define an extended input data vector that includes both $\boldsymbol{y}_k$ and $\boldsymbol{\xi}_k$ components as

$$
\boldsymbol{w}_k =
\begin{bmatrix}
y(kT_s + (N_F + p - 1)T_p) \\
y(kT_s + (N_F + p - 2)T_p) \\
\vdots \\
y(kT_s + N_F T_p) \\
\cdots\cdots\cdots\cdots\cdots\cdots \\
y(kT_s + (N_F - 1)T_p) \\
y(kT_s + (N_F - 2)T_p) \\
\vdots \\
y(kT_s + T_p) \\
y(kT_s) \\
\cdots\cdots\cdots\cdots\cdots\cdots \\
\overline{a}_k \\
\cdots\cdots\cdots\cdots\cdots\cdots \\
\overline{a}_{k-1} \\
\vdots \\
\overline{a}_{k-N_B}
\end{bmatrix}
\tag{5.65}
$$

$\boldsymbol{S}$ and $\boldsymbol{T}$ are matrices to exchange elements of $\boldsymbol{w}_k$ as

$$
\boldsymbol{S}\boldsymbol{w}_k = \begin{bmatrix} \boldsymbol{\xi}_k \\ \cdots \\ \boldsymbol{y}_k \end{bmatrix}
\tag{5.66}
$$

$$
\boldsymbol{Q}\boldsymbol{w}_k = \begin{bmatrix} \boldsymbol{y}_{k+1} \\ \cdots \\ \boldsymbol{\rho}_k \end{bmatrix}
\tag{5.67}
$$

Because $\boldsymbol{S}$ and $\boldsymbol{Q}$ have only a single 1 in each column or row, and because the rest of the elements are all 0, they have the following relationships:

$$
\boldsymbol{S}^{-1} = \boldsymbol{S}^T
\tag{5.68}
$$

$$
\boldsymbol{Q}^{-1} = \boldsymbol{Q}^T
\tag{5.69}
$$

In the fast Kalman algorithm, $\boldsymbol{\xi}_k$ corresponds to the forward-estimated value of $\boldsymbol{y}_k$ as

$$
\boldsymbol{\xi}_k = -\boldsymbol{A}_{k-1}^T \boldsymbol{y}_{k-1} + \boldsymbol{\varepsilon}_k
\tag{5.70}
$$

where $-\boldsymbol{A}_{k-1}$ is the $N_{total} \times p$-dimensional forward-estimation coefficient matrix and $\boldsymbol{\varepsilon}_k$ is the p-row forward-estimation error vector. In the same manner, $\boldsymbol{\rho}_k$ corresponds to the backward-estimated value of $\boldsymbol{y}_{k+1}$ given by

$$\boldsymbol{\rho}_k = -\boldsymbol{D}_{k-1}^T \boldsymbol{y}_{k-1} + \boldsymbol{\eta}_k \qquad (5.71)$$

where $-\boldsymbol{D}_{k-1}$ is the $N_{total} \times p$-dimensional backward-estimation coefficient matrix and $\boldsymbol{\eta}_k$ is the p-row backward-estimation error vector. $\boldsymbol{A}_{k-1}$ and $\boldsymbol{D}_{k-1}$ are recursively updated to minimize the estimation errors given by

$$E_f = \sum_{k=1}^{n} \lambda^{n-k} \left| \boldsymbol{\xi}_k + \boldsymbol{A}_{k-1}^T \boldsymbol{y}_k \right|^2 \qquad (5.72)$$

$$E_b = \sum_{k=1}^{n} \lambda^{n-k} \left| \boldsymbol{\rho}_k + \boldsymbol{D}_{k-1}^T \boldsymbol{y}_{k+1} \right|^2 \qquad (5.73)$$

The equalizer output and the estimation error are given by

$$z_n = \boldsymbol{h}_{n-1}^T \boldsymbol{y}_n \qquad (5.74)$$

$$e_n = \bar{a}_n - z_n = \bar{a}_n - \boldsymbol{h}_{n-1}^T \boldsymbol{y}_n \qquad (5.75)$$

and the tap gain vector is updated every T_s second as

$$\boldsymbol{h}_n = \boldsymbol{h}_{n-1} + \boldsymbol{k}_n e_n \qquad (5.76)$$

$$\boldsymbol{k}_n = \boldsymbol{R}_{1,n}^{-1} \boldsymbol{y}_n^* \qquad (5.77)$$

where $\boldsymbol{R}_{1,n}$ is the covariance matrix of the tap data vector, expressed as

$$\boldsymbol{R}_{1,n} = \lambda \boldsymbol{R}_{1,n-1} + \boldsymbol{y}_n^* \boldsymbol{y}_n^T$$
$$= \sum_{k=1}^{n} \lambda^{n-k} \boldsymbol{y}_n^* \boldsymbol{y}_n^T + \delta \boldsymbol{I} \qquad (5.78)$$

and λ is a forgetting factor. In the same manner, we can obtain the covariance matrix of $\boldsymbol{w}_n$ as

$$\boldsymbol{W}_n = \lambda \boldsymbol{W}_{n-1} + \boldsymbol{w}_n^* \boldsymbol{w}_n^T \qquad (5.79)$$

and the extended Kalman gain can be defined as

$$\boldsymbol{k}_n^E = \boldsymbol{W}_n^{-1} \boldsymbol{w}_n^*$$
$$= \boldsymbol{S}^T (\boldsymbol{S} \boldsymbol{W}_n \boldsymbol{S}^T)^{-1} \boldsymbol{S} \boldsymbol{w}_n^* \qquad (5.80)$$

In equation (5.80), $\boldsymbol{S} \boldsymbol{W}_n \boldsymbol{S}^T$ can be modified as

$$SW_nS^T$$

$$= S\left[\sum_{k=0}^{n}\lambda^{n-k}w_k^*w_k^T\right]S^T$$

$$=\begin{bmatrix}\sum_{k=0}^{n}\lambda^{n-k}\xi_k^*\xi_k^T & \sum_{k=0}^{n}\lambda^{n-k}\xi_k^*y_k^T \\ \sum_{k=0}^{n}\lambda^{n-k}y_k^*\xi_k^T & \sum_{k=0}^{n}\lambda^{n-k}y_k^*y_k^T\end{bmatrix}$$

$$=\begin{bmatrix}b_n & B_n^{*T} \\ B_n & R_n\end{bmatrix}$$

(5.81)

where

$$b_n = \lambda b_{n-1} + \xi_n^*\xi_n^T \tag{5.82}$$

$$B_n = \lambda B_{n-1} + y_n^*\xi_n^T \tag{5.83}$$

One of the most important points in the fast Kalman algorithm is how to efficiently calculate the inverse of a matrix. First, let's calculate the inverse of SW_nS^T.

When we apply an identical equality given by

$$\begin{bmatrix}A & B \\ C & D\end{bmatrix}^{-1} = \begin{bmatrix}K^{-1} & -K^{-1}BD^{-1} \\ -D^{-1}CK^{-1} & D^{-1}+D^{-1}CK^{-1}BD^{-1}\end{bmatrix} \tag{5.84}$$

$$K = A - BD^{-1}C \tag{5.85}$$

we can calculate the inverse of equation (5.81) as

$$\left(SW_nS^T\right)^{-1}$$

$$=\begin{bmatrix}E_n^{-1} & -E_n^{-1}B_n^{*T}R_n^{-1} \\ -R_n^{-1}B_nE_n^{-1} & R_n^{-1}+R_n^{-1}B_nE_n^{-1}B_n^{*T}R_n^{-1}\end{bmatrix} \tag{5.86}$$

where

$$E_n = b_n - B_n^{*T}R_n^{-1}B_n \tag{5.87}$$

Because A_n has to minimize E_f given by equation (5.72), it is given by

$$A_n = -R_n^{-1}B_n \tag{5.88}$$

Consequently, equation (5.87) can be modified as

$$E_n = b_n + B_n^{*T} A_n \tag{5.89}$$

and $(SW_nS^T)^{-1}$ is given by

$$(SW_nS^T)^{-1}$$
$$= \begin{bmatrix} E_n^{-1} & E_n^{-1}A_n^{*T} \\ A_nE_n^{-1} & R_n^{-1} + A_nE_n^{-1}A_n^{*T} \end{bmatrix} \tag{5.90}$$

When equations (5.77) and (5.90) are substituted into equation (5.80), we can calculate that

$$k_n^E = S^T\left(SW_nS^T\right)^{-1}Sw_n^*$$

$$= S^T\left(SW_nS^T\right)^{-1}\begin{bmatrix} \xi_n^* \\ \cdots \\ y_n^* \end{bmatrix}$$

$$= S^T\begin{bmatrix} E_n^{-1}\varepsilon_n'^* \\ \cdots\cdots\cdots\cdots \\ k_n + A_nE_n^{-1}\varepsilon_n'^* \end{bmatrix} \tag{5.91}$$

where

$$\varepsilon_n' = \xi_n + A_n^T y_n \tag{5.92}$$

E_n can be modified using equations (5.78), (5.82), (5.83), (5.88), and (5.92) as

$$E_n = \lambda E_{n-1} + \varepsilon_n'^* \varepsilon_n^T \tag{5.93}$$

Using equations (5.70), (5.77), (5.78), (5.83), and (5.88), A_n can be expressed as

$$A_n = A_{n-1} - k_n\varepsilon_n^T \tag{5.94}$$

Although equation (5.91) is much simpler than equation (5.80), there still remains an inverse matrix (E_n^{-1}) to be calculated. To calculate E_n^{-1}, we can apply the following identical equality:

$$(A + BC)^{-1} = A^{-1} - A^{-1}B(I_m + CA^{-1}B)^{-1}CA^{-1} \tag{5.95}$$

where A, B, and C are matrices of $n \times n$, $n \times m$, and $m \times n$, and $|A| \neq 0$. Using equation (5.95), we can figure the inverse matrix of $E_n^{-1} (= P_n)$ as

$$E_n^{-1} = P_n = \frac{1}{\lambda}\left[P_{n-1} - k_n^s\varepsilon_n^T P_{n-1}\right] \tag{5.96}$$

where

$$k_n^s = \frac{1}{\lambda + v_n} P_{n-1} \varepsilon_n^{'*}$$

(5.97)

$$v_n = \varepsilon_n^T P_{n-1} \varepsilon_n^{'*}$$

(5.98)

Using equations (5.96)–(5.98), we can calculate

$$P_n \varepsilon_n^{'*} = k_n^s$$

(5.99)

thereby modifying equation (5.91) as

$$k_n^E = S^T \begin{bmatrix} k_n^s \\ \cdots\cdots\cdots \\ k_n + A_n k_n^s \end{bmatrix}$$

(5.100)

In the same manner, we can also calculate the following equations:

$$Q w_n^* = (Q W_n Q^T) Q k_n^E$$

(5.101)

$$Q W_n Q^T = Q \left[\sum_{k=0}^{n} \lambda^{n-k} w_k^* w_k^T \right] Q^T$$

$$= \begin{bmatrix} \displaystyle\sum_{k=0}^{n} \lambda^{n-k} y_{n+1}^* y_{n+1}^T & \displaystyle\sum_{k=0}^{n} \lambda^{n-k} y_{n+1}^* \rho_n^T \\ \displaystyle\sum_{k=0}^{n} \lambda^{n-k} \rho_n^* y_{n+1}^T & \displaystyle\sum_{k=0}^{n} \lambda^{n-k} \rho_{n+1}^* \rho_{n+1}^T \end{bmatrix}$$

$$= \begin{bmatrix} R_{n+1} & J_{n+1} \\ J_{n+1}^{*T} & f_n \end{bmatrix},$$

(5.102)

where

$$f_n = \lambda f_{n-1} + \rho_n^* \rho_n^T$$

(5.103)

$$J_n = \lambda J_{n-1} + y_{n+1}^* \rho_n^T$$

(5.104)

When we define m_n and μ_n as

$$Q k_n^E = \begin{bmatrix} m_n \\ \cdots \\ \mu_n \end{bmatrix}$$

(5.105)

we can modify equation (5.101) as

$$Qw_n^* = \begin{bmatrix} R_{n+1} & J_n \\ J_n^{*T} & f_n \end{bmatrix} \begin{bmatrix} m_n \\ \mu_n \end{bmatrix}$$
$$= \begin{bmatrix} R_{n+1}m_n + J_n\mu_n \\ J_n^{*T}m_n + f_n\mu_n \end{bmatrix}$$

(5.106)

When we compare equations (5.67) and (5.106), we can calculate the following equation.

$$y_{n+1}^* = R_{n+1}m_n + J_n\mu_n$$

(5.107)

Because D_n is controlled to minimize equation (5.73), it is given by Wiener-Hopf's equation as

$$D_n = -R_{n+1}^{-1}J_n$$

(5.108)

Moreover, using equations (5.78), (5.104), and (5.106), we can calculate

$$D_n - D_{n-1} = -k_{n+1}\eta_n^T$$

(5.109)

$$\eta_n^T = \rho_n^T + D_{n-1}^T y_{n+1}$$

(5.110)

As a result, we can modify equation (5.107) as

$$m_n - D_n m_n = R_{n+1}^{-1}y_{n+1}^*$$

(5.111)

Using equations (5.77), (5.108), and (5.111), we can calculate this result.

$$k_{n+1} = \left[m_n - D_{n-1}m_n \right]\left[1 - \eta_n^T\mu_n \right]^{-1}$$

(5.112)

With these results, the fast Kalman algorithm can be summarized as follows:

☞ **Initialization of vectors and matrices**

$$k_1 = y_0 = c_0 = 0 \ (N - \text{row vector})$$

(5.113)

$$A_0 = D_0 = 0 \ (N - \text{row}, q - \text{column matrices})$$

(5.114)

$$k_0^s = 0 \ (q - \text{row vector})$$

(5.115)

$$P_0 = 0 \ (q - \text{row}, q - \text{column matrix})$$

(5.116)

☞ **Tap gain update**

$$\varepsilon_n = \xi_n + A_{n-1}^T y_{n-1}$$

(5.117)

$$A_n = A_{n-1} - k_n \varepsilon_n^T \tag{5.118}$$

$$\varepsilon_n^{'} = \xi_n + A_n^T y_n \tag{5.119}$$

$$k_n^s = P_{n-1} \varepsilon_n^{'*} \left[\lambda + \varepsilon_n^T P_{n-1} \varepsilon_n^{'*} \right]^{-1} \tag{5.120}$$

$$P_n = \frac{1}{\lambda} \left[P_{n-1} - k_n^s \varepsilon_n^T P_{n-1} \right] \tag{5.121}$$

$$\begin{bmatrix} m_n \\ \cdots \\ \mu_n \end{bmatrix} = QS^T \begin{bmatrix} k_n^s \\ \cdots\cdots\cdots\cdots \\ k_n + A_n k_n^s \end{bmatrix} \tag{5.122}$$

$$\eta_n^T = \rho_n^T + D_{n-1}^T y_{n+1} \tag{5.123}$$

$$k_{n+1} = \left[m_n - D_{n-1} m_n \right] \left[1 - \eta_n^T \mu_n \right]^{-1} \tag{5.124}$$

$$D_n = D_{n-1} - k_{n+1} \eta_n^T \tag{5.125}$$

When $n < N_F/p$, because $\rho_n = 0$, equations (5.123)–(5.125) can be simplified as

$$k_{n+1} = m_n \tag{5.126}$$

Using the calculated Kalman gain k_{n+1}, the tap gain vector is updated as

$$e_n = \bar{a}_n - h_n^T y_n \tag{5.127}$$

$$h_{n+1} = h_n + e_n k_{n+1} \tag{5.128}$$

5.4.5.1 Number of Computations for the Fast Kalman Algorithm   The number of multiplications per symbol for the fast Kalman algorithm is given by

$$C_{FK}(N_{total}) = \begin{cases} 12N_{total} + 28qN_{total} + 14q^2 + 12q + 8 & (n \geq N_F / p) \\ 8N_{total} + 16qN_{total} + 14q^2 + 8q + 4 & (n < N_F / p) \end{cases} \tag{5.129}$$

Figure 5.14 shows the number of multiplications per symbol versus R_s for $\tau_{max} = 10$, 25, and 50 μs, where FF filter of the DFE is a $T_s/2$-spaced filter, which corresponds to q = 3 in the fast Kalman algorithm. When this performance is compared to Figure 5.13, we find that, when N_{total} is small (symbol rate is low), the RLS algorithm requires a lower number of multiplications. On the other hand, when N_{total} is large, the fast Kalman algorithm requires less computation. The crossing point is $N_{total} = 14$.

5.4.5.2 Performance Comparison between RLS And Fast Kalman Algorithms

Figure 5.15 shows performance comparison of the initial acquisition process between RLS and fast Kalman algorithms, and Figure 5.16 shows a BER performance compar-

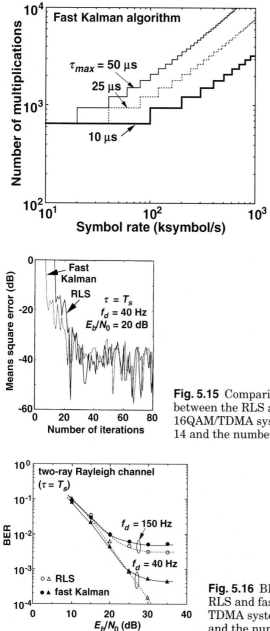

Fig. 5.14 Number of multiplications per symbol versus R_s for $\tau_{max} = 10, 25,$ and 50 µs in the case of the fast Kalman algorithm, where FF filter of the DFE is a $T_s/2$-spaced filter.

Fig. 5.15 Comparison of the tap gain convergence performance between the RLS and fast Kalman algorithms in the case of 16QAM/TDMA systems, where the number of FF taps (N_F) is 14 and the number of the FB taps (N_B) is 6.

Fig. 5.16 BER performance comparison between the RLS and fast Kalman algorithms in the case of 16QAM/TDMA systems, where the number of FF taps (N_F) is 14 and the number of the FB taps (N_B) is 6.

ison between these two algorithms. In these figures, the 16QAM/TDMA system is employed, the number of FF taps (N_F) is 14, and the number of FB taps (N_B) is 6. These results show that both algorithms give almost the same performance. Therefore, we can select either the RLS or fast Kalman algorithms considering only the required number of computations.

5.4.6 Tap Gain Interpolation Techniques

Figure 5.17 shows a burst format and basic concept of the tap gain interpolation technique [5-9; 5-15]. First, the initial tap gain acquisition is carried out at both preamble and postamble using the RLS algorithm. Because a known training sequence is feedbacked to the FB taps during the initial acquisition process, the tap gain vector is rapidly converged to the optimum value within a very short time period [5.3].

Given that the optimum tap gain at the end of preamble and postamble is $\boldsymbol{h}_{pre}$ and $\boldsymbol{h}_{post}$, the tap gain vector for the k-th information symbol is given by linearly interpolating $\boldsymbol{h}_{pre}$ and $\boldsymbol{h}_{post}$ as

$$\boldsymbol{h}(k) = \frac{k}{N_T} \boldsymbol{h}_{post} + \frac{N_T - k}{N_T} \boldsymbol{h}_{pre} \tag{5.130}$$

$$N_T = N_D + N_{post} \tag{5.131}$$

When several training sequences are embedded in each burst, for example, in preamble, midamble, and postamble, we can cope with faster channel variation because higher-order interpolation is applicable. However, as discussed in chapter 4, first- or second-order interpolations give very accurate channel estimation performance provided that the training sequence period is appropriately selected. Moreover, higher-order interpolation requires more computation accuracy of DSP. Therefore, we will discuss only first-order interpolation next.

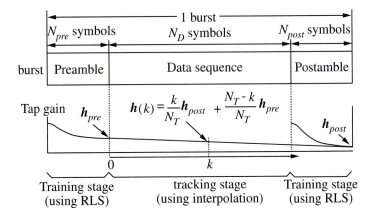

Fig. 5.17 Burst format and the concept of the tap gain interpolation technique.

5.4.6.1 Performance Comparison between First-order and Zeroth-order Interpolation   First of all, let's compare the performance of the first-order interpolation and zeroth-order interpolation in which $\boldsymbol{h}_{pre}$ is used for any information symbols during each burst. Figure 5.18 shows the BER versus f_d for a QPSK/TDMA system using zeroth-order and first-order interpolation techniques under two-ray Rayleigh fading conditions with D/U = 0 dB, $\tau = T_s$, and $E_b/N_0 = 20$ dB; where $N_{pre} = N_{post} = 26$ and $N_d = 200$ are employed [5-9]. The figure shows that the first-order interpolation has four times more maximum Doppler frequency immunity than the zeroth-order interpolation.

5.4.6.2 Optimization of N_{pre} and N_{post}   To improve BER performance, larger N_{pre} and N_{post} are preferable because it would guarantee better tap gain acquisition performance during the initial acquisition process. However, larger N_{pre} and N_{post} degrade TDMA frame efficiency because these symbols do not carry any information. Therefore, N_{pre} and N_{post} have to be optimized.

Figure 5.19 shows BER versus N_{pre} performance, where $f_d = 40$ Hz and $N_d = 200$ [5-9]. This figure shows that the optimum N_{pre} increases with E_b/N_0 because less residual distortion by tap gain misadjustment is required with increasing E_b/N_0. When we want to improve BER performance in the range of $10^{-3} \le \text{BER} \le 10^{-2}$, $N_{pre} = 26$ is optimum.

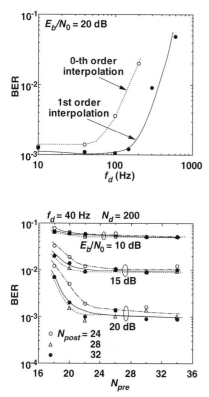

Fig. 5.18 BER versus f_d for a QPSK/TDMA system using zeroth-order and first-order interpolation techniques under two-ray Rayleigh fading conditions with D/U = 0 dB, $\tau = T_s$ and $E_b/N_0 = 20$ dB; where $N_{pre} = N_{post} = 26$ and $N_d = 200$ are employed (from Ref. 5-9, © Institute of Electrical and Electronic Engineers, 1991).

Fig. 5.19 BER versus N_{pre} performance, where $f_d = 40$ Hz and $N_d = 200$ (from Ref. 5-9, © Institute of Electrical and Electronic Engineers, 1991).

Figure 5.20 shows BER versus N_{post} performance, where $f_d = 40$ Hz, $N_{pre} = 26$, and $N_d = 200$ [5-9]. As in the case of N_{pre} optimization, the optimum N_{post} increases with E_b/N_0. When we optimize the BER in the range of $10^{-3} \leq$ BER $\leq 10^{-2}$, the optimum N_{post} is 26 as shown in Figure 5.20.

5.4.6.3 Performance Comparison between DFE with Interpolation and DFE without Interpolation
We also have to evaluate the performance difference between the RLS algorithm with interpolation and without interpolation. Figure 5.21 shows BER performance of these two equalizing algorithms for QPSK under two-ray Rayleigh fading conditions with D/U = 0 dB and $\tau = T_s$. The figure shows that although the performance for the DFE with interpolation is a little bit degraded in the BER $< 10^{-3}$ region, these two performances are almost the same in the important regions ($10^{-3} \leq$ BER $\leq 10^{-2}$).

When we apply the interpolation technique, the average number of multiplications per information symbol is given by

$$C_{INT} = (N_{pre} + N_{post} - 2M_1)(7N_{total}^2 + 12N_{total}) / N_d + 4N_{total} \qquad (5.132)$$

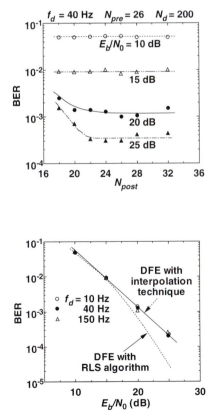

Fig. 5.20 BER versus N_{post} performance, where $f_d = 40$ Hz, $N_{pre} = 26$ and $N_d = 200$ (from Ref. 5-9, © Institute of Electrical and Electronic Engineers, 1991).

Fig. 5.21 BER performance of DFE with and without interpolation for QPSK/TDMA under the two-ray Rayleigh model conditions with D/U = 0 dB and $\tau = T_s$ (from Ref. 5-9, © Institute of Electrical and Electronic Engineers, 1991).

where $M_1 = N_F/p + N_B$. When we compare the C_{RLS} given by equation (5.60) and C_{INT}, we find that the number of multiplications is reduced to

$$C_{INT} / C_{RLS}$$
$$= (N_{pre} + N_{post} - 2M_1) / N_d + 4N_{total} / (7N_{total}^2 + 12N_{total})$$

(5.133)

For example, when $N_{pre} = N_{post} = 20$, $N_d = 200$, $N_F = 14$, $N_B = 6$, and $p = 2$, C_{INT}/C_{RLS} is approximately 15%. Therefore, the tap gain interpolation is a very effective technique to reduce computation complexity without degrading the performance too much.

5.4.6.4 BER Performance with a Parameter of Delay Time of the Delayed Wave

In the previous discussion, we have assumed a two-ray Rayleigh model with a delay time of the delayed wave (τ_{max}) of T_s. Figure 5.22 shows BER performance with a parameter of τ_{max}, where $f_d = 150$ Hz. To compensate for τ_{max} of up to $6T_s$, $N_F = 14$ and $N_B = 6$ are employed. The results show that the performance between the two is almost the same when $\tau_{max} \leq 3T_s$. On the other hand, it gets worse with increasing τ_{max} because the lower number of FF taps are effectively combined in the case of larger τ_{max}. The bidirectional equalizer is a very powerful technique we can use to overcome this problem.

5.4.7 Bidirectional DFE

The bidirectional DFE (BDDFE) proposed by Suzuki [5-11] is very effective in improving the BER performance under nonminimum phase conditions. Figure 5.23 shows burst structure and equalizing procedure of the BDDFE. In the forward equalizing process of BDDFE, a training sequence embedded in the preamble is used for the initial tap gain acquisition. Next, in the backward equalizing process, a training sequence embedded in the postamble is used for the initial tap gain acquisition. After the forward and backward equalizing is finished, the less distorted data are selected.

In the BDDFE, there are two types of equalizing direction selection processes.

☞ **Symbol-by-symbol selection,** in which magnitude of the equalizing error is compared for both equalizing directions at each symbol, and the less distorted

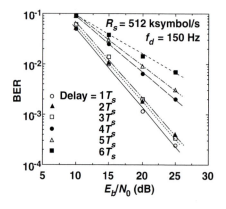

Fig. 5.22 BER performance with a parameter of τ_{max}, where $f_d = 150$ Hz, and $N_F = 14$ and $N_B = 6$ are employed.

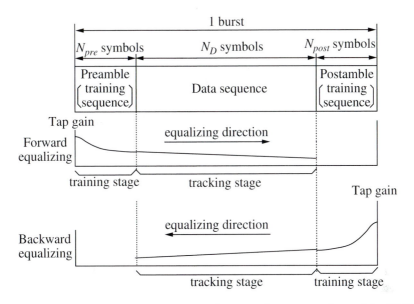

Fig. 5.23 Configuration of the burst structure and its equalizing procedure of the BDDFE.

symbol is selected. When $f_d T_B$ (T_B is the burst length) is large, this selection gives better performance than the burst-by-burst selection rule because it has a good tracking ability to the channel variation.

☞ **Burst-by-burst selection**, in which magnitude of the equalizing error is accumulated over the whole burst in each equalizing direction, and the equalizing direction with the smaller accumulated value is selected as the equalizing direction for the burst. This selection gives better performance when $f_d T_B$ is small because it has a noise-suppression effect.

We can select either one according to the channel variation speed and burst length.

As we have discussed, in the case of DFE, the FB taps are used to equalize signals in the minimum phase conditions, and the FF taps are used in the nonminimum phase conditions. In the case of the BDDFE, because nonminimum phase conditions in forward equalizing are converted to the minimum phase conditions in backward equalizing, most of the equalizing is done using FB taps. As a result, we can reduce the number of FF taps.

Another advantage of the BDDFE is that it can easily be combined with the tap gain interpolation techniques because its burst format is the same as that for the tap gain interpolation scheme [5-15]. Figure 5.24 shows the equalizing process for the BDDFE with tap gain interpolation. In forward equalizing, initial tap gain acquisition is carried out at both preamble and postamble, as shown in Figure 5.24 where $\boldsymbol{h}_{pre}^F$ and $\boldsymbol{h}_{post}^F$ are the optimum tap gain vectors obtained at the end of preamble and postamble in forward equalizing. The optimum tap gain vector at k-th information symbol is then obtained by interpolating $\boldsymbol{h}_{pre}^F$ and $\boldsymbol{h}_{post}^F$ as

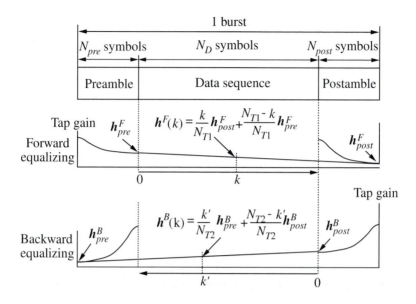

Fig. 5.24 Equalizing process for the BDDFE with tap gain interpolation.

$$\boldsymbol{h}^F(k) = \frac{k}{N_{T1}}\boldsymbol{h}^F_{post} + \frac{N_{T1} - k}{N_{T1}}\boldsymbol{h}^F_{pre} \tag{5.134}$$

$$N_{T1} = N_D + N_{post} \tag{5.135}$$

In backward equalizing, the same process is carried out as shown in Figure 5.24. In this case, the optimum tap gain vector for k'-th symbol is given by

$$\boldsymbol{h}^B(k) = \frac{k'}{N_{T2}}\boldsymbol{h}^B_{pre} + \frac{N_{T2} - k'}{N_{T2}}\boldsymbol{h}^B_{post} \tag{5.136}$$

$$N_{T2} = N_D + N_{pre} \tag{5.137}$$

where $\boldsymbol{h}^B_{post}$ and $\boldsymbol{h}^B_{pre}$ are the optimum tap gain vectors at the end of postamble and preamble in backward equalizing and the relationship between k and k' is given by

$$k' = N_D + 1 - k \tag{5.138}$$

After forward and backward equalizing is finished, the more reliable equalizing direction is selected symbol by symbol or burst by burst.

5.4.7.1 Reference Symbol Timing Control during Backward Equalizing In the BDDFE, reference symbol timing during backward equalizing is very important to effectively operate the BDDFE [5-15]. Figure 5.25 shows the relationship between the input signals in the FF filter, decision-feedback symbols in the FB filter, and the refer-

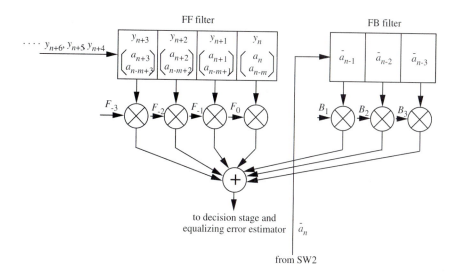

Fig. 5.25 Relationship between the input signals in the FF filter, decision-feedbacked symbols in the FB filter, and the reference symbol during backward equalizing.

ence symbol during backward equalizing. In this figure, the upper symbol in the bracket of each FF tap corresponds to the first-arrived path component, and the lower one corresponds to the delayed path component. Let's assume that the channel impulse response is given by

$$c(t) = c_0 \delta(t) + c_m \delta(t - mT_s) \tag{5.139}$$

and the tap spacing of the FF filter is equal to a symbol duration to simplify our discussion. In this case, the received baseband signal at $t = nT_s$ (y_n) includes transmitted symbols of a_n and a_{n-m}. Moreover, $\{y_n\}$ is fed to the BDDFE in the reverse order of reception—i.e., $\{y_n, y_{n-1}, y_{n-2}, \dots\}$—during backward equalizing. As a result, each FF tap includes transmitted symbols as shown in Figure 5.25. To effectively operate BDDFE, the first-arrived path (upper symbol in the bracket of each FF tap) has to be exactly canceled out by using FB taps during the backward equalizing process if the channel is in nonminimum phase conditions. In this case, the FB taps should include one of the symbol components in the upper bracket of the FF taps (a_{n-3}, a_{n-2}, a_{n-1}, or a_n in Figure 5.25) to cancel out the first-arrived path. However, if the reference symbol is synchronized to the timing of a_n, FB taps include a_{n+1}, a_{n+2}, and a_{n+3}, which means that there is no transmitted symbol component in the FB taps to cancel out the first-arrived path in the FF taps. As a result, backward equalizing does not work at all.

To solve this problem, it is effective to delay the reference symbol timing by DT_s. First, let's assume that $D = 3$.

Tables 5.1, 5.2, and 5.3 show transmitted symbol components included in each tap in the case of $(m, D)=(1, 3)$, $(2, 3)$ and $(3, 3)$. Because $D = 3$ is selected, the BDDFE operates to pick up a_{n-3} in backward equalizing. When $m = 1$, the pair (a_{n-2}, a_{n-3}) in the F_{-3}-tap is combined with a_{n-2} in the B_1-tap to pick up a_{n-3} component in the F_{-3}-tap. In

Table 5.1 Transmitted symbol components included in each tap ($m = 1$, $D = 3$).

tap	F_{-3}	F_{-2}	F_{-1}	F_0	B_1	B_2	B_3
transmitted symbol components included in each tap	a_{n-3}	a_{n-2}	a_{n-1}	a_n	a_{n-2}	a_{n-1}	a_n
	a_{n-4}	a_{n-3}	a_{n-2}	a_{n-1}			

Table 5.2 Transmitted symbol components included in each tap ($m = 2$, $D = 3$).

tap	F_{-3}	F_{-2}	F_{-1}	F_0	B_1	B_2	B_3
transmitted symbol components included in each tap	a_{n-3}	a_{n-2}	a_{n-1}	a_n	a_{n-2}	a_{n-1}	a_n
	a_{n-5}	a_{n-4}	a_{n-3}	a_{n-2}			

Table 5.3 Transmitted symbol components included in each tap ($m = 3$, $D = 3$).

tap	F_{-3}	F_{-2}	F_{-1}	F_0	B_1	B_2	B_3
transmitted symbol components included in each tap	a_{n-3}	a_{n-2}	a_{n-1}	a_n	a_{n-2}	a_{n-1}	a_n
	a_{n-6}	a_{n-5}	a_{n-4}	a_{n-3}			

the same manner, F_{-1}-tap and B_2-tap are used in the case of $m = 2$, and F_0 and B_3 are used in the case of $m = 3$. When $m \geq 4$, backward equalizing is useless because there is no a_{n-3} component in the FF taps. Moreover, we can see that F_{-3} is not used at all when $D = 3$ and $1 \leq m \leq 3$. This means that, when the maximum delay time of the delayed wave is mT_s, the optimum number of FF taps is m for the T_s-spaced BDDFE and mp taps for the fractional-spaced BDDFE.

Moreover, when $m = 1$, only forward equalizing can sufficiently compensate for the effect of the delayed wave if the number of FF taps is sufficient as shown in Figure 5.22. Therefore, we can further reduce the number of FF taps to $p(m - 1)$.

Figure 5.26 shows the configuration of the BDDFE, where tap spacing of the FF filter is $T_s/2$ ($p = 2$). During the forward equalizing process, switch 2 (SW2) is switched to 1. On the other hand, SW2 is switched to 2 during backward equalizing by delaying the reference signal timing by DT_s second.

Figure 5.27 shows BER versus delay time of the delayed wave (τ) with parameters of (N_F, D). In Figure 5.27, $N_F = 14$ and $N_F = 10$ correspond to the optimum number of FF taps to compensate for $\tau_{max} = 6T_s$ in the case of DFE and BDDFE, respectively. When we compare the performance for these two cases, we find that $N_F = 10$ gives a better performance because the lower number of FF taps gives better initial acquisition performance. When we reduce N_F to $N_F = 8$, on the other hand, the performance at $t = 2T_s$ is degraded. When D is decreased, BER performance is degraded at $\tau > DT_s$.

Figure 5.28 shows BER performance of QPSK with BDDFE under two-ray Rayleigh fading conditions, where (N_F, D) = (8, 5), $p = 2$, $N_B = 6$, and $\tau_{max} = 3T_s$. In this figure, BER performance of QPSK with DFE at $f_d = 160$ Hz is also shown as a reference.

We can find from this figure that the BDDFE can improve BER performance by about 2.5 dB owing to backward equalizing during nonminimum phase conditions.

5.4.7.2 Application of DFE and BDDFE to 16QAM The application of M-ary QAM is very efficient not only for increasing system capacity but also for reducing the number of computations of the DFE and BDDFE. However, one of the most serious

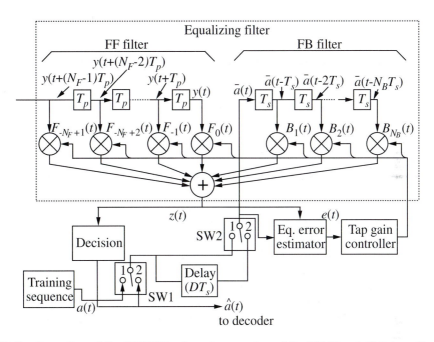

Fig. 5.26 Configuration of the BDDFE, where tap spacing of the FF filter is $T_s/2$ ($p = 2$).

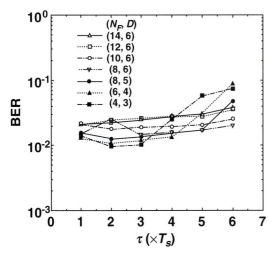

Fig. 5.27 BER versus delay time of the delayed wave (τ) with parameters of (N_F, D).

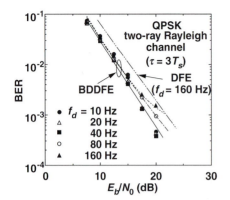

Fig. 5.28 BER performance of QPSK with BDDFE under two-ray Rayleigh fading channels, where $(N_F, D) = (8, 5), p = 2, N_B = 6$, and $\tau_{max} = 3T_s$.

problems in applying BDDFE to M-ary QAM is that a modulation with a higher modulation level is more sensitive to the error propagation effect. There are two types of error propagation effects in the DFE and BDDFE. In the first type, the feedback symbol with error diverges the tap gain vector from the optimum value. This effect was shown in Figure 5.12. Fortunately, the tap gain interpolation technique is free from this effect. Another error propagation effect is that the feedback symbol with error distorts the equalized output. When we apply the tap gain interpolation technique, it will be the main phenomenon of the error propagation effect.

Figure 5.29 shows BER performances of the BDDFE with perfect feedback and decision feedback for QPSK and 16QAM, where the tap gain interpolation technique is applied and the two-ray Rayleigh model with $f_d = 150$ Hz and $\tau = T_s$ is assumed. In the case of QPSK, degradation due to the error propagation effect is very small—less than 1 dB. On the other hand, such degradation becomes more than 2 dB in the case of 16QAM. Furthermore, M-ary QAM is more sensitive to noise due to its shorter minimum signal distance. Therefore, we have to combine some other techniques to improve receiver sensitivity if we want to apply 16QAM to wideband wireless systems. One method for this purpose is, of course, BDDFE as we have discussed before. Figure 5.30 shows BER performance of 16QAM with both DFE and BDDFE under two-ray Rayleigh fading conditions, where $f_d = 150$ Hz, $(N_F, D) = (8, 5)$ and $N_B = 6$. As

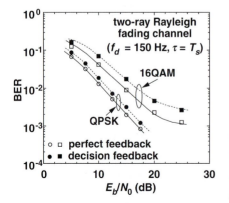

Fig. 5.29 BER performances of the BDDFE with perfect feedback and decision feedback for QPSK and 16QAM, where the tap gain interpolation technique is applied and the two-ray Rayleigh model with $f_d = 150$ Hz and $\tau = T_s$ is assumed.

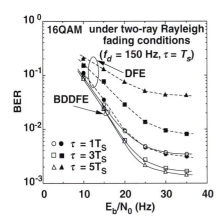

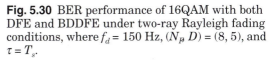

Fig. 5.30 BER performance of 16QAM with both DFE and BDDFE under two-ray Rayleigh fading conditions, where $f_d = 150$ Hz, $(N_B, D) = (8, 5)$, and $\tau = T_s$.

shown in this figure, BDDFE is very effective in improving BER performance for 16QAM systems, especially for longer τ.

5.4.7.3 Space Diversity with BDDFE Using Predecision Scheme for Equalizing Direction and Diversity Branch Selection

Although BDDFE is an effective technique for improving BER performance of wideband 16QAM, further BER performance improvement could be necessary in some cases. Space diversity combined with BDDFE or DFE is a very popular and effective technique for improving BER performance under frequency-selective fading conditions [5-10; 5-16; 5-17; 5-18]. However, when BDDFE and space diversity are merely combined, it requires twice as much computation as a single branch receiver. A predecision scheme for equalizing the direction of the BDDFE and the diversity branch is a technique for solving this problem.

Figure 5.31 shows the concept of a predecision scheme for the BDDFE equalizing direction [5-15]. This scheme operates according to the following procedure:

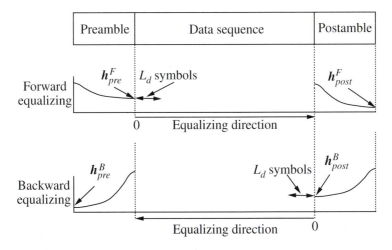

Fig. 5.31 Concept of a predecision scheme for the BDDFE equalizing direction.

1. Initial acquisition for forward equalizing is carried out at the preamble, and the optimum tap gain vector $\boldsymbol{h}_{pre}^{F}$ is calculated at the end of preamble.

2. L_d information symbols next to the preamble are equalized in the forward direction using the tap gain vector $\boldsymbol{h}_{pre}^{F}$. At the same time, the equalizing error $e_F(i)$ ($1 \leq i \leq L_d$) for each symbol is calculated. After L_d symbols are equalized, $|e_F(i)|^2$ is accumulated from $i = 1$ to L_d.

3. The initial acquisition for backward equalizing is carried out at the postamble.

4. L_d information symbols next to the postamble are equalized in the backward direction using the tap gain vector $\boldsymbol{h}_{post}^{B}$. At the same time, the equalizing error $e_B(i)$ ($1 \leq i \leq L_d$) for each symbol is calculated, and the accumulated $|e_B(i)|^2$ from $i = 1$ to L_d is obtained.

5. Steps 1–4 are carried out for all the branch signals.

6. Combination of the equalizing direction and diversity branch with the smallest accumulated error is selected.

7. Equalization using tap gain interpolation is carried out for all the information symbols in the burst.

Because this scheme equalizes only a part of the data to select the equalizing direction of BDDFE and the diversity branch, the amount of computation can be reduced. Figure 5.32 shows BER versus L_d performance of 16QAM with BDDFE and space diversity using the predecision scheme. When L_d is smaller than 40 symbols, BER performance is degraded because incorrect selection might occur. On the other hand, when L_d is larger than around 60 symbols, the BER performance is also degraded because channel condition is changing during the L_d symbol duration. As a result, $40 \leq L_d \leq 60$ is considered optimum.

Figure 5.33 shows BER performances of 16QAM with BDDFE and space diversity using the predecision scheme for equalizing direction and diversity branch, where $L_d = 40$ symbols, symbol rate is 512 ksymbol/s, and $f_d = 150$ Hz. When we compare Figures 5.30 and 5.33, we find that BDDFE and space diversity can improve BER performance by about 5 dB at BER = 10^{-2}, and it can also drastically reduce irreducible BER at high E_b/N_0 region.

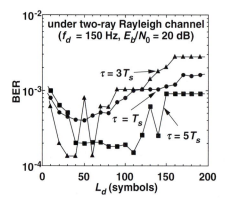

Fig. 5.32 BER versus L_d performance of 16QAM with BDDFE and space diversity using the predecision scheme.

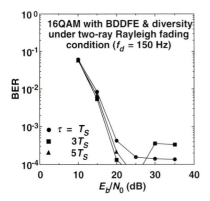

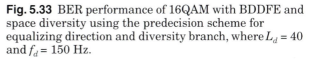

Fig. 5.33 BER performance of 16QAM with BDDFE and space diversity using the predecision scheme for equalizing direction and diversity branch, where $L_d = 40$ and $f_d = 150$ Hz.

Now let's discuss the number of computations necessary for BDDFE with space diversity using the predecision scheme. Table 5.4 shows equations for the number of multiplications for various types of receivers including the adaptive equalizer and/or space diversity, where $\lambda = 1.0$ is employed and the predecision scheme is employed for BDDFE and BDDFE with space diversity schemes, and Table 5.5 shows an example of the number of multiplications to get one symbol output for each receiver type. As we can calculate from this table, the number of multiplications for BDDFE with interpolation and a predecision scheme is only 16% in comparison with the conventional DFE. When we further apply space diversity to the BDDFE scheme, the number of computations is increased by 64% in comparison with the BDDFE scheme without diversity. However, this number is still 27% of computations compared with the conventional DFE without diversity. When the number of computations is compared to that of the conventional DFE with space diversity, it is only 14%. Therefore, tap gain interpolation and a predecision scheme are very effective in reducing the number of computations without degrading the BER performance too much.

Table 5.4 Average number of multiplications to obtain one symbol output.

Types	Number of Multiplications per Symbol
DFE	$(N_{pre} + N_D - N_B)R(N_{total})/N_D$
DFE with space diverisity	$2(N_{pre} + N_D - N_B)R(N_{total})/N_D + 4$
DFE with interpolation	$(2N_{pre} - N_s - N_B)R(N_{total})/N_D + 4N_{total}$
DFE with interpolation and diversity	$2(2N_{pre} - N_s - N_B)R(N_{total})/N_D + 8N_{total} + 4$
BDDFE with interpolation	$[2\{(N_{pre} - N_B)R(N_{total}) + L_d(4N_{total} + 2)\} + (N_{pre} - N_s)R(N_{total})]/N_D + 4N_{total}$
BDDFE with interpolation and diversity	$[4\{(N_{pre} - N_B)R(N_{total}) + L_d(4N_{total} + 2)\} + (N_{pre} - N_s)R(N_{total})]/N_D + 4N_{total}$

Table 5.5 Number of multiplications to get one symbol output for each receiver type.

Types	N_{pre}	N_D	N_{total}	N_B	N_s	Number of Multiplications
DFE	24	200	20	6	13	3208
DFE with diversity	24	200	20	6	13	6420
DFE with interpolation	30	200	20	6	13	622
DFE with interpolation and diversity	30	200	20	6	13	1248
BDDFE with interpolation	30	200	14	6	10	537
BDDFE with interpolation and diversity	30	200	14	6	10	882

In Table 5.5, $N_s = N_B + N_F/p$, $p = 2$, $N_{pre} = N_{post}$ and

$$R(N_{total}) = \begin{cases} 7N_{total}^2 + 12N_{total}^2 & (\lambda \neq 1.0) \\ 6N_{total}^2 + 12N_{total} & (\lambda = 1.0) \end{cases} \qquad (5.140)$$

5.5 MAXIMUM LIKELIHOOD SEQUENCE ESTIMATION

5.5.1 Configuration of the MLSE

Figure 5.34 shows a configuration of the MLSE [5-19; 5-20; 5-21]. It consists of a channel estimator to estimate channel impulse response $c(t)$, a matched filter that maximizes the SNR of the Viterbi decoder input, and a Viterbi decoder that estimates the transmitted data sequence using the Viterbi algorithm.

The received baseband signal is given by

$$y(t) = \sum_{i=-\infty}^{\infty} a_i c(t - iT_s) \qquad (5.141)$$

First, in the MLSE receiver, the channel impulse response $c(t)$ is measured using the training sequence in each burst. When the channel variation is slow, we can assume that $c(t)$ is constant during each burst. On the other hand, when the channel varies

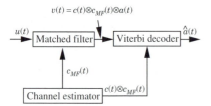

Fig. 5.34 Configuration of an MLSE.

very fast, we have to estimate the channel variation simultaneously with the sequence estimation.

Let us assume that $c(t)$ is constant during each burst. In this case, the best location of the training sequence is the center of each burst (midamble type). When the symbol timing is $t = nT_s$, and $y(nT_s)$ is expressed as y_n, the MLSE receiver estimates the transmitted data sequence $\{\alpha_n\}$ that maximizes the conditional probability $p(\{y_n\} \mid \{\alpha_n\})$ using the Viterbi algorithm. When the observation interval I is defined as

$$\left\{ t \in I, \ (n - N + 1)T_s \le t \le nT_s \right\} \qquad (5.142)$$

the conditional probability is given by

$$p(\{y_n\} \mid \{\alpha_n\}) = A \exp\left\{ -\frac{1}{4N_0} \left[\sum_{nT_s \in I} 2\,\mathrm{Re}(\alpha_n^* v_n) - \sum_{iT_s \in I} \sum_{kT_s \in I} \alpha_i^* s_{i-k} \alpha_k \right] \right\} \qquad (5.143)$$

$$v_n = c_{MF}(t) \otimes y(t)\Big|_{t = nT_s}$$
$$= \sum_{m=-\infty}^{\infty} a_{n-m} s_m + n_n \qquad (5.144)$$

$$s_m = s_{-m}^* = c_{MF}(t) \otimes c(t)\Big|_{t = mT_s} \qquad (5.145)$$

where $c_{MF}(t)$ is the impulse response of the matched filter given by

$$c_{MF}(t) = c^*(-t) \qquad (5.146)$$

and A is a constant that satisfies

$$\int_{-\infty}^{\infty} A \cdot p(\{y_n\} \mid \{\alpha_n\}) dy_n = 1 \qquad (5.147)$$

Equation (5.143) means that the maximum $p(\{y_n\} \mid \{\alpha_n\})$ is obtained by maximizing the metric given by

$$J_I(\{\alpha_n\}) = \sum_{nT_s \in I} 2\,\mathrm{Re}(\alpha_n^* v_n) - \sum_{iT_s} \sum_{kT_s} \alpha_i^* s_{i-k} \alpha_k \qquad (5.148)$$

When the maximum delay time to be compensated for is LT_s, equation (5.148) can be expressed as

$$J_k(\sigma_k) = J_{k-1}(\sigma_{k-1}) + BR(\sigma_{k-1}, \sigma_k) \qquad (5.149)$$

$$BR(\sigma_{k-1}, \sigma_k) = 2\,\mathrm{Re}(\alpha_k^* v_k) - \left\{ \alpha_k^* s_0 \alpha_k + 2\,\mathrm{Re}\left(\alpha_k^* \sum_{m=1}^{L} s_m \alpha_{k-m} \right) \right\} \qquad (5.150)$$

where

$$\sigma_{k-1} = (\alpha_{k-L},\ \alpha_{k-L+1},\ \cdots,\ \alpha_{k-1}) \tag{5.151}$$

is a state at $t = (k-1)T_s$ and $BR(\sigma_{k-1}, \sigma_k)$ is called the *branch metric*. When the modulation level is M, the number of states is M^L.

Assume the state σ^i at $t = kT_s$ is σ_k^i. Each state has M incoming paths. Among them, only one path with the maximum metric will finally survive. As a result, the survived path history at the state σ_k^i can be determined by

$$H_k(\sigma_k^i) = [\alpha_k,\ \alpha_{k-1}, \cdots \alpha_{k-L}, \cdots]_i \tag{5.152}$$

Consequently, the metric given by equation (5.149) can be modified as

$$J_k\left(\sigma_k^i\right) = \max_{\sigma_{k-1}^j \to \sigma_k^i} \left\{ J_{k-1}\left(\sigma_{k-1}^j\right) + BR\left(\sigma_{k-1}^j,\ \sigma_k^i\right) \right\} \tag{5.153}$$

where

$$\max_{\sigma_{k-1}^j \to \sigma_k^i} \{\cdot\} \tag{5.154}$$

means to select a path from a state σ_{k-1}^j that has the maximum value of $\{\cdot\}$. Path history at σ_k^i is also updated as

$$H_k\left(\sigma_k^i\right) = \left[\alpha_k : H_{k-1}\left(\sigma_{k-1}^j\right)\right] \tag{5.155}$$

When we recursively calculate equations (5.150), (5.153), and (5.155), we can obtain the optimum data sequence. In the burst signal transmission, we usually add several tail symbols to merge the survival path to a certain state.

5.5.2 Channel Estimation Scheme

The most popular channel estimation scheme is to embed a PN sequence in each burst, and then take the correlation between the received signal and the transmitted PN sequence. Its principle is the same as that for the CDM-type channel sounding technique (discussed in chapter 4). The difference is that the PN sequence used for the channel estimation in the MLSE receiver is relatively short due to the limitation of the frame efficiency.

Another channel estimation technique is the RLS algorithm [5-22]. Let's define the channel impulse response at $t = nT_s$ as

$$c(nT_s) = \sum_{i=0}^{L} c_{i,n} \delta(t - iT_s) \tag{5.156}$$

the impulse response vector at $t = nT_s$ as

$$\boldsymbol{c}_n = \begin{bmatrix} c_{0,n}, & c_{1,n}, & \cdots, & c_{L,n} \end{bmatrix}^T \qquad (5.157)$$

and the training sequence vector at $t = nT_s$ as

$$\boldsymbol{a}_n = \begin{bmatrix} a_n, & a_{n-1}, & \cdots, & a_{n-L} \end{bmatrix}^T \qquad (5.158)$$

The optimum $\boldsymbol{c}_n$ can be obtained using the RLS algorithm as

$$\boldsymbol{k}_n = \boldsymbol{P}_{n-1}\boldsymbol{a}_n(\boldsymbol{a}_n^{*T}\boldsymbol{P}_{n-1}\boldsymbol{a}_n + \lambda v)^{-1} \qquad (5.159)$$

$$\boldsymbol{P}_n = \frac{1}{\lambda}\left(\boldsymbol{P}_{n-1} - \boldsymbol{k}_n\boldsymbol{a}_n^{*T}\boldsymbol{P}_{n-1}\right) \qquad (5.160)$$

$$e_n = y_n - \boldsymbol{c}_{n-1}^T\boldsymbol{a}_n \qquad (5.161)$$

$$\boldsymbol{c}_n = \boldsymbol{c}_{n-1} + e_n\boldsymbol{k}_n \qquad (5.162)$$

In equations (5.159) and (5.160), because a_n is a known symbol sequence and these equations do not include $\boldsymbol{c}_n$, we can preliminarily obtain $\boldsymbol{k}_n$ and $\boldsymbol{P}_n$ if $\boldsymbol{P}_0$ is initially determined. As a result, we can obtain the optimum $\boldsymbol{c}_n$ by calculating only equations (5.161) and (5.162), which means that the number of multiplications necessary for this channel estimation is only $8(L + 1)$ for each iteration.

5.5.3 Estimation for Time-Varying Channel Impulse Response

In the previous discussions, we have assumed that the channel impulse response is constant in each burst. However, this is not true when terminal mobility becomes high. Figure 5.35 shows BER performance of QPSK with an MLSE receiver under two-ray Rayleigh fading conditions, where delay time of the delayed wave is one symbol duration, $f_d = 100$ Hz, symbol rate is 256 ksymbol/s, and the channel impulse response is estimated using 20 symbols of preamble. The obtained impulse response is used during all the following information symbols in the burst. As shown in this figure, the BER

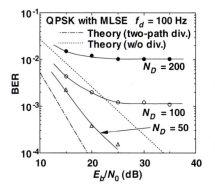

Fig. 5.35 BER performance of QPSK with an MLSE receiver under two-ray Rayleigh fading conditions, where delay time of the delayed wave is one symbol duration, $f_d = 100$ Hz, symbol rate is 256 ksymbol/s, and the channel impulse response is estimated at the 20-symbol of preamble (from Ref. 5-22, © Institute of Electronics, Information and Communication Engineers, 1990).

performance is degraded when N_D is increasing because the channel characteristics are not constant. Only when $N_D = 50$, can we obtain acceptable BER performance.

One of the solutions to this problem is the interpolation-type channel impulse response estimation scheme using preamble and postamble [5-22]. In this case, channel impulse response is measured at both preamble and postamble. When the obtained channel impulse responses are expressed as $\boldsymbol{c}_{pre}$ and $\boldsymbol{c}_{post}$, the channel impulse response for k-th information symbol is given by

$$\boldsymbol{c}(k) = \frac{k}{N_T}\boldsymbol{c}_{post} + \frac{N_T - k}{N_T}\boldsymbol{c}_{pre} \tag{5.163}$$

Figure 5.36 shows BER performance of a Viterbi equalizer using the interpolation-type channel impulse response estimator, where the channel condition is the same as that for Figure 5.35. When we compare Figure 5.35 and Figure 5.36, we can find that the interpolation-type impulse response estimator can keep up with eight times faster channel variation in comparison with the holding-type channel estimator. Therefore, although the interpolation type requires twice as many training symbols, it can accept eight times longer burst length thereby achieving higher frame efficiency than the holding type channel estimator.

5.5.4 Delayed Decision Feedback Sequence Estimation (DDFSE)

One of the most serious problems for the Viterbi equalizer is that the number of computations is exponentially increasing with the number of modulation levels M and the number of delayed paths L. When we evaluate the number of computations by the number of multiplications as in the case of DFE or BDDFE, it is given by

$$C_{MLSE} = (2L + 11)M^{L+1} \tag{5.164}$$

Figure 5.37 shows C_{MLSE} versus symbol rate (R_s) in the case of BPSK and QPSK, where $\tau_{max} = 10$ μs. As shown in this figure, C_{MLSE} for QPSK is approximately 10 times more than that for BPSK when R_s is in the range of 100–500 ksymbol/s. Therefore, how to reduce the number of computations is a very important issue for the Viterbi equalizer.

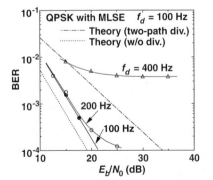

Fig. 5.36 BER performance of QPSK with the Viterbi equalizer using an interpolation-type channel impulse response estimator, where delay time of the delayed wave is one symbol duration, $f_d = 100$ Hz, symbol rate is 256 ksymbol/s, and the channel impulse response is estimated at the 20-symbol of preamble (from Ref. 5-22, © Institute of Electronics, Information and Communication Engineers, 1990).

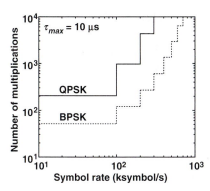

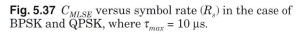

Fig. 5.37 C_{MLSE} versus symbol rate (R_s) in the case of BPSK and QPSK, where $\tau_{max} = 10$ µs.

One method of reducing the number of computations is the reduction of L. However, if we just reduce L, the BER performance is degraded due to residual distortion. The DDFSE is one of the effective schemes for reducing the number of computations [5-23].

In the DDFSE algorithm, the delayed path to be considered in the Viterbi algorithm is limited to the m-symbol duration, where $m \le L$. As a result, the state at $t = (k - 1)T_s$ can be expressed as

$$\xi_k = \left(\alpha_{k-m+1},\ \alpha_{k-m+2},\ \cdots,\ \alpha_k\right)\ \left(0 \le m \le L\right) \tag{5.165}$$

In this case, the path history for the state ξ_{k-1} is given by

$$H_k(\xi_k^i) = \left[\alpha_k,\ \alpha_{k-1},\ \cdots\alpha_{k-L},\ \cdots\right]_i \tag{5.166}$$

and the metric for the state ξ_k^i is given by

$$J_k(\xi_k^i) = \max_{\xi_{k-1}^j \to \xi_k^i}\left\{J_{k-1}(\xi_{k-1}^j) + BR(\xi_{k-1}^j,\ \xi_k^i)\right\} \tag{5.167}$$

where the branch metric $BR(\xi_{k-1}^j,\ \xi_k^i)$ is given by

$$BR(\xi_{k-1}^j,\ \xi_k^i)$$
$$= 2\,\mathrm{Re}\left(\alpha_k^* v_k\right) - \alpha_k^* s_0 \alpha_k - 2\,\mathrm{Re}\left(\alpha_k^* \sum_{l=1}^{m} s_l \alpha_{k-l}\right) - 2\,\mathrm{Re}\left(\alpha_k^* \sum_{l=m+1}^{L} s_l \hat{\alpha}_{k-l}\right) \tag{5.168}$$

In equation (5.168), the first three terms corresponds to the metric calculation of the 2^m-state Viterbi algorithm, and the fourth term cancels out the residual intersymbol interference due to the restriction of the number of delayed paths used in the Viterbi algorithm. To calculate the fourth term in equation (5.168), α_k in the path history $H_k(\xi_{k-1}^j)$ is employed. Path history at the state ξ_k^i is updated as

$$H_k\left(\xi_k^i\right) = \left[\alpha_k : H_{k-1}^j\left(\xi_{k-1}^j\right)\right] \tag{5.169}$$

In the DDFSE algorithm, the number of computations is reduced by $M^{-(L-m)}$ because the number of states is reduced to M^m instead of M^L.

Figure 5.38 (b) shows an example of the BER performance of QPSK with DDFSE ($m = 2, L = 5$) under the average power delay profile characteristics as shown in Figure 5.38 (a), where all the paths are independently Rayleigh faded with the maximum Doppler frequency of f_d = 100 Hz [5-22]. The first two paths are combined by the Viterbi algorithm, whereas the third one is canceled out by using the path history data. First of all, when we employ the 2^m-state Viterbi algorithm, the BER performance is extremely degraded due to the residual delayed paths. On the other hand, when we employ DDFSE, we can achieve almost the same performance as that shown in Figure 5.36. Therefore, the DDFSE algorithm is considered to be very effective for reducing the number of computations without significantly degrading the BER performance.

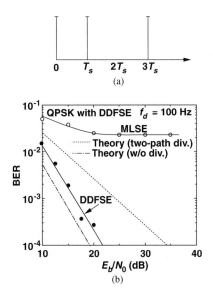

Fig. 5.38 (a) Average delay profile of the BER performance evaluation, (b) BER performance of QPSK with DDFSE ($m = 2$ and $L = 5$), where f_d = 100 Hz (from Ref. 5-22, © Institute of Electronics, Information and Communication Engineers, 1990).

5.6 How Do We Select DFE or Viterbi Equalizer?

When we compare the BER performance of the DFE and Viterbi equalizer, the Viterbi equalizer gives better performance than the DFE because all of the path is combined by the maximal ratio combining in the case of the Viterbi equalizer, whereas selection combining is carried out in the case of the DFE. On the other hand, from the viewpoint of algorithm complexity, DFE is much simpler when the number of delayed paths to be compensated for becomes large or when we employ a higher modulation level. Therefore, we have to take into account symbol rate and the maximum delay time of the delayed path to be compensated for when we select the adaptive equalizing techniques.

Figure 5.39 shows which of the two is more suitable, considering both modulation level and symbol rate. As shown in Figure 5.38, in the case of low-symbol-rate systems like PDC and IS-54, although both the Viterbi equalizer and DFE are applicable, the Viterbi equalizer is considered to be more suitable because it gives better receiver sensitivity. In the case of GSM, although the symbol rate is much higher than that for PDC and IS-54, the number of computations for the Viterbi algorithm is still tolerable because it employs a 1bit/symbol modulation scheme [5-22]. Therefore, the Viterbi equalizer is fairly often applied to the GSM systems. However, if we want to employ a high spectral efficient modulation scheme with a high symbol rate—for example, QPSK or 16QAM with its symbol rate of several hundred ksymbol/s—DFE is considered to be more suitable from the viewpoint of algorithm complexity.

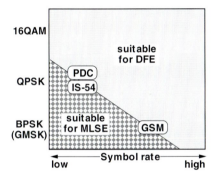

Fig. 5.39 Preferable choice of DFE and MLSE in terms of modulation scheme and symbol rate.

5.7 ADAPTIVE ARRAY ANTENNA

5.7.1 Basic Configuration of the Adaptive Array

The adaptive array antenna, which consists of plural omnidirectional antennas, is an adaptive directional antenna that controls its directivity by adjusting the complex weight of each antenna element according to the propagation path conditions [5-24; 5-25; 5-26; 5-27; 5-28]. Figure 5.40 shows a configuration of the adaptive array antenna consisting of four omnidirectional antennas with antenna spacing of d. When a signal comes from the incoming direction of θ, the received signal of each antenna element can be expressed as follows:

$$s_{R1}(t) = h_1(t)\exp\left(-j\frac{2\pi}{\lambda}\frac{d}{\sqrt{2}}\cos\theta\right)s_{R0}(t) \text{ (for antenna \#1)} \qquad (5.170a)$$

$$s_{R2}(t) = h_2(t)\exp\left(j\frac{2\pi}{\lambda}\frac{d}{\sqrt{2}}\sin\theta\right)s_{R0}(t) \text{ (for antenna \#2)} \qquad (5.170b)$$

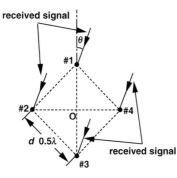

Fig. 5.40 Configuration of the adaptive array antenna, where it consists of four omnidirectional antennas with antenna spacing of d.

$$s_{R3}(t) = h_3(t) \exp\left(j \frac{2\pi}{\lambda} \frac{d}{\sqrt{2}} \cos\theta \right) s_{R0}(t) \text{ (for antenna \#3)} \tag{5.170c}$$

$$s_{R4}(t) = h_4(t) \exp\left(-j \frac{2\pi}{\lambda} \frac{d}{\sqrt{2}} \sin\theta \right) s_{R0}(t) \text{ (for antenna \#4)} \tag{5.170d}$$

where $s_{R0}(t)$ represents the received signal at the point O in Figure 5.40 and $h_1(t)$, $h_2(t)$, $h_3(t)$, and $h_4(t)$ represent the complex weights for antenna #1, #2, #3, and #4, respectively. As a result, the combined received signal is given by

$$s_R(t) = \left[h_1(t) e^{-j\frac{2\pi}{\lambda}\frac{d}{\sqrt{2}}\cos\theta} + h_2(t) e^{j\frac{2\pi}{\lambda}\frac{d}{\sqrt{2}}\sin\theta} \right.$$
$$\left. + h_3(t) e^{j\frac{2\pi}{\lambda}\frac{d}{\sqrt{2}}\cos\theta} + h_4(t) e^{-j\frac{2\pi}{\lambda}\frac{d}{\sqrt{2}}\sin\theta} \right] s_{R0}(t) \tag{5.171}$$

$$= H(t;\theta) s_{R0}(t)$$

$$H(t;\theta) = h_1(t) e^{-j\frac{2\pi}{\lambda}\frac{d}{\sqrt{2}}\cos\theta} + h_2(t) e^{j\frac{2\pi}{\lambda}\frac{d}{\sqrt{2}}\sin\theta}$$
$$+ h_3(t) e^{j\frac{2\pi}{\lambda}\frac{d}{\sqrt{2}}\cos\theta} + h_4(t) e^{-j\frac{2\pi}{\lambda}\frac{d}{\sqrt{2}}\sin\theta} \tag{5.172}$$

where $H(t;\theta)$ represents directivity of the adaptive array antenna. As shown in equation (5.172), $H(t;\theta)$ can be controlled by adjusting $h_1(t)$ to $h_4(t)$.

Figure 5.41 shows examples of directivity of the adaptive array antenna in the case of (a) $h_1(t) = h_2(t) = h_3(t) = h_4(t) = 1.0$ and (b) $h_1(t) = 0.8 - j0.48$, $h_2(t) = 0.707 + j0.706$, $h_3(t) = -0.6 + j0.8$, and $h_4(t) = -0.8 - j0.6$. As shown in these figures, directivity of the adaptive array antenna can flexibly be controlled by adjusting the complex weights $h_i(t)$ ($1 \le i \le 4$). Using this feature, we can apply the adaptive array antenna as an interference canceler or an anti-frequency-selective fading technique.

When we employ the adaptive array antenna as an interference canceler, we can suppress the interference by reducing $|H(t;\theta)|$ for the incident angle of the interference signals [5-28; 5-29]. On the other hand, when we employ it as an anti-frequency-selective fading technique, its operation greatly depends on the channel conditions [5-24; 5-30]. In the case of a two-ray Rayleigh channel and minimum phase conditions, the adaptive array creates null in the direction of the delayed path, and, in the case of nonminimum phase conditions, it creates null in the direction of the direct path.

However, it is not practical to search all the signal paths and to distinguish which are the desired and undesired paths. Therefore, the adaptive array antenna employs an indirect means to create antenna beam directivity, i.e., to maximize the signal-to-distortion power ratio after combining.

Figure 5.42 shows the configaration of the adaptive array systems. Although this figure represents the conceptual configuration of the adaptive array, this configuration, as it is, is not practical because signal processing at the IF band requires very complicated hardware.

The practical receiver configuration of the adaptive array is investigated in studies by Ohgane et al. [5-31; 5-32], in which antenna beam directivity is equivalently controlled at the baseband (zero IF) using digital signal processing techniques. Figure 5.43 shows the configuration of the adaptive array antenna system using baseband

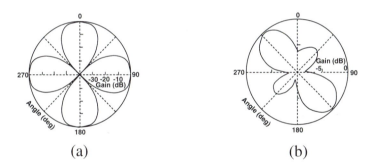

(a)   (b)

Fig. 5.41 Examples of directivity of the adaptive array antenna: (a) $h_1(t) = h_2(t) = h_3(t) = h_4(t) = 1.0$; (b) $h_1(t) = 0.8 - j0.48$, $h_2(t) = 0.707 + j0.706$, $h_3(t) = -0.6 + j0.8$, and $h_4(t) = -0.8 - j0.6$.

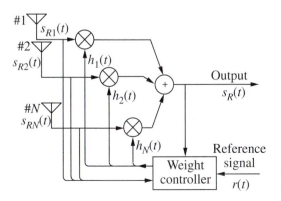

Fig. 5.42 Configuration of the adaptive array systems.

digital signal processing techniques. The received signal in each antenna is first down-converted to the IF band, and the signal level is controlled to a proper level to guarantee sufficient dynamic range for baseband digital signal processing. At this stage, a common-control signal level is fed back to the AGC amplifier for each antenna branch because the power ratio of the received signal between two arbitrary antenna elements at the AGC output has to be the same as that at the AGC input as we discussed in chapter 4, section 4.4.3. In Ref. [5-31], the AGC is controlled to keep the maximum power level among all branch signals constant.

The AGC output is then quadrature-demodulated to obtain the baseband signals of each branch. Assume the baseband signal for i-th antenna element is $s_{Bi}(t)$. These baseband signals are then weighted and summed at the adaptive directive controller to equivalently create antenna directivity of the adaptive array antenna at the baseband. At this stage, the combined output signal is given by

$$s_B(t) = \sum_{i=1}^{N} h_i(t) s_{Bi}(t) \tag{5.173}$$

where $s_{Bi}(t)$ is the band-limited received signal expressed as an equivalent low-pass system.

The combined output given by equation (5.173) is then compared to the reference signal to caclulate distortion included in $s_B(t)$. At this stage, two criteria are used to assess distortion in the $s_B(t)$—the LMS algorithm and the constant modulus algorithm (CMA).

5.7.2 LMS Algorithm

In this method, first the difference between $s_B(t)$ and the ideal (reference) signal waveform ($r(t)$) is obtained as follows:

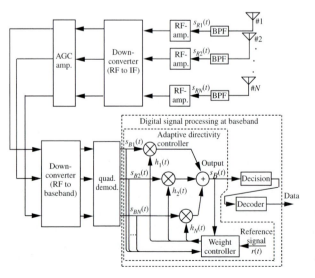

Fig. 5.43 Configuration of the adaptive array antenna system using baseband digital signal processing techniques.

$$
\begin{aligned}
e(t) &= e_I(t) + j \cdot e_Q(t) \\
&= r(t) - s_B(t) \\
&= r(t) - \sum_{i=1}^{N} h_i(t) s_{Bi}(t)
\end{aligned}
\tag{5.174}
$$

Then the weight for each antenna element is controlled to minimize deviation of $e(t)$. One of the effective means of obtaining the optimum weights is the RLS algorithm. In this case, the input signal vector at $t = nT_s$ is given as

$$
\boldsymbol{y}_n = \left[s_{B1}(nT_s), \ s_{B2}(nT_s), \ \ldots, \ s_{BN}(nT_s) \right]^T
\tag{5.175}
$$

The weight vector at $t = nT_s$ is given as

$$
\boldsymbol{h}_n = \left[h_1(nT_s), \ h_2(nT_s), \ \ldots, \ h_N(nT_s) \right]^T
\tag{5.176}
$$

As a result, the output of the adaptive array $s_{Bn} = s_B(nT_s)$ and its error $e_n = e(nT_s)$ are given by

$$
s_{Bn} = \boldsymbol{h}_n^T \boldsymbol{y}_n
\tag{5.177}
$$

$$
e_n = r_n - s_{Bn} = r_n \boldsymbol{h}_n^T \boldsymbol{y}_n
\tag{5.178}
$$

and the optimum weight vector can be updated as

$$
\boldsymbol{P}_0 = \delta^{-1} \boldsymbol{I} \ (\boldsymbol{I}: \text{unit matrix})
\tag{5.179}
$$

$$
\boldsymbol{k}_n = \boldsymbol{P}_{n-1} \boldsymbol{y}_n (\boldsymbol{y}_n^{*T} \boldsymbol{P}_{n-1} \boldsymbol{y}_n + \lambda v)^{-1}
\tag{5.180}
$$

$$
\boldsymbol{h}_n = \boldsymbol{h}_{n-1} + e_n \boldsymbol{k}_n
\tag{5.181}
$$

$$
\boldsymbol{P}_n = (\boldsymbol{P}_{n-1} - \boldsymbol{k}_n \boldsymbol{y}_n^{*T} \boldsymbol{P}_{n-1}) \lambda^{-1}
\tag{5.182}
$$

When r_n is synchronized to the timing of the first-arrived path, the directivity is controlled to pick up the first-arrived path and to suppress the delayed paths. On the other hand, when r_n is synchronized to the timing of one of the delayed path, the directivity is controlled to pick up the path and to suppress the rest of the paths. Therefore, if we apply the block demodulation technique and carry out weight control for plural synchronization timings, we can pick up several path signals and can achieve path diversity combining [5-33].

This algorithm is the same as the DFE algorithm. However, the number of elements in $\boldsymbol{y}_n$ and $\boldsymbol{h}_n$ is quite different. In the case of the DFE, the number of elements in $\boldsymbol{y}_n$ and $\boldsymbol{h}_n$ depends on the maximum delay time of the delayed wave. When the maximum delay time increases, we have to increase the number of taps. On the other hand, in the case of the adaptive array antenna, the number of elements in $\boldsymbol{y}_n$ and $\boldsymbol{h}_n$

is determined not by the maximum delay time but by the number of undesired paths as well as the angle resolution of the desired and undesired paths. For example, when undesired paths are coming from different angles of the desired path, we can suffi- ciently suppress the undesired paths. On the other hand, when the desired and undes- ired paths are coming from almost the same angle, we cannot suppress the undesired signals. Fortunately, the desired signal is coming from various directions, and the delayed signals are coming from limited directions in most cases of the frequency- selective fading environments; we can expect a great effect to mitigate the multipath fading effect even though we employ a small number of antenna elements. Ref. [5-31] shows that, even though $N = 4$, the adaptive array antenna works very effectively in the actual propagation path conditions to compensate for frequency-selective fading regardless of the maximum delay time.

When we compare this number with the number of taps for the DFE, we can find that the adaptive array antenna requires a lower number of computations, especially in the multipath fading environments with a relatively long delayed wave. For exam- ple, when we assume the maximum delay time of 10 µs, and the symbol rate of 500 ksymbol/s, the number of multiplications for DFE is 2,127 using equation (5.60). On the other hand, in the case of an adaptive array antenna with $N = 4$, the number of multiplications is 160, which is only 7.5% of the computations in comparisons with that for the DFE.

One of the important issues for the adaptive array antenna using the LMS algo- rithm is how to obtain the reference (known) signal. However, when we apply it to the TDMA system in which a reference sequence is embedded in the preamble, midamble, or postamble, we can employ almost the same initial weight acquisition process as that for the DFE.

5.7.3 Constant Modulus Algorithm

The CMA is another weight-control technique for the adaptive array antenna [5- 34]. In the CMA, the weight vector is controlled to keep the envelope of the combined signal to be constant. Therefore, this algorithm is very suitable when the transmitted signal is modulated by a constant envelope modulation, such as GMSK, MSK, and any other FM schemes.

In the CMA, the weight vector is controlled to minimize the following criteria:

$$J_{pq}(t) = E\left[\left||s_B(t)|^p - 1\right|^q\right]$$
(5.183)

There are several recursive algorithms to obtain the optimum weight vector that min- imizes equation (5.183). One of the simplest algorithms is the gradient algorithm given by the following equation:

$$\boldsymbol{h}_n = \boldsymbol{h}_{n-1} - \mu e_n \boldsymbol{y}_n^*$$
(5.184)

where

$$e_n = s_{Bn} - \frac{s_{Bn}}{|s_{Bn}|} \tag{5.185}$$

is used in the case of $(p, q) = (1, 2)$, and

$$e_n = s_{Bn}\left(|s_{Bn}|^2 - 1\right) \tag{5.186}$$

is used in the case of $(p, q) = (2, 2)$. μ is a constant that determines the weight vector convergence speed as well as the residual estimation error of the weight vector. When we employ a larger μ, although the convergence speed becomes faster, its estimation error becomes larger. On the other hand, when we employ a smaller μ, although the estimation error of weight vector becomes smaller, the weight vector convergence speed becomes slower.

In the case of the DFE, although μ is optimized, the convergence speed is too slow to keep up with the fading variations. Therefore, we usually employ the RLS algorithm instead of the LMS algorithm. On the other hand, in the case of CMA, we do not have to synchronize the weight vector update timing with the symbol timing. Rather we can update the weight vector with its period of a fraction of a symbol duration ($T_s/p, p \geq 2$), provided that the envelope of the transmitted signal is constant. From this viewpoint, the LMS algorithm is also effective for the CMA. Ref. [5-35] shows the results of the optimization of μ as well as the optimum combination of p and q. The results show that $p = 1, q = 2$, and $\mu = 0.01$ can sufficiently compensate for the frequency-selective fading ($f_d = 150$ Hz) in the case of 256 kbit/s GMSK/TDMA systems.

5.7.4 Comparison between LMS and CMA Algorithms

As we have discussed before, LMS and CMA are mainly investigated as the weight control algorithms for the adaptive array antenna. Because these two algorithms have quite different features, we will compare their features.

5.7.4.1 Reference Signal for Weight Vector Adaptation
In the LMS algorithm in which mean square error is used as a cost function, both the envelope and phase distortion due to fading can be compensated for using a reference signal. A known training sequence embedded in each burst is often employed as a reference in the case of TDMA systems. On the other hand, the CMA does not require any reference signals for weight vector updating because only the envelope information is used for the cost function of this algorithm. As a result, the CMA is considered to have a potential to achieve higher frame efficiency than the LMS algorithm. In this case, however, we have to offset residual phase error which is not compensated for by the CMA.

5.7.4.2 Path Diversity Effect
In the LMS algorithm, the reference training sequence must be generated with the timing coincident with the strongest path. For this purpose, we have to obtain the delay profile; we also have to decide to which timing the reference sequence has to be synchronized. When we select several reference timings and carry out array signal combining for each timing using block demodulation tech-

niques, we can combine them by either selection combining or maximal ratio combining.

In contrast, the CMA automatically selects one or several multipaths to be combined. In most cases, the CMA captures the strongest path. As a result, the CMA inherently has a selection combining path diversity effect by using no delay profile estimator.

5.7.4.3 Synchronization Circuit Complexity In the LMS algorithm, we have to take frame and symbol timing synchronization before the adaptive array combining because the LMS algorithm requires this timing information. On the other hand, in the CMA, owing to the constancy of the transmitted signal envelope, we do not have to synchronously sample the received signal with the symbol and frame timings. This feature has two meanings. The first one is that we can arbitrarily determine the weight vector updating period regardless of the symbol rate. Another one is that we can put frame and symbol timing synchronization stages after the adaptive directivity control stage, which means that we can take more reliable frame and symbol timing synchronization using the frequency-selective fading-compensated signal.

5.7.4.4 Applicable Modulation Schemes The LMS algorithm is basically applicable to any modulation scheme. On the other hand, the application of the CMA is limited to only the constant envelope modulation scheme, such as GMSK. Moreover, the output of the adaptive array includes phase ambiguity in the case of the CMA because no phase information is fed back to the weight controller. To remove the effect of the phase ambiguity, we have to embed reference symbols or employ differential encoding and differential decoding, which could degrade the BER performance.

5.7.5 Field Experimental Result of the CMA Adaptive Array

A field trial of the adaptive array antenna was conducted by Ohgane et al. using the 1.5-GHz band [5-31; 5-32]. Table 5.6 shows specifications of the experimental systems shown in Ref. [5-31; 5-32].

Table 5.6 Specifications of the experimental CMA adaptive array system.

Item	Specification
Carrier frequency	1431.5 MHz
Tx power	5 watts (W) (37 dBm)
Tx antenna	omnidirectional height: 30 m
Modulation/demodulation scheme	GMSK with coherent detection
Access/multiplexing scheme	24-channel TDM
User bit rate/channel	8 kbit/s
Bit rate for air interface	256 kbit/s
Adaptive array	4-ary CMA adaptive array height: 2.2 m

Figure 5.44 shows the area map for the field experiments. The field trial was carried out in central Tokyo, Japan. Figure 5.45 shows delay-angle spectra at point P1. As shown in this figure, relatively long paths with delay time of up to 10 μs are observed. Figure 5.46 shows the measured directivity of the adaptive array antenna at the point

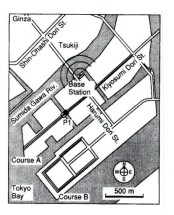

Fig. 5.44 Map central Tokyo, Japan, location of the field experiments (from Ref. 5-32, © Institute of Electrical and Electronic Engineers, 1993).

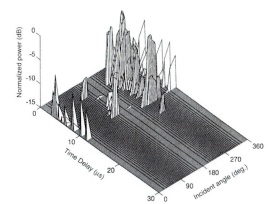
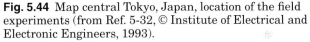

Fig. 5.45 Delay-angle spectra at point P1 (from Ref. 5-32, © Institute of Electrical and Electronic Engineers, 1993).

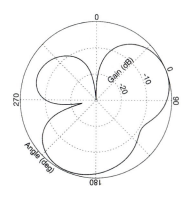
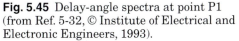

Fig. 5.46 Antenna beam directivity of the adaptive array antenna at point P1 (from Ref. 5-32, © Institute of Electrical and Electronic Engineers, 1993).

P1. As we can see in this figure, nulls are directed to approximately 0, 120, and 260 degrees, and the main beam is directed toward 65 degrees. When we multiply this directivity with the delay-angle spectra shown in Figure 5.45, we can obtain the delay-angle spectra at point P1 as shown in Figure 5.47. This figure shows that long delayed paths located around 0 degree and 260 degrees are clearly suppressed by the adaptive array.

Figure 5.48 shows BER performance of the CMA adaptive array measured along course A. In this figure, BER performance with a single omnidirectional antenna is also shown as a reference. When we employ a single omnidirectional antenna, irreducible BER of 7.0×10^{-3} occurs at high E_b/N_0 regions. On the other hand, when we employ the CMA adaptive array, BER = 10^{-2} can be achieved at E_b/N_0 = 13 dB, which is 22 dB lower than that for the single omnidirectional antenna system. With these results, the CMA adaptive array is considered to be one of the effective anti-frequency-selective fading techniques for land mobile communication systems.

5.8 RAKE DIVERSITY FOR SPREAD-SPECTRUM SIGNALS

Let's assume that the channel impulse response is given by

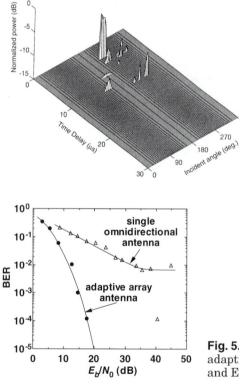

Fig. 5.47 Delay-angle spectra at point P1 after the array pattern is multiplied (from Ref. 5-32, © Institute of Electrical and Electronic Engineers, 1993).

Fig. 5.48 BER performance of GMSK with CMA adaptive array (from Ref. 5-32, © Institute of Electrical and Electronic Engineers, 1993).

$$c(t) = \sum_{i=0}^{L-1} c_i \delta(t - iT_c) \tag{5.187}$$

where L is the number of paths to be combined. When we transmit a pilot channel associated with the traffic channel in the case of CDMA, we can measure the channel impulse response as we discussed in chapter 4. Let the measured impulse response be

$$\hat{c}(t) = \sum_{i=0}^{L-1} \hat{c}_i \delta(t - iT_c) \tag{5.188}$$

As we discussed in chapter 4, section 4.3.3, when we multiply the complex conjugate of $\hat{c}_m$ and the complex conjugate of the m-chip delayed spreading code with the received baseband signal of the traffic channel $u^T(t)$, we can obtain the transmitted baseband signal included in the m-th path as

$$v_m(t) = |c_m|^2 a(t) \tag{5.189}$$

Therefore, when this process is carried out for all the delayed paths and they are added up, we can achieve path diversity with maximal ratio combining as [5-36], which is called Rake diversity.

$$z(t) = \sum_{i=0}^{L-1} v_m(t) = \sum_{i=0}^{L-1} |c_m|^2 a(t) \tag{5.190}$$

Let's calculate BER performance of path diversity in a case where all the paths have the same average power, as in

$$E\left[|c_1|^2\right] = E\left[|c_2|^2\right] = \cdots = E\left[|c_L|^2\right] \tag{5.191}$$

In this case, p.d.f. of each path is given by

$$p(\gamma) = \frac{1}{(L-1)!} \frac{\gamma^{L-1}}{(\gamma_0/L)^{L_1}} e^{-\frac{\gamma}{(\gamma_0/L)}} \tag{5.192}$$

When BER performance under AWGN conditions is given by $a \cdot \mathrm{erfc}\left(\sqrt{b\gamma}\right)$, BER performance under a frequency-selective fading channel represented by m-path Rayleigh model is given by

$$P_{path}(\gamma_0)$$
$$= a\left[1 - \frac{1}{\sqrt{1+1/(b\gamma_0/L)}} \sum_{k=0}^{L-1} \frac{(2k-1)!!}{(2k)!!} \frac{1}{(b\gamma_0/L+1)^k}\right] \tag{5.193}$$

The difference between equations (5.193) and (4.97) is that γ_0/L instead of γ_0 is substituted in equation (5.193). This is because the average power of each path is given by C/L, where C is the total received signal power.

In the same manner, when BER performance under AWGN conditions is given by $a \cdot e^{-b\gamma}$, BER performance under a frequency-selective fading channel represented by L-path Rayleigh model is given by

$$P_{path}(\gamma_0) = a \left[\frac{(L-1)!}{(1 + b\gamma_0/L)^L} \right] \tag{5.194}$$

REFERENCES

5-1. Sampei, S., "Development of Japanese adaptive equalizing technology toward high bit rate data transmission in land mobile communications," IEICE Trans., Vol. E74, No. 6, pp. 1512–21, June 1991.

5-2. Sampei, S. and Kinoshita, N., "A method for rejecting adjacent channel interference using an adaptive equalizer," Rev. of the Communications Res. Lab., Vol. 37, No. 1, pp. 75–84, February 1991.

5-3. Godard, D., "Channel equalization using a Kalman filter for fast data transmission," IBM J. Res. Develop., Vol. 18, No. 3, pp. 267–73, May 1974.

5-4. Muller, S., "Least squares algorithms for adaptive equalizer," Bell Syst. Tech. J., Vol. 60, No. 8, pp. 1905–25, October 1981.

5-5. Falconer, D. D. and Ljung, L., "Application of fast Kalman estimation to adaptive equalization," IEEE Trans. Commun., Vol. COM-26, No. 10, pp. 1439–46, October 1978.

5-6. Namiki, J., "Block demodulation for short radio packet," Trans. IECE of Japan, Vol. 67-B, No. 1, pp. 54–61, January 1984.

5-7. Nakajima, M. and Sampei, S., "Performance of a decision feedback equalizer under frequency selective fading in land mobile communications," Trans. IEICE (B-II), Vol. J72-B-II, No. 10, pp. 515–23, October 1989.

5-8. Suzuki, H. and Fukawa, K., "Dynamic performance analysis on RLS adaptive equalizers for mobile radio transmission," Trans. IEICE (B-II), Vol. J76-B-II, No. 4, pp. 189–201, April 1993.

5-9. Sampei, S., "Computation reduction of decision feedback equalizer using interpolation for land mobile communications," GLOBECO'M91 (Phoenix, Arizona), pp. 521–25, November 1991.

5-10. Lo, N. W. K., Falconer, D. D. and Sheikh, A. U. H., "Adaptive equalizer and diversity combining for mobile radio using interpolated channel estimates," IEEE Trans. Veh. Technol.,Vol. 40, No. 3, pp. 636–45, August 1991.

5-11. Suzuki, H., "Performance of a new adaptive diversity-equalization for digital mobile radio," Electron. Letters, Vol. 26, No. 10, pp. 627–28, May 1990.

5-12. Higashi, A. and Suzuki, H., "Dual-mode equalization for digital mobile radio," Trans. IEICE (B-II), Vol. J74-B-II, No. 3, pp. 91–100, March 1991.

5-13. Kamio, Y. and Sampei, S., "Performance of reduced complexity DFE using bidirectional equalizing in land mobile communications," 42nd IEEE Veh. Tech. Conf. (Denver, Colorado), pp. 372–75, May 1992.

5-14. Ariyavisitakul, S., "Equalization of a hard-limited slowly-fading multipath signal using a phase equalizer with a time-reversal structure," 40th IEEE Veh. Tech. Conf. (Orlando, Florida), pp. 520–26, May 1990.

5-15. Nagayasu, T., Sampei, S. and Kamio, Y., "Complexity reduction and performance improvement of a decision feedback equalizer for 16QAM in land mobile communications," IEEE Trans. Veh. Technol., Vol. 44, No. 3, pp. 570–78, August 1995.

5-16. Iida, M. and Sakaniwa, K., "Frequency selective compensation technology of digital 16QAM for microcellular mobile radio communication system," 42nd IEEE Veh. Tech. Conf. (Denver, Colorado), pp. 662–65, May 1992.

5-17. Teshima, I., Asano, M., Kamata, Y., Urabe, K. and Sasaoka, H., "Equalizer development for wideband 16QAM/TDMA in land mobile communications," PIMRC'94 (The Hague, Netherlands), pp. 225–31, September 1994.

5-18. Balaban, P. and Salz, J., "Dual diversity combining and equalization in digital cellular mobile radio," IEEE Trans. Veh. Technol., Vol. 40, No. 2, pp. 342–54, May 1991.

5-19. Forney, G. D. Jr., "Maximum-likelihood sequence estimation of digital sequences in the presence of intersymbol interference," IEEE Trans. Inf. Theory, Vol. IT-18, No. 3, pp. 363–78, May 1972.

5-20. Ungerboeck, G., "Adaptive maximum-likelihood receiver for carrier-modulated data-transmission systems," IEEE Trans. Commun. ,Vol. COM-22, No. 5, pp. 624–36, May 1974.

5-21. D'Avella, R., Moreno, L. and Agostino, M. S., "An adaptive MLSE receiver for TDMA digital mobile radio," IEEE J. Sel. Areas Commun., Vol. 7, No. 1, pp. 122–29, January 1989.

5-22. Okada, M. and Sampei, S., "Performance of DDFSE with interpolation type channel estimator under land mobile frequency selective fading channels," Trans. IEICE, Vol. J73-B-II, No. 11, pp. 727–35, November 1990.

5-23. Duel, A. and Heegard, C., "Delayed decision feedback sequence estimation," IEEE Trans. Commun., Vol. 37, No. 5, pp. 428–36, May 1989.

5-24. Ogawa, Y., Ohmiya, M. and Itoh, K., "An LMS adaptive array for multipath fading reduction," IEEE Trans. Aerosp. & Electron. Syst., Vol. 23, pp. 17–23, January 1987.

5-25. Widraw, B. and Stearns, S. D., Adaptive signal processing, Prentice Hall, Englewood Cliffs, 1985.

5-26. Monzingo, R. A. and Miller T. W., Introduction to adaptive arrays, John Wiley & Sons, New York, 1980.

5-27. Shan, T. J. and Kailath, T., "Adaptive beam-forming for coherent signals and interferences," IEEE Trans. Acoust. Speech & Signal Processing, Vol. ASSP-33, No. 3, pp. 527–36, June 1985.

5-28. Riegler, R. L. and Compton, R. T. Jr., "An adaptive array for interference rejection," Proc. IEEE, Vol. 61, No. 6, pp. 748–58, June 1973.

5-29. Ohgane, T., "Spectral efficiency improvement by base station antenna pattern control for land mobile cellular systems," IEICE Trans. Commun., Vol. E77-B, No. 5, pp. 598–605, May 1994.

5-30. Ogawa, Y., Ohmiya, M. and Itoh, K., "Fading equalization using an adaptive antenna for high-speed digital mobile communications," Proc. ISAP'89, pp. 857–60, August1989.

5-31. Ohgane, T., Shimura, T., Matsuzawa, N. and Sasaoka, H., "An implementation of a CMA adaptive array for a high speed GMSK transmission in mobile communications," IEEE Trans. Veh. Technol., Vol. 42, No. 3, pp. 282–88, August 1993.

5-32. Ohgane, T., Matsuzawa, N., Shimura, T., Mizuno, M. and Sasaoka, H., "BER performance of CMA adaptive array for high-speed GMSK mobile communication—a description of measurements in central Tokyo," IEEE Trans. Veh. Technol., Vol. 42, No. 4, pp. 484–90, November 1993.

5-33. Ogawa, Y., Yokohata, K. and Itoh, K., "Spatial-domain path-diversity using an adaptive array for mobile communications," 4th IEEE ICUPC (Tokyo, Japan), pp. 600–4, November 1995.

5-34. Treichler, J. R. and Larimore, M. G., "New processing techniques based on the constant adaptive algorithm," IEEE Trans. Acoust. Speech & Signal Process., Vol. 33, No. 4, pp. 420–31, April 1985.

5-35. Ohgane, T., "Characteristics of CMA adaptive array for selective fading compensation in digital land mobile communications," Electron. and Commun. in Japan, Scripta Technica Inc., Part 1, Vol. 74, No. 9, pp. 43–53, September 1991.

5-36. Abeta, S., Sampei, S. and Morinaga, N., "DS/CDMA coherent detection system with a suppressed pilot channel," GLOBECOM'94 (San Francisco, California), pp. 1622–26, November 1994.

Error Control Techniques

As discussed in previous chapters, propagation path conditions for wireless communication systems are very severe—path loss with respect to distance is much larger than path loss for fixed wireless links, and BER performance is severely degraded by multipath fading. Although the anti-fading techniques discussed in chapters 4 and 5 are very effective in improving receiver sensitivity for voice transmissions, they are not sufficient to satisfy link quality requirements for nonvoice data transmission services.

Error control techniques have a very long history, and there is much written on this subject [6-1; 6-2]. However, these books are not complete enough for engineers in the field of wireless mobile communication systems because they are missing a systematic point of view; for example, what the characteristics are of an error sequence of wireless mobile communication systems, how to evaluate burst errors, and how to design error control techniques for wireless communication systems considering such burst errors.

Therefore, this chapter will discuss features of the error sequence in wireless mobile communication systems, how to analyze statistics of the bit error, and how to design error control techniques.

6.1 OUTLINE OF ERROR CONTROL TECHNIQUES

Error control techniques that intend to reduce the probability of bit errors are very effective in achieving high-quality data transmission in wireless communication systems. To detect or correct errors, we add some redundant bits to the source information by using an encoding rule that maximizes the error detection or error correction abilities. Such encoding is called *channel coding*.

There are mainly two methods involved in error control techniques. One is the automatic repeat request (ARQ) and the other is forward error correction (FEC). In ARQ systems, the receiver simply detects whether the transmitted data block includes errors or not. When errors are detected, a retransmission request is transmitted to the sender [6-3]. In this system, we can achieve error-free conditions if we can accept long transmission delay time. Therefore, ARQ is effective for data, facsimile, or still-image transmissions rather than voice or moving-picture transmissions. On the other hand, FEC techniques are effective in improving transmission quality of voice or moving-picture transmissions in which retransmission is impossible.

Figure 6.1 shows a configuration of the transmitter and receiver using both ARQ and FEC techniques. The binary source information sequence is cyclic redundancy code (CRC) encoded for the ARQ and then FEC encoded. In wireless communication channels, we usually insert an interleaver after the FEC encoder because the bit error sequence is characterized by burst rather than random due to fading. The interleaved sequence is then fed to the modulator. At the receiver, after the demodulation process, deinterleaver, error correction, and error detection are carried out. When the received message includes no error, an acknowledgment (ACK) is sent back to the message sender. On the other hand, when an error is detected in the message, nonacknowledgment (NACK) is returned to request retransmission.

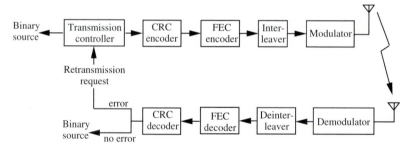

Fig. 6.1 Configuration of the transmitter and receiver using both ARQ and FEC techniques.

6.2 DIGITAL CHANNEL MODEL

Let's assume that an n-bit data sequence

$$\boldsymbol{a} = [a_{n-1},\ a_{n-2},\ ...,\ a_2,\ a_1,\ a_0] \tag{6.1}$$

is transmitted, and the corresponding received data sequence is given by

$$\boldsymbol{b} = [b_{n-1},\ b_{n-2},\ ...,\ b_2,\ b_1,\ b_0] \tag{6.2}$$

When there is no error in the received sequence

$$b_k = a_k, \quad (k = n-1, \; n-2, \; \dots, 2, \; 1, \; 0) \tag{6.3}$$

is satisfied. On the other hand, when an error occurred at the k-th bit, the received bit b_k is given by

$$b_k = a_k \oplus 1 \tag{6.4}$$

where $\oplus$ denotes a modulo-2 addition. Therefore, we can express the relationship between the transmitted and received bit sequences as

$$b_k = a_k \oplus e_k \tag{6.5}$$

where $\{e_n\}$ is called the *bit error sequence* in which, $e_k = 0$ means no error and $e_k = 1$ means error. Figure 6.2 shows this relationship, which is called the *digital radio channel model*.

    When the cause of the error is only AWGN, a bit error has occurred randomly. We call this error sequence the *random error sequence*. On the other hand, when we transmit data over a fading channel in which the received signal level is time variant, the error sequence becomes bursty. We call such a sequence a *burst error sequence* [6-4].

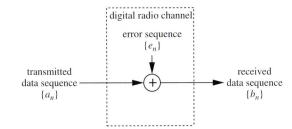

Fig. 6.2 Digital radio channel model.

6.3 CONCEPT OF ERROR DETECTION AND ERROR CORRECTION

First, we will discuss the basic concept of error detection and error correction schemes in the case of (7, 4) Hamming code. (7, 4) Hamming code encodes 4-bit source data $[a_3, a_2, a_1, a_0]$ to the 7-bit codeword $[b_6, b_5, b_4, b_3, b_2, b_1, b_0]$ using the following encoding rule:

$$
\begin{aligned}
b_6 &= a_3 \\
b_5 &= a_2 \\
b_4 &= a_1 \\
b_3 &= a_0 \\
b_2 &= b_6 \oplus b_5 \oplus b_4 \qquad = a_3 \oplus a_2 \oplus a_1 \\
b_1 &= \qquad b_5 \oplus b_4 \oplus b_3 = \qquad a_2 \oplus a_1 \oplus a_0 \\
b_0 &= b_6 \oplus b_5 \qquad \oplus b_3 = a_3 \oplus a_2 \qquad \oplus a_0
\end{aligned}
\tag{6.6}
$$

When we apply this rule, the relationship between the 4-bit source data and the 7-bit codeword is that shown in Table 6.1.

Minimum Hamming distance of a code (d_{min}) is usually used as criteria to evaluate error detection or error correction ability. Hamming distance, $d(\boldsymbol{u}, \boldsymbol{v})$, between two codewords $\boldsymbol{u}$ and $\boldsymbol{v}$ is defined as the number of different bits; the minimum Hamming distance is defined as the smallest value of $d(\boldsymbol{u}, \boldsymbol{v})$. In the case of $(7, 4)$ Hamming code, we find that minimum Hamming distance of the 4-bit source data is $d_{min} = 1$, whereas the minimum Hamming distance after coding is $d_{min} = 3$. In this case, when the received codeword includes a 1-bit error, the Hamming distance between the transmitted codeword and the received codeword is 1. On the other hand, Hamming distance between the received codeword and a codeword other than the transmitted one is still more than 1. Therefore, we can correctly detect the transmitted codeword when there is only 1 error bit in a codeword. On the other hand, when there are 2 error bits in a codeword, Hamming distance between the received codeword and one of the codewords other than the transmitted one could be shorter than that between the transmitted and the received codewords. Therefore, we cannot correct 2-bit errors. However, we can detect 2-bit errors because the received codeword with a 2-bit error is still different from any one of the codewords.

Table 6.1 Relationship between the source data and the codeword.

4-Bit Source Data	7-Bit Codeword
0000	0000000
0001	0001011
0010	0010110
0011	0011101
0100	0100111
0101	0101100
0110	0110001
0111	0111010
1000	1000101
1001	1001110
1010	1010011
1011	1011000
1100	1100010
1101	1101001
1110	1110100
1111	1111111

Generally, when the minimum Hamming distance is d_{min}, we can detect $(d_{min}-1)$-bit error and can correct $\text{Int}[(d_{min}-1)/2]$-bit error, where $\text{Int}[x]$ means the integer part of x.

6.4 EVALUATION CRITERIA FOR ERROR SEQUENCE CHARACTERISTICS

For the evaluation of the performance of error detection or error correction schemes, error sequence characteristics are very important. For example, error correction schemes greatly depend not only on the BER but also on whether the error sequence is bursty or not [6-5]. Among the many criteria for the evaluation of burst errors, the following three criteria are frequently used.

☞ **Error-free length:** the number of successive nonerror bits

☞ **Error run length:** the number of successive error bits

☞ **Burst length:** a sequence of error bits preceded and followed by error-free runs of at least n_0 bits

Let's assume that the number of events for k-bit of error-free length, k-bit of error run length, and k-bit of burst length are $e_f(k)$, $e_r(k)$, and $e_b(k)$, respectively. When the received bit error sequence is given by

$$[000000101101110000010110101000000] \tag{6.7}$$

and $n_0 = 3$ bits is assumed for burst length, $e_f(k)$, $e_r(k)$, and $e_b(k)$, are calculated as follows:

Error run length
$$e_r(1) = 4, e_r(2) = 2, e_r(3) = 1, e_r(4) = 0, e_r(k \geq 5) = 0$$
Error-free length
$$e_f(1) = 5, e_f(2) = 0, e_f(3) = 0, e_f(4) = 0, e_f(5) = 2, e_f(6) = 1, e_f(k \geq 7) = 0$$
Burst length
$$e_b(8) = 2, e_b(1 \leq k \leq 7) = e_b(k \geq 9) = 0$$

Among them, burst length is the most popular measure to evaluate burst error characteristics because burst length is very closely related to the performance of the FEC in comparison with the other statistics [6-6; 6-7]. Therefore, we will discuss burst length as a measure for burst error characteristics.

For the evaluation of burst length statistics, we usually use cumulative distribution of burst length given by

$$E(k) = \sum_{l=1}^{k} e_b(k) \tag{6.8}$$

Figure 6.3 shows the relationship between instantaneous E_b/N_0 and instantaneous BER for 64 ksymbol/s QPSK, where $f_d = 1$ Hz and instantaneous BER is averaged over

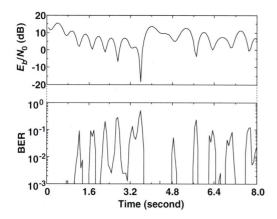

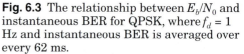

Fig. 6.3 The relationship between E_b/N_0 and instantaneous BER for QPSK, where $f_d = 1$ Hz and instantaneous BER is averaged over every 62 ms.

every 62 ms. As shown in this figure, when the received signal level is faded, burst errors occur.

Figure 6.4 shows cumulative distribution of burst length with a parameter of E_b/N_0 for 64 ksymbol/s QPSK under flat Rayleigh fading environments with $f_d = 1$ Hz, where $n_0 = 20$ bits is employed. When we employ FEC, knowing how to prevent a long burst error is very important. From this viewpoint, cumulative distribution at longer burst lengths should be as high as possible. When we observe Figure 6.4, we can see that longer burst length is more probable when E_b/N_0 becomes small because fade duration becomes long.

Figure 6.5 shows an example of the field experimental results of the burst length distribution for 1.5-GHz-band pilot symbol-aided 16QAM/FDMA modem, where $n_0 = 50$ bits is employed, the received signal level is in the range of −100.0 to −98.0 dBm, and vehicular speed is classified into three ranges—15.0–25.0, 25.0–35.0, and 35.0–45.0 kilometers per hour (km/h) [6-8]. In this figure, a longer burst is getting more probable with vehicular speed (with f_d) for the same reason cited for Figure 6.4.

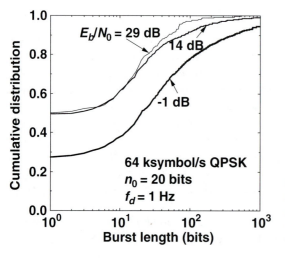

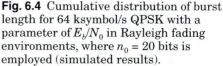

Fig. 6.4 Cumulative distribution of burst length for 64 ksymbol/s QPSK with a parameter of E_b/N_0 in Rayleigh fading environments, where $n_0 = 20$ bits is employed (simulated results).

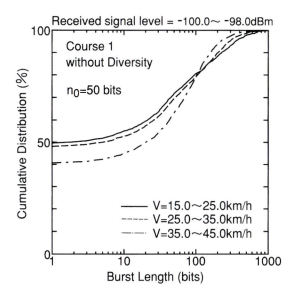

Fig. 6.5 Cumulative distribution of the burst length of 1.5-GHz-band pilot symbol-aided 16QAM/FDMA modem obtained by field experiments, where $n_0 = 50$ bits is employed (from Ref. 6-8, © Communications Research Laboratory, 1993).

6.5 INTERLEAVING FOR RANDOMIZATION OF BURST ERRORS

As discussed previously, the bit error sequence is bursty in wireless mobile communication systems. On the other hand, the error-correcting code is ineffective or, rather, it could degrade the BER performance in the case of burst errors. For example, when we apply (7, 4) Hamming code, although we would be correcting 1-bit error in each 7-bit block, we could be producing extra errors when there are 2 or more bit errors in each 7-bit block. Therefore, knowing ways to randomize a burst error sequence is very important for the design of FEC for wireless mobile communication systems.

Interleaving is an effective means of randomizing burst errors [6-6; 6-7]. We usually insert an interleaver using an m-row and n-column matrix after the FEC encoder in the transmitter, as was shown in Figure 6.1. Figure 6.6 shows an example of interleaving and deinterleaving processes using a 4-row, 7-column matrix. In the interleaving process, the coded data sequence is written row-wise, and the written data is read out column-wise. In the case of Figure 6.6, the transmitted data sequence is given by (1, 8, 15, 22, 2, 9, 16, 23, 3, 10, 17, 24, ..., 7, 14, 21, 28).

At the receiver, the deinterleaver is inserted before the FEC decoder as shown in Figure 6.1. At this stage, the received data sequence is written column-wise and is read out row-wise. As a result, a burst error of length $l = mb$ is broken up into m burst of length b. In the case of Figure 6.6, four consecutive errors are distributed into 4 bursts of length 1.

Let's assume that four consecutive bits (15, 22, 2, 9) are errored. For example, when (7, 4) Hamming code is applied, such a burst error cannot be corrected if inter-

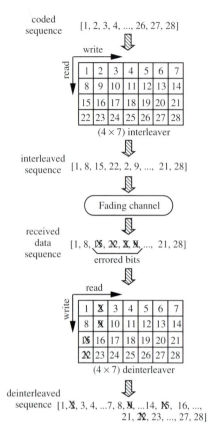

Fig. 6.6 Interleaving and deinterleaving processes using 4-row, 7-column matrix.

leaving and deinterleaving are not employed. On the other hand, when interleaving and deinterleaving are employed, (7,4) Hamming code can correct errors exactly.

Figure 6.7 shows cumulative distributions of burst length for 64 ksymbol/s QPSK with an interleaver and without an interleaver, where f_d = 1 Hz, average BER = 10^{-2}, and n_0 = 20 bits. The performance at BER = 10^{-2} under AWGN channel is also

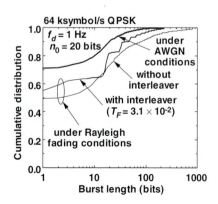

Fig. 6.7 Cumulative distributions of burst length of 64 ksymbol/s QPSK with interleaver and without interleaver, where f_d = 1 Hz, average BER = 10^{-2}, and n_0 = 20 bits (simulated results).

shown in this figure as a reference. When an interleaver is not employed, a burst length of more than 100 bits occurs with its probability of approximately 10%. On the other hand, if an interleaver with its frame length of $T_F = 3.1 \times 10^{-2}$ is employed, we can reduce the probability of a long burst, where $T_F = mnT_s$ (T_s is a symbol duration) for an m-row, n-column interleaver. By using the interleaver, the curve for cumulative distribution moves toward the curve for the AWGN channel, and the probability for a burst length of more than 100 bits becomes less than 5%. However, the position of the curve for interleaved sequence depends on the frame length of the interleaver (T_F). When T_F is small, the curve locates near that for a noninterleaved sequence. On the other hand, when T_F is increased, the curve gets closer to that for the AWGN channel.

6.6 ERROR DETECTION CODE

Let's assume a k-bit data sequence expressed as

$$[a_{k-1}, \ a_{k-2}, \ \dots \ , a_2, \ a_1, \ a_0] \tag{6.9}$$

When we add some redundant bits at the end of this sequence to increase minimum Hamming distance, we can detect errors in this sequence. The simplest error detection scheme is *parity check*. In this method, a check bit called *parity* given by

$$p = a_{k-1} \oplus a_{k-2} \oplus \cdots \oplus a_2 \oplus a_1 \oplus a_0 \tag{6.10}$$

is attached after the k-bit data sequence. Therefore, the obtained codeword can be expressed as

$$[a_{k-1}, \ a_{k-2}, \ \dots \ , a_2, \ a_1, \ a_0, \ p] \tag{6.11}$$

When we sum up all the bits in this codeword on the modulo-2 basis, the result becomes 0. In other words, when the result is not 0, we can detect error occurrence. Although the parity check is not a very powerful error detection scheme, it is sometimes used for error detection because its algorithm is very simple.

A more powerful and popular error detection scheme is the CRC. Let's assume that k-bit of source data expressed as the code polynomial of

$$a(x) = a_{k-1}x^{k-1} + a_{k-2}x^{k-2} + \cdots + a_2x^2 + a_1x + a_0 \tag{6.12}$$

is CRC encoded using the $(n\text{-}k)$-th order generator polynomial of $G(x)$. To obtain the n-bit of the codeword, k-bit of the source data is first upper-shifted by $(n\text{-}k)$ bit. As a result, the upper-shifted source bit is given by

$$\begin{aligned} b_1(x) &= b_{n-1}x^{n-1} + b_{n-2}x^{n-2} + \cdots + b_{n-k}x^{n-k} \\ &= a_{k-1}x^{n-1} + a_{k-2}x^{n-2} + \cdots + a_0x^{n-k} \\ &= x^{n-k}a(x) \end{aligned} \tag{6.13}$$

Next, $b_1(x)$ is divided by an $(n\text{-}k)$-th order generator polynomial $G(x)$. When $Q(x)$ is the quotient and $R(x)$ is the reminder on the modulo-2 basis, $b_1(x)$ can be expressed as

$$b_1(x) = Q(x)G(x) + R(x) \tag{6.14}$$

This $R(x)$ is then mapped onto the lower $(n\text{-}k)$ bit of the CRC-encoded data. As a result, the code polynomial of the encoded bit sequence is given by

$$\begin{aligned} b(x) &= b_1(x) + R(x) \\ &= x^{n-k}a(x) + R(x) \\ &= Q(x)G(x) + 2R(x) \\ &= Q(x)G(x) \end{aligned} \tag{6.15}$$

This equation means that the code polynomial of the CRC-encoded data can be divided by $G(x)$.

The code polynomial of the received codeword is

$$\hat{b}(x) = \hat{b}_{n-1}x^{n-1} + \hat{b}_{n-2}x^{n-2} + \cdots + \hat{b}_2x^2 + \hat{b}_1x + \hat{b}_0 \tag{6.16}$$

When there is no error

$$\hat{b}_i = b_i, \ (0 \le i \le n-1) \tag{6.17}$$

is satisfied. On the other hand, when an error occurs at the i-th bit, $\hat{b}_i$ is given by

$$\hat{b}_i = b_i \oplus 1 \tag{6.18}$$

Therefore, the code polynomial of the received data sequence can be expressed as

$$\begin{aligned} \hat{b}(x) &= b(x) + e(x) \\ &= Q(x)G(x) + e(x) \end{aligned} \tag{6.19}$$

where $e(x)$ is the polynominal of the error sequence expressed as

$$e(x) = e_{n-1}x^{n-1} + e_{n-2}x^{n-2} + \cdots + e_ix^i + \cdots + e_1x + e_0 \tag{6.20}$$

$$e_i = \begin{cases} 0, & \text{no error occurred} \\ 1, & \text{error occurred} \end{cases} \tag{6.21}$$

When no error occurred ($e(x) = 0$), $\hat{b}(x)$ can be divided by $G(x)$. Therefore, it is desirable to reduce the error miss-detection probability (probability that $e(x)$ can be divided by $G(x)$ even if an error occurred) to achieve reliable error detection performance.

One of the popular codes is the 16-bit CRC recommended by the International Telecommunication Union Telecommunication Section (ITU-T), whose generator polynomial is given by

$$G(x) = x^{16} + x^{12} + x^5 + 1$$

$$= (x + 1)(x^{15} + x^{14} + x^{13} + x^{12} + x^4 + x^3 + x^2 + x + 1)$$

(6.22)

$G(x)$ has two factors, $(x + 1)$ and $(x^{15} + x^{14} + x^{13} + x^{12} + x^4 + x^3 + x^2 + x + 1)$. They are both irreducible polynomials, and their periods are 1 and 32767, respectively. First of all, when $\hat{b}(x)$ is divided by $(x + 1)$, its reminder becomes

$$e(x) = e_{n-1} \oplus e_{n-2} \oplus \cdots \oplus e_2 \oplus e_1 \oplus e_0$$

(6.23)

This process is exactly the same as that used for parity check. Therefore, we can detect an odd number of errors using this code.

Next, a generator polynominal of $(x^{15} + x^{14} + x^{13} + x^{12} + x^4 + x^3 + x^2 + x + 1)$ produces a code set having a minimum Hamming distance of 4. Therefore, we can detect random errors of 3 bits or less. At the same time, because the order of $G(x)$ is 16, we can perfectly detect burst errors of 16 bits or less. Furthermore, we can detect 17 bits of burst errors with a probability of $(1-2^{-15})$ and more than 17 bits of burst errors with a probability of $(1-2^{-16})$.

6.6.1 CRC Encoder Circuit

Figure 6.8 (a) shows a configuration of the CRC encoder, where D means a 1-bit delay element, such as D-flip flop (D-FF). First, SW_1, SW_2, and SW_3 are connected to 1, and the information bit sequence is fed to the encoder. During this process, the information bit sequence is also output as an information part of the codeword. When the last information bit is input, the content in the D-FFs corresponds to the 16-bit parity check bits. Therefore, SW1, SW2, and SW3 are switched to 2, and these parity bits are shifted out.

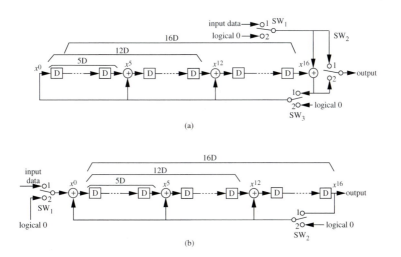

Fig. 6.8 Configurations of CRC encoder and decoder.

6.6.2 CRC Decoder Circuit

Figure 6.8 (b) shows a configuration of the CRC decoder. As we have discussed before, we can detect errors by dividing the transmitted data sequence by the generator polynomial. First, SW_1 and SW_2 are connected to 1, and the received bit sequence is fed to the CRC decoder. Just after all the received bits are input, SW_1 and SW_2 are switched to 2, and 16 bits of the reminder are shifted out. When all these 16 bits are 0, we will decide that there is no error in the received sequence. Otherwise, we will detect that errors occurred.

6.7 FORWARD ERROR CORRECTION USING CYCLIC CODE

Cyclic code is a very popular linear block code because it allows us to easily implement the encoder and decoder using linear shift registers. Moreover, encoding and decoding operations of the cyclic codes are based on the algebraic structure in the Galois field with q elements $GF(q)$.

A cyclic code has a property such that any cyclic shift of a codeword produces another codeword. For example, when

$$v_a(x) = v_{n-1}x^{n-1} + v_{n-2}x^{n-2} + \cdots + v_2x^2 + v_1x + v_0 \tag{6.24}$$

is one of the codewords, a 1-bit-shifted code

$$v_b(x) = v_{n-2}x^{n-1} + v_{n-3}x^{n-2} + \cdots + v_1x^2 + v_0x + v_{n-1} \tag{6.25}$$

is another codeword. Because x represents a 1-bit shift and because the code should be cyclically shifted, the following should be satisfied:

$$
\begin{aligned}
v_b(x) &= xv_a(x) \\
&= v_{n-2}x^{n-1} + v_{n-3}x^{n-2} + \cdots + v_1x^2 + v_0x + v_{n-1}x^n
\end{aligned} \tag{6.26}
$$

As a result, n-bit cyclic codewords should satisfy the following equation:

$$x^n + 1 = 0 \tag{6.27}$$

where these calculations are on the modulo-2 basis.

Next, we have to find the generator polynomial $G(x)$ for the encoding. Because x is one of the roots for equation (6.27), $G(x)$ should be one of the factors of $x^n + 1$.

Let's assume $n = 7$ as an example. $x^7 + 1$ can be factorized as follows:

$$x^7 + 1 = (x^3 + x + 1)(x^3 + x^2 + 1)(x + 1) \tag{6.28}$$

The generator polynomial $G(x)$ should satisfy the following requirements:

☞ $G(x)$ is irreducible.
☞ Root of $G(x)$ should satisfy the following requirement to guarantee that l-bit-shifted code ($l < n$) becomes another codeword.

$$\begin{cases} \alpha^l \neq 1 & for \ \ l < n \\ a^l = 1 & for \ \ l = n \end{cases} \tag{6.29}$$

Among $x^3 + x + 1, x^3 + x^2 + 1$, and $x + 1$, $G_1(x) = x^3 + x + 1$ and $G_2(x) = x^3 + x^2 + 1$ satisfy these two requirements. When α is a root for $G_1(x) = 0$, $G_2(1/\alpha) = 0$ is also satisfied. Therefore, we can employ either one as the generator polynomial. In general, the relationship between the order of generator polynomial p and the number of bits in a codeword n and source bits in the codeword k is given by

$$n = 2^p - 1 \tag{6.30a}$$

$$k = n - p \tag{6.30b}$$

Popular generator polynomials for each p are shown in Table 6.2 [6-9]. For example, when the order of generator polynomial is fifth, the generator polynomial is given by $G(x) = x^5 + x^2 + 1$. In this case, (n, k) is given by $(31, 26)$.

Table 6.2 Generator polynomials on GF(2).

p	$G(x)$	p	$G(x)$
1	$x + 1$	17	$x^{17} + x^3 + 1$
2	$x^2 + x + 1$	18	$x^{18} + x^7 + 1$
3	$x^3 + x + 1$	19	$x^{19} + x^5 + x^2 + x + 1$
4	$x^4 + x + 1$	20	$x^{20} + x^3 + 1$
5	$x^5 + x^2 + 1$	21	$x^{21} + x^2 + 1$
6	$x^6 + x + 1$	22	$x^{22} + x + 1$
7	$x^7 + x + 1$	23	$x^{23} + x^5 + 1$
8	$x^8 + x^4 + x^3 + x^2 + 1$	24	$x^{24} + x^7 + x^2 + x + 1$
9	$x^9 + x^4 + 1$	25	$x^{25} + x^3 + 1$
10	$x^{10} + x^3 + 1$	26	$x^{26} + x^6 + x^2 + x + 1$
11	$x^{11} + x^2 + 1$	27	$x^{27} + x^5 + x^2 + x + 1$
12	$x^{12} + x^6 + x^4 + x + 1$	28	$x^{28} + x^3 + 1$
13	$x^{13} + x^4 + x^3 + x + 1$	29	$x^{29} + x^2 + 1$
14	$x^{14} + x^{10} + x^6 + x + 1$	30	$x^{30} + x^{23} + x^2 + x + 1$
15	$x^{15} + x + 1$	31	$x^{31} + x^3 + 1$
16	$x^{16} + x^{12} + x^3 + x + 1$	32	$x^{32} + x^{22} + x^2 + x + 1$

6.8 BCH CODE

6.8.1 Encoding for Single-Bit Error-Correcting BCH Code

Single-bit error-correcting BCH code is a linear cyclic code and one of the simplest error-correcting codes. Let's consider (n,k) BCH code, where n is the number of coded bits and k is the number of information bits.

The encoding process is the same as that used for the error detection code

$$b(x) = x^{n-k}a(x) + R(x) \tag{6.31}$$

where
    $a(x)$ =   code polynomial of the source bit sequence
    $b(x)$ =   code polynomial of the coded sequence
    $R(x)$ =   reminder after $x^{n-k}a(x)$ is divided by an $(n-k)$-th order generator polynomial $G(x)$

As discussed, the order of $G(x)$ is $p = \log_2(n + 1) = n - k$. Therefore, k is given by

$$k = n - \log_2(n + 1) \tag{6.32}$$

Table 6.3 shows the relationship between n and k, as well as the corresponding generator polynomials for single-bit error-correcting BCH codes.

6.8.2 Decoding for Single-Bit Error-Correcting BCH Code

When there is no error in the received codeword, the received code polynomial $\hat{b}(x)$ is divisible by the generator polynomial $G(x)$. However, when i-th bit is in error, the received code polynomial becomes

$$\begin{aligned} \hat{b}(x) &= b(x) + x^i \\ &= G(x)Q(x) + x^i \end{aligned} \tag{6.33}$$

As discussed earlier in this chapter, $e(x) = x^i \, (0 \le i \le n - 1)$ is not divisible by the $G(x)$ because $x^i \ne 1$ for any i in the range of $0 \le i \le n - 1$. As a result, x^i can be expressed

Table 6.3 Generator polynomials for (n,k)-single-bit error-correcting BCH codes.

n	k	G(x)
7	4	$x^3 + x + 1$
15	11	$x^4 + x + 1$
31	26	$x^5 + x^2 + 1$
63	57	$x^6 + x + 1$
127	120	$x^7 + x + 1$

by the linear combination of x^m $(0 \leq m \leq n - k - 1)$, which corresponds to the reminder of $e(x)$ divided by $G(x)$. This $(n - k - 1)$-th order polynomial is called the *syndrome*. For example, in the case of (7, 4) and (15, 11) BCH codes, the relationship between $e(x)$ and the syndrome is given by Table 6.4 and Table 6.5.

For example, when the calculated syndrome is $x^3 + x + 1$ in the case of (15, 11) BCH code, we can see from Table 6.5 that the corresponding $e(x)$ is x^7. As a result, we can detect the 7-th bit is in error, and we can correct the 7-th bit by adding 1 to the bit on the modulo-2 basis.

6.8.3 Two-Bit Error-Correcting BCH Code

As discussed before, the error polynomial is non-zero when errors are included in the received codeword. Let's assume that i-th and k-th bits of the received codeword are in error. In this case, the error sequence polynomial is expressed as

$$e(x) = x^i + x^k \tag{6.34}$$

Because there are two unknown values i and k in this equation, we need two equations to figure them out.

Table 6.4 Relationship between $e(x)$ and the syndrome for (7, 4) BCH code.

$e(x)$	Syndrome	$e(x)$	Syndrome
x^0	1	x^4	$x^2 + x$
x^1	x	x^5	$x^2 + x + 1$
x^2	x^2	x^6	$x^2 + 1$
x^3	$x + 1$		

Table 6.5 Relationship between $e(x)$ and the syndrome for (15,11) BCH code.

$e(x)$	Syndrome	$e(x)$	Syndrome
x^0	1	x^8	$x^2 + 1$
x	x	x^9	$x^3 + x$
x^2	x^2	x^{10}	$x^2 + x + 1$
x^3	x^3	x^{11}	$x^3 + x^2 + x$
x^4	$x + 1$	x^{12}	$x^3 + x^2 + x + 1$
x^5	$x^2 + x$	x^{13}	$x^3 + x^2 + 1$
x^6	$x^3 + x^2$	x^{14}	$x^3 + 1$
x^7	$x^3 + x + 1$		

Let's assume a cyclic code with $n = 15$. $x^{15} + 1$ can be factorized into

$$x^{15} + 1 = (x^4 + x + 1)(x^4 + x^3 + x^2 + x + 1)(x^2 + x + 1)(x + 1)(x^4 + x^3 + 1) \quad (6.35)$$

When α is defined as a root of $x^4 + x + 1 = 0$, we can further factorize each factor in equation (6.35) as follows:

$x^4 + x + 1$ factorized into $(x - \alpha)(x - \alpha^2)(x - \alpha^4)(x - \alpha^8)$

$x^4 + x^3 + x^2 + x + 1$ factorized into $(x - \alpha^3)(x - \alpha^6)(x - \alpha^9)(x - \alpha^{12})$

$x^2 + x + 1$ factorized into $(x - \alpha^5)(x - \alpha^{10})$

$x^4 + x^3 + 1$ factorized into $(x - \alpha^7)(x - \alpha^{11})(x - \alpha^{13})(x - \alpha^{14})$

In these factorizations, we can find that α and α^3 are roots of different equations. Therefore, we can expect to obtain two independent syndrome equations when we define the generator polynomial $G(x)$ as

$$G(x) = (x^4 + x + 1)(x^4 + x^3 + x^2 + x + 1) \quad (6.36)$$

In this case, the number of parity bits is 8 bits because $G(x)$ is the 8-th order poly-nominal. Therefore, this code can be expressed as $(n, k) = (15, 7)$. Because the encod-ing process is the same with the other cyclic code, we will discuss only the decoding process.

6.8.3.1 Basic Calculation for Decoding Process   Before we discuss the decod-ing process, let's summarize basic calculations for the decoding process. First, we define α as a root of $G(x)$. When $n = 15$, α is a root of $G(x) = x^4 + x + 1$. Therefore, the following equation is satisfied:

$$\alpha^4 = \alpha + 1 \quad (6.37)$$

It means that α^i ($0 \le i \le n - 1$) can be represented by a linear combination of $1, \alpha, \alpha^2,$ and a^3. For example, α^7 can be expressed as

$$\alpha^7 = \alpha^4 \alpha^3 = (\alpha + 1)\alpha^3 = \alpha^3 + \alpha + 1 \quad (6.38)$$

This equation can also be expressed by 4-bit data as $(1, 0, 1, 1)$. Table 6.6 shows the linear combination of $1, \alpha, \alpha^2,$ and a^3 (polynominal expression) as well as the 4-bit data (binary) expression for α^i ($0 \le i \le 14$) in the case of $\alpha^4 = \alpha + 1$, where $(1, \alpha, ..., \alpha^{14})$ are the elements of $GF(2^4)$.

Addition of of these elements is defined as the modulo-2 addition of each bit in the binary notation. For example, the addition of $\alpha^6 = (1100)$ and $\alpha^8 = (0101)$ is obtained by

$$\alpha^6 + \alpha^8 = [1100] + [0101] = [1001] = \alpha^{14} \quad (6.39)$$

Table 6.7 shows the addition table for $GF(16)$ as an example [6-10]. On the other hand, multiplication of α^i and α^k over $GF(16)$ is given by Equation 6.40.

Table 6.6 Polynomial and binary expressions for powers of α.

Exponential Notation	Polynomial Notation	Binary Notation	Exponential Notation	Polynomial Notation	Binary Notation
0	0	0000	α^7	$\alpha^3 + \alpha + 1$	1011
1	1	0001	α^8	$\alpha^2 + 1$	0101
α	α	0010	α^9	$\alpha^3 + \alpha$	1010
α^2	α^2	0100	α^{10}	$\alpha^2 + \alpha + 1$	0111
α^3	α^3	1000	α^{11}	$\alpha^3 + \alpha^2 + \alpha$	1110
α^4	$\alpha + 1$	0011	α^{12}	$\alpha^3 + \alpha^2 + \alpha + 1$	1111
α^5	$\alpha^2 + \alpha$	0110	α^{13}	$\alpha^3 + \alpha^2 + 1$	1101
α^6	$\alpha^3 + \alpha^2$	1100	α^{14}	$\alpha^3 + 1$	1001

Table 6.7 Addition table for $GF(16)$ [6-10].

+	0	1	α	α^2	α^3	α^4	α^5	α^6	α^7	α^8	α^9	α^{10}	α^{11}	α^{12}	α^{13}	α^{14}
0	0	1	α	α^2	α^3	α^4	α^5	α^6	α^7	α^8	α^9	α^{10}	α^{11}	α^{12}	α^{13}	α^{14}
1	1	0	α^4	α^8	α^{14}	α	α^{10}	α^{13}	α^9	α^2	α^7	α^5	α^{12}	α^{11}	α^6	α^3
α	α	α^4	0	α^5	α^9	1	α^2	α^{11}	α^{14}	α^{10}	α^3	α^8	α^6	α^{13}	α^{12}	α^7
α^2	α^2	α^8	α^5	0	α^6	α^{10}	α	α^3	α^{12}	1	α^{11}	α^4	α^9	α^7	α^{14}	α^{13}
α^3	α^3	α^{14}	α^9	α^6	0	α^7	α^{11}	α^2	α^4	α^{13}	α	α^{12}	α^5	α^{10}	α^8	1
α^4	α^4	α	1	α^{10}	α^7	0	α^8	α^{12}	α^3	α^5	α^{14}	α^2	α^{13}	α^6	α^{11}	α^9
α^5	α^5	α^{10}	α^2	α	α^{11}	α^8	0	α^9	α^{13}	α^4	α^6	1	α^3	α^{14}	α^7	α^{12}
α^6	α^6	α^{13}	a^{11}	α^3	α^2	α^{12}	α^9	0	α^{10}	α^{14}	α^5	α^7	α	α^4	1	α^8
α^7	α^7	α^9	α^{14}	α^{12}	α^4	α^3	α^{13}	α^{10}	0	α^{11}	1	α^6	α^8	α^2	α^5	α
α^8	α^8	α^2	α^{10}	1	α^{13}	α^5	α^4	α^{14}	α^{11}	0	α^{12}	α	α^7	α^9	α^3	α^6
α^9	α^9	α^7	α^3	α^{11}	α	α^{14}	α^6	α^5	1	α^{12}	0	α^{13}	α^2	α^8	α^{10}	α^4
α^{10}	α^{10}	α^5	α^8	α^4	α^{12}	α^2	1	α^7	α^6	α	α^{13}	0	α^{14}	α^3	α^9	α^{11}
α^{11}	α^{11}	α^{12}	α^6	α^9	α^5	α^{13}	α^3	α	α^8	α^7	α^2	α^{14}	0	1	α^4	α^{10}
α^{12}	α^{12}	α^{11}	α^{13}	α^7	α^{10}	α^6	α^{14}	α^4	α^2	α^9	α^8	α^3	1	0	α	α^5
α^{13}	α^{13}	α^6	α^{12}	α^{14}	α^8	α^{11}	α^7	1	α^5	α^3	α^{10}	α^9	α^4	α	0	α^2
α^{14}	α^{14}	α^3	α^7	α^{13}	1	α^9	α^{12}	α^8	α	α^6	α^4	α^{11}	α^{10}	α^5	α^2	0

$$\alpha^i \cdot \alpha^k = \alpha^{(i+k \mod 15)} \tag{6.40}$$

When the number of bits in the codeword is n, all the elements belong to the $GF(n + 1)$. For example, when $n = 31$, all the elements belong to $GF(32)$.

6.8.3.2 Decoding Process of Two-Bit Error-Correcting BCH Code   The received codeword is first divided by $G(x)$ to obtain the reminder $R(x)$. When i-th and k-th bits in the received codeword are in error, $R(x)$ is given by

$$R(x) = e(x) = x^i + x^k \tag{6.41}$$

As in the case of single error-correcting code, the source codeword is divisible by $(x^4 + x + 1)$. Therefore, when i-th and k-th bits are in error, a syndrome given by

$$S_0 = R(\alpha) = \alpha^i + \alpha^k \tag{6.42}$$

is obtained. At the same time, the transmitted codeword is also divisible by $(x^4 + x^3 + x^2 + x + 1)$. Therefore, we can obtain another syndrome

$$S_1 = R(\alpha^3) = \alpha^{3i} + \alpha^{3k} = (\alpha^3)^i + (\alpha^3)^k \tag{6.43}$$

when i-th and k-th bits are errored.
Because the received codeword is expressed as

$$\hat{b}(x) = Q(x)G(x) + R(x) \tag{6.44}$$

and $G(\alpha) = 0$, syndromes S_0 and S_1 can also be expressed as

$$S_0 = \hat{b}(\alpha) = R(\alpha) = \alpha^i + \alpha^k \tag{6.45}$$

$$S_1 = b(\alpha^3) = R(\alpha^3) = \alpha^{3i} + \alpha^{3k} \tag{6.46}$$

S_0 and S_1 can be modified as

$$S_0^2 = (\alpha^i + \alpha^k)^2 = (\alpha^i)^2 + (\alpha^k)^2 \tag{6.47}$$

$$S_1 = (\alpha^i)^3 + (\alpha^k)^3$$
$$= (\alpha^i + \alpha^k)\{(\alpha^i)^2 + \alpha^i\alpha^k + (\alpha^k)^2\} \tag{6.48}$$
$$= S_0(S_0^2 + \alpha^i\alpha^k)$$

Therefore, α^i and α^k are given as the roots of the following equations:

$$(x + \alpha^i)(x + \alpha^k) = x^2 + (\alpha^i + \alpha^k)x + \alpha^i\alpha^k$$
$$= x^2 + S_0x + [(S_1/S_0) + S_0^2] = 0 \tag{6.49}$$

When we substitute α^0 to α^{14} into equation (6.49) and we check whether the result is 0 or not, we can find out the roots of equation (6.49). A summary of the decoding process is

☞ Calculate syndromes S_0 and S_1 given by equations (6.45) and (6.46).
☞ Calculate $(S_1/S_0) + S_2^0$ for equation (6.49).
☞ Find out the roots of equation (6.49) by substituting α^0 to α^{14} into equation (6.49).
☞ Add 1 to the error bit on the modulo-2 basis.

In this process, addition, multiplication, and division for $GF(n + 1)$ can be achieved by using digital signal processors (DSPs) and read-only memory (ROM) look up tables. Therefore, we can flexibly implement any type of BCH decoder using DSP techniques.

E X A M P L E   6 – 1

The 7-bit source data is $(1, 1, 0, 1, 0, 0, 1)$.

1. Encode this data by using a generator polynomial of $G(x) = x^8 + x^7 + x^6 + x^4 + 1$.

2. When the received codeword is $(1, 1, 0, 1, 1, 0, 1, 1, 1, 0, 1, 0, 0, 0, 0)$, find the error positions and correct them.

Solution
1. The source data polynomial is expressed as

$$a(x) = x^6 + x^5 + x^3 + x^0 \tag{6.50}$$

When $x^8 a(x)$ is divided by $G(x)$, the reminder is given by

$$R(x) = x^6 + x^4 \tag{6.51}$$

Therefore, the polynomial of the codeword is given by

$$b(x) = x^{14} + x^{13} + x^{11} + x^8 + x^6 + x^4 \tag{6.52}$$

2. The polynominal of the received codeword is given by

$$\hat{b}(x) = x^{14} + x^{13} + x^{11} + x^{10} + x^8 + x^7 + x^6 + x^4 \tag{6.53}$$

Therefore, S_0 and S_1 are given by

$$S_0 = \alpha^{14} + \alpha^{13} + \alpha^{11} + \alpha^{10} + \alpha^8 + \alpha^7 + \alpha^6 + \alpha^4 \tag{6.54}$$

$$S_1 = \alpha^{42} + \alpha^{39} + \alpha^{33} + \alpha^{30} + \alpha^{24} + \alpha^{21} + \alpha^{18} + \alpha^{12} \tag{6.55}$$

Using the addition table shown in Table 6.7 and $\alpha^{15} = 1$, S_0 and S_1 can be simplified as $S_0 = \alpha^6$ and $S_1 = \alpha^{13}$. When we substitute these values into equation (6.49), we can obtain the following equation:

$$x^2 + \alpha^6 x + \left[\alpha^{13}/\alpha^6 + (\alpha^6)\right] = x^2 + \alpha^6 x + \alpha^2 = 0 \qquad (6.56)$$

When α^0 to α^{14} are substituted into this equation, we find that α^7 and α^{11} are the roots of this equation. This means that 7-th and 11-th bits in the received codeword are in error. When we add $e(x) = x^{11} + x^7$ to the received codeword, we can obtain the same codeword as the transmitted one.

6.8.4 Three-Bit Error-Correcting BCH Code

In general, when the generator polynomial $G(x)$ is defined as the least common multiple of the minimal polynomials having $2t$ consecutive powers of α—α, a^2, ..., α^{2t}—its Hamming distance becomes $2t + 1$. As a result, such code can correct up to t-bit errors in each block. For example, $G(x) = x^4 + x + 1$ can correct only 1 bit because its roots are α, α^2, and α^4. In the case of $G(x) = (x^4 + x + 1)(x^4 + x^3 + x^2 + x + 1)$, it can correct 2 bits because α, a^2, a^3, and α^4 are included as the roots of $G(x)$.

Therefore, we can extend this idea to more powerful error-correcting code. For example, we can obtain 3-bit error-correcting code when we further multiply a polynominal having α^5 as a root. As discussed earlier, $(x^2 + x + 1)$ has α^5 as its root. Therefore, we can obtain the generator polynomial of 3-bit-correcting BCH code with $n = 15$ as

$$G(x) = (x^4 + x + 1)(x^4 + x^3 + x^2 + x + 1)(x^2 + x + 1) \qquad (6.57)$$

In this case, we have to calculate the following three syndromes:

$$S_0 = \alpha^i + \alpha^k + \alpha^m \qquad (6.58)$$

$$S_1 = \alpha^{3i} + \alpha^{3k} + \alpha^{3m} \qquad (6.59)$$

$$S_2 = \alpha^{5i} + \alpha^{5k} + \alpha^{5m} \qquad (6.60)$$

The error bit position can be obtained by substituting α^0 for α^{14} into the following equation to find out α^i that satisfies the equation.

$$\begin{aligned} x^3 + S_0 x^2 + \left[(S_2 + S_1 S_0^2)/(S_0^3 + S_1)\right]x \\ + \left[(S_0 S_4 + S_1^2 + S_0^6 + S_0^3 S_1)/(S_0^3 + S_1)\right] = 0 \end{aligned} \qquad (6.61)$$

6.8.5 More than Three-Bit Error-Correcting BCH Code

With almost the same process for the 2-bit or 3-bit error-correcting code, we can implement 4-bit or more error-correcting BCH codes. In the encoding process, we can implement t-bit error-correcting code if the generator polynomial has roots of α, α^2, α^3, α^4, ..., α^{2t-1}, α^{2t}.

Decoding for 4-bit or more error-correcting codes can be implemented using Peterson's method as follows [6-11]. First, the following syndromes are calculated:

$$S_0' = \hat{b}(\alpha) = e(\alpha)$$
$$S_1' = \hat{b}(\alpha^2) = e(\alpha^2)$$
$$S_2' = b(\alpha^3) = e(\alpha^3)$$
$$S_3' = b(\alpha^4) = e(\alpha^4)$$

(6.62)

.

.

.

$$S_{2t-1}' = b(\alpha^{2t}) = e(\alpha^{2t})$$

Then, coefficients of the error locator polynomial, given by

$$\sigma(x) = (x - e^{n_1})(x - e^{n_2}) \cdots (x - e^{n_t})$$
$$= x^t + \sigma_1 x^{t-1} + \sigma_2 x^{t-2} + \cdots + \sigma_{t-1} x + \sigma_t$$

(6.63)

are calculated as follows:

$$S_0' + \sigma_1 = 0$$
$$S_1' + S_0'\sigma_1 + 2\sigma_2 = 0$$
$$S_2' + S_1'\sigma_1 + S_0'\sigma_2 + 3\sigma_3 = 0$$
$$S_3' + S_2'\sigma_1 + S_1'\sigma_2 + S_0'\sigma_3 + 4\sigma_4 = 0$$
$$S_4' + S_3'\sigma_1 + S_2'\sigma_2 + S_1'\sigma_3 + S_0'\sigma_4 + 5\sigma_5 = 0$$

(6.64)

.

.

.

Then, the roots of equation (6.63) are obtained by substituting α, α^2,..., α^n into the equation and checking whether the equations are satisfied or not.

Table 6.8 shows generator polynomials of (n, k, t) BCH code, where n, k, and t are the number of bits of the codeword, the number of information bits in the codeword, and the number of correctable bits, respectively [6-9]. In this table, $G(x)$ is expressed by the octal notation. For example, $G(x)$ for (15, 11, 1) code is shown as 23 (octal notation) in Table 6.8. Because 23 expressed by the octal notation corresponds to 10011 in the binary notation, the generator polynomial is given by $G(x) = x^4 + x + 1$.

In the case of $t \geq 2$, $G(x)$ is given by multiplying the corresponding factors and all the other factors written above it in the table. For example, in the case of (127, 120, 1), its generator polynomial is given by $x^7 + x + 1$ (203 in octal notation). In the case of (127, 113, 2), $x^7 + x^5 + x^3 + x + 1$ (253 in octal notation) is shown in this table. Therefore, the generator polynomial is given by $G(x) = (x^7 + x + 1)(x^7 + x^5 + x^3 + x + 1)$. In the same

Table 6.8 Generator polynomial for (n, k, t) BCH code [6-9].

(n, k, t)	G(x)	(n, k, t)	G(x)	(n, k, t)	G(x)
(7, 4, 1)	13	(127, 120, 1)	203	(255, 207, 6)	×747
		(127, 113, 2)	×253	(255, 199, 7)	×453
(15, 11, 1)	23	(127, 106, 3)	×217	(255, 191, 8)	×727
(15, 7, 2)	×37	(127, 99, 4)	×375	(255, 187, 9)	×23
(15, 5, 3)	×7	(127, 92, 5)	×271	(255, 179, 10)	×545
		(127, 85, 6)	×211	(255, 171, 11)	×613
(31, 26, 1)	45	(127, 78, 7)	×345	(255, 163, 12)	×543
(31, 21, 2)	×75	(127, 71, 9)	×277	(255, 155, 13)	×433
(31, 16, 3)	×67	(127, 64, 10)	×357	(255, 147, 14)	×477
(31, 11, 5)	×57	(127, 57, 11)	×313	(255, 139, 15)	×615
(31, 6, 7)	×73	(127, 50, 13)	×247	(255, 131, 18)	×455
		(127, 43, 14)	×367	(255, 123, 19)	×537
(63, 57, 1)	103	(127, 36, 15)	×221	(255, 115, 21)	×771
(63, 51, 2)	×127	(127, 29, 21)	×325	(255, 107, 22)	×703
(63, 45, 3)	×147	(127, 22, 23)	×323	(255, 99, 23)	×471
(63, 39, 4)	×111	(127, 15, 27)	×361	(255, 91, 25)	×651
(63, 36, 5)	×15	(127, 8, 31)	×235	(255, 87, 26)	×37
(63, 30, 6)	×155			(255, 79, 27)	×607
(63, 24, 7)	×133	(255, 247, 1)	435	(255, 71, 29)	×661
(63, 18, 10)	×165	(255, 239, 2)	×567	(255, 63, 30)	×515
(63, 16, 11)	×7	(255, 231, 3)	×763	(255, 55, 31)	×717
(63, 10, 13)	×163	(255, 223, 4)	×551	(255, 47, 42)	×735
(63, 7, 15)	×13	(255, 215, 5)	×675	(255, 45, 43)	×7

manner, $G(x)$ for (127, 106, 3) is given by $G(x) = (x^7 + x + 1)(x^7 + x^5 + x^3 + x + 1)(x^7 + x^3 + x^2 + x + 1)$.

6.8.6 Modified Code

In the original BCH code, n and k are given by $n = 2^m - 1$ ($m = 2, 3, 4, 5, ...$) and $k = 2^m - m$. In practical systems, however, preferable numbers of n and k—especially n—could be different from these numbers due to parameters of air interfaces and so on. Among the many modified codes, we will discuss shortened code and extended code.

6.8.6.1 Shortened Code Figure 6.9 shows encoding process of the shortened BCH code. This process is summarized as follows [6-1]:

☞ Source data sequence is divided into $(k\text{-}l)$-bit.

☞ Add l-bit zeros after the $(k\text{-}l)$-bit source data.

☞ Encode k-bit of source and zero data to make n-bit-coded data.

☞ Remove l-bit zeros from the coded data to make $(n\text{-}l)$-bit codeword.

With this process, we can make $(n\text{-}l, k\text{-}l, t)$ shortened BCH code.

On the other hand, we can decode this shortened code with the inverse operation of the encoding process summarized as follows:

☞ Insert l-bits of zeros between $(k\text{-}l)$-bit of source bits and $(n\text{-}l)$-bit of check bits to make (n, k, t) BCH code

☞ Decode zero-inserted code using the same process as the BCH decoding.

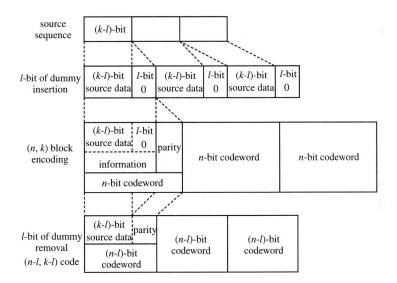

Fig. 6.9 Encoding process of the shortened BCH code.

E X A M P L E   6 – 2

When the source data sequence is (11010001), obtain single-bit error-correcting, shortened BCH code using $G(x) = x^4 + x + 1$.

Solution

$G(x)$ is the primitive polynomial, and the irreducible factors of $x^{15} + 1 = 0$. Therefore, we can encode the data sequence by (15, 11, 1) single-bit error-correcting BCH code. Because (11010001) is an 8-bit source data, we can apply 3-bit shortened (12, 8, 1) code. First, 3 bits of zeros are inserted after the source data to make 11-bit data (11010001000). Then this sequence is decoded using $G(x)$ to obtain the codeword of (110100010000010). In this codeword, 9-th to 11-th bits from the left are the dummy bits. Therefore, these 3 bits are removed. As a result, we can obtain the shortened (12, 8, 1) BCH-encoded codeword of (110100010010).

6.8.6.2 Extended Code   The extended code is obtained by adding l bits of parity to the (n, k, t) BCH code to make $(n + l, k, t)$ BCH code [6-1]. One of the popular extended BCH codes is the 1-bit extended code.

Let's assume that the BCH-encoded codeword is given by $(b_{n-1}, b_{n-2}, ..., b_2, b_1, b_0)$. In the 1-bit extended code, a parity given by

$$p_1 = b_{n-1} \oplus b_{n-2} \oplus \cdots \oplus b_2 \oplus b_1 \oplus b_0 \tag{6.65}$$

is added at the lowest bit in the codeword. As a result, the 1-bit extended codeword is given by $(b_{n-1}, b_{n-2}, ..., b_2, b_1, b_0, p_1)$.

When the minimum Hamming distance before the extension is $d_{min} = 2t + 1$, the parity bit addition results in the minimum Hamming distance of $d_{min} = 2t + 2$. Therefore, such extension can increase error detection ability from $2t$ bits to $(2t + 1)$ bits, although error correction ability is still t bits.

6.9 REED-SOLOMON CODE

6.9.1 Encoding Process

Reed-Solomon (RS) code is an extension of BCH code from binary to nonbinary code [6-10]. In the case of BCH code, all the coefficients of the polynomials are the elements of $GF(2)$—0 or 1. On the other hand, coefficients of generator polynomials for Reed-Solomon code are the elements of $GF(2^m)$. Therefore, each coefficient can be also expressed as the m-bit binary notation.

In the Reed-Solomon encoder, the binary data sequence is converted to m-bit parallel data to produce a symbol according to this m-bit data. In this case, the symbol sequence polynomial can be expressed as

$$a(x) = a_{k-1}x^{k-1} + a_{k-2}x^{k-2} + \cdots + a_2x^2 + a_1x + a_0 \tag{6.66}$$

where $(a_{k-1}, a_{k-2}, ..., a_1, a_0)$ are elements of $GF(2^m)$, and they represent m bits of data.

E X A M P L E   6 – 3

Let's assume that the source data sequence is (1101000101011111). Express this sequence in the polynomial expression of $GF(16)$.

Solution
The elements of $GF(16)$ are expressed as $(0, \alpha^0, \alpha, \alpha^2, \alpha^3, ..., \alpha^{13}, \alpha^{14})$, where α should satisfy $\alpha^4 + \alpha + 1 = 0$. Therefore, these elements can also be expressed as the linear combination

of $(\alpha^0, \alpha, \alpha^2, \alpha^3)$, i.e., 4-bit data. First, the source data sequence is blocked into 4-bit data. As a result, the blocked symbol sequence becomes (1101), (0001), (0101), and (1111). From Table 6.6, (1101), (0001), (0101), and (1111) correspond to α^{13}, α^0, α^8, and α^{12}, respectively. Therefore, the source data sequence can be expressed as

$$a(x) = \alpha^{13}x^3 + \alpha^0 x^2 + \alpha^8 x + \alpha^{12} \tag{6.67}$$

As we have discussed before, α, α^2, α^3, α^4, ..., α^{2t-1}, and α^{2t} should be the roots of the generator polynomial $G(x)$ for t-bit error-correcting binary BCH code. At the same time, coefficients of $G(x)$ should be 0 or 1 in the case of binary BCH code. As a result, the order of $G(x)$ is generally much higher than $2t$. On the other hand, coefficients for $G(x)$ are the elements of $GF(2^m)$ in the case of Reed-Solomon code. Therefore, $G(x)$ for Reed-Solomon code can be given exactly by the degree-$2t$ polynomial as

$$G(x) = (x - \alpha)(x - \alpha^2)(x - \alpha^3)(x - \alpha^4) \cdots (x - \alpha^{2t-1})(x - \alpha^{2t}) \tag{6.68}$$

As in the case of binary BCH code, the number of symbols in the codeword is given by $n = 2^m - 1$. When $G(x)$ is the degree-$2t$ polynomial, the number of source symbols in the codeword of Reed-Solomon code is given by $k = n - 2t$.

E X A M P L E   6 – 4

1. Calculate $G(x)$ for 3-symbol error-correcting Reed-Solomon code.

2. Encode $(\alpha^{10}, \alpha^2, 0, 0, 0, 0, 0, 0, \alpha^9)$.

Solution
1. $G(x)$ for $t = 3$ is given by

$$G(x) = (x + \alpha)(x + \alpha^2)(x + \alpha^3)(x + \alpha^4)(x + \alpha^5)(x + \alpha^6)$$
$$= x^6 + \alpha^{10}x^5 + \alpha^{14}x^4 + \alpha^4 x^3 + \alpha^6 x^2 + \alpha^9 x + \alpha^6 \tag{6.69}$$

2. The source symbol sequence can be expressed as

$$a(x) = \alpha^{10}x^8 + \alpha^2 x^7 + \alpha^9 \tag{6.70}$$

When $x^6 a(x)$ is divided by $G(x)$, its reminder is given by

$$R(x) = \alpha^{14}x^5 + \alpha^3 x^4 + \alpha^2 x^3 + \alpha^7 x^2 + \alpha^3 x + \alpha^{13} \tag{6.71}$$

As a result, the codeword polynomial is given by

$$b(x) = \alpha^{10}x^{14} + \alpha^2 x^{13} + \alpha^9 x^6$$
$$+ \alpha^{14}x^5 + \alpha^3 x^4 + \alpha^2 x^3 + \alpha^7 x^2 + \alpha^3 x + \alpha^{13} \tag{6.72}$$

6.9.2 Decoding for Single-Symbol Error-Correcting Code

Let's assume that the polynomials of the transmitted and received codewords are given by

$$b(x) = b_{n-1}x^{n-1} + b_{n-2}x^{n-2} + \cdots + b_2x^2 + b_1x + b_0 \qquad (6.73)$$

$$\hat{b}(x) = \hat{b}_{n-1}x^{n-1} + \hat{b}_{n-2}x^{n-2} + \cdots + \hat{b}_2x^2 + \hat{b}_1x + \hat{b}_0 \qquad (6.74)$$

where

$$\hat{b}_i = b_i + e_i \qquad (6.75)$$

Although e_i is 0 or 1 in the binary BCH case, it takes one of the values in $GF(2^m)$ in the case of Reed-Solomon code. As a result, the polynomial of the received codeword can be expressed as

$$\hat{b}(x) = b(x) + e(x) \qquad (6.76)$$

$$e(x) = e_{n-1}x^{n-1} + e_{n-2}x^{n-2} + \cdots + e_2x^2 + e_1x + e_0 \qquad (6.77)$$

In the case of single-symbol error-correcting Reed-Solomon code, syndromes are given by

$$S_0 = \hat{b}(\alpha) = e(\alpha) \qquad (6.78)$$

$$S_1 = \hat{b}(\alpha^2) = e(\alpha^2) \qquad (6.79)$$

When no error occurs, $S_0 = S_1 = 0$ can be obtained. On the other hand, when i-th symbol is in error, the syndromes are given by

$$S_0 = e_i\alpha^i \qquad (6.80)$$

$$S_1 = e_i\alpha^{2i} \qquad (6.81)$$

Using these two equations, we can obtain the error symbol location (α^i) and the error value (e_i) as follows:

$$\alpha^i = S_1/S_0 \qquad (6.82)$$

$$e_i = S_0^2/S_1 \qquad (6.83)$$

Then we can correct the error symbol by subtracting e_i from the i-th symbol in the received codeword.

6.9.3 Decoding for Two-Symbol Error-Correcting Code

Because the generator polynomial for 2-symbol error-correcting RS code has the degree of 4, the codeword includes $(n - 4)$-symbol of the source symbol. Therefore, it can be expressed as the $(n, n - 4, 2)$ RS code.

Let's assume that i-th and k-th bits are in error. To correct 2-symbol errors, the following syndromes are used:

$$S_0 = \hat{b}(\alpha) = e(\alpha) = e_i \alpha^i + e_k \alpha^k \qquad (6.84)$$

$$S_1 = \hat{b}(\alpha^2) = e(\alpha^2) = e_i \alpha^{2i} + e_k \alpha^{2k} \qquad (6.85)$$

$$S_2 = \hat{b}(\alpha^3) = e(\alpha^3) = e_i \alpha^{3i} + e_k \alpha^{3k} \qquad (6.86)$$

$$S_3 = \hat{b}(\alpha^4) = e(\alpha^4) = e_i \alpha^{4i} + e_k \alpha^{4k} \qquad (6.87)$$

Let's define Y_1 as the error value and V_1 as the error location of the first error, and Y_2 and V_2 are the same for the second error. When the first error occurs at the i-th symbol with its error value of e_i and the second one occurs at the k-th symbol with its error value of e_j, we can express $Y_1 = e_i$, $V_1 = \alpha^i$, $Y_2 = e_k$, and $V_2 = \alpha^k$. Using these values, we can express syndromes as

$$S_m = e_i (\alpha^i)^{m+1} + e_k (\alpha^k)^{m+1} = \sum_{u=1}^{2} Y_u V_u^{m+1} \quad (m = 0,\ 1,\ 2,\ 3) \qquad (6.88)$$

Using these equations, we will obtain Y_1, Y_2, V_1, and V_2.

First, we will define the error locator polynomial as

$$\sigma(x) = (x + V_1)(x + V_2) = x^2 + \sigma_1 x + \sigma_2 \qquad (6.89)$$

Because V_u ($u = 1$ or 2) is a root of this equation, the following equation has to be satisfied.

$$V_u^2 + \sigma_1 V_u + \sigma_2 = 0 \qquad (6.90)$$

When we multiply $Y_u V_u^{1+i}$ on both sides of equation (6.90) and sum up equations for $u = 1$ and 2, we can obtain the following equation:

$$\sum_{u=1}^{2} \left(Y_u V_u^{3+i} + \sigma_1 Y_u V_u^{2+i} + \sigma_2 Y_u V_u^{1+i} \right) = S_{i+2} + \sigma_1 S_{i+1} + \sigma_2 S_i = 0 \qquad (6.91)$$

where $i = 0, 1, 2, 3$. Using equation (6.91) for $i = 0$ and 1, we can obtain

$$\begin{bmatrix} S_1 & S_0 \\ S_2 & S_1 \end{bmatrix} \begin{bmatrix} \sigma_1 \\ \sigma_2 \end{bmatrix} = \begin{bmatrix} S_2 \\ S_3 \end{bmatrix} \qquad (6.92)$$

As a result, σ_1 and σ_2 are given by

$$\sigma_1 = (S_1 S_2 + S_0 S_3) / \Delta_2 \qquad (6.93)$$

$$\sigma_2 = (S_1 S_3 + S_2^2) / \Delta_2 \qquad (6.94)$$

where

$$\Delta_2 = S_1^2 + S_0 S_2 \qquad (6.95)$$

The symbol error locations V_1 and V_2 can be obtained by substituting α^0 to α^{14} into equation (6.90) and finding out a value for α^i that satisfies equation (6.90).

When the calculated V_1 and V_2 are substituted into equation (6.88) for $m = 0$ and 1, we get the following equation:

$$\begin{bmatrix} V_1 & V_2 \\ V_1^2 & V_2^2 \end{bmatrix} \begin{bmatrix} Y_1 \\ Y_2 \end{bmatrix} = \begin{bmatrix} S_0 \\ S_1 \end{bmatrix} \qquad (6.96)$$

As a result, we can calculate the error values Y_1 and Y_2 as

$$Y_1 = (V_2^2 S_0 + V_2 S_1) / (V_1 V_2^2 + V_1^2 V_2) \qquad (6.97)$$

$$Y_2 = (V_1^2 S_0 + V_1 S_1) / (V_1 V_2^2 + V_1^2 V_2) \qquad (6.98)$$

When we add these values to the corresponding symbols of the received codeword, we can correct 2-symbol errors.

When only one error occurred, Δ_2 in equations (6.93) and (6.94) is 0. In this case, we cannot calculate σ_1 and σ_2. To prevent this situation, we have to check whether or not $\Delta_2 = 0$ just after the calculation of the syndromes. When $\Delta_2 \neq 0$, we can calculate V_1, V_2, Y_1, and Y_2 using the above-mentioned process. On the other hand, when $\Delta_2 = 0$, we decide that only a single-symbol error occurred, and we will do the decoding process for the single-symbol error-correcting code.

6.9.4 Decoding for Three-Symbol Error-Correcting Code

The decoding process for a 3-symbol error-correcting RS code is almost the same as that for a 2-symbol error-correcting code.

Let's assume that the error symbol sequence polynomial is given by

$$e(x) = e_i x^i + e_k x^k + e_m x^m \qquad (6.99)$$

First, the following syndromes are calculated:

$$S_j = e_i(\alpha^i)^{j+1} + e_k(\alpha^k)^{j+1} + e_m(\alpha^m)^{j+1} = \sum_{u=1}^{3} Y_u V_u^{1+j} \qquad (6.100)$$

where Y_1, Y_2, and Y_3 are the error locations of the first, second, and third errors, and V_1, V_2, and V_3 are the error values of those errors.

The error locator polynomial for a 3-symbol error-correcting code is given by

$$\sigma(x) = (x + V_1)(x + V_2)(x + V_3) = x^3 + \sigma_1 x^2 + \sigma_2 x + \sigma_3 \tag{6.101}$$

When V_u ($u = 1, 2,$ or 3) is substituted into equation (6.101), the result is 0. Therefore, when we substitute V_u in equation (6.101), multiply $Y_u V_u^{1+j}$ on both sides, and sum up equations for $u = 1, 2,$ and 3, we get the following equation:

$$\sum_{u=1}^{3} \left(Y_u V^{4+j} + \sigma_1 Y_u V_u^{3+j} + \sigma_2 Y_u V_u^{2+j} + \sigma_3 Y_u V_u^{1+j} \right) \tag{6.102}$$
$$= S_{j+3} + \sigma_1 S_{j+2} + \sigma_2 S_{j+1} + \sigma_3 S_j = 0,$$

where $j = 0, 1, 2, 3, 4, 5$. Using equation for $j = 0, 1, 2$, we can obtain

$$\begin{bmatrix} S_2 & S_1 & S_0 \\ S_3 & S_2 & S_1 \\ S_4 & S_3 & S_2 \end{bmatrix} \begin{bmatrix} \sigma_1 \\ \sigma_2 \\ \sigma_3 \end{bmatrix} = \begin{bmatrix} S_3 \\ S_4 \\ S_5 \end{bmatrix} \tag{6.103}$$

As a result, we can calculate σ_1, σ_2, and σ_3 as

$$\sigma_1 = (S_1^2 S_5 + S_2^2 S_3 + S_3^2 S_1 + S_0 S_3 S_4 + S_0 S_2 S_5 + S_1 S_2 S_4) / \Delta_3 \tag{6.104}$$

$$\sigma_2 = (S_2^2 S_4 + S_3^2 S_2 + S_4^2 S_0 + S_0 S_3 S_5 + S_1 S_2 S_5 + S_1 S_3 S_4) / \Delta_3 \tag{6.105}$$

$$\sigma_3 = (S_2^2 S_5 + S_3^3 + S_4^2 S_1 + S_1 S_3 S_5) / \Delta_3 \tag{6.106}$$

where

$$\Delta_3 = S_2^3 + S_0 S_3^2 + S_1^2 S_4 + S_0 S_2 S_4 \tag{6.107}$$

When $\Delta_3 = 0$, it means the number of symbol errors are two or less. In this case, we have to apply the decoding process for 2-symbol or less error-correcting code. On the other hand, when $\Delta_3 \neq 0$, symbol error locations V_1, V_2, and V_3 can be obtained by substituting α^0 to α^{14} into equation (6.101), and we can find α^i that satisfies equation (6.101).

When we substitute V_1, V_2, and V_3 in equation (6.100) for $j = 0, 1,$ and 2, we can obtain the following equation:

$$\begin{bmatrix} V_1 & V_2 & V_3 \\ V_1^2 & V_2^2 & V_3^2 \\ V_1^3 & V_2^3 & V_3^3 \end{bmatrix} \begin{bmatrix} Y_1 \\ Y_2 \\ Y_3 \end{bmatrix} = \begin{bmatrix} S_0 \\ S_1 \\ S_2 \end{bmatrix} \tag{6.108}$$

When we solve this equation, we can obtain Y_1, Y_2, and Y_3 as

$$Y_1 = [S_0 V_2 V_3 + S_1 (V_2 + V_3) + S_2] / [V_1 (V_1 + V_2)(V_1 + V_3)] \tag{6.109}$$

$$Y_2 = [S_0 V_1 V_3 + S_1(V_1 + V_3) + S_2]/[V_2(V_2 + V_3)(V_1 + V_2)] \qquad (6.110)$$

$$Y_3 = [S_0 V_1 V_2 + S_1(V_1 + V_2) + S_2]/[V_3(V_2 + V_3)(V_1 + V_3)] \qquad (6.111)$$

When we add these values to the corresponding symbols of the received codeword, we can correct 3-symbol errors.

E X A M P L E   6 – 5

When source data are encoded using $(15, 9, 3)$ RS code, decode the following received codewords:

1. $(\alpha^{10}, \alpha^2, 0, 0, 0, \alpha^3, 0, 0, \alpha^9, \alpha^{14}, \alpha^3, \alpha^9, \alpha^7, \alpha^3, \alpha^{11})$
2. $(\alpha^1, \alpha^2, 0, 0, 0, 0, 0, 0, \alpha^{12}, \alpha^{14}, \alpha^3, \alpha^2, \alpha^7, \alpha^3, \alpha^{13})$
3. $(\alpha^{10}, \alpha^2, 0, 0, 1, 0, 0, 0, \alpha^9, \alpha^{14}, \alpha^3, \alpha^2, \alpha^7, \alpha^3, \alpha^{13})$
4. $(\alpha^{10}, \alpha^2, 0, 0, 0, 0, 0, 0, \alpha^9, \alpha^{14}, \alpha^3, \alpha^2, \alpha^7, \alpha^3, \alpha^{13})$

Solution
1. $(\alpha^{10}, \alpha^2, 0, 0, 0, \alpha^3, 0, 0, \alpha^9, \alpha^{14}, \alpha^3, \alpha^9, \alpha^7, \alpha^3, \alpha^{11})$

As the received codeword polynomial is expressed as

$$\hat{b}(x) = \alpha^{10}x^{14} + \alpha^2 x^{13} + \alpha^3 x^9 + \alpha^9 x^6 + \alpha^{14}x^5$$
$$+ \alpha^3 x^4 + \alpha^9 x^3 + \alpha^7 x^2 + \alpha^3 x + \alpha^{11} \qquad (6.112)$$

the syndromes are given by $S_0 = \alpha^8$, $S_1 = \alpha^7$, $S_2 = \alpha^2$, $S_3 = \alpha^6$, $S_4 = \alpha^8$, and $S_5 = \alpha^8$. First, Δ_3 given by equation (6.107) is obtained as $\Delta_3 = \alpha^{14}$. Therefore, the received codeword includes three errored symbols.

Coefficients for the error locator polynomial are obtained as $\sigma_1 = \alpha^4$, $\sigma_2 = \alpha^{13}$, and $\sigma_3 = \alpha^{12}$ using equations (6.104–7). As a result, we can obtain the error locator polynomial as

$$\sigma(x) = x^3 + \alpha^4 x^2 + \alpha^{13}x + \alpha^{12} = (x+1)(x+\alpha^3)(x+\alpha^9) \qquad (6.113)$$

which shows that $V_1 = 1$, $V_2 = \alpha^3$, and $V_3 = \alpha^9$.

Next, using equations (6.109–11), we can obtain the error values as $Y_1 = \alpha^4$, $Y_2 = \alpha^{11}$, and $Y_3 = \alpha^3$, which means that the error symbol sequence polynomial is expressed as

$$e(x) = \alpha^3 x^9 + \alpha^{11}x^3 + \alpha^4 \qquad (6.114)$$

When we add $e(x)$ to the received codeword polynomial, we can obtain the error-corrected codeword polynomial as

$$b(x) = \hat{b}(x) + e(x)$$
$$= \alpha^{10}x^{14} + \alpha^2 x^{13} + (\alpha^3 + \alpha^3)x^9 + \alpha^9 x^6 + \alpha^{14}x^5$$
$$+ \alpha^3 x^4 + (\alpha^9 + \alpha^{11})x^3 + \alpha^7 x^2 + \alpha^3 x + (\alpha^{11} + \alpha^4) \qquad (6.115)$$
$$= \alpha^{10}x^{14} + \alpha^2 x^{13} + \alpha^9 x^6 + \alpha^{14}x^5$$
$$+ \alpha^3 x^4 + \alpha^2 x^3 + \alpha^7 x^2 + \alpha^3 x + \alpha^{13}$$

2. $(\alpha^1, \alpha^2, 0, 0, 0, 0, 0, 0, \alpha^{12}, \alpha^{14}, \alpha^3, \alpha^2, \alpha^7, \alpha^3, \alpha^{13})$

As the received codeword polynomial is expressed as

$$\hat{b}(x) = \alpha x^{14} + \alpha^2 x^{13} + \alpha^{12} x^6 + \alpha^{14} x^5 \tag{6.116}$$
$$+ \alpha^3 x^4 + \alpha^2 x^3 + \alpha^7 x^2 + \alpha^3 x + \alpha^{13}$$

the syndromes are given by $S_0 = \alpha$, $S_1 = \alpha^9$, $S_2 = \alpha^3$, $S_3 = \alpha^{10}$, $S_4 = \alpha^{13}$, and $S_5 = \alpha^{13}$. In this case, $\Delta_3 = 0$. Therefore, the received codeword includes two or less error symbols. When we calculate Δ_2 given by equation (6.95), we can obtain $\Delta_2 = \alpha^7$. Therefore, we can find that the received codeword includes two symbol errors.

The coefficients for the error locator polynomial are obtained as $\sigma_1 = \alpha^8$ and $\sigma_2 = \alpha^5$ using equations (6.93–95). As a result, we can obtain the error locator polynomial as

$$\sigma(x) = x^2 + \alpha^8 x + \alpha^5 = (x + \alpha^6)(x + \alpha^{14}) \tag{6.117}$$

which shows that $V_1 = \alpha^6$ and $V_2 = \alpha^{14}$.

Next, using equations (6.97–98), we can obtain the error values as $Y_1 = \alpha^8$ and $Y_2 = \alpha^8$, which means that the error symbol sequence polynomial is expressed as

$$e(x) = \alpha^8 x^{14} + \alpha^8 x^6 \tag{6.118}$$

When we add $e(x)$ to the received codeword polynomial, we can obtain the error-corrected codeword polynomial as

$$
\begin{aligned}
b(x) &= \hat{b}(x) + e(x) \\
&= (\alpha + \alpha^8) x^{14} + \alpha^2 x^{13} + (\alpha^{12} + \alpha^8) x^6 + \alpha^{14} x^5 \\
&\quad + \alpha^3 x^4 + \alpha^2 x^3 + \alpha^7 x^2 + \alpha^3 x + \alpha^{13} \\
&= \alpha^{10} x^{14} + \alpha^2 x^{13} + \alpha^9 x^6 + \alpha^{14} x^5 \\
&\quad + \alpha^3 x^4 + \alpha^2 x^3 + \alpha^7 x^2 + \alpha^3 x + \alpha^{13}
\end{aligned}
\tag{6.119}
$$

3. $(\alpha^{10}, \alpha^2, 0, 0, 1, 0, 0, 0, \alpha^9, \alpha^{14}, \alpha^3, \alpha^2, \alpha^7, \alpha^3, \alpha^{13})$

The received codeword polynomial is expressed as

$$\hat{b}(x) = \alpha^{10} x^{14} + \alpha^2 x^{13} + x^{10} + \alpha^9 x^6 \tag{6.120}$$
$$+ \alpha^{14} x^5 + \alpha^3 x^4 + \alpha^2 x^3 + \alpha^3 x + \alpha^{13}$$

and the syndromes are obtained as $S_0 = \alpha^{10}$, $S_1 = \alpha^5$, $S_2 = 1$, $S_3 = \alpha^{10}$, $S_4 = \alpha^5$, and $S_5 = 1$. When we calculate Δ_3 and Δ_2, we find that both of them are 0. Therefore, we find that the received codeword includes a single symbol error. Using equations (6.80) and (6.81), we find that the 10-th symbol is in error and its value is 1—i.e., the error symbol polynomial is expressed as

$$e(x) = x^{10} \tag{6.121}$$

Therefore, the transmitted codeword sequence can be obtained as

$$b(x) = \hat{b}(x) + e(x)$$
$$= \alpha^{10}x^{14} + \alpha^2 x^{13} + (1+1)x^{10} + \alpha^9 x^6$$
$$+ \alpha^{14}x^5 + \alpha^3 x^4 + \alpha^2 x^3 + \alpha^3 x + \alpha^{13} \tag{6.122}$$
$$= \alpha^{10}x^{14} + \alpha^2 x^{13} + \alpha^9 x^6 + \alpha^{14}x^5 + \alpha^3 x^4$$
$$+ \alpha^2 x^3 + \alpha^3 x + \alpha^{13}$$

4. $(\alpha^{10}, \alpha^2, 0, 0, 0, 0, 0, 0, \alpha^9, \alpha^{14}, \alpha^3, \alpha^2, \alpha^7, \alpha^3, \alpha^{13})$

When we calculate the syndromes, we find that all the syndromes are 0. Therefore, there is no error in this codeword.

6.10 BER PERFORMANCE OF BLOCK CODE

6.10.1 BER Performance of BCH Code

Figure 6.10 shows computer-simulated results of BER versus C/N_0 performance for BCH-coded QPSK systems under AWGN conditions. As we can see from this figure, when the transmitted data sequence is encoded to keep its symbol rate constant, BER is improved by one to two order of magnitude in the case of (15, 11, 1) BCH code. When we employ (15, 7, 2) BCH code, we can further improve BER by one more order of magnitude. However, we have to reduce the source bit rate by 11/15 or 7/15 in the case of (15, 11, 1) code or (15, 7, 2), respectively, when we keep the symbol rate constant.

To evaluate the BER performance, including this bit rate reduction effect, we usually evaluate BER versus E_b/N_0 performance, where E_b is defined as the received signal level divided by the source bit rate. Figure 6.11 shows BER versus E_b/N_0 performance of the same BCH-coded QPSK system under AWGN conditions. The required E_b/N_0 difference between uncoded QPSK and coded QPSK to achieve a certain BER is called *coding gain*. As shown in this figure, coding gain for (15, 11, 1) code and (15, 7, 2) code is almost the same. At BER = 10^{-3}, the coding gain is about 1 dB. When the

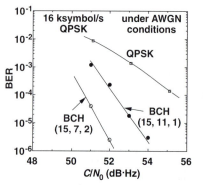

Fig. 6.10 BER versus C/N_0 performance of BCH-coded 16-ksymbol/s QPSK system under AWGN conditions (simulated results).

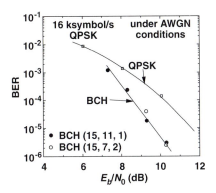

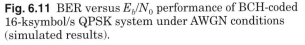

Fig. 6.11 BER versus E_b/N_0 performance of BCH-coded 16-ksymbol/s QPSK system under AWGN conditions (simulated results).

required BER is smaller, larger coding gain is obtained. For example, when the required BER is 10^{-5}, the coding gain becomes about 2 dB.

When BCH-encoded QPSK is operated under flat Ralyeigh fading conditions, we have to employ an appropriate size of interleaver because we cannot obtain sufficient coding gain unless the size is sufficiently large. To evaluate the maximum ability of BCH code, let us obtain the BER performance under flat Rayleigh fading conditions in the case of an interleaver with infinite frame length ($T_F=\infty$).

Figure 6.12 shows simulated results of BER performance for BCH-encoded QPSK with an infinite size of interleaver under flat Rayleigh fading conditions. As we find from this figure, when the interleaver frame length is infinite, coding gain at small BER is very high, whereas that at high BER is very small, or rather the performance after coding is worse than that for an uncoded case. Table 6.9 shows the coding gain for both (15, 11, 1) and (15, 7, 2) BCH codes at BER = 10^{-3} and 10^{-5}. As shown in this table, (15, 11, 1) code gives a higher coding gain for BER = 10^{-3}, and (15, 7, 2) gives a higher coding gain for BER = 10^{-5}. Therefore, when we select a channel coding scheme, we have to take the required BER into account. Moreover, when we compare Figure 6.11 and Figure 6.12, we find that the BCH code is more effective under Rayleigh fading conditions. Therefore, BCH code is applied to some wireless communication systems, such as PDC, which will be discussed in chapter 11.

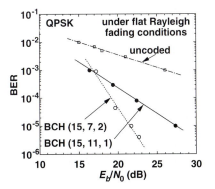

Fig. 6.12 BER performance of BCH-encoded QPSK with infinite size-interleaver under flat Rayleigh fading conditions (simulated results).

Table 6.9 Coding gain for BCH code with a parameter of required BER.

Required BER	Coding Gain	
	(15, 11, 1) BCH Code	(15, 7, 2) BCH Code
BER = 10^{-3}	9.5 dB	8.8 dB
BER = 10^{-5}	18.5 dB	24.5 dB

6.10.2 BER Performance of Reed-Solomon Code

Figure 6.13 shows simulated results of BER versus C/N_0 performance, and Figure 6.14 shows BER versus E_b/N_0 performance of the same system under AWGN conditions. We find from this figure that the coding gain of (15, 13, 1) RS code for BER = 10^{-3} is 0.8 dB and that for BER = 10^{-5} is 2.0 dB. When we employ (15, 11, 2) RS code, coding gain for BER = 10^{-3} is 1.0 dB and that for BER = 10^{-5} is 2.5 dB. These results show that RS code can achieve higher coding gain than BCH code at higher required BER.

Figure 6.15 shows BER performance of 64-ksymbol/s RS-encoded QPSK with an infinite-size interleaver under flat Rayleigh fading conditions, and Table 6.10 shows coding gains for (15, 13, 1) and (15, 11, 2) RS codes for BER = 10^{-3} and 10^{-5}. We can

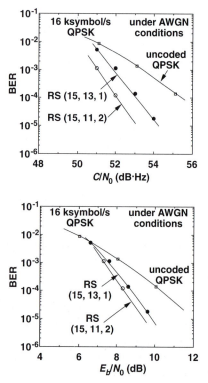

Fig. 6.13 BER versus C/N_0 performance of RS-coded 16-ksymbol/s QPSK system under AWGN conditions (simulated results).

Fig. 6.14 BER versus E_b/N_0 performance of RS-coded 16-ksymbol/s QPSK system under AWGN conditions (simulated results).

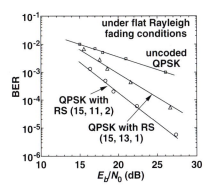

Fig. 6.15 BER performance of RS-encoded QPSK with infinite-size interleaver under flat Rayleigh fading conditions (simulated results).

Table 6.10 Coding gain for RS code with a parameter of required BER.

	Coding Gain	
Required BER	**(15, 13, 1) RS Code**	**(15, 11, 2) RS Code**
BER = 10^{-3}	6.5 dB	9.4 dB
BER = 10^{-5}	14.3 dB	21.8 dB

find from these results that even though we employ (15, 13, 1) RS code in which redundancy is only 2/15, we can obtain high coding gain. When we employ (15, 11, 2) RS code, we can achieve higher coding gain than that for (15, 11, 1) BCH code although its redundancy is the same. For example, (15, 11, 2) RS code can achieve approximately a 3-dB higher coding gain than (15, 11, 1) BCH code at BER = 10^{-5}. Therefore, RS code is considered to be more powerful FEC than the BCH code under both AWGN and flat Rayleigh fading conditions.

6.11 FORWARD ERROR CORRECTION USING CONVOLUTIONAL CODE

6.11.1 Convolutional Encoder

Convolutional code is a very powerful channel-coding scheme for wireless mobile communication systems [6-12]. First, we will discuss the process of the convolutional encoder with a coding rate of 1/2 as an example.

Figure 6.16 shows the configuration of the convolutional encoder with a coding rate of 1/2 with its constraint length (K) of 3 [6-13]. The convolutional encoder consists of shift registers and two modulo-2 adders connected to some of the shift registers. The performance of the convolutional code is determined by two parameters—coding rate (r) and constraint length (K). The input data sequence is fed to the K-stage shift register, and two output data are calculated using the contents of K-stage shift registers. The generator polynomials determine the encoding process. For the convolutional encoder shown in Figure 6.16, two coded bits are produced when one source bit is input. In this case, the generator polynomials determine the encoding rule.

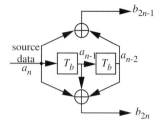

Fig. 6.16 Configuration of the convolutional encoder with a coding rate of 1/2 with $K = 3$.

Let's assume the generator polynomials for $r = 1/2$ code are expressed as

$$G_1(x) = g_{K-1}^1 x^{K-1} + g_{K-2}^1 x^{K-2} + \cdots + g_2^1 x^2 + g_1^1 x + g_0^1 \tag{6.123}$$

$$G_2(x) = g_{K-1}^2 x^{K-1} + g_{K-2}^2 x^{K-2} + \cdots + g_2^2 x^2 + g_1^2 x + g_0^2 \tag{6.124}$$

where x means a 1-bit timing delay, $G_1(x)$ is the generator polynomial for the first encoded bit and $G_2(x)$ is for the second bit, and g_m^n means whether or not an m-bit delayed source bit is added on the modulo-2 basis to obtain n-th encoded bit. For example, in the case of the encoder given by Figure 6.16, the generator polynomials are given by

$$G_1(x) = x^2 + x + 1 \tag{6.125}$$

$$G_2(x) = x^2 + 1 \tag{6.126}$$

When n-th source bit (a_n) is fed to the shift register, these two generator polynomials determine the relationship between the source bit a_n and the 2-bit codeword (b_{2n-1}, b_{2n}) as

$$b_{2n-1} = a_{n-2} \oplus a_{n-1} \oplus a_n \tag{6.127}$$

$$b_{2n} = a_{n-2} \qquad \oplus a_n \tag{6.128}$$

As shown in equations (6.127) and (6.128), the codeword is determined by the previous source data (a_{n-1}, a_{n-2}) and the input data a_n. Now, let's define the state as $\sigma_n = (a_{n-1}, a_{n-2})$. In the case of $K = 3$, each state takes one of the four states—$\sigma^0 = (0, 0)$, $\sigma^1 = (0, 1)$, $\sigma^2 = (1, 0)$, and $\sigma^3 = (1, 1)$. When $(n + 1)$-th source bit (a_{n+1}) is fed to the shift register, all the data in the shift registers are shifted to the right and the state is changed from $\sigma_n = (a_{n-1}, a_{n-2})$ to $\sigma_{n+1} = (a_n, a_{n-1})$. At the same time, a 2-bit codeword (b_{2n+1}, b_{2n+2}) is generated according to equations (6.127) and (6.128).

Table 6.11 summarizes the relationships between the input data, old and new states, and the output codeword.

There are two schematic diagrams that represent the relationship given by Table 6.11—the state diagram and trellis diagram. Figure 6.17 shows the state diagram representation of Table 6.11, where solid lines mean that the input (a_{n+1}) is logical 1 and the dashed lines mean logical 0. Moreover, the 2-bit data written in each line repre-

Table 6.11 State transitions of the encoder.

Old state σ_n	Input a_{n+1}	New State σ_{n+1}	Codeword (b_{2n+1}, b_{2n+2})
00	0	00	00
	1	10	11
01	0	00	11
	1	10	00
10	0	01	10
	1	11	01
11	0	01	01
	1	11	10

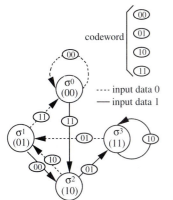

Fig. 6.17 State diagram representation for the state transition given by Table 6.11.

sents the output codeword. Although this diagram is very suitable for understanding the state transition between two arbitrary states, it is not suitable for expressing a state transition sequence for a certain period.

The trellis diagram is very suitable for expressing the state transition sequence. Figure 6.18 shows the trellis diagram that represents relationships given by Table 6.11. In this figure, solid lines mean that the input is logical 1 and dashed lines mean logical 0. Moreover, 2-bit data written near each line indicate the output codeword, and σ_n^j means σ^j for n-th timing.

E X A M P L E   6 – 6

When $G_1(x)$ and $G_2(x)$ are given by equations (6.125) and (6.126), and input data sequence is given by $(1, 0, 1, 1, 0, 0, 0, 1)$

1. Obtain the encoded data sequence.

2. Find the state transition sequence.

3. Draw the state transition sequence using the trellis diagram.

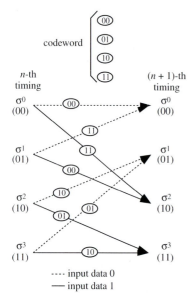

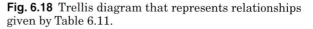

Fig. 6.18 Trellis diagram that represents relationships given by Table 6.11.

Solution

The initial state is assumed to be 00. At each time slot, input, state, and output are given as shown in Table 6.12.

1. This table means that the coded output sequence is given by

$(1, 1, 1, 0, 0, 0, 0, 1, 0, 1, 1, 1, 0, 0, 1, 1)$.

2. This table also shows that the state transition sequence is given by

$(\sigma^0, \sigma^2, \sigma^1, \sigma^2, \sigma^3, \sigma^1, \sigma^0, \sigma^0, \sigma^2)$.

3. Figure 6.19 shows the trellis state diagram for $n = 0$ to 8. The thick lines show the obtained state transition sequence.

When we can calculate a state transition, say, from σ^i to σ^j, we can obtain the corresponding input as shown in Figure 6.18. For example, when the state transition sequence is obtained as $(\sigma^0, \sigma^2, \sigma^1, \sigma^2, \sigma^3, \sigma^1, \sigma^0, \sigma^0, \sigma^2)$, we can detect the source bit sequence as $(1, 0, 1, 1, 0, 0, 0, 1)$.

For the convolutional encoder design, we have to select generator polynomials that maximize the minimum Hamming distance. Table 6.13 shows the optimum generator polynomials for $r = 1/2$ convolutional encoder, and Table 6.14 shows the optimum generator polynomials for $r = 1/3$ [6-13]. For example, when $K = 7$ and $r = 1/3$, the first to third bits of the n-th codeword are given by

$$b_{3n-2} = a_{n-6} \oplus a_{n-5} \oplus a_{n-3} \oplus a_{n-2} \oplus a_n \qquad (6.129)$$

$$b_{3n-1} = a_{n-6} \oplus a_{n-4} \oplus a_{n-1} \oplus a_n \qquad (6.130)$$

$$b_{3n} = a_{n-6} \oplus a_{n-4} \oplus a_{n-3} \oplus a_{n-2} \oplus a_{n-1} \oplus a_n \qquad (6.131)$$

Table 6.12 Relationship between input, state, and output for $r = 1/2$ and $K = 3$ convolutional encode

Time	Input	State	Output
0		00	
1	1	10	11
2	0	01	10
3	1	10	00
4	1	11	01
5	0	01	01
6	0	00	11
7	0	00	00
8	1	10	11

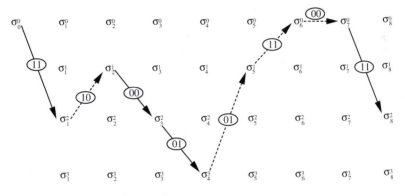

Fig. 6.19 Trellis state diagram of example 6.6.

6.11.2 Maximum Likelihood Sequence Estimation with Hard Decision Decoding

The MLSE estimates the transmitted data sequence $\{\alpha_n\}$ that maximizes the conditional probability $p(\{y_n\} \mid \{\alpha_n\})$. Because the Viterbi algorithm is employed to achieve MLSE in most cases, the MLSE is also called *Viterbi decoding* [6-14]. When we conduct the MLSE using the detected binary data, we call it *hard decision decoding*. On the other hand, when we conduct it using demodulated baseband signals, we call it *soft decision decoding*.

First, let's discuss the hard decision decoding algorithm for the $r = 1/2$ and $K = 3$ convolutional encoder given by Figure 6.16. As shown in Figure 6.18, there are eight possible state transitions. In the case of hard decision decoding, the branch metric for a state transition from σ_n^i to σ_{n+1}^j is defined by the Hamming distance between the codeword produced by the state transition β_{n+1} and the received codeword $\hat{b}_{n+1}$ as

$$BR(\sigma_n^i, \ \sigma_{n+1}^j) = d_H(\hat{\boldsymbol{b}}_n, \ \boldsymbol{\beta}_n)$$
$$= (b_{2n+1} \oplus \beta_{2n+1}) + (b_{2n+2} \oplus \beta_{2n+2}) \tag{6.132}$$

where

$$\hat{\boldsymbol{b}}_{n+1} = (\hat{b}_{2n+1}, \ \hat{b}_{2n+2}) \tag{6.133}$$

$$\boldsymbol{\beta}_{n+1} = (\beta_{2n+1}, \ \beta_{2n+2}) \tag{6.134}$$

In the case of hard decision decoding, $\{\alpha_n\}$ that maximizes $p(\{y_n\} \mid \{\alpha_n\})$ corresponds to a sequence having a minimum accumulated branch metric. This accumulated value is called the *metric*.

Let us define the metric for σ_n^i as $J_n(\sigma_n^i)$. When we observe one of the states for the $(n + 1)$-th time slot in Figure 6.18, we find that each state has two incoming paths. Between them, only one path with a smaller metric (accumulated Hamming distance) will finally survive. Therefore, $J_{n+1}(\sigma_{n+1}^i)$ is given by

$$J_{n+1}(\sigma_{n+1}^j) = \min_{\sigma_n^i \to \sigma_{n+1}^j} \left\{ J_n(\sigma_n^i) + BR(\sigma_n^i, \ \sigma_{n+1}^j) \right\} \tag{6.135}$$

At each stage, equation (6.135) is calculated to obtain the metric for each state. At the same time, the path history that represents the information bit sequence of the survivor path for each state is also updated as

$$H_{n+1}\left(\sigma_{n+1}^i\right) = \left[H_n\left(\sigma_n^j\right), \ \alpha_{n+1}\right] \tag{6.136}$$

where $H_n(\sigma_n^i) = (\alpha_1, \alpha_2, ..., \alpha_n)$ represents the data sequence of the survival path for the state σ_n^i. When such a metric and path history are recursively calculated, we can obtain the optimum data sequence based on MLSE.

Let the source data sequence be $(1, 0, 1, 1, 0, 0)$, where the last 2 bits are tail bits. When this sequence is encoded using the $K = 3$ and $r = 1/2$ convolutional encoder, the corresponding coded sequence is given by $(1, 1, 1, 0, 0, 0, 1, 0, 0, 1, 1, 1)$. Let us assume that the third bit is received in error. In this case, the received data sequence is given by $(1, 1, 0, 0, 0, 0, 1, 0, 0, 1, 1, 1)$.

Figure 6.20 shows the first three trellis levels, where the initial state is assumed to be σ_0^0. As shown in this figure, the state is changed to σ_1^0 when the input data is logical 0, and it produces the codeword of 00. On the other hand, it is changed to σ_1^2 when the input is logical 1, and it produces the output codeword of 11. Because the received codeword is 11, the branch metric for each state transition is given by $BR(\sigma_0^0, \sigma_1^0) = 2$ and $BR(\sigma_0^0, \sigma_1^2) = 0$. Therefore, the metric for each state is given by $J(\sigma_1^0) = 2$ and $J(\sigma_1^2) = 0$, and the corresponding path histories are given by $H_1(\sigma_1^0) = (0)$ and $H_1(\sigma_1^2) = (1)$.

At the second trellis level, there are four possible state transitions—$(\sigma_1^0$ to $\sigma_2^0)$, $(\sigma_1^0$ to $\sigma_2^2)$, $(\sigma_1^2$ to $\sigma_2^1)$, and $(\sigma_1^2$ to $\sigma_2^3)$. Because the received codeword is 00, the branch metric for each state transition is given by $BR(\sigma_1^0$ to $\sigma_2^0) = 0$, $BR(\sigma_1^0$ to $\sigma_2^2) = 2$, $BR(\sigma_1^2$ to $\sigma_2^1) = 1$,

Table 6.13 Optimum generator polynominals for $r = 1/2$.

K	$G_1(x)$	$G_2(x)$
3	$x^2 + 1$	$x^2 + x + 1$
4	$x^3 + x + 1$	$x^3 + x^2 + x + 1$
5	$x^4 + x^3 + 1$	$x^4 + x^2 + x + 1$
6	$x^5 + x^4 + x^2 + 1$	$x^5 + x^3 + x^2 + x + 1$
7	$x^6 + x^5 + x^3 + x^2 + 1$	$x^6 + x^3 + x^2 + x + 1$
8	$x^7 + x^6 + x^5 + x^2 + 1$	$x^7 + x^4 + x^3 + x^2 + x + 1$
9	$x^8 + x^4 + x^3 + x^2 + 1$	$x^8 + x^7 + x^5 + x^3 + x^2 + x + 1$
10	$x^9 + x^8 + x^7 + x^5 + x^4 + x^3 + 1$	$x^9 + x^7 + x^4 + x^3 + x + 1$
11	$x^{10} + x^8 + x^7 + x^6 + x^4 + x^3 + 1$	$x^{10} + x^6 + x^5 + x^3 + x^2 + x + 1$
12	$x^{11} + x^9 + x^8 + x^7 + x^5 + x^4 + 1$	$x^{11} + x^{10} + x^7 + x^5 + x^4 + x^3 + x^2 + 1$
13	$x^{12} + x^{11} + x^9 + x^8 + x^6 + x^4 + 1$	$x^{12} + x^8 + x^7 + x^5 + x^4 + x^3 + x^2 + x + 1$
14	$x^{13} + x^{11} + x^{10} + x^9 + x^8 + x^6 + x^5 + x^4 + 1$	$x^{13} + x^{12} + x^9 + x^7 + x^4 + x^3 + x^2 + 1$

Table 6.14 Optimum generator polynominals for $r = 1/3$.

K	$G_1(x)$	$G_2(x)$	$G_3(x)$
3	$x^2 + 1$	$x^2 + x + 1$	$x^2 + x + 1$
4	$x^3 + x^2 + 1$	$x^3 + x + 1$	$x^3 + x^2 + x + 1$
5	$x^4 + x^2 + 1$	$x^4 + x^3 + x + 1$	$x^4 + x^3 + x^2 + x + 1$
6	$x^5 + x^4 + x^3 + 1$	$x^5 + x^4 + x^2 + 1$	$x^5 + x^3 + x^2 + x + 1$
7	$x^6 + x^5 + x^3 + x^2 + 1$	$x^6 + x^4 + x + 1$	$x^6 + x^4 + x^3 + x^2 + x + 1$
8	$x^7 + x^5 + x^3 + 1$	$x^7 + x^4 + x^3 + x + 1$	$x^7 + x^4 + x^3 + x^2 + x + 1$
9	$x^8 + x^7 + x^6 + x^5 + x^3 + x^2 + 1$	$x^8 + x^7 + x^4 + x^3 + x + 1$	$x^8 + x^5 + x^2 + x + 1$
10	$x^9 + x^8 + x^7 + x^6 + x^3 + 1$	$x^9 + x^7 + x^5 + x^4 + x^3 + x^2 + 1$	$x^9 + x^8 + x^6 + x^5 + x^2 + x + 1$
11	$x^{10} + x^8 + x^7 + x^5 + x^4 + x^3 + 1$	$x^{10} + x^7 + x^6 + x^5 + x^3 + x^2 + 1$	$x^{10} + x^8 + x^7 + x^6 + x^5 + x^4 + x^3 + x + 1$
12	$x^{11} + x^{10} + x^9 + x^7 + x^6 + x^5 + x^4 + x^3 + 1$	$x^{11} + x^{10} + x^7 + x^5 + x^4 + x^3 + x^2 + 1$	$x^{11} + x^9 + x^7 + x^6 + x^4 + x + 1$
13	$x^{12} + x^{11} + x^9 + x^8 + x^6 + x^4 + 1$	$x^{12} + x^{10} + x^{10} + x^9 + x^8 + x^7 + x^5 + x^4 + 1$	$x^{12} + x^8 + x^7 + x^5 + x^4 + x^3 + x^2 + x + 1$
14	$x^{13} + x^{11} + x^{10} + x^9 + x^8 + x^6 + x^5 + x^4 + 1$	$x^{13} + x^9 + x^8 + x^6 + x^5 + x^4 + x^2 + x + 1$	$x^{13} + x^{12} + x^{10} + x^9 + x^7 + x^4 + x^3 + x^2 + x + 1$

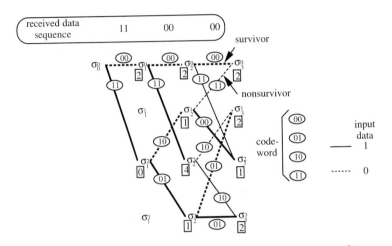

Fig. 6.20 First three trellis levels, where the initial state is assumed to be σ_0^0.

and $BR(\sigma_1^2$ to $\sigma_2^3) = 1$. Therefore, the corresponding metrics and path histories are given by

$$J(\sigma_2^0) = J(\sigma_1^0) + BR(\sigma_1^0, \ \sigma_2^0) = 2 + 0 = 2 \tag{6.137}$$

$$J(\sigma_2^1) = J(\sigma_1^2) + BR(\sigma_1^2, \ \sigma_2^1) = 0 + 1 = 1 \tag{6.138}$$

$$J(\sigma_2^2) = J(\sigma_1^0) + BR(\sigma_1^0, \ \sigma_2^2) = 2 + 2 = 4 \tag{6.139}$$

$$J(\sigma_2^3) = J(\sigma_1^2) + BR(\sigma_1^2, \ \sigma_2^3) = 0 + 1 = 1 \tag{6.140}$$

and $H_2(\sigma_2^0) = (0,0)$, $H_2(\sigma_2^1) = (1,0)$, $H_2(\sigma_2^2) = (0,1)$, $H_2(\sigma_2^3) = (1, 1)$.

At the third level, there are two incoming paths at each state as shown in Figure 6.20. Among them, only a path having a smaller metric will finally survive. Therefore, we will compare metrics for two incoming paths at each state and select one path having the smaller metric. When the incoming metrics have the same value, we can select either one. In this manner, we can obtain the metric and path history for each state of the third stage as follows:

$J_3(\sigma_3^0) = 2, J_3(\sigma_3^1) = 2, J_3(\sigma_3^2) = 1, J_3(\sigma_3^3) = 2$
$H_3(\sigma_3^0) = (0, 0, 0), H_3(\sigma_3^1) = (1, 1, 0), H_3(\sigma_3^2) = (1, 0, 1), H_3(\sigma_3^3) = (1, 1, 1)$.

In the same manner, we can obtain the metric and path history for each state at the fourth trellis level as follows:

$J_4(\sigma_4^0) = 3, J_4(\sigma_4^1) = 1, J_4(\sigma_4^2) = 3, J_4(\sigma_4^3) = 1$
$H_4(\sigma_4^0) = (0, 0, 0, 0), H_4(\sigma_4^1) = (1, 0 , 1, 0), H_4(\sigma_4^2) = (1, 1, 0, 1), H_4(\sigma_4^3) = (1, 0, 1, 1)$

At the fifth level, we already know that the input is logical 0 because it is a tail bit. Therefore we have to calculate the metric and path histories for only the states of σ_5^0 and σ_5^1 because all the paths relating to the input of logical 0 will reach to these two states. The obtained results are as follows:

$$J_5(\sigma_5^0) = 2, J_5(\sigma_5^1) = 1$$

$$H_5(\sigma_5^0) = (1, 0, 1, 0, 0), H_5(\sigma_5^1) = (1, 0, 1, 1, 0)$$

At the sixth level, we know that this input is also logical 0. When the input is logical 0, both states σ_5^0 and σ_5^1 will reach to σ_6^0. The metric and path history for σ_6^0 are given by $J_6(\sigma_6^0) = 1$ and $H_6(\sigma_6^0) = (1, 0, 1, 1, 0, 0)$. As a result, we can obtain the transmitted data sequence by $H_6(\sigma_6^0)$.

6.11.3 Maximum Likelihood Sequence Estimation with Soft Decision Decoding

In the previous section, we discussed hard decision decoding in which the hard-limited signal is used for the metric calculation. Now, let's extend our discussion to soft decision decoding. The difference between hard decision and soft decision is whether or not some analog signal information is used in the metric calculation [6-10].

Let's remember the definition of the likelihood sequence estimation again. The definition is to estimate a data sequence $\{\alpha_n\}$ that maximizes the conditional probability density function $p(\{y_n\} | \{\alpha_n\})$. When we transmit a BPSK signal over an AWGN channel, the received signal is coherently detected by using an ideal carrier reference, and the observation period is assumed to be $1 \le n \le N$; the probability density function is given by

$$p\big(\{y_n\}|\{\alpha_n\}\big) = \left(\frac{1}{\sqrt{2\pi\sigma_n^2}}\right)^N \prod_{n=1}^{N} \exp\left(-\frac{\big(y_n - S(\alpha_n)\big)^2}{2\sigma_n^2}\right) \tag{6.141}$$

where σ_n^2 is the noise power and $S(\alpha_n)$ is the complex baseband signal corresponding to the information bit α_n. Because this function is monotonously increasing with $-(y_n - S(\alpha_n))^2$, we usually use the logarithmic likelihood function given by

$$\log\big(p(\{y_n\}|\{\alpha_n\})\big) = -\frac{1}{2P_n} \sum_{n=1}^{N} \big(y_n - S(\alpha_n)\big)^2 - \frac{N}{2} \log\big(2\pi\sigma_n^2\big) \tag{6.142}$$

instead of the likelihood function given by equation (6.141). Because we can remove a constant value or a constant factor in the logarithmic likelihood function without loss of generality, we usually use the following equation as a branch metric:

$$BR(\sigma_n^i, \ \sigma_{n+1}^j) = \sum_{l=1}^{Ra} \big(y_{Ra(n-1)+l} - S(\alpha_{Ra(n-1)+l})\big)^2 \tag{6.143}$$

where Ra is the number of bits in a coded symbol. When we employ equation (6.143) as a branch metric, the path history with the minimum metric gives the optimum data sequence.

6.11.4 Decoding for BPSK with Absolute Phase Coherent Detection

In the case of convolutionally encoded BPSK with $r = 1/2$ and $K = 3$, when the received signal sequence is $(y_1, y_2, y_3, y_4, y_5, y_6, y_7, y_8) = (1.2, 0.6, -0.3, 0.1, \cdots)$, where the initial state is σ_0^0, the metric calculation is carried out this way.

When the input data is logical 0, the state is changed to σ_1^0, producing the code-word of 00 . Because $S(0) = -1$ in the case of BPSK, the corresponding baseband signal is $(-1, -1)$. On the other hand, when the input data is logical 1, the state is changed to σ_1^2 producing the output codeword of 11. This codeword corresponds to the baseband signal of $(1, 1)$ because $S(1) = 1$.

The received signal corresponding to the first codeword is $(1.2, 0.6)$. Therefore, the metric for each state transition is given by

$$J(\sigma_1^0) = BR(\sigma_0^0, \ \sigma_1^0) = (1.2 - (-1))^2 + (0.6 - (-1))^2 = 7.4 \tag{6.144}$$

$$J(\sigma_1^2) = BR(\sigma_0^0, \ \sigma_1^2) = (1.2 - 1)^2 + (0.6 - 1)^2 = 0.2 \tag{6.145}$$

In the same manner, the metric for each state at the second codeword is given by

$$J(\sigma_2^0) = J(\sigma_1^0) + BR(\sigma_1^0, \ \sigma_2^0) = 7.4 + (-0.3 + 1)^2 + (0.1 + 1)^2 = 9.1 \tag{6.146}$$

$$J(\sigma_2^1) = J(\sigma_1^2) + BR(\sigma_1^2, \ \sigma_2^1) = 0.2 + (-0.3 + 1)^2 + (0.1 + 1)^2 = 1.9 \tag{6.147}$$

$$J(\sigma_2^2) = J(\sigma_1^0) + BR(\sigma_1^0, \ \sigma_2^2) = 7.4 + (-0.3 - 1)^2 + (0.1 - 1)^2 = 9.9 \tag{6.148}$$

$$J(\sigma_2^3) = J(\sigma_1^2) + BR(\sigma_1^2, \ \sigma_2^3) = 0.2 + (-0.3 - 1)^2 + (0.1 - 1)^2 = 2.7 \tag{6.149}$$

In the preceding discussions, we have assumed AWGN channels. In practice, however, the channel is a fading channel rather than an AWGN channel. In such channel conditions, the branch metric has to be weighted by a factor representing its reliability. Thus, the following branch metric is used instead of equation (6.143).

$$BR(\sigma_n^i, \ \sigma_{n+1}^j) = \sum_{l=1}^{Ra} \left(y_{Ra(n-1)+l} - w_{Ra(n-1)+l}S(\alpha_{Ra(n-1)+l})\right)^2 \tag{6.150}$$

where w_i is a weighting factor. In the case of flat Rayleigh fading, w_i is given by the envelope level of i-th received signal $(w_i = r_i)$. In this equation, the AWGN channel condition can be expressed as $w_i = 1$ for any i. When we include the envelope level in the branch metric calculation, we have to add an envelope level detector before the demodulator because we usually remove envelope fluctuation by using a hard-limiter or an AGC amplifier before demodulation to keep a sufficient dynamic range at the demodulator. The pilot channel multiplexing schemes discussed in chapter 5 are methods of detecting the envelope level.

Figure 6.21 shows BER performance of convolutionally encoded QPSK with both hard decision Viterbi decoding and soft decision Viterbi decoding, where $r = 1/2, K = 7$, and Gray mapping is employed for QPSK modulation. As shown in this figure, soft decision decoding can achieve about 2 dB higher coding gain than hard decision decoding.

6.11.5 Branch Metric for BPSK with Differential Detection

When differential detection is employed as a demodulation scheme, the branch metric calculation is given by

$$BR(\sigma_n^i, \ \sigma_{n+1}^j) = \sum_{l=1}^{Ra} \left(z_{Ra(n-1)+l} - w_{Ra(n-1)+l} S(\alpha_{Ra(n-1)+l}) \right)^2 \tag{6.151}$$

where

$$z_i = y_{i-1}^* y_i \tag{6.152}$$

$$w_i = r_{i-1} \cdot r_i \tag{6.153}$$

6.11.6 Branch Metric for QPSK with Coherent Detection

In the case of QPSK, a serial source data sequence is converted to 2-bit parallel data, and each of them are fed to I-ch and Q-ch of the QPSK modulator. When I-ch and Q-ch of the received baseband signal at $t = nT_s$ are expressed as x_{In} and x_{Qn}, when the envelope level at $t = nT_s$ is r_n, and when Gray encoding with absolute phase coherent detection is employed, the input signal sequence for the Viterbi decoder is given by parallel-to-serial conversion of the received complex baseband signal as

$$y_{2n-1} = x_{In} \tag{6.154}$$

$$y_{2n} = x_{Qn} \tag{6.155}$$

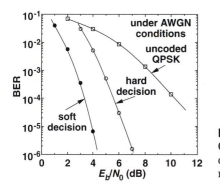

Fig. 6.21 BER performance of convolutionally encoded QPSK with both hard decision Viterbi decoding and soft decision Viterbi decoding, where r = 1/2, K = 7, and Gray mapping is employed for QPSK modulation.

where a symbol duration of sequence $\{y_n\}$ is $T_s/2$ and the corresponding envelope for the weight in the soft decision is given by

$$w_{2n-1} = w_{2n} = r_n \tag{6.156}$$

Using these data, the branch metric is given by

$$BR(\sigma_n^i, \; \sigma_{n+1}^j) = \sum_{l=1}^{Ra} \left(y_{Ra(n-1)+l} - w_{Ra(n-1)+l}S(\alpha_{Ra(n-1)+l})\right)^2 \tag{6.157}$$

6.11.7 Branch Metric for QPSK with Differential Detection

When baseband signal mapping for differential encoding with differential detection is given by Table 3.2, the input signal sequence of the Viterbi decoder is given by

$$z_{2n-1} = \mathrm{Im}\left[y_{i-1}^* y_i \exp(j\pi/4)\right] \tag{6.158}$$

$$z_{2n} = \mathrm{Re}\left[y_{i-1}^* y_i \exp(j\pi/4)\right] \tag{6.159}$$

and the branch metric is given by

$$BR(\sigma_n^i, \; \sigma_{n+1}^j) = \sum_{l=1}^{Ra} \left(z_{Ra(n-1)+l} - w_{Ra(n-1)+l}S(\alpha_{Ra(n-1)+l})\right)^2 \tag{6.160}$$

where

$$S(1) = -1 \tag{6.161}$$

$$S(0) = 1 \tag{6.162}$$

$$w_{2i-1} = w_{2i} = r_{i-1} \cdot r_i \tag{6.163}$$

6.11.8 Soft Decision Viterbi Decoding for *M*-ary QAM

In the case of 2^m-ary QAM, a symbol includes m bits of data. On the other hand, a convolutional encoder with $r = 1/R_a$ includes R_a bits in a codeword.

The branch metric for soft decision is determined by the Euclidean distance between the received baseband signal and the candidates of the transmitted symbols. When the number of bits in both a codeword and the number of bits in a QAM symbol are the same, we can easily obtain the branch metric. However, it is not always satisfied. For example, when $r = 1/3$ convolutionally encoded data is transmitted by using 16QAM, although the bit sequence of the codeword is blocked every 3 bits, every 4 bits of this sequence is mapped onto the 16QAM baseband signal. Therefore, how to obtain the QAM symbol independent branch metric is a problem of combining 2^m-ary QAM and convolutional encoding [6-15].

Figure 6.22 shows a constellation of 16QAM with Gray code. Let us assume that (a_1, a_2, a_3, a_4) represents 4 bits of data included in each symbol. When $a_1 = 0$, the transmitted symbol is $S_1, S_2, S_3, S_4, S_5, S_6, S_7$, or S_8. On the other hand, when $a_1 = 1$, the transmitted symbol is $S_9, S_{10}, S_{11}, S_{12}, S_{13}, S_{14}, S_{15}$, or S_{16}. In the following, we will call a symbol candidate group for $a_1 = 0$ as G_{10} and that for $a_1 = 1$ as G_{11} as shown in Table 6.15. Similarly, we can define G_{i0} for $a_i = 0$ and G_{i1} for $a_i = 1$ as shown in Tables 6.16, 6.17, and 6.18.

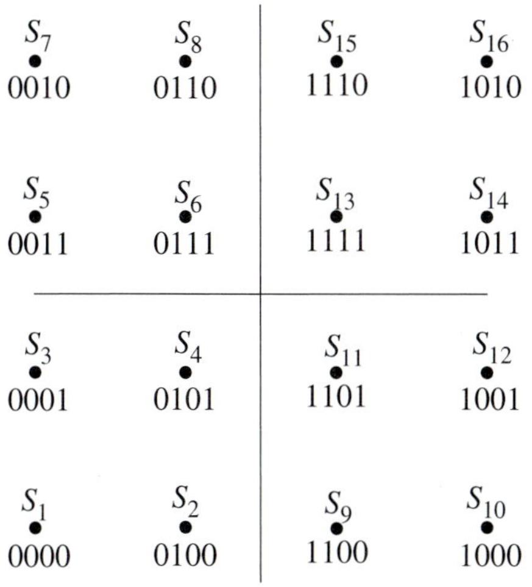

S_7	S_8	S_{15}	S_{16}
•	•	•	•
0010	0110	1110	1010

S_5	S_6	S_{13}	S_{14}
•	•	•	•
0011	0111	1111	1011

S_3	S_4	S_{11}	S_{12}
•	•	•	•
0001	0101	1101	1001

S_1	S_2	S_9	S_{10}
•	•	•	•
0000	0100	1100	1000

Fig. 6.22 Constellation of 16QAM with Gray code.

Table 6.15 QAM symbol grouping with respect to a_1.

G_{10}	S_1	S_2	S_3	S_4	S_5	S_6	S_7	S_8
G_{11}	S_9	S_{10}	S_{11}	S_{12}	S_{13}	S_{14}	S_{15}	S_{16}

Table 6.16 QAM symbol grouping with respect to a_2.

G_{20}	S_1	S_3	S_5	S_7	S_{10}	S_{12}	S_{14}	S_{16}
G_{21}	S_2	S_4	S_6	S_8	S_9	S_{11}	S_{13}	S_{15}

Table 6.17 QAM symbol grouping with respect to a_3.

G_{30}	S_1	S_2	S_3	S_4	S_9	S_{10}	S_{11}	S_{12}
G_{31}	S_5	S_6	S_7	S_8	S_{13}	S_{14}	S_{15}	S_{16}

Table 6.18 QAM symbol grouping with respect to a_4.

G_{40}	S_1	S_2	S_7	S_8	S_9	S_{10}	S_{15}	S_{16}
G_{41}	S_3	S_4	S_5	S_6	S_{11}	S_{12}	S_{13}	S_{14}

When we define 4 bits of data at $t = nT_s$ as $\boldsymbol{a}_n = (a_{n1}, a_{n2}, a_{n3}, a_{n4})$, the optimum received sequence $\{\boldsymbol{a}_n\}$ is the one that maximizes the following conditional probability:

$$p(y_n | \boldsymbol{a}_n) = \prod_{i=1}^{N} p(y_i | \boldsymbol{a}_i) \tag{6.164}$$

where

$$p(y_i | \boldsymbol{a}_i) = \prod_{l=1}^{4} p(y_i | a_{il}) \tag{6.165}$$

and $p(y_i | a_{il})$ corresponds to the p.d.f. of l-th bit in i-th symbol. Therefore, when we calculate $p(y_i | a_{il} = 0)$ and $p(y_i | a_{il} = 1)$, we get the branch metric for l-th bit in i-th symbol.

When $a_{i1} = 0$ is transmitted, the transmitted symbol belongs to G_{10} in Table 6.15. Therefore, $p(y_i | a_{i1} = 0)$ is given by

$$\begin{aligned}
p(y_i | a_{i0} = 0) &= \frac{1}{8} \big\{ p(y_i | S_1) + p(y_i | S_2) + p(y_i | S_3) + p(y_i | S_4) \\
&\quad + p(y_i | S_5) + p(y_i | S_6) + p(y_i | S_7) + p(y_i | S_8) \big\} \\
&= \frac{1}{8} \frac{1}{2\pi\sigma_n^2} \sum_{k=1}^{8} \exp\left(-\frac{\left(\mathrm{Re}[y_i] - \mathrm{Re}[S_k]\right)^2 + \left(\mathrm{Im}[y_i] - \mathrm{Im}[S_k]\right)^2}{2\sigma_n^2} \right)
\end{aligned} \tag{6.166}$$

where σ_n^2 is the noise power, and where the selection probability of each symbol is assumed to be equal.

Practically, $p(y_i | a_{i0} = 0)$ is dominated by the Euclidean distance between the received symbol and the closest symbol candidate. Therefore, $p(y_i | a_{i0} = 0)$ can be approximated by

$$\begin{aligned}
&p(y_i | a_{i1} = 0) \\
&\cong \underset{S_k \in G_{10}}{\mathrm{Max}} \left[\frac{1}{8} \frac{1}{2\pi\sigma_n^2} \exp\left(-\frac{\left(\mathrm{Re}[y_i] - \mathrm{Re}[S_k]\right)^2 + \left(\mathrm{Im}[y_i] - \mathrm{Im}[S_k]\right)^2}{2\sigma_n^2} \right) \right]
\end{aligned} \tag{6.167}$$

In the same manner, $p(y_i | a_{il} = 0)$ and $p(y_i | a_{il} = 1)$ can be calculated as follows:

$$\begin{aligned}
&p(y_i | a_{il} = 0) \\
&\cong \underset{S_k \in G_{l0}}{\mathrm{Max}} \left[\frac{1}{8} \frac{1}{2\pi\sigma_n^2} \exp\left(-\frac{\left(\mathrm{Re}[y_i] - \mathrm{Re}[S_k]\right)^2 + \left(\mathrm{Im}[y_i] - \mathrm{Im}[S_k]\right)^2}{2\sigma_n^2} \right) \right]
\end{aligned} \tag{6.168}$$

$$p(y_i|a_{il}=1)$$

$$\cong \underset{S_k \in G_{l1}}{\text{Max}} \left[\frac{1}{8} \frac{1}{2\pi\sigma_n^2} \exp\left(-\frac{\left(\text{Re}[y_i]-\text{Re}[S_k]\right)^2 + \left(\text{Im}[y_i]-\text{Im}[S_k]\right)^2}{2\sigma_n^2} \right) \right] \qquad (6.169)$$

Therefore, the branch metric of l-th bit in i-th symbol both for $a_{il} = 0$ and for $a_{il} = 1$ [6-15] are given by

$$BR(a_{il}=0) = \underset{S_k \in G_{l0}}{\text{Min}} \left[\left(\text{Re}[y_i]-\text{Re}[S_k]\right)^2 + \left(\text{Im}[y_i]-\text{Im}[S_k]\right)^2 \right] \qquad (6.170)$$

$$BR(a_{il}=1) = \underset{S_k \in G_{l1}}{\text{Min}} \left[\left(\text{Re}[y_i]-\text{Re}[S_k]\right)^2 + \left(\text{Im}[y_i]-\text{Im}[S_k]\right)^2 \right] \qquad (6.171)$$

Figure 6.23 shows BER performance of convolutionally encoded QPSK, 16QAM, 64QAM, and 256QAM with (a) hard decision Viterbi decoding and (b) soft decision Viterbi decoding, where $r = 1/2$ and $K = 7$ are employed in the convolutional encoder. As we can see from this figure, the above-mentioned soft decision can achieve higher coding gain in comparison with the hard decision decoding, and the improvement is more remarkable with an increasing modulation level.

6.11.9 BER Performance of Convolutional Code under Flat Rayleigh Fading Conditions

Interleaving is a necessary technique for obtaining sufficient coding gain even for the convolutional code. As we have discussed before, an infinite interleaver gives

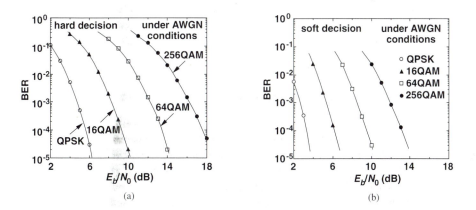

Fig. 6.23 BER performance of convolutionally encoded QPSK, 16QAM, 64QAM, and 256QAM with (a) hard decision Viterbi decoding and (b) soft decision Viterbi decoding, where $r = 1/2$ and $K = 7$ are employed in the convolutional encoder.

the lower limit of the BER. Therefore, we will discuss the BER performance of a convolutionally encoded QPSK system with an infinite-size interleaver.

Figure 6.24 shows the BER performance of convolutionally encoded QPSK with soft decision Viterbi decoding under flat Rayleigh fading conditions, where $r = 1/2$ and $K = 7$ are employed, and the interleaver size is assumed to be infinite. When we compare this performance with that for BCH shown in Figure 6.12 and that for RS shown in Figure 6.15, we find that convolutional encoding can obtain higher coding gain than BCH and RS encoding. The main cause of this coding gain difference is owing to the soft decision decoding for the convolutional encoding. Although soft decision decoding requires more computation, it is not a problem at present due to the recent development of DSP and LSI technologies. Therefore, convolutional encoding with Viterbi decoding is applied to various wireless communication systems, such as GSM and IS-95.

6.11.10 Punctured Code

Convolutional code with its coding rate of k/M and constraint length of K can be implemented by the following process [6-13]:

☞ k-bit information sequence is inserted into K-stage shift register every kT_b second, where T_b is a bit duration.

☞ M-bit of encoded symbol is obtained every kT_b second by connecting modulo-2 adders and shift register stages determined by the generator polynomials.

Because the state transition occurs when k-bit of information is fed to the shift register, there exists k incoming path at each state. This means that the number of calculations for Viterbi decoding exponentially increases with k.

Punctured code is a very effective means of simplifying the decoding process for rate $(n - 1)/n$ convolutional encoding with Viterbi decoding [6-16; 6-17; 6-18]. Figure 6.25 shows a configuration of the punctured code encoder. The information bit sequence is encoded by a rate $1/M$ convolutional encoder, and it is blocked every $M(n - 1)$-bit. In Figure 6.25, the relationship between information bits and coded bits is also shown, where a_k is the information bit, and b_{ki} is the i-th encoded bits at $t = kT_b$. Next,

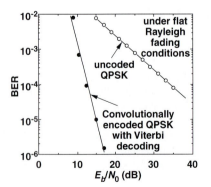

Fig. 6.24 BER performance of convolutionally encoded QPSK with soft decision Viterbi decoding under flat Rayleigh fading conditions, where $r = 1/2$ and $K = 7$ are employed and interleaver size is assumed to be infinite.

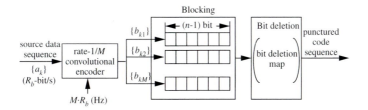

Fig. 6.25 Configuration of the punctured code encoder.

$M(n-1)$–n bits in the $M(n-1)$-bit in the blocked bits are periodically deleted to make n-bit of the punctured code. Table 6.19 shows a bit deletion map for punctured coding using a $r = 1/2$ and $K = 7$ convolutional encoder, where 1 means the transmitted bit and 0 means the deleted bit [6-17].

Table 6.19 Bit deletion map for punctured coding using a rate 1/2 convolutional encoder.

Coding Rate	Block Size (n)	Optimum Bit Deletion Map; 1 = Transmitted Bit; 0 = Deleted Bit	Minimum Free Distance
1/2	2	1 1	10
2/3	3	1 1 1 0	6
3/4	4	1 1 0 1 0 1	5
4/5	5	1 1 1 1 1 0 0 0	4
5/6	6	1 1 0 1 0 1 0 1 0 1	4
6/7	7	1 1 1 0 1 0 1 0 0 1 0 1	3
7/8	8	1 1 1 1 0 1 0 1 0 0 0 1 0 1	3

E X A M P L E   6 – 7

When the information bit sequence is given by (1, 0, 1, 1, 0, 0, 0, 1), and the rate 4/5 punctured convolutional encoder is employed, obtain the transmitted data sequence, where generator polynomials of the original convolutional encoder are given by $G_1(x) = x^2 + x + 1$ and $G_2(x) = x^2 + 1$.

Solution
First, the transmitted bit sequence after the rate 1/2 convolutional encoder is given as shown in Table 6.20.

From Table 6.19, a rate 4/5 corresponds to the case of $n = 5$. Therefore, the coded bit sequence are blocked every 8 bits. As a result, data in Table 6.20 are divided into two groups, $k = 1\text{--}4$ and $k = 5\text{--}8$. Table 6.21 shows the obtained coded data sequence; X means the punctured bit.

Therefore, the transmitted bit sequence is given by $(1, 1, 1, 0, 0, 0, 1, 1, 0, 1)$.

Table 6.20 The obtained coded data sequence including bit deletion.

k	1	2	3	4	5	6	7	8
b_{k1}	1	1	0	0	0	1	0	1
b_{k2}	1	X	X	X	1	X	X	X

Table 6.21 The obtained coded data sequence.

k	1	2	3	4	5	6	7	8
b_{k1}	1	1	0	0	0	1	0	1
b_{k2}	1	0	0	1	1	1	0	1

In the decoding procedure, metric calculation is achieved based on either hard or soft decision Viterbi decoding similar to that for the original rate of $1/M$ decoding after inserting dummy bits into positions of the deleted bits. As a result, the dummy-inserted sequence is given by $(1, 1, 1, X, 0, X, 0, X, 0, 1, 1, X, 0, X, 1, X)$. In the metric calculation, the branch metric for the punctured bit is set to 0 because whether the punctured bit is logical 0 or logical 1 is equiprobable.

Figure 6.26 shows BER performance of convolutionally encoded QPSK with $r = 1/2$ and $K = 7$ punctured code under AWGN conditions. When we employ a lower coding rate, we can obtain a higher coding gain. On the other hand, when we employ a higher rate code, although BER performance is not good at the higher BER region, we can obtain sufficient coding gain at low BER regions.

Figure 6.27 shows BER performance of convolutionally encoded QPSK with $r = 1/2$ and $K = 7$ punctured code under flat Rayleigh fading conditions on the assumption

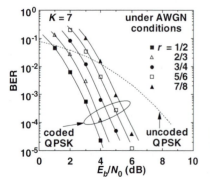

Fig. 6.26 BER performance of convolutionally encoded QPSK with punctured code under AWGN conditions, where $r = 1/2$ and $K = 7$ are employed in the convolutional encoder.

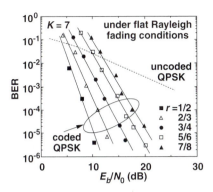

Fig. 6.27 BER performance of convolutionally encoded QPSK with punctured code under flat Rayleigh fading conditions, where r = 1/2 and K = 7 are employed in the convolutional encoder, and an infinite size of bit interleaver is assumed.

that interleaver size is infinite so that the burst error is perfectly randomized. Although the BER performance is getting degraded with an increasing coding rate, $r = 7/8$ still gives better BER performance at BER $< 10^{-2}$. On the other hand, in the higher region of BER $> 10^{-2}$, because we cannot obtain a coding gain if we employ a higher coding rate, we have to employ a lower coding gain. Therefore, we have to carefully select the coding rate considering the required BER as well as the acceptable bandwidth.

6.12 AUTOMATIC REPEAT REQUEST

6.12.1 Configuration of ARQ System

Figure 6.28 shows a configuration of a two-way radio system using ARQ [6-19]. The transmitted message is first stored in the data buffer and divided into a certain size of data blocks. After each block is channel encoded and frame formatted, it is fed to the modulator and then transmitted. At the receiver, after the received signal is demodulated and decoded, the received data is evaluated, whether or not the message is correctly received. When the message is correctly received, ACK is returned via the return link. On the other hand, when the message includes errors, NACK is returned.

For message transmission, there are mainly five protocols.

☞ Stop-and-wait (SAW) ARQ

☞ Go-back-N (GBN) ARQ

☞ Selective-repeat (SR) ARQ

☞ Type I hybrid ARQ

☞ Type II hybrid ARQ

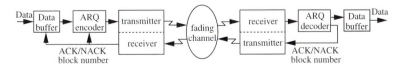

Fig. 6.28 Configuration of a two-way radio system using ARQ.

We will discuss features of each protocol.

6.12.2 SAW ARQ Protocol

Figure 6.29 shows the SAW ARQ protocol. In this protocol, n-th data block is transmitted just after ACK for the $(n-1)$-th data block is received. When NACK for $(n-1)$-th block is returned, the transmitter sends the $(n-1)$-th block again. In this figure, the second and fourth blocks are in error.

Performance of the ARQ is determined by the error miss-detection probability. Let's assume that P_c is the probability that no error is included in the received signal, P_d is the probability that the error detection code can detect errors in a block, and P_e is the probability that the error detection code cannot detect errors. These probabilities satisfy

$$P_c + P_d + P_e = 1 \tag{6.172}$$

In the SAW ARQ algorithm, ACK is returned when the error detection code decoder does not detect errors. Therefore, probability to return ACK is given by

$$P_{ACK} = P_c + P_e \tag{6.173}$$

Because P_c is usually much larger than P_e, P_{ACK} is given by

$$P_{ACK} = P_c = (1-P_b)^n \tag{6.174}$$

where n is the number of bits in a block and P_b is the BER of the channel. On the other hand, probability to return NACK is given by

$$P_{NACK} = P_d = 1 - P_{ACK} \tag{6.175}$$

Let's discuss throughput performance of the SAW ARQ algorithm under the condition that k is the number of information bits in a block, D is the round-trip propagation delay time, and R_b is the bit rate. When a data block is transmitted, its ACK is received after DR_b bit duration, and no data is transmitted during this period as shown in Figure 6.29. Therefore, one data block transmission takes $(1 + DR_b/n)$-block

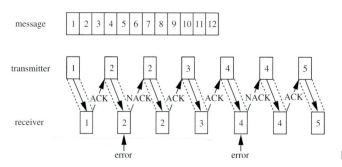

Fig. 6.29 SAW ARQ protocol.

duration. As a result, the average transmission time in terms of a block duration is given by

$$
\begin{aligned}
T_{SAW} &= \left(1 + DR_b/n\right)P_{ACK} + 2\left(1 + DR_b/n\right)P_{ACK}(1 - P_{ACK}) \\
&\quad + 3\left(1 + DR_b/n\right)P_{ACK}(1 - P_{ACK})^2 + \cdots \\
&= \frac{1 + DR_b/n}{P_{ACK}}
\end{aligned}
\tag{6.176}
$$

As a result, throughput for SAW ARQ is given by

$$
\eta_{SAW} = \frac{1}{T_{SAW}}\left(\frac{k}{n}\right) = \frac{P_{ACK}}{1 + DR_b/n}\left(\frac{k}{n}\right)
\tag{6.177}
$$

6.12.3 GBN ARQ Protocol

Figure 6.30 shows protocol of GBN ARQ, where the second and sixth blocks are in error. In this algorithm, the data blocks are transmitted regardless of the ACK or NACK signals. When NACK for the $(n - N)$-th block is returned, the data blocks are transmitted from the $(n - N)$-th block. In the case of Figure 6.30, NACK for the second block is received just after the fourth block is transmitted. Therefore, the transmitter retransmits from the second block again even though the third and fourth blocks have been correctly received.

Because one block error results in retransmission for the errored block as well as the following $(N - 1)$ blocks, the average transmission time in terms of a block duration is given by

$$
\begin{aligned}
T_{GBN} &= 1 \cdot P_{ACK} + (N + 1)P_{ACK}(1 - P_{ACK}) \\
&\quad + (2N + 1)P_{ACK}(1 - P_{ACK})^2 + \cdots \\
&= 1 + \frac{N(1 - P_{ACK})}{P_{ACK}}
\end{aligned}
\tag{6.178}
$$

As a result, throughput performance for GBN ARQ is given by

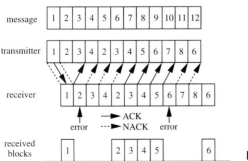

Fig. 6.30 GBN ARQ protocol.

$$\eta_{GBN} = \frac{1}{T_{GBN}}\left(\frac{k}{n}\right) = \frac{P_{ACK}}{P_{ACK} + (1 - P_{ACK})N}\left(\frac{k}{n}\right) \qquad (6.179)$$

6.12.4 SR ARQ Protocol

In the SR ARQ protocol, only the errored blocks are retransmitted. Figure 6.31 shows protocol of SR ARQ, where the second and sixth blocks are in error. In this protocol, we have to temporarily store all the received blocks in the buffer to reorder the received blocks. In Figure 6.31, all the correctly received data blocks are stored in buffer #1. In this case, because the second block is correctly received after the fourth block, the third to fifth blocks are transferred to buffer #2 after the second block is correctly received as shown in this figure.

Similar to GBN, we can obtain the average transmission time in terms of a block duration as

$$\begin{aligned} T_{SR} &= 1 \cdot P_{ACK} + 2P_{ACK}(1 - P_{ACK}) \\ &\quad + 3P_{ACK}(1 - P_{ACK})^2 + \cdots \\ &= \frac{1}{P_{ACK}} \end{aligned} \qquad (6.180)$$

Using this result, we can obtain its throughput as

$$\eta_{SR} = \frac{1}{T_{SR}}\left(\frac{k}{n}\right) = \left(\frac{k}{n}\right)P_{ACK} \qquad (6.181)$$

6.12.5 Comparison between SAW, GBN, and SR

Figure 6.32 shows transmission efficiency versus channel BER performance for SAW, GBN, and SR algorithms under AWGN conditions, where transmission efficiency is defined as the inverse of the average transmission time in terms of a block

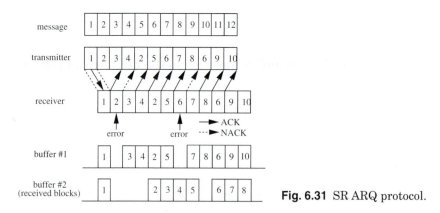

Fig. 6.31 SR ARQ protocol.

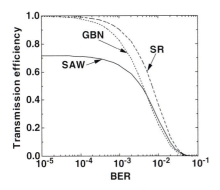

Fig. 6.32 Transmission efficiency versus channel BER performance for SAW, GBN, and SR algorithms under AWGN conditions, where transmission efficiency is defined as the inverse of the average transmission time in terms of a block duration ($1/T_{SAW}$, $1/T_{GBN}$, and $1/T_{SR}$), $D = 20$ μs, $R_b = 2$ Mbit/s, $n = 100$ bits, and N for GBN is 2.

duration ($1/T_{SAW}$, $1/T_{GBN}$, and $1/T_{SR}$), $D = 20$ μs, $R_b = 2$ Mbit/s, $n = 100$ bits, and N for GBN is 2. In the case of SAW, throughput at low BER is limited by the round-trip propagation delay time as shown in equation (6.177). In the case of GBN, its throughput is higher than that for SAW at lower BER. On the other hand, throughput for GBN is worse than that for SAW at higher BER because N blocks of data are discarded by 1-bit error, and this probability increases with BER. As a result, SR can achieve the highest throughput performance among these three algorithms because it is not affected by the round-trip propagation delay time or N.

Because the SAW algorithm is simple and applicable to the half-duplex systems, it is applied to various data transmission systems. However, its throughput is relatively low in comparison with the other algorithms when round-trip propagation delay time is long. Therefore, SAW is effective only for short-range wireless communication systems in which hardware simplicity is required.

In the case of SR, although SR can achieve the highest throughput, a relatively large-size buffer (buffer #1 in Figure 6.31) is required when the channel condition is very bad. Several methods are proposed to overcome this problem, such as the limitation of the number of retransmissions [6-20]. Hybrid ARQ in which FEC and ARQ are combined is another technique to overcome this problem.

6.12.6 Type I Hybrid ARQ

In the type I hybrid ARQ protocol, after each transmitted data block is CRC encoded, the coded block is further encoded by an FEC code. In the receiver, after the FEC is decoded, error detection decoding is carried out. Because the receiver is more robust to fading as well as interference and noise owing to the FEC, we can improve throughput when the received signal level is low. However, when the received signal level is high, its throughput is lower than that for the SR ARQ due to redundant bits for FEC.

6.12.7 Type II Hybrid ARQ

In the type II hybrid ARQ protocol, after each transmitted data block is encoded by an error-detecting code such as CRC, the encoded sequence is further encoded by an FEC code. Let's call the CRC-encoded data block the *original sequence* and the redundant (parity) bits after FEC encoding the *parity sequence*. In this case, as the FEC code, we usually employ an invertible code by which the original sequence can be regenerated by the parity sequence [6-19].

Let's consider an (n, k) block code whose generator polynomial $G(x)$, source message polynomial $u(x)$, and codeword polynomial $w(x)$ are given by

$$G(x) = x^{n-k} + g_{n-k-1}x^{n-k-1} + \cdots + g_1 x + g_0 \tag{6.182}$$

$$u(x) = u_{k-1}x^{k-1} + u_{k-2}x^{k-2} + \cdots + u_1 x + u_0 \tag{6.183}$$

$$w(x) = w_{n-1}x^{n-1} + w_{n-2}x^{n-2} + \cdots + w_1 x + w_0 \tag{6.184}$$

and $u_1(t)$ and $u_2(x)$ are two distinct source messages.

When we divide $x^{n-k}u_1(x)$ and $x^{n-k}u_2(x)$ by $G(x)$, respectively, we can obtain the following relationships:

$$x^{n-k}u_1(x) = a_1(x)G(x) + b_1(x) \tag{6.185}$$

$$x^{n-k}u_2(x) = a_2(x)G(x) + b_2(x) \tag{6.186}$$

where $a_1(x)$ and $b_1(x)$ are quotient and remainder for $x^{n-k}u_1(x)$, and $a_2(x)$ and $b_2(x)$ are those for $x^{n-k}u_2(x)$.

In the case of invertible code, $b_1(x) \neq b_2(x)$ should be satisfied when $u_1(t)$ and $u_2(x)$ are not identical. Suppose $b_1(x) = b_2(x)$. When we add equations (6.185) and (6.186), we can obtain

$$\left[u_1(x) + u_2(x)\right]x^{n-k} = \left[a_1(x) + a_2(x)\right]G(x) \tag{6.187}$$

Because the order of $G(x)$ and x^{n-k} are the same, $u_1(t) + u_2(x)$ must be divisible by $G(x)$.

When $k > n - k$ ($2k > n$) is satisfied, $u_1(t) + u_2(x)$ is divisible by $G(x)$. In this case, $b_1(x) = b_2(x)$ could be satisfied even though $u_1(x) \neq u_2(x)$. On the other hand, in the case of $2k \leq n$, $u_1(t) + u_2(x)$ is not divisible, which means that, when $u_1(x)$ and $u_2(x)$ are different source data, its parity is not identical. Therefore, we can generate invertible code when its coding rate is lower than or equal to 1/2 [6-19].

In the type II hybrid ARQ, the original sequence is first transmitted. When ACK for the original sequence is returned, the protocol is the same as that for SR ARQ. On the other hand, when NACK is received, its parity sequence is transmitted in the retransmission process. In this case, the original sequence is first regenerated from the parity sequence (code inversion process), and error detection decoding is carried out. When the parity sequence is found to include no error, ACK is returned. On the other hand, when the parity sequence also includes errors, both the detected original and parity sequences are combined, FEC decoded, and checked to see whether the FEC-decoded sequence includes errors or not. When the result includes no errors, ACK is returned. On the other hand, when the result still includes errors, we will go back to the original sequence transmission process again.

We might have a question as to why the code inversion process is carried out for the parity sequence before combining original and parity sequences for FEC decoding. Let's assume that the channel condition for the original sequence transmission is very bad—BER is 0.1, for example—and that the channel condition for the parity sequence is error free due to the time-varying nature of the wireless communication channel. In such a case, although original and parity sequences are combined and FEC decoding is carried out, the result would include errors because too many errors are included in the combined sequence. On the other hand, if the channel condition for parity transmission is very good, we can correctly obtain the original sequence by inverting the parity sequence. Therefore, the code inversion process would be necessary in the type II hybrid ARQ protocol.

Although we have discussed block codes for hybrid ARQ schemes, we can also apply convolutional codes including rate-compatible codes and punctured codes [6-21; 6-22; 6-23; 6-24].

6.12.8 Performance of ARQ under Flat Rayleigh Fading Conditions

Although there are many papers on the performance of ARQ in the Rayleigh fading channel, we will show you an experimental result using the GBN ARQ as an example. Figure 6.33 shows configuration of the experiment setup [6-25]. In this system, 9-stage PN code is fed to the GBN ARQ encoder to construct a data block, and the data block is fed to the GMSK modulator, where the transmitted message is encoded by 15-stage CRC, the bit rate is 16 kbit/s, and a 3-dB bandwidth normalized by a bit duration for the premodulation filter is $B_b T_b = 0.25$. After this signal is distorted by flat Rayleigh fading by the fading simulator, it is received by the GMSK demodulator with a coherent detector. The regenerated data sequence obtained at the GMSK demodulator is then fed to the ARQ decoder to detect whether the message includes errors and returns ACK or NACK to the transmitter. In this system, the ARQ encoder and decoder are constructed by DSPs, and the returned link is wire-connected.

Figure 6.34 shows transmission efficiency versus BER for a GBN ARQ system under flat Rayleigh fading conditions, where a frame consists of 128 bits of information and 15 bits of CRC check bits, and $N = 2$. First, we can find that the transmission efficiency under flat Rayleigh fading conditions with smaller f_d is higher than that under AWGN channel conditions because long error-free length is more probable

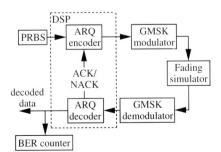

Fig. 6.33 Configuration of the experiment setup for a GBN ARQ system (from Ref. 6-25, © Communications Research Laboratory, 1991).

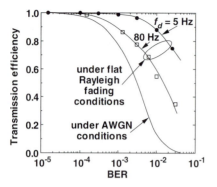

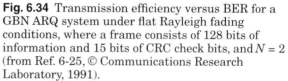

Fig. 6.34 Transmission efficiency versus BER for a GBN ARQ system under flat Rayleigh fading conditions, where a frame consists of 128 bits of information and 15 bits of CRC check bits, and $N = 2$ (from Ref. 6-25, © Communications Research Laboratory, 1991).

under flat Rayleigh fading conditions with smaller f_d than under AWGN channel conditions owing to burstness of the error sequence.

Figure 6.35 shows transmission efficiency versus f_d with a parameter of E_b/N_0 under flat Rayleigh fading conditions, where a frame consists of 128 bits of information and 15 bits of CRC check bits, and $N = 2$. As shown in this figure, when f_d increases, transmission efficiency decreases because error-free length becomes short with increasing f_d. Moreover, we can find that transmission efficiency depends on E_b/N_0.

Figure 6.36 shows transmission efficiency versus E_b/N_0 with a parameter of a frame length under flat Rayleigh fading conditions, where $f_d = 5$ Hz and $N = 2$. As

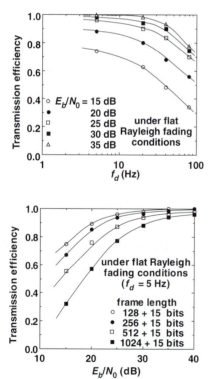

Fig. 6.35 Transmission efficiency versus f_d with a parameter of E_b/N_0 under flat Rayleigh fading conditions, where a frame consists of 128 bits of information and 15 bits of CRC check bits, and $N = 2$ (from Ref. 6-25, © Communications Research Laboratory, 1991).

Fig. 6.36 Transmission efficiency versus E_b/N_0 with a parameter of a frame length under flat Rayleigh fading conditions, where $f_d = 5$ Hz and $N = 2$ (from Ref. 6-25, © Communications Research Laboratory, 1991).

shown in this figure, higher E_b/N_0 can achieve higher transmission efficiency because BER becomes small at high E_b/N_0. Moreover, we can find that longer frame length degrades transmission efficiency performance because the block error rate increases with a frame length.

REFERENCES

6-1. Clark, G. C. Jr. and Cain, J. B., *Error-correction coding for digital communications*, Plenum, New York, 1981.

6-2. Pless, V., *Introduction to the theory of error-correcting codes* (2d Edition), John Wiley & Sons, New York, 1989.

6-3. Lin, S., Costello, D. J., Jr. and Miller, M. J., "Automatic-repeat-request error-control schemes," IEEE Commun. Mag., Vol. 22, No. 12, pp. 5–15, December 1984.

6-4. Proakis, J. G., *Digital communications* (2d Edition), McGraw-Hill, New York, 1989.

6-5. Mumter, M. and Wolf, J., "Predicted performance of error-control over real channel," IEEE Trans. Inf. Theory, Vol. IT-14, No. 5, pp. 640–50, September 1968.

6-6. Ohtani, K. and Ohmori, H., "Distribution of burst error lengths in Rayleigh fading radio channels," Electron. Letters, Vol. 16, No. 23, pp. 889–91, November 1980.

6-7. Sampei, S. and Kamio, Y., "Performance of FEC with interleaving in digital land mobile communications," Trans. IECE of Japan, Vol. E68, No. 10, pp. 651–52, October 1985.

6-8. Sampei, S., Kamio, Y. and Sasaoka, H., "Field experiments on pilot symbol aided 16QAM for land mobile communication systems," Review of the Communications Research Laboratory, Vol. 39, No. 2, pp. 83–93, June 1993.

6-9. Peterson, W. W. and Weldon, E. J., *Error-correcting code*, M.I.T. Press, Cambridge, Massachusetts, 1972.

6-10. Steele, R. (Editor), *Mobile radio communications*, Pentech Press, London, 1992.

6-11. Peterson, W. W., "Encoding and error correction procedures for the Bose-Chaudhuri codes," IRE Trans. Inform. Theory, Vol. IT-6, pp. 459–70, September 1960.

6-12. Viterbi, A. J., "Convolutional codes and their performance in communication systems," IEEE Trans. Commun., Vol. COM-19, No. 5, pp. 835–48, October 1971.

6-13. Bargava, N., Haccoun, D., Matyas, R. and Nuspl, P. P., *Digital communications by satellite*, John Wiley & Sons, New York, 1981.

6-14. Viterbi, A. J., "Error bounds for convolutional codes and an asymptotically optimum decoding algorithm," IEEE Trans. Inform. Theory, Vol. IT-13, No. 2, pp. 260–69, April 1967.

6-15. Matsuoka, H., Sampei, S. and Morinaga, N., "Adaptive modulation system with punctured convolutional code for high quality personal communication systems," IEICE Trans. Commun., Vol. E79-B, No. 3, pp. 328–34, March 1996.

6-16. Trumpis, B. D. and McAdam, P. L., "Performance of convolutional codes on burst noise channels," IEEE National Telecommun. Conf. (Los Angeles, California), pp. 36.3.1–36.3.14, December 1977.

6-17. Yasuda, Y., Kashiki, K. and Hirata, Y., "High-rate punctured convolutional codes for soft decision Viterbi decoding," IEEE Trans. Commun., Vol. COM-32, No. 3, pp. 315–19, March 1984.

6-18. Cain, J. B. Clark, G. C. and Geist, J. M., "Punctured convolutional codes of rate $(n-1)/n$ and simplified maximum likelihood decoding," IEEE Trans. Info. Theory, Vol. IT-25, No. 1, pp. 97–100, January 1979.

6-19. Lin, S. and Costello, J. Jr., *Error control coding, fundamental and applications*, Prentice Hall, Englewood Cliffs, New Jersey, 1983.

6-20. Yang, Q. and Bhargava, V. K., "Delay and coding gain analysis of a truncated Type-II hybrid ARQ protocol," IEEE Trans. Veh. Technol., Vol. 42, No. 1, pp. 22–31, February 1993.

6-21. Aridhi, S. and Despins, C. L., "Performance analysis of Type-I and Type-II hybrid ARQ protocols using concatenated codes in a DS-CDMA Rayleigh fading channel," 4th IEEE ICUPC (Tokyo, Japan), pp. 748–52, November 1995.

6-22. Hagenauger, J., "Rate-compatible punctured convolutional code (RCPC codes) and their applications," IEEE Trans. Commun., Vol. 36, No. 4, pp. 389–400, April 1988.

6-23. Kallel, S. and Haccoun, D., "Generalized Type-II hybrid ARQ scheme using punctured convolutional coding," IEEE Trans. Commun., Vol. 38, No. 11, pp. 1938–46, November 1990.

6-24. Lou, H. and Cheung, A. S., "Performance of punctured channel codes with ARQ for multimedia transmission in Rayleigh fading channels," 46th IEEE Veh. Tech. Conf. (Atlanta, Georgia), pp. 282–86, April 1996.

6-25. Kamio, Y., "Characteristics of GBN-ARQ in digital land mobile communications," Review of the Communications Research Laboratory, Vol. 37, No. 1, pp. 167–73, February 1991.

CHAPTER 7

Access and Duplex Techniques

In this chapter we will discuss access and duplex techniques for wireless communication systems. Due to the recent development of wireless communication technologies, we can apply any access scheme—including frequency division multiple access (FDMA), time division multiple access (TDMA), and code division multiple access (CDMA)—to wireless communication systems. Moreover, we can also apply either frequency division duplex (FDD) or time division duplex (TDD) as a duplex scheme. As a result, we have a lot of debate about which is better—TDMA or CDMA, FDD or TDD. The answer to this depends on the combined techniques, such as the modulation scheme, anti-fading techniques, forward error correction, and so on, as well as the requirements of services, such as the coverage area, capacity, traffic, and type of information. Thus, this chapter discusses only the key techniques specific for each access and duplex scheme. Design strategy for TDMA and CDMA will be discussed in chapters 15 and 16.

7.1 FREQUENCY DIVISION MULTIPLE ACCESS

FDMA is the most popular multiple access scheme for land mobile communication systems because it can easily discriminate channels by filters in the frequency domain. Actually, all analog cellular systems in the world use FDMA.

Figure 7.1(a) shows the spectrum of FDMA systems. The assigned system bandwidth is divided into bands with its bandwidth of W_{ch}. Between adjacent channels, there is a guard space to prevent spectrum overlapping due to carrier frequency insta-

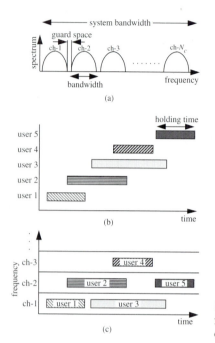

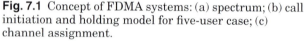

Fig. 7.1 Concept of FDMA systems: (a) spectrum; (b) call initiation and holding model for five-user case; (c) channel assignment.

bility. When each user sends a call request to the BS, the BS assigns one of the unused channels to the user, and the channel is used exclusively by that user during a call. However, when the call is terminated, the channel is reassigned to a different user. When five users initiate and hold calls as shown in Figure 7.1(b), an example of channel assignment might be that shown in Figure 7.1(c).

The most important advantage of FDMA is that its hardware is simple because it can easily discriminate channels only by filters. It does not require any synchronization or timing control, which are essential for TDMA or CDMA systems. Furthermore, as for the anti-fading techniques, we have to consider only flat fading because the bandwidth of each channel in the FDMA is sufficiently narrow.

On the other hand, system capacity for FDMA systems greatly depends on the instability of carrier frequency. Let's assume that we consider only voice transmission. When the bit rate of voice codec is R_b (bit/s), modulation level is M, and roll-off factor is α, its bandwidth per channel (W_{ch}) is given by

$$W_{ch} = \frac{(1+\alpha)R_b}{\log_2 M} + 2\delta f_c \qquad (7.1)$$

where δ means carrier frequency instability, which is specified as around 3×10^{-6} for the terminal in Japan, and $2\delta f_c$ means necessary guard space to prevent spectrum overlapping due to carrier frequency instability. In the case of large-zone systems, the system capacity can be expressed as the number of voice channels per unit bandwidth. When the unit bandwidth is selected as 1 MHz, the system capacity (number of channels per 1 MHz bandwidth) is given by

$$N_{ch} = \frac{10^6}{W_{ch}} \tag{7.2}$$

When the first term in equation (7.1) is much larger than the second term—when R_b is large or M is small—carrier frequency instability has little influence on system capacity. However, when low-bit-rate voice codec or high-spectral efficient modulation are applied to the FDMA systems, the first and second terms in equation (7.1) become comparable; thereby system capacity is limited by carrier frequency instability.

Figure 7.2 shows N_{ch} versus R_b with a parameter of δ for (a) 1 bit/symbol modulation ($M = 2$), (b) 2 bit/symbol modulation ($M = 4$), and (c) 4 bit/symbol modulation ($M =$

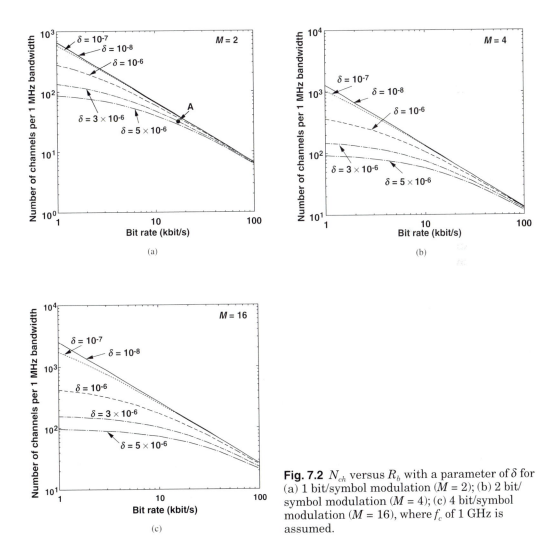

(a)

(b)

(c)

Fig. 7.2 N_{ch} versus R_b with a parameter of δ for (a) 1 bit/symbol modulation ($M = 2$); (b) 2 bit/symbol modulation ($M = 4$); (c) 4 bit/symbol modulation ($M = 16$), where f_c of 1 GHz is assumed.

16), where f_c of 1 GHz is assumed. As shown in this figure, the system capacity is limited by the carrier frequency instability at lower R_b, and it is more remarkable when the modulation level becomes large.

During the 1970s to early 1980s, R_b = 16 kbit/s and M = 2 were popular values. In this case, carrier frequency instability had almost no impact on the capacity, as shown in Figure 7.2 (point A corresponds to this case). Since the mid-1980s, a lot of studies were started on the development of low-bit-rate voice codec and spectral efficient modulation schemes. From Figure 7.2(b) and (c), we find that N_{ch} for $\delta = 3 \times 10^{-6}$ is about 50% as much as that for $\delta = 10^{-8}$ at R_b = 8 kbit/s and M = 4, and it is about 33% as much at R_b = 8 kbit/s and M = 16. These results mean that carrier frequency instability of less than 10^{-7} is necessary to improve spectral efficiency for FDMA systems using low-bit-rate voice codec and high spectral efficiency modulation scheme.

Furthermore, FDMA systems have several other problems.

☞ Intermodulation interference increases with the number of carriers.

☞ Variable rate transmission is difficult because such a terminal has to prepare a lot of modems. For the same reason, composite transmission of voice and non-voice data is also difficult.

☞ High Q-value for the transmitter and receiver filters is required to guarantee high channel selectivity.

7.2 TIME DIVISION MULTIPLE ACCESS

7.2.1 Historical Overview of TDMA for Land Mobile Communication Systems

TDMA is a method that enables users to access the assigned bandwidth on a time basis—each channel occupies whole the system bandwidth, but it occupies only a fraction of the time, called *slot*, on a periodic basis. TDMA was first investigated for satellite communication systems in the late 1960s to solve the disadvantages of FDMA systems that we discussed previously. The first commercial TDMA satellite system was operated in Canada in 1976.

Investigation into the application of TDMA to land mobile communication systems was started in around 1980 by Kinoshita et al., who applied TDMA to user access links [7-1]. At that time, unfortunately, TDMA was not considered to be suitable for land mobile communications because of its complicated timing controls. However, in 1982, the CEPT organized the GSM committee and started to make specifications for the pan-European digital cellular system, in which TDMA was included as a candidate access scheme. Finally, TDMA was selected as the access scheme for the GSM system [7-2] in 1988. As a result of the extensive studies on TDMA in this committee, TDMA technologies for land mobile communications were highly developed. As a result, TDMA is now a very popular access scheme for land mobile communication systems, and it is applied to many second-generation cellular systems and cordless phone systems all over the world [7-2; 7-3; 7-4; 7-5; 7-6].

7.2.2 Basic Concept of TDMA

Figure 7.3 shows a concept of TDMA operation. In this system, one frame consists of N_{ch} slots, and its frame length is T_f second. In the uplink, each terminal transmits information using an assigned slot in each frame. Therefore, each terminal has to transmit its slot exactly in the assigned slot timing to prevent signal collisions. A simple way to prevent such collisions is to prepare a guard time that is sufficiently long to buffer the propagation path length difference. Another way to prevent burst collision is to use the time alignment technique. Because this technique can reduce guard time, most of the TDMA land mobile communication systems apply this technique.

On the other hand, in the downlink, all the slot signals are transmitted by the BS. Therefore, the transmitted signal is not a TDMA signal but a TDM signal.

Figure 7.4 shows the relationship between the source information signal and the TDMA signal. Let's assume that the bit rate of the source signal is R_b kbit/s, one TDMA frame consists of N_{ch} slots, its frame length is T_f second, and the source infor-

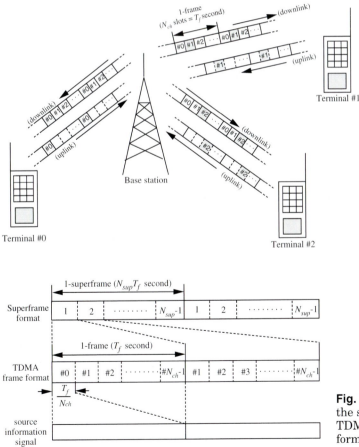

Fig. 7.3 Concept of TDMA operation.

Fig. 7.4 Relationship between the source information signal, TDMA frame, and superframe format.

mation sequence is transmitted using one of the slots in each frame. When this signal is transmitted via a TDMA channel, the source information sequence is first blocked every T_f second. In this case, each block contains $R_b T_f$ bits. Because these data are transmitted during a slot duration of T_f/N_{ch} second, the bit rate for the TDMA signal should be N_{ch} times higher than that of the source information sequence. In other words, the transmitted signal is compressed in the time domain. In practical systems, because there are 15–25% of the redundant bits in each slot, the bit rate for the TDMA signal is given by $N_{ch} R_b/F_{eff}$, where F_{eff} is frame efficiency given by

$$F_{eff} = 1 - \frac{\text{total number of redundant symbols in each slot}}{\text{total number of symbols in each slot}} \tag{7.3}$$

When this TDMA signal is received, the demodulated data is expanded to the source bit rate. In the TDMA systems, we also employ a superframe format to multiplex several types of channels in a slot. Figure 7.4 also shows an N_{sup}-frame multiplexed superframe format. We will discuss the superframe in more detail later in this chapter.

Figure 7.5 shows (a) spectrum of TDMA systems, (b) a model for call from five users, and (c) slot (channel) assignment, where 3-channel TDMA is assumed. As discussed above, occupied bandwidth for N_{ch} TDMA is about N_{ch} times wider than that of the FDMA system. When a call request is received at the BS as shown in Figure 7.5(b), the BS assigns one of the unused slots to the terminal as shown in Figure 7.5(c).

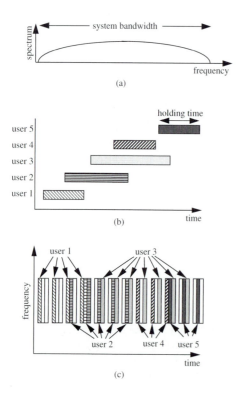

Fig. 7.5 Concept of TDMA systems: (a) spectrum of TDMA systems; (b) a call initiation and holding model for five-user case; (c) slot (channel) assignment, where 3-channel TDMA is assumed.

7.2.3 Frame and Slot Format for TDMA Systems

Figure 7.6 shows an example of the frame and slot format of a 6-channel TDMA system. Each frame consists of 6 slots. In each slot, unique carrier phase synchronization words (CRSW), symbol timing synchronization words (STSW), and slot timing synchronization words (SLSW) are included in the preamble followed by the data symbols. Ramp symbols are used to smooth the waveform of each burst, thereby suppressing out-of-band spectrum. Guard time is used to prevent slot collision due to the difference in propagation path distance for each of the terminals. When the cell radius is R, the necessary guard time is given by $G = 2R/c$, where c is the speed of light. For example, when $R = 1$ km, the guard time should be longer than 6.67 μs.

Although unique words are allocated in the preamble in Figure 7.6, we can also allocate these words in the middle part of a slot (midamble) or in the last part of a slot (postamble). Or we can distribute them throughout the preamble, midamble, and postamble. Figure 7.7 shows slot formats for (a) unique words in preamble, (b) unique words in midamble, and (c) unique words in preamble and postamble. One big advantage for allocating unique words in the preamble is that we can sequentially demodulate a slot signal. On the other hand, when we allocate unique words in the midamble or distribute them in the preamble and postamble, we can improve tracking ability of anti-fading techniques as discussed in chapters 4 and 5, although we have to store all the waveforms of a slot in the memory and have to demodulate by using block demod-

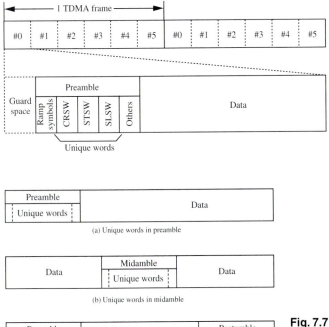

Fig. 7.6 An example of frame and slot format of a 6-channel TDMA system.

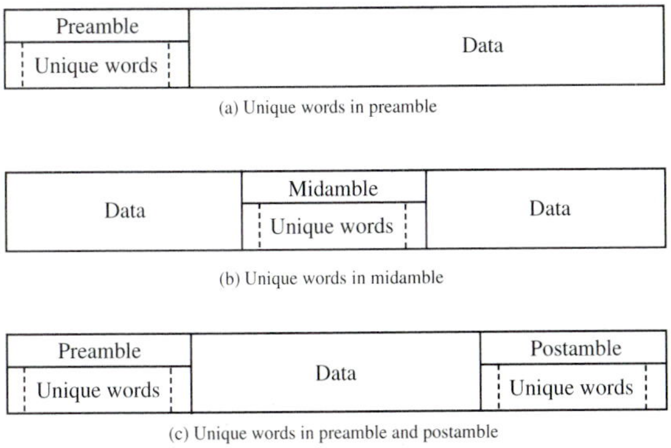

Fig. 7.7 Slot formats: (a) unique words in preamble; (b) unique words in midamble; (c) unique words in preamble and postamble.

ulation techniques. Therefore, we have to carefully design unique word allocation considering the robustness-to-fading variation and hardware complexity. Sequential demodulation and block demodulation will be detailed in a later section.

7.2.4 Superframe Format

We sometimes employ a superframe format to achieve accurate time synchronization between the BS and terminals. Figure 7.8 shows an example of a superframe format in which a superframe consists of N_{sup} frames (N_{sup} = 7 in Figure 7.8). In this case, one superframe length is $N_{sup}T_f$. When a terminal takes superframe and symbol timing synchronization, and each terminal has a time base with an instability of less than $N_{sup}T_f$, the terminal can synchronize to BS timing with its resolution of a symbol duration (T_s).

Another advantage of a superframe format is that we can multiplex low-bit-rate control information in the time domain using a superframe format. In Figure 7.8, a broadcast control channel (BCCH), three paging channels (PCH) #0–#2, and three signaling control channels (SCCH) #0–#2 are multiplexed on the superframe. The BCCH is allocated at the first frame. PCHs and SCCHs are alternately allocated after the BCCH. In this case, the average bit rate for each channel is $R_b/7$.

7.2.5 Sequential Demodulation and Block Demodulation

In the TDMA systems, carrier phase synchronization, symbol timing synchronization, and slot timing synchronization should be used in each burst. In conventional-type TDMA systems such as TDMA satellite systems, the CRSW, STSW, and SLSW are included in the beginning of each burst (preamble) as shown in Figure 7.6. When a terminal receives a signal, synchronizations of the carrier, symbol timing, and slot timing are used in this order, and the demodulation process is started after these synchronizations are finished. Thus, a lot of preamble symbols are required for this type of demodulation, thereby reducing the frame efficiency.

Block demodulation is an alternative scheme to solve this problem [7-7]. Figure 7.9 shows a slot format for block demodulation. In this scheme, the received signal is first analog-to-digital (A/D) converted and stored in memory. Because we can use carrier phase and symbol timing synchronization with all the stored received signals as

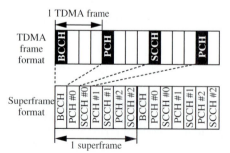

Fig. 7.8 An example of a superframe format.

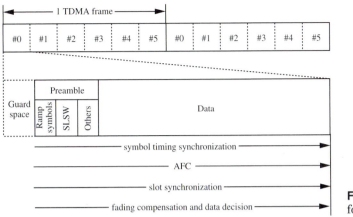

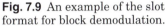

Fig. 7.9 An example of the slot format for block demodulation.

shown in Figure 7.9, we can remove the CRSW and STSW from preamble, thereby improving frame efficiency. Furthermore, we can improve carrier and symbol timing synchronization performances because we can use all of the burst signal for synchronization.

Because this scheme requires an A/D converter and memory to store the whole slot signal, its hardware could be a bit complicated in comparison with the hardware for sequential demodulation. However, such circuit complexity problems are getting less serious due to the recent device and digital signal processing technologies. Actually, when we apply the adaptive equalizer to land mobile communications as in the case of GSM or IS-54 systems, the A/D converter and memory is indispensable. Therefore, the block demodulation technique is becoming a more and more popular and important technique for achieving high frame efficiency as well as for improving synchronization performances for TDMA systems.

7.2.6 Frame, Superframe, and Slot Synchronization

7.2.6.1 Frame Synchronization   In a TDMA-based digital land mobile communication system, each slot in a frame has an SLSW to establish frame synchronization. Figure 7.10 outlines the frame synchronization circuit in which correlation between the received baseband signal and the SLSW is taken. As shown in Figure 7.10, we can obtain correlation peaks at every frame timing. But, we also obtain randomly appearing correlation peaks due to false detection at the same time. Although some correlation peaks due to false detection can be removed by the threshold detector, we cannot remove such peaks in their entirety. Therefore, forward and backward protection circuits follow the threshold detector to remove such random peaks as shown in Figure 7.10 [7-8].

In the initial acquisition process for the frame synchronization, frame synchronization is taken when consecutive N_B peaks with an interval of a frame duration (T_f) are detected. This process is called N_B-stage backward protection. Even after frame synchronization is taken, the correlation peak by the SLSW could not be detected due

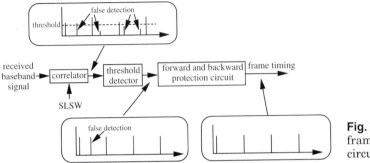

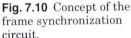

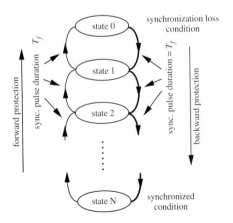

Fig. 7.10 Concept of the frame synchronization circuit.

to noise and fading. To prevent such synchronization loss, synchronization loss is detected only when consecutive N_F peaks at the frame timing cannot be detected. This process is called N_F-stage forward protection. Figure 7.11 shows a state diagram of the forward and backward protection circuit, where N_F and N_B are assumed to be (= N).

In the previous discussion, we assumed that there is only one correlation peak in each slot if there are no false-detection peaks. This is true when the channel is under AWGN or flat Rayleigh fading conditions. In actual propagation path conditions, however, there could be several correlation peaks around the SLSW due to delayed paths. Figure 7.12 shows an example of the frame-timing synchronization circuit operation, where its propagation path consists of three paths that are independently Rayleigh faded. As shown in Figure 7.12, there are three correlation peaks at each SLSW location as well as at the false-detection points. Let's call each correlation peak cluster around the SLSW and false-detection positions a *correlation peak cluster*. When the correlator output is fed to the threshold detector, several peaks among each cluster can be detected. Of course, the number of detected peaks in each cluster depends on the threshold level. As a result, the correlation peak interval has jitters.

Moreover, to which path the frame timing should be locked depends on the fading compensation techniques. When we employ an adaptive equalizer, we have to lock frame timing to the first path. On the other hand, when we employ a flat Rayleigh fading compensation technique, we have to lock frame timing to the strongest path.

Fig. 7.11 An example of a state diagram of the forward and backward protection circuit.

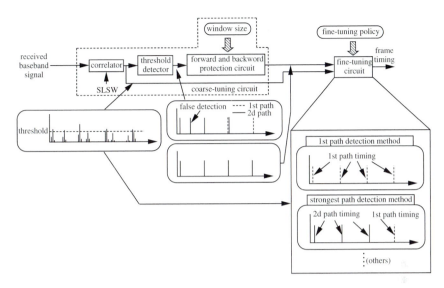

Fig. 7.12 An example of frame-timing synchronization circuit operation, where its propagation path consists of three paths that are independently Rayleigh faded.

To cope with such requirements, it is convenient to divide the frame-timing synchronization circuit into coarse-tuning and fine-tuning circuits as shown in Figure 7.12 [7-9]. In the coarse-tuning stage, the average timing for the SLSW correlation peak cluster is detected with an ambiguity of $\pm dT_s$, where d is a positive integer, T_s is a symbol duration, and this ambiguous region width $(2dT_s)$ is called a *detection window*. In this case, frame synchronization is taken when consecutive N_B peaks with an interval of $T_f \pm dT_s$ are detected. In the forward protection procedure, synchronization loss is detected only when consecutive N_F peaks at the frame timing with timing ambiguity of $\pm dT_s$ cannot be detected.

In the fine-tuning process, the delay profile around the coarsely detected frame timing is obtained again, and the optimum frame timing suitable for the applied fading compensation is detected.

7.2.6.2 Superframe Synchronization

When a superframe configuration is employed in a system, we also have to take superframe synchronization. There are several methods of identifying the first slot in each superframe.

- ☞ Assign a special SLSW to the first frame in each superframe.
- ☞ Map a special logical channel (for example, BCCH) onto the first frame in each superframe for the control channel-multiplexed downlink carrier.
- ☞ Embed a superframe synchronization counter number in the downlink slot.
- ☞ Inform the superframe counter number from the BS to the terminal during the initial channel allocation process.

7.2.6.3 Slot Synchronization   When carrier synchronization and symbol timing synchronization are taken and the received signal is demodulated, we have to achieve slot synchronization to know data symbol position in each slot. For this purpose, an SLSW is embedded in each slot. At the receiver, the correlation between the received signal and the SLSW are taken, and the correlation peak position corresponding to the SLSW position is searched in each slot.

The SLSW detection performance is determined by the SLSW miss-detection probability and the SLSW false-detection probability. The SLSW miss-detection probability is the probability that the correlation value is lower than a certain threshold. When the number of symbols in the SLSW is N symbols, the channel BER is P_e, correlation is hard decision detected, and the threshold value of correlation is ε, the SLSW miss-detection probability is given by

$$P_M = \sum_{i=\varepsilon+1}^{N} P_e^i (1-P_e)^{N-i} \tag{7.4}$$

Equation (7.4) means that we can reduce the SLSW miss-detection probability by increasing ε. However, the SLSW false detection is more probable when we increase ε. For example, when a burst consists of N-bit SLSW and N-bit data, the SLSW false-detection probability is given by

$$P_F = \frac{1}{2^N} \sum_{k=0}^{\varepsilon} {}_N C_k \tag{7.5}$$

When we compare equations (7.4) and (7.5), we can easily find that P_F is much higher than P_N, and that it cannot be reduced much by increasing N. When the number of bits in the data section is increased, such false detection becomes more probable. These results mean that only the optimization of ε is unable to reduce both the SLSW miss-detection and false-detection probabilities. Therefore, we usually apply some other techniques to reduce the SLSW false-detection probability.

When the transmitted signal is a TDM signal as in the case of a downlink or a TDMA signal with a relatively long holding time, we can reduce both probabilities by introducing forward and backward protection techniques or windowing techniques. On the other hand, when the transmitted signal is a short burst as in the case of a common access channel, we cannot apply such protection or windowing techniques. There are several techniques for reducing the SLSW miss-detection and false-detection probabilities in the case of short-burst transmissions.

One such technique is to embed plural SLSWs in each burst. When K SLSWs are used in each slot, although miss-detection probability of SLSW is increased by K times, its false-detection probability is given by P_F^K, where P_F is the false-detection probability for a single SLSW case, and the codeword length for the SLSWs is assumed to be the same. This means that the false-detection probability can be drastically reduced by using plural SLSWs. Of course, we can assign SLSWs having different codeword lengths in the slot.

Another technique is to conduct error detection for the information data part after the SLSW position is detected. The CRC check is one of the most popular error detection schemes. In this case, only when no error is detected by the CRC check is the detected SLSW position decided to be correct.

The personal digital cellular (PDC) system, which will be discussed in chapter 11, applies these two techniques to achieve slot synchronization for the common access channel in the uplink. In this system, an SLSW and a color code (a known codeword) are detected, and the CRC check is carried out to decide the SLSW synchronization.

7.3 CODE DIVISION MULTIPLE ACCESS

7.3.1 Historic Overview of CDMA

Although study on CDMA was started long time ago, the most important event for the application of CDMA to land mobile communications is the Fourteenth Plenary Assembly of the International Radio Consultative Committee (CCIR) held in June 1978 in Kyoto, Japan. This conference looked at spared spectrum techniques as a promising key technology for high-capacity land mobile communication systems [7-10]. Since this meeting, many studies have been made on the application of spread spectrum (SS) to land mobile communication [7-11; 7-12; 7-13]. The Radio Research Laboratory, Ministry of Posts and Telecommunications (MPT) of Japan (presently, Communications Research Laboratory, MPT of Japan), developed experimental systems of the direct sequence spread spectrum (DS/SS) and frequency hopping (FH) systems in 1981 and conducted laboratory and field experiments during 1981–1986 [7-14; 7-15; 7-16; 7-17]. As a result, these experiments confirmed that

☞ A rake receiver can effectively combine direct and delayed paths under frequency-selective fading conditions [7-14].

☞ Fast FH (FFH) is robust to frequency-selective fading, and it is more remarkable when FEC is effectively combined with FFH [7-15; 7-16; 7-17].

However, such developments were not strong driving forces in promoting the use of SS technologies in land mobile communications because

☞ It was premature for the device technology to make a fast power control device, which is essential to mitigate the near-far problem for DS/SS systems.

☞ It was difficult to produce an FFH synthesizer.

☞ System capacity using SS technology was estimated to be low in comparison with the existing analog FM cellular systems [7-18].

As a result, most of the land mobile communication engineers terminated their studies on CDMA systems by the mid-1980s.

However, in the late 1980s, Qualcomm Inc. proposed a DS/CDMA system that achieved a 10- to 20-fold increase in system capacity over the Advanced Mobile Phone Systems (AMPS) [7-19]. Because a 10-fold increase in system capacity sounds very attractive for cellular systems as well as future personal communication systems, many engineers restarted the development of CDMA systems.

When Qualcomm Inc. proposed a DS/CDMA system, most people doubted a 10- to 20-fold system capacity increase, because SS systems were not considered effective in increasing system capacity [7-18]. The key factors of the proposed system that would lead to such high system capacity are

☞ Fast power control large-scale integration (LSI) with its dynamic range of more than 80 dB was developed, which is a solution for the near-far problem as well as an improvement in receiver sensitivity.

☞ Convolutional code with low coding rate and orthogonal code are effectively combined with the DS/SS technique to improve receiver sensitivity.

☞ Soft and softer handover, which are kinds of macro-diversity techniques, are employed to improve receiver sensitivity as well as to suppress short-duration peaks of co-channel interference (CCI) due to compensation of deep fades.

☞ Variable-rate voice codec is employed to achieve voice activation, thereby reducing CCI.

☞ Sector cell is employed to reduce incoming CCI.

Therefore, the high capacity of the Qualcomm system can be attributed to the effective and efficient integration of the DS/SS technique and the other various technologies according to the following strategies:

☞ Removal of the causes of limited system capacity. Fast power control, variable-rate voice codec, and sector cell are part of this strategy.

☞ Improvement of receiver sensitivity to enhance system capacity. Low-rate convolutional code and soft/softer handover are part of this strategy.

In addition to these technological advances, the proposal of the DS/CDMA was supported by the rapidly growing device technology, especially VLSI and DSP technologies. When we compare central processing unit (CPU) devices in 1980 to what was available in 1990, we find that there was only a 16-bit CPU in 1980, but there was already a 64-bit CPU in 1990, which is 50 times faster than a 16-bit CPU. The same thing happened with memory capacity. In 1990, memory capacity was about 50 times larger than it was in 1980. As for DSP, DSP did not exist in 1980, but there were floating-point DSPs in 1990. DSP technology is a very powerful driving force for the application of TDMA and CDMA to practical cellular systems because a DSP chip can achieve very complicated signal processing, such as the adaptive equalizer, Rake receiver, voice codec, and so on.

With this technological and historical background, the U.S. Federal Communications Commission (FCC) authorized the application of SS to instruments using the industrial, scientific, and medical (ISM) band and to amateur and police radio in 1985 [7-20]. Then, in 1993, TIA approved the DS/CDMA system proposed by Qualcomm Inc. as an interim standard of digital cellular systems [7-21].

7.3.2 Basic Concept of CDMA

Figure 7.13 shows a transmitter and receiver configuration for CDMA systems. Let's assume that the source signal bit rate is R_b bit/s. In the transmitter, this source signal is multiplied by a preliminary assigned code (code #1 in Figure 7.13) with a chip rate of R_c chip/s followed by the quadrature modulator. As a result, the transmitted signal bandwidth becomes $G_p = R_c/R_b$ times as wide as that for the source signal. Because all the terminals use the same carrier frequency and chip rate (bandwidth), all the transmitted signals other than the desired signal are regarded as CCI signals.

At the CDMA receiver, the desired signal can be picked up by taking correlation between the received signal and a code used at the transmitter (code #1 in Figure 7.13). When the spreading code sequence for the receiver and that used at the transmitter are synchronized, the resultant signal spectrum becomes the same as that for the source signal. On the other hand, the signal bandwidth of the interference signals still remains the same bandwidth even after taking correlation. As a result, the carrier-to-CCI power ratio (C/I_c) for the correlator output is improved by $10\log_{10}G$ (dB) in comparison with that for the correlator input as shown in the following equation:

$$\left(C/I_c\right)_{out} \ [\text{dB}] = G \ [\text{dB}] + \left(C/I_c\right)_{in} [\text{dB}] \tag{7.6}$$

When the propagation path condition is considered to be a frequency-selective fading condition, and delay time of the delayed wave is longer than a chip duration, we can pick up the first path component if spreading code timing is synchronized to the first path because correlation between the first path and the other paths becomes negligibly small owing to the low autocorrelation value for the delay time of longer

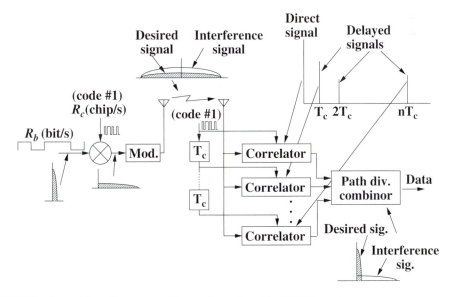

Fig. 7.13 Configuration of a transmitter and receiver for CDMA systems.

than a chip duration. On the other hand, when the spreading code timing is synchronized to one of the delayed path timings, we can pick up the delayed path component. Therefore, we can achieve path diversity when we prepare the spreading code synchronized to each path timing as we have discussed in chapter 5.

Figure 7.14(a) shows the spectrum of a CDMA system. The most distinct feature of CDMA systems is that all the terminals share the whole system bandwidth, and each terminal signal is discriminated by the code. When each user sends a call request to the BS, the BS assigns one of the spreading codes to the user. When five users initiate and hold calls as shown in Figure 7.14(b), time and frequency are occupied as shown in Figure 7.14(c).

7.3.3 Near-Far Problem and Fast Power Control

In DS/CDMA systems, although the received power level of each traffic channel is kept at the same level in the downlink, they are not equal in the uplink because of the near-far problem. When all of the terminals transmit signals with the same transmission power, the received signal level is strong from a near-in mobile unit, whereas it is weak from a far-end mobile unit due to propagation path loss. In this case, when the received signal level of a terminal is 10 times higher than that of the other terminals, for example, this terminal is equivalently transmitting 10 channels, although it

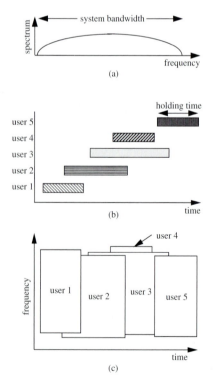

Fig. 7.14 Concept of a CDMA system: (a) spectrum of a CDMA system; (b) a call initiation and holding model for five-user case; (c) channel assignment to each user.

is actually transmitting data for only one channel. As a result, such received power imbalance causes a severe reduction in the system capacity.

To prevent such capacity reduction, fast power control that adjusts all the received signal levels to be equal at the BS is essential in the uplink. Figure 7.15 outlines this concept. When a terminal is located near the fringe area as in the case of terminal #2 in Figure 7.15, its transmission power is increased. On the other hand, when a terminal is located near the BS as in the case of terminal #1 in Figure 7.15, the transmission power is lowered. With this process, all the received power from the terminals are controlled to be almost equal at the BS. However, there is one thing we have to bear in mind—although all of the received signal level is controlled to be equal within the cell (cell A in Figure 7.15), the transmission power from a terminal at the fringe (terminal #2 in Figure 7.15) could be a strong CCI for cell site B because fading between terminal #2 and cell site A and fading between terminal #2 and cell site B are independently changed. As a result, perfect power control could result in increasing intercell CCI, although it helps to reduce intracell interference [7-22; 7-23].

7.3.3.1 Strategies to Cope with Wide Dynamic Range of Power Control Requirements—Open- and Closed-Loop Controls   In uplink CDMA systems, a very wide dynamic range of transmitter power control is required. First of all, path loss due to distance and shadowing (log-normal fading) would change the received signal level about 60 dB or more. This dynamic range becomes large with the coverage area. Moreover, it is rapidly changing in the range of −30 to +10 dB from the average signal level due to multipath fading. As a result, dynamic range of the received signal level including rapid variation becomes more than 100 dB.

The IS-95 system employs two strategies to cope with such wide dynamic range of the power control requirements—reduction of the dynamic range of rapid fading and two-stage power control.

Reduction of the Dynamic Range of Rapid Fading.   To reduce dynamic range of the rapid fading variation, the IS-95 systems employ space diversity and path diversity using a Rake receiver. First, let's consider only flat Rayleigh fading conditions. When two-branch antenna diversity with maximal ratio combining is employed, 99.99% value of the received signal level (probability that 99.99% of the received signal level is higher than this level) is −18 dB from the rms received signal level while it

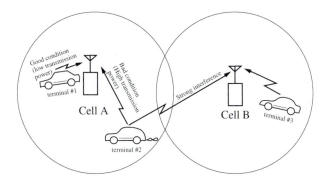

Fig. 7.15 Concept of fast power control.

is −40 dB in the case of single-branch reception. This means that the dynamic range is reduced by 22 dB due to the two-branch diversity reception.

When the channel is in frequency-selective fading conditions, we can further reduce the dynamic range. In frequency-selective fading environments, if the delay time of the delayed path is longer than a chip duration, we can resolve and combine the first path and the delayed paths on the maximal ratio-combining basis by using a Rake receiver. In other words, if we can resolve L paths having the same average power level and employing two-branch diversity, we can achieve $2L$-path diversity with maximal ratio combining. Because of these diversity effects, the dynamic range of the rapid fading variation is reduced to around 10 dB.

Another technique for dynamic range reduction is the soft handoff, which is a macro-diversity technique. Because the transmitter power of the terminal at the fringe is received by plural BSs, and because they are combined by selective combining diversity, the dynamic range due to path loss and shadowing can further be reduced.

Two-Stage Power Control Technique.   The IS-95 system employs a two-stage power control—open-loop and closed-loop controls. In the open-loop control, path loss due to distance and shadowing is measured in the downlink, and the expected average path loss is compensated for using the measured path loss in the downlink because such path loss has a high correlation between the downlink and the uplink, even in FDD systems. Using this open-loop control, the dynamic range of the fading variation is controlled to several dB, which corresponds to the expected dynamic range of the fast fading variation after space diversity and path diversity.

In the closed-loop control, deviation of the received signal level is controlled to a constant level with its deviation of about 1 dB. However, this standard deviation has a big impact on the capacity of CDMA systems. For example, when the standard deviation is 2 dB, there is a capacity reduction of 20–50% [7-24; 7-25; 7-26]. In IS-95, the transmitter power is updated every 1.25 ms to cope with a very fast fading variation to reduce this standard deviation to around 1–2 dB [7-27]. Other than the received signal level control scheme, C/I_c-based transmitter power control [7-28] and open-loop forward power control using TDD [7-29] have also been proposed as power control schemes.

7.3.4 Application of Low-Coding-Rate FEC

In FD/TDMA systems like GSM and PDC, a low-rate channel encoder is not preferable because redundancy caused by coding requires a wider bandwidth, thereby degrading spectral efficiency in terms of frequency. On the other hand, in the case of CDMA systems, bandwidth expansion due to a low-rate channel encoder is not a problem because the SS occupies a much wider bandwidth. In this case, the question is which coding rate of the channel encoder is appropriate, and how do we optimize the coding rate and processing gain? To consider this question, we will assume two fading environments—flat Rayleigh fading and frequency-selective fading.

Figure 7.16 shows the configuration of the DS/CDMA modulator including FEC. First let's assume that the channel is in a flat Rayleigh fading environment, the

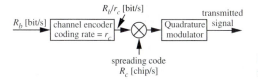

Fig. 7.16 Configuration of the DS/CDMA modulator including FEC.

source bit rate is R_b (bit/s), the coding rate is r_c, and the chip rate is R_c. When the source bit is encoded, the coded bit rate is given by R_b/r_c [bit/s]. This coded sequence is then multiplied by a spreading code.

At the receiver, when C/I_c before despreading is given by $(C/I_c)_{rec}$, C/I_c after despreading is given by

$$
\begin{aligned}
\left(C/I_c\right)_{desp}[\text{dB}] &= \left(C/I_c\right)_{rec}[\text{dB}] + 10\log_{10}\left(r_c R_c / R_b\right)[\text{dB}] \\
&= \left(C/I_c\right)_{rec}[\text{dB}] + G[\text{dB}] + 10\log_{10}(r_c)[\text{dB}]
\end{aligned}
\tag{7.7}
$$

where G is the total processing gain given by R_c/R_b. Equation (7.7) means that $(C/I_c)_{desp}$ is degraded by $-10\log10(r_c)$ dB in comparison with the noncoding case ($r_c = 1$). On the other hand, we can obtain channel coding gain (G_c). Consequently, we can improve receiver sensitivity by $G_c - 10\log(r_c)$ (dB). This means that the performance is determined by the trade-off between the coding gain of the channel encoder and the reduction of the processing gain [7-30; 7-31]. Especially when a power control technique is employed, we can expect a large coding gain because of the reduction in burst error length owing to the power control. Therefore, a convolutional encoder with its coding rate of 1/2 and 1/3 is employed in the IS-95 system.

In frequency-selective fading environments, we also have to take the resolution of the multipath into account. Under frequency-selective fading conditions, all the paths have to be resolved by a despreading process to effectively achieve path diversity. Let's assume a two-ray Rayleigh model, delay time of the delayed path of more than a chip duration, and signal level of both paths being the same. In this case, when the spreading code synchronized to the first path is multiplied to the received signal, the residual distortion due to the power of the second path is suppressed by $10\log_{10}(r_c G)$ dB. Therefore, this residual distortion should be sufficiently small to efficiently obtain the path diversity effect at the Rake receiver. In other words, when the required C/I_c is Γ dB

$$
\Gamma \ll 10\log_{10}(r_c G)
\tag{7.8}
$$

has to be satisfied. For example, when $\Gamma = 7$ dB and $r_c = 0.5$, $G > 100$ is necessary. Therefore, selection of the coding gain is determined by the trade-off between the coding gain, the processing gain, and the resolution of multipath discrimination.

7.3.5 Soft and Softer Handover

Because DS/CDMA uses the same carrier in every cell, each terminal can receive downlink signals from all the nearby BSs. For the same reason, each BS can receive uplink signals not only from the terminals within the cell but also from the terminal outside the cell. This means that DS/CDMA systems can employ a macro-diversity using neighboring BSs. In the DS/CDMA system, such macro-diversity is combined with the handover process, and it is called *soft handover*.

Let's assume that the downlink signal includes both traffic channels and a pilot channel and that each BS employs the same spreading code with a different initial phase. When the initial phase of the spreading code for each BS is appropriately shifted, each terminal can resolve the delay profile of each BS. Using the obtained delay profiles of all the BSs nearby, soft handoff is carried out.

Figure 7.17 shows an outline of the soft handover process. Let's assume that a mobile terminal is moving from BS #1 to BS #2. Each terminal measures the received signal level from both BS #1 and BS #2. In this case, when the largest L-path is combined at the Rake receiver, the total power of the largest L-path from each BS is measured. Then, the received signal levels are compared with two threshold levels—the add threshold and the drop threshold. When the signal level from a BS is higher than the add threshold and a signal from another BS is lower than the drop threshold, the terminal has a connection with only one of the BSs with the higher received signal level. On the other hand, when one of the received signal levels becomes weaker than the add threshold, the terminal sends a handover request, and soft handover is initiated. During the soft handover, the MS receives the downlink signal from both BS #1 and #2. At the same time, the uplink signal of the terminal is received by both BSs. The soft handoff process is continued until one of the received signal levels goes down below the drop threshold. Although the received signal level goes down below the drop threshold, the BS with the low received signal level is disconnected only when the received signal level is continuously below the drop threshold for a certain time period (time margin) to prevent a ping-pong effect.

When each cell is divided into sectors and a terminal is moving from one sector to another sector of the same cell site, we can apply the same handover scheme. This is called *softer handover*.

The effects of soft and softer handover are as follows:

☞ Improvement of transmission quality of a terminal at the fringe area

☞ Reduction of the required dynamic range of the power control

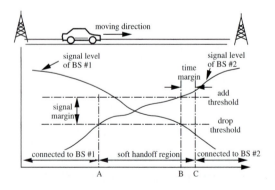

Fig. 7.17 Outline of the soft handover process.

☞ Suppression of the transmission power of the terminal, thereby saving battery lifetime

☞ Suppression of the short-duration peak of the intercell interference, thereby reducing total CCI

On the other hand, the soft handover could reduce system capacity if a sufficient amount of spreading codes are not prepared for the downlink.

7.3.6 Synchronization Specific for CDMA

In CDMA systems, the following synchronization is required.

☞ Chip timing synchronization
☞ Spreading code synchronization
☞ Frame synchronization
☞ Carrier synchronization
☞ Fading variation tracking

Chip timing synchronization techniques will be discussed in chapter 8.

Many methods have been proposed for spreading code synchronization, such as the delay lock loop [7-32] and the coherent matched filter [7-33]. The matched filter is the most popular scheme for code synchronization.

Figure 7.18 shows the configuration of the matched filter. When chip timing synchronization has already been taken before spreading code synchronization, we can employ the configuration shown in Figure 7.18(a). Let's assume that the PN sequence with its code length of N_p is employed, the amplitude of each chip is 1 or -1, and a chip duration is T_c. The sequence can be expressed as the following equation:

$$\beta(t) = \sum_{l=0}^{N_p-1} \beta_l \delta(t - lT_c) \qquad (7.9)$$

and the tap vector and tap gain vector of the matched filter are given by

$$\boldsymbol{y}_n = \left[y(nT_c), \ y\big((n-1)T_c\big), \ \dots \ , y\big((n-N_p+1)T_c\big) \right]^T \qquad (7.10)$$

$$\boldsymbol{h}_n = \left[\beta_{N_p-1}, \ \beta_{N_p-2}, \ \dots \ , \beta_2, \ \beta_1, \ \beta_0 \right]^T \qquad (7.11)$$

When all the components stored in the tap vector are the same as the corresponding tap gain, the output of the matched filter given by

$$z_n = \boldsymbol{h}_n^{*T} \boldsymbol{y}_n \qquad (7.12)$$

becomes N_p. On the other hand, when the tap vector is not the same as the tap gain vector, z_n becomes -1 due to the autocorrelation characteristics of the PN code. There-

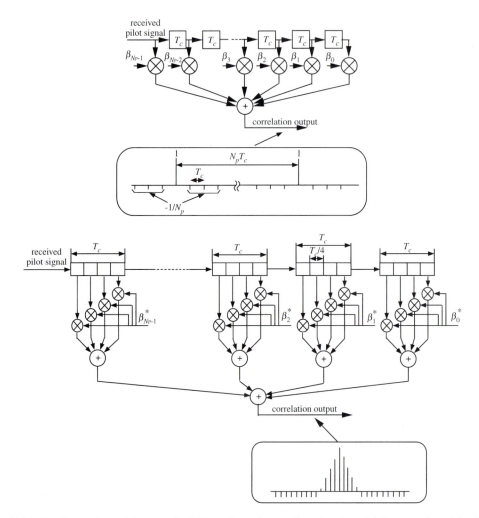

Fig. 7.18 Configuration of the matched filter for code synchronization: (a) T_c-spaced matched filter; (b) oversampling-type matched filter.

fore, when z_n is normalized by N_p, z_n becomes 1 only when the tap vector and the tap gain vector are the same; otherwise, it is $-1/N_p$. Because the spreading code is repeated with a period of N_p-chip duration as shown in Figure 7.18(a), we can detect the synchronization timing by searching the regularly appearing peak timing of the matched filter output.

However, we can employ Figure 7.18(a)'s type of spreading code synchronization circuit only when chip timing synchronization is taken before this synchronization. Unfortunately, this is impossible because of low C/I_c before despreading of the DS/CDMA systems. Therefore, we have to employ an oversampling technique to take spreading code synchronization. Figure 7.18(b) is the spreading code synchronization circuit using the oversampling technique, where its sampling rate is four times the chip rate as an example. In this case, the tap vector and tap gain vector of the matched filter are given by

$$
\boldsymbol{y}_n = \left[y(nT_c),\ y\left(nT_c - \frac{T_c}{4}\right),\ y\left(nT_c - \frac{T_c}{2}\right),\ y\left(nT_c - \frac{3}{4}T_c\right), \right.
$$

$$
\left. y((n-1)T_c),\ \dots\ , y\left((n-N_p+1)T_c - \frac{3}{4}T_c\right) \right]^T
\tag{7.13}
$$

$$
\boldsymbol{h}_n = \left[\beta_{N_p-1},\ \beta_{N_p-1},\ \beta_{N_p-1},\ \beta_{N_p-1},\ \beta_{N_p-2},\ \dots\ , \right.
$$

$$
\left. \beta_1,\ \beta_1,\ \beta_1,\ \beta_1,\ \beta_0,\ \beta_0,\ \beta_0,\ \beta_0 \right]^T
\tag{7.14}
$$

and the output of the matched filter is given by equation (7.12). The waveform of the correlator output is shown in Figure 7.18(b). As shown in this figure, the optimum sampling point corresponds to chip timing. Therefore, we can take chip timing synchronization if we search the regularly appearing local maximum point with its period of a chip duration. This chip timing synchronization scheme will be detailed in chapter 8.

7.4 COMPARISON OF FDMA, TDMA, AND CDMA

Table 7.1 shows a comparison of the features of FDMA, TDMA, and CDMA systems. From the viewpoint of system configuration, FDMA is the simplest access scheme of the three. However, it is not suitable for achieving high-capacity voice transmission systems using a low-bit-rate voice codec and spectral efficient modulation schemes because it requires very high stability of the oscillator. Moreover, variable transmission rate control is very difficult in the case of FDMA because it requires K-set of modems to achieve variable transmission rate control from R_b (bit/s) to KR_b (bit/s). As a result, no second-generation cellular system applies the FDMA scheme at present. Furthermore, it is very difficult for FDMA systems to monitor the received signal level of the adjacent cells for channel reassignment or handover processes.

When we apply TDMA, although we can mitigate the requirement for carrier frequency stability and achieve variable transmission rate control using a modem, we need a highly accurate slot, frame, or superframe synchronization. Moreover, we have to develop anti-frequency-selective fading techniques if the number of slots in each frame (N_{ch}) is large. Furthermore, the transmitter amplifier should be operated at K times higher peak power than the average power. Fortunately, we can solve these problems at present thanks to extensive developments in timing control techniques, adaptive equalizing techniques, and high-power-efficient power amplifier techniques. Another important advantage of TDMA systems is that we can measure the received signal level of adjacent cells during idle time slots. Such received signal level measurement is very effective for the handover process as well as for achieving C/I_c-based dynamic channel assignment.

Table 7.1 Comparison of the features of FDMA, TDMA, and CDMA systems.

	FDMA	**TDMA**	**CDMA**
Timing control	not required	required	required
Carrier frequency stability	high stability is required	low stability is acceptable if large number of channels are multiplexed	low stability is acceptable if chip rate is sufficiently high
Near-far problem	not affected	not affected	fast power control is required
Peak/average power ratio	1	K	1
Variable transmission rate	difficult	easy	easy
Anti-multipath fading techniques	- diversity - high coding rate FEC	- diversity - high coding rate FEC - adaptive equalizer (if N_{ch} is large)	- Rake diversity - low coding rate FEC - fast power control
Received signal level monitoring	difficult	easy	easy
Suitable zone radius	any size is OK	any size is OK (time alignment required)	large size is not suitable

In the case of CDMA, the most serious problem is the near-far problem as we have discussed before. The near-far problem is now solved by fast power control techniques. In addition to mitigating the near-far problem, the fast power control technique is also effective for improving receiver sensitivity because it makes the received signal level constant. Moreover, the following CDMA-specific techniques can further improve receiver sensitivity.

☞ Low-coding-rate FEC is applicable.

☞ Peak power is the same as the average power.

☞ Soft and softer handover is applicable.

Therefore, CDMA has the potential to achieve lower power consumption than TDMA or FDMA provided that very accurate power control is applicable, which will be discussed in chapter 16.

Another advantage for CDMA is that we can easily compensate for frequency-selective fading by using the path diversity technique. Furthermore, we can easily monitor the received signal level of adjacent cells just by changing a reference code at the correlator for the channel delay profile monitor because all the BSs use the same carrier frequency and chip rate. This feature is actively applied to the soft handover process.

On the other hand, in the case of CDMA, smaller zone radius is preferable for power control because larger zones require a wider dynamic range of the power control. Even in the case of TDMA, a larger zone radius requires longer guard time or accurate time alignment if smaller guard time is necessary. On the other hand, zone radius is limited only by the requirement of the transmitter power in the case of FDMA.

7.5 FREQUENCY DIVISION DUPLEX

FDD is the most popular duplex scheme for two-way radio communication systems because it can easily discriminate between uplink and downlink signals by filters. Actually, most of the land mobile communication systems other than the DECT and PHS employ FDD.

Figure 7.19 shows an example of spectrum allocation and the modem configuration of FDD systems. In the FDD systems, a different frequency band with its bandwidth of W_{sys} is employed for uplink and downlink. Moreover, transmission and reception are carried out through the same antenna. Therefore, a duplexer that discriminates the spectrum for uplink and downlink is inserted in both the BS and the terminal. In this case, the carrier frequency spacing should be sufficiently large from the hardware implementation point of view because shorter carrier spacing requires higher Q-value for the duplex filter. In PDC systems, 130 MHz is used for an 800- to 900-MHz band and 48 MHz is used for a 1.5-GHz band.

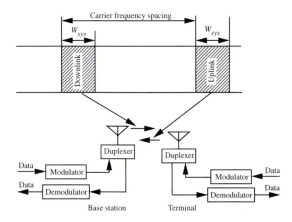

Fig. 7.19 An example of spectrum allocation and the modem configuration of FDD systems.

7.6 TIME DIVISION DUPLEX

TDD is another duplex scheme for two-way radio systems. In this scheme, both the BS and terminal transmit a signal over the same radio frequency channel but at different segments in time. Figure 7.20 shows an example of spectrum allocation and the modem configuration of TDD systems. In TDD systems, the uplink and downlink

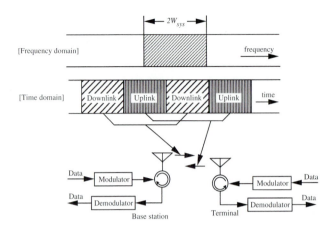

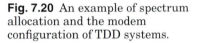

Fig. 7.20 An example of spectrum allocation and the modem configuration of TDD systems.

alternately use the same spectrum. Because each signal has to transmit data during half a period for FDD systems, the occupied bandwidth for each link is twice as wide as that for FDD systems, although the total bandwidth for FDD and TDD are the same bandwidth.

One of the most important features of the TDD system is that it does not require a duplexer that occupies a relatively large mass in the FDD modem because uplink and downlink signals are discriminated in the time domain. However, the TDD system requires guard space or time alignment as in the case of TDMA.

7.7 COMPARISON OF FDD AND TDD

Table 7.2 shows a comparison of the features for FDD and TDD systems. The most important feature of the FDD system is that it does not require any timing synchronization. This advantage is more important if the coverage area for each BS becomes large because a larger zone radius requires a larger dynamic range of the time alignment or a longer guard space. Moreover, FDD is more robust to delay spread because TDD requires twice as much symbol rate as FDD.

On the other hand, TDD does not require an RF duplexer which occupies a large amount of volume of the modem. Moreover, spectrum management will be more flexible if we employ TDD because we do not have to prepare a pair of spectra as in the case of FDD. Especially, it is a very important advantage for systems using discontinuous radio spectrum as in the case of the FPLMTS in which a radio spectrum of 1885–2025 MHz (140 MHz) and 2110–2200 MHz (90 MHz) are assigned and available on a worldwide basis by the ITU WARC-92 [7-34]. Furthermore, TDD is more suitable for channel condition monitoring because reciprocity of the propagation path characteristics [7-35] is satisfied if time difference between the transmission and reception time slots is very small. This feature is now actively used for the transmitter (BS) diversity in the PHS systems.

Table 7.2 Comparison between FDD and TDD systems.

Items	FDD system	TDD system
Required total bandwidth	same as TDD	same as FDD
Symbol rate	R_s	$2R_s$
Duplexer	necessary	not necessary
Flexibility of radio resource management	a pair of spectrums required	flexible
Immunity to multipath fading	more robust	less robust
Requirement to synchronization	no synchronization specific for FDD required	uplink and downlink timing synchronization required
Requirement to zone radius	applicable to either small cell or large cell systems	preferable to smaller cell systems
Reciprocity between uplink and downlink channels	not satisfied	satisfied for the desired signal
Transmission diversity	impossible	possible
Direct communication between terminals	possible	possible (easy)

REFERENCES

7-1. Kinoshita, K., Hata, M. and Hirade, K., "Digital mobile telephone system using TD/FDMA scheme," IEEE Trans. Veh. Technol., Vol. VT-31, pp. 153–57, November 1982.

7-2. Mallinder, B. J. T., "An overview of the GSM system," Proc. 3rd Nordic Seminar on DLMRC '88, pp. 1–4, September 1988.

7-3. DECT, "Digital European cordless telecommunication system–common interface specifications," Code RES-3(89), DECT, 1989.

7-4. EIA/TIA, "Cellular system: Dual-mode mobile station–base station compatibility standard," IS-54, Project 2215, Washington D.C., December 1989.

7-5. RCR, "Digital cellular telecommunication system," RCR STD-27, April 1991.

7-6. RCR, "Personal handy phone system," RCR STD-28, May 1993.

7-7. Namiki, J., "Block demodulation for short radio packet," *Electronics and Communication in Japan*, Vol. 67-B, No. 5, pp. 47–56, 1984 (Translated to English from Trans. IECE, Vol. 67-B, No. 1, pp. 54–61, January 1984).

7-8. Feher, K., ed., *Digital communications, satellite / earth station engineering*, Prentice-Hall Inc., Englewood Cliffs, New Jersey, 1983.

7-9. Ue, T., Sampei, S. and Morinaga, N., "Symbol rate and modulation level controlled adaptive modulation/TDMA/TDD for personal communication systems," 45th IEEE Veh. Tech. Conf. (Chicago, Illinois), pp. 306–10, July 1995.

7-10. CCIR Study Programme 18B/1, 1978.

7-11. Yue, O.C., "Spread spectrum mobile radio, 1977–1982," IEEE Trans. Veh. Technol., Vol. VT-32, No. 1, pp. 98–105, February 1983.

7-12. Goodman, D. J., Henry, P. S. and Prabhu, V. K., "Frequency-hopped multilevel FSK for mobile radio," Bell Syst. Tech. J., Vol. 59, No. 7, pp. 1257–75, September 1980.

7-13. Cook, C. E., Ellersick, F. W., Milstein, L. B. and Schilling, eds., *Spread spectrum communications*, IEEE Press, 1983.

7-14. Moriyama, E., Ikegami, S. and Kadokawa, Y., "Configuration and basic performance of direct-sequence spread-spectrum land mobile communication equipment," Rev. Radio Res. Lab., Vol. 30, No. 155, pp. 131–57, June 1984.

7-15. Mizuno, M., "Randomization effect of errors by means of frequency hopping techniques in a fading channel," IEEE Trans. Commun., Vol. COM-30, No. 5, pp. 1052–56, May 1982.

7-16. Mizuno, M., Nogami, H. and Kadokawa, Y., "Configuration and basic performance of SSFH land mobile communication equipment," Rev. Radio Res. Lab., Vol. 30, No. 154, pp. 49–60, June 1984.

7-17. Mizuno, M., Moriyama, E. and Saruwatari, T., "Diversity improvement in FH-MFSK land mobile commradio," ISAP'85 (Kyoto, Japan), pp. 569–72, August 1985.

7-18. Turin, G. L., "The effect of multipath and fading on the performance of direct-sequence CDMA systems," IEEE Trans. Veh. Technol., Vol. VT-33, No. 3, pp. 213–19, August 1984.

7-19. Gilhousen, K. S., Jacobs, I. M., Padovani, R., Viterbi, A. J., Weaver, L. A. Jr. and Wheatley, C. E. III, "On the capacity of a cellular CDMA system," IEEE Trans. Veh. Technol., Vol. 40, No. 2, pp. 303–12, May 1991.

7-20. Newman, D. B. Jr., "Communication and the law, FCC authorizes spread spectrum," *IEEE Communications Magazine*, Vol. 24, No. 7, pp. 46–47, July 1986.

7-21. TIA/EIA, "Mobile station—base station compatibility standard for dual mode wideband spread spectrum cellular system," TIA/EIA IS-95, 1993.

7-22. Ariyavisitakul, S., "Autonomous SIR-based power control for TDMA system," IEEE GLOBECOM'93 (Houston, Texas), pp. 307–10, November 1993.

7-23. Abeta, S., Sampei, S. and Morinaga, N., "Adaptive coding rate and processing gain control for cellular DS/CDMA systems," 4th ICUPC (Tokyo, Japan), pp. 241–45, November 1995.

7-24. Cameron, R. and Woerner, B. D., "An analysis of CDMA with imperfect power control," 42d IEEE Veh. Tech. Conf. (Colorado, Denver), pp. 977–80, May 1992.

7-25. Falciasecca, G., Gaiani, E., Missiroli, M., Muratore, F., Palestini, V. and Riva, G., "Influence of propagation parameters on cellular CDMA capacity and effects of imperfect power control," IEEE 2d International Symposium on Spread Spectrum Techniques and Applications (Yokohama, Japan), November 1992.

7-26. Viterbi, A. J. and Viterbi, A. M.., "Erlang capacity of a power controlled CDMA system," IEEE J. Select. Areas Commun., Vol. 11, No. 6, pp. 892–900, June 1993.

7-27. Viterbi, A. J., Viterbi, A. M. and Zehavi, E., "Performance of power controlled wideband terrestrial digital communication," IEEE Trans. Commun., Vol. 41, No. 4, pp. 559–69, April 1993.

7-28. Ariyavisitakul, S. and Chang, L. F., "Signal and interference statistics of a CDMA system with feedback power control," IEEE Trans. Commun., Vol. 41, No. 11, pp. 1626–34, November 1993.

7-29. Hayashi, M., Miya, K., Kato, O. and Homma, K., "CDMA/TDD cellular systems utilizing a base-station-based diversity scheme," 45th IEEE Veh. Tech. Conf. (Chicago, Ilinois), pp. 799–803, July 1995.

7-30. Viterbi, A. J., "Spread spectrum communications–myths and realities," *IEEE Communication Magazine*, Vol. 17, No. 3, pp. 11–18, May 1979.

7-31. Viterbi, A. J., "Very low rate convolutional codes for maximum theoretical performance of spread-spectrum multiple-access channels," J. Select. Areas Commun., Vol. 8, No. 4, pp. 641–49, May 1990.

7-32. Spilker, J. J. Jr., *Digital communications by satellite*, Prentice-Hall, Inc., Englewood Cliffs, New Jersey, 1977.

7-33. Hamamoto, N., Suzuki, R., Nishiyama, I., Miura, R. and Nishigaki, T., "PN-SS equipment for satellite communications with data demodulation by matched filters," Trans. IECE of Japan, Vol. J69-B, No. 11, pp. 1540–47, November 1986.

7-34. Resolution ITU-R 17, "Integration of future land mobile telecommunication systems (FPLMTS) into existing networks," 1993.

7-35. Lee, W. C. Y., *Mobile communication design fundamentals* (2d edition), John Wiley & Sons, Inc. (New York), 1993.

CHAPTER 8

Synchronization for Digital Wireless Transmission Systems

In digital wireless transmission systems, there are many synchronization circuits, especially in the receiver. Synchronization commonly used for both TDMA or CDMA systems includes

☞ Symbol timing synchronization (chip timing synchronization for CDMA)

☞ Carrier frequency synchronization (offset frequency compensation)

☞ Carrier phase synchronization

☞ Frame synchronization

In the case of CDMA, spreading code synchronization for DS/CDMA or frequency hopping pattern synchronization for FFH will also be required. Among these synchronization schemes, carrier phase synchronization techniques have already been discussed in chapters 4 and 5, and frame synchronization techniques have been discussed in chapter 7. Therefore, this chapter will discuss symbol timing synchronization, including chip timing synchronization for DS/CDMA and offset frequency compensation techniques.

As for the demodulation scheme, we can apply both sequential demodulation and block demodulation. Of these schemes, block demodulation is more frequently applied to the terminals for practical systems such as PDC, GSM, and IS-95. That and the fact that we already have many sources explaining conventional synchronization circuits for sequential demodulation systems leads me to discuss only synchronization schemes specific for block demodulation systems in this chapter.

8.1 SYMBOL TIMING SYNCHRONIZATION FOR TDMA SYSTEMS

Symbol timing synchronization is the most important circuit for receivers because recently developed AFC, fading compensation, and so on are performed using the baseband signal sampled at the symbol timing. Therefore, the symbol timing synchronization circuit has to be robust to frequency offset, DC offset, fading, and noise. Moreover, fast acquisition capability is also required from the viewpoint of high transmission efficiency and spectral efficiency.

8.1.1 Classification of the Symbol Timing Synchronization Schemes

Figure 8.1 shows classification of the symbol timing synchronization schemes. Symbol timing synchronization schemes are categorized into decision-directed and nondecision-directed schemes [8-1; 8-2]. In both cases, the optimum sampling timing can be obtained by the maximum likelihood estimations. Theoretically, the decision-directed methods give better jitter performance provided that the carrier phase and frame timing are already synchronized. However, the symbol timing synchronizations have to be taken before any other synchronizations if we would like to apply baseband signal processing-type carrier and frame synchronizations. Therefore, we will discuss only nondecision-directed methods in these sections.

8.1.2 Maximum Likelihood Estimation

Let's assume that the received baseband signal transmitted over a flat Rayleigh fading channel is expressed as

$$y(t) = re^{j\theta}s(t; \tau) + n(t) \tag{8.1}$$

$$s(t; \tau) = \sum_{i=-\infty}^{\infty} a_i u(t - iT_s - \tau) \tag{8.2}$$

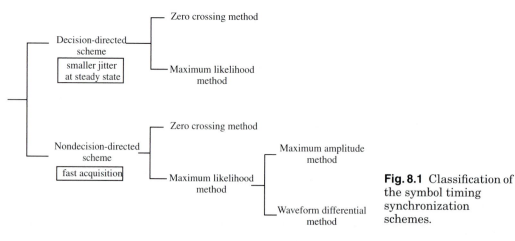

Fig. 8.1 Classification of the symbol timing synchronization schemes.

where T_s is the symbol duration, τ is the symbol timing, $u(t)$ is the wavewform for each symbol, and r and θ are the amplitude and phase distortion by flat Rayleigh fading.

When the noise component is AWGN, the likelihood function to estimate the optimum timing τ is given by [8-1; 8-2]

$$L(\tau) = \exp\left[-\frac{1}{4N_0}\int_{T_0}\left|y(t) - re^{j\theta}s(t;\tau)\right|^2\right] \tag{8.3}$$

where T_0 is the observation interval for the estimation of τ. Therefore, the optimum timing is the value that gives the maximum $L(\tau)$. As we can easily find, the maximization of $L(\tau)$ is equivalent to the maximization of the following equation called the *log-likelihood function*.

$$\Lambda(\tau) = \int_{T_0} re^{-j\theta}y(t)s^*(t;\tau)dt$$

$$= \int_{T_0} re^{-j\theta}y(t)\sum_{i=-\infty}^{\infty}a_i^*g(t - iT_s - \tau)dt \tag{8.4}$$

$$= \sum_{i=-\infty}^{\infty}a_i^*v_i(\tau)$$

where

$$v_i(\tau) = v_{Ii}(\tau) + j \cdot v_{Qi}$$

$$= \sum_{i=-\infty}^{\infty}(re^{-j\theta})q_i(\tau) \tag{8.5}$$

$$q_i(\tau) = \int_{T_0} y(t)g(t - iT_s - \tau)dt \tag{8.6}$$

$q_i(\tau)$ is the sampled value of the matched filter output at $t = iT_s + \tau$, where its input signal is $y(t)$ and its impulse response is $g(-t)$.

When we assume QPSK as a modulation scheme with its transmitted symbol of

$$a_i = a_{Ii} + ja_{Qi} \tag{8.7}$$

we can modify the log-likelihood function as

$$\Lambda(\tau) = \sum_{i=-\infty}^{\infty}a_{Ii}v_{Ii}(\tau) + \sum_{i=-\infty}^{\infty}a_{Qi}v_{Qi}(\tau) \tag{8.8}$$

Equation (8.8) gives the optimum sampling timing for the decision-directed scheme.

For the nondecision-directed scheme, the likelihood function is obtained by averaging the likelihood function over the data values (over and a_{Ii} and a_{Qi}). The likelihood function corresponding to equation (8.8) is given by

$$L(\tau) = \prod_{i=-\infty}^{\infty} \exp\left[a_{Ii}v_{Ii}(\tau)\right] \prod_{i=-\infty}^{\infty} \exp\left[a_{Qi}v_{Qi}(\tau)\right] \qquad (8.9)$$

Let's assume that the modulation scheme is QPSK and

$$\Pr[a_{Ii} = 1] = \Pr[a_{Ii} = -1] = \frac{1}{2} \qquad (8.10a)$$

$$\Pr[a_{Qi} = 1] = \Pr[a_{Qi} = -1] = \frac{1}{2} \qquad (8.10b)$$

The expected value of $\exp[a_{Ii}v_{Ii}(t)]$ is obtained as

$$E\left[\exp\left(a_{Ii}v_{Ii}(\tau)\right)\right] = \frac{1}{2}\exp\left(-v_{Ii}(\tau)\right) + \frac{1}{2}\exp\left(v_{Ii}(\tau)\right)$$
$$= \cosh\left(v_{Ii}(\tau)\right) \qquad (8.11)$$

By substituting equation (8.11) into equation (8.9), we can obtain the likelihood function as

$$L(\tau) = \prod_{i=-\infty}^{\infty} \cosh(v_{Ii}) \prod_{i=-\infty}^{\infty} \cosh(v_{Qi}) \qquad (8.12)$$

By taking the natural logarithm of equation (8.12), the log-likelihood function is obtained as

$$\Lambda(\tau) = \sum_{i=-\infty}^{\infty} \ln \cosh(v_{Ii}) + \sum_{i=-\infty}^{\infty} \ln \cosh(v_{Qi}) \qquad (8.13)$$

Because $\ln\cosh(x)$ can be approximated by $x^2/2$ for small x, we can obtain a very simple approximated log-likelihood function as

$$\Lambda(\tau) = \sum_{i=-\infty}^{\infty} v_{Ii}^2 + v_{Qi}^2$$
$$= \sum_{i=-\infty}^{\infty} |v_i|^2 \qquad (8.14)$$
$$= r^2 \sum_{i=-\infty}^{\infty} |q_i(\tau)|^2$$

In most of the wireless communication systems, fading variation is considered to be much slower than the symbol rate. In such a case, we can consider that the envelope variation due to fading is almost constant during observation period T_0. Therefore, we can further simplify the log-likelihood function as

$$\Lambda(\tau) = \sum_{i=-\infty}^{\infty} |q_i(\tau)|^2 \qquad (8.15)$$

Equations (8.14) and (8.15) can be interpreted that the average location of the maximum amplitude position during one symbol gives the optimum sampling point based on the maximum likelihood estimation regardless of the average received signal level. In other words, the optimum sampling point is the maximum eye opening point. Moreover, $\Lambda(\tau)$ is not affected by any phase distortion as we can observe in equation (8.14).

This scheme can also be applied to the multilevel modulation, such as M-ary QAM.

8.1.2.1 Maximum Amplitude Method (MAM)

Because the approximated log-likelihood function searches the average maximum point during a symbol period, we will call the method using equation (8.15) the *maximum amplitude method* (MAM) here [8-3].

Let's assume that the sampling frequency of the A/D converter is

$$f_{AD} = M / T_s + f_{off} = Mf_s + f_{off} \qquad (8.16)$$

where f_s is a symbol transmission rate and f_{AD} is not perfectly synchronized to the symbol rate. Therefore, there is a frequency offset (f_{off}) in equation (8.16). The effect of f_{off} depends on the burst length in TDMA systems or the frame length in CDMA systems. When a TDMA burst or a CDMA frame includes N_{total} symbols and instability of the oscillation frequency (f_{off}/f_{AD}) is 3.0×10^{-6}, timing offset at the end of the burst or frame is given by

$$T_{off} / T_{AD} = 3.0 \times 10^{-6} N_{total} \qquad (8.17)$$

where T_{off} is the timing offset at the end of the burst or frame and T_{AD} is the sampling period (= $1/f_{AD}$). Equation (8.17) shows that, when N_{total} is less than 1,000, we can ignore the effect of f_{off} because T_{off}/T_{AD} is negligibly small (less than 0.01). Actually, N_{total} is up to 500 in most cases. Therefore, we will ignore the effect of f_{off} here.

Let's assume that the sampling timing for the A/D converter is

$$t_{k,m} = kT_s + (m / M)T_s \qquad (8.18a)$$

where

$$k = 0, \ 1, \ 2, \ .... \qquad (8.18b)$$

$$m = 0, \ 1, \ 2, \ ...., \ M - 1 \qquad (8.18c)$$

In this case, the instantaneous log-likelihood function at $t = t_{k,m}$ is given by

$$\sigma_{MAM}(k,m) = \left| v_k \left(\frac{mT_s}{M} \right) \right|^2 \qquad (8.19)$$

and the log-likelihood function is given by

$$\Lambda_{MAM}(m) = \sum_{k=N_b}^{N_b-1+N_a} \sigma_{MAM}(k,m) \qquad (8.20)$$

In this equation, this accumulation is carried out from N_b-th symbol to $(N_b - 1 + N_a)$-th symbol, where N_a is the number of symbols to be accumulated. The optimum sampling point m_a is then obtained by searching m which gives the maximum value of $\Lambda_{MAM}(m)$.

Figure 8.2 shows the configuration of symbol timing synchronization circuit using MAM. Symbol timing is detected by block estimation using a major portion of each burst to reduce error due to noise. At first, $\sigma_{MAM}(k, m)$ is calculated at the instantaneous likelihood calculator (ILC), and the calculated value is fed to the M-cyclic accumulator. In the M-cyclic accumulator, there are M accumulators and they are accessed in circular order with a period of M to accumulate $\sigma_{MAM}(k, m)$. After the accumulation of N_a symbols, the index of the accumulator counting the maximum value is determined as the sampling point at the sampling timing selector.

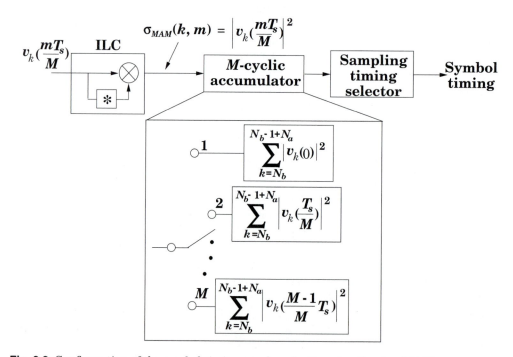

Fig. 8.2 Configuration of the symbol timing synchronization circuit using MAM.

8.1.2.2 Wave Differential Method (WDM)   The MAM operates to search the average location of the maximum eye opening points. This means that the optimum sampling timing corresponds to the average zero-slope position. The WDM is based on this idea [8-4].

Log-likelihood function for WDM is given by

$$\Lambda_{WDM}(m) = \sum_{k=N_b}^{N_b-1+N_a} \sigma_{WDM}(k,m) \tag{8.21}$$

$$\sigma_{WDM}(k,m) = \left| v_k\left(\frac{(m+1)T_s}{M} - \frac{(m-1)T_s}{M}\right) \right|^2 \tag{8.22}$$

After accumulation over N_a symbols, the index of the accumulator counting the minimum value is determined as the sampling timing at the sampling timing selector. Thus, the WDM can be implemented by changing the function of ILC from $\sigma_{MAM}(k, m)$ to $\sigma_{WDM}(k, m)$, and the decision logic of the sampling timing selector from maximum detection to minimum detection in Figure 8.2.

8.1.3 Zero-Crossing Method (ZCM)

One of the most popular symbol timing synchronization schemes is the ZCM [8-5; 8-6]. This scheme is very frequently used because it is easy to implement using digital logic devices, such as the digital phase-locked loop. For the application of ZCM to 16QAM, however, jitter performance of ZCM is degraded compared with that for BPSK or QPSK because jitter for 16QAM at zero-crossing points is larger.

One scheme for the ZCM is proposed by Gardner [8-6] in which the optimum sampling timing is given by searching the timing (m) that minimizes the following likelihood function.

$$\Lambda_G(m) = \sum_{k=N_b}^{N_b-1+N_a} \sigma_G(k,m) \tag{8.23}$$

$$\sigma_G(k,m) = \left| v_k\left(\frac{m-M/2}{M}T_s\right)\left\{ v_k\left(\frac{m}{M}T_s\right) - v_{k-1}\left(\frac{m}{M}T_s\right)\right\} \right|^2 \tag{8.24}$$

In the case of BPSK, when two consecutive symbols include the same data (these two symbols have the same polarity), $v_k(m/M) - v_{k-1}(m/M) = 0$ is satisfied if m is the sampling timing. On the other hand, when $v_k(m/M)$ and $v_{k-1}(m/M)$ have opposite polarity with each other, $v_k((m-M/2)T_s/M)$ becomes 0 if m is the sampling timing. Therefore, the optimum sampling timing (m) is obtained by searching m that minimizes equation (8.24).

8.1.4 Jitter Performance

Symbol timing synchronization has to be robust to various types of distortion, such as noise, fading, and frequency offset. First of all, let's discuss timing jitter performance against noise under static conditions.

Figure 8.3 shows computer-simulated results of the rms jitter versus the number of accumulated symbols (N_a) for ZCM, WDM, and MAM under AWGN conditions for (a) BPSK and (b) 16QAM; the parameters used in these simulation are

- ☞ Modulation scheme: BPSK, 16QAM
- ☞ Tx and Rx filters: root roll-off filter (roll-off factor is 0.5)
- ☞ A/D over-sampling rate: $M = 32$

and rms jitter normalized by a symbol duration (T_s) is defined as

$$\tau_{rms} / T_s = \frac{1}{T_s} \sqrt{\sum_{i=0}^{M-1} (\tau_i - \tau_{opt})^2 p(\tau_i)} \tag{8.25}$$

where $p(\tau_i)$ is the probability for the selected sampling timing of τ_i and τ_{opt} is the optimum sampling timing.

When the modulation scheme is BPSK, all the methods show almost the same performance at $E_b/N_0 = 10$ dB. However, when $E_b/N_0 = 2$ dB, rms jitter performance of ZCM is degraded in comparison with that for WDM and MAM because the SNR around the zero crossing point is extremely low in the low E_b/N_0 case. In the case of 16QAM, on the other hand, the performance of the ZCM is much worse than that for WDM and MAM at both $E_b/N_0 = 2$ dB and 10 dB because jitter around the zero crossing point is larger than that of BPSK.

When we design symbol timing synchronization circuits, we have to define tolerance of timing jitter for each modulation scheme. Figure 8.4 shows BER versus normalized rms jitter performances of (a) BPSK and (b) 16QAM. From Figure 8.4(a), we

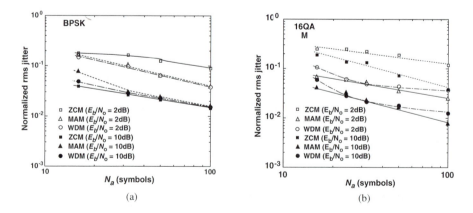

Fig. 8.3 Computer-simulated results of the rms jitter versus the number of accumulated symbols (N_a) for ZCM, WDM, and MAM under AWGN conditions for (a) BPSK and (b) 16QAM.

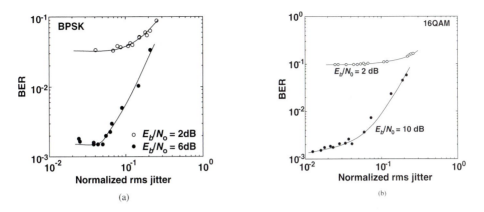

Fig. 8.4 BER versus normalized rms jitter performance of (a) BPSK and (b) 16QAM.

can find that BER is not degraded when the normalized rms jitter is less than 5.0×10^{-2} for $E_b/N_0 = 6.0$ dB, and it is less than 1.0×10^{-1} for $E_b/N_0 = 2.0$ dB. These results and the data in Figure 8.3(a) show that $N_a = 25$ is necessary for MAM and WDM and that $N_a = 70$ is necessary for ZCM in the case of BPSK. In the same manner, $N_a = 25$ for MAM and WDM and $N_a = 150$ for ZCM are required for 16QAM, as shown in Figure 8.3(b) and Figure 8.4(b). In the practical TDMA systems, one burst consists of around 100 symbols. Therefore, MAM and WDM have sufficient jitter performances for such systems. On the other hand, the jitter performance of ZCM could sometimes be insufficient if a burst length is shorter than 100 symbols in duration.

As shown in Figure 8.3 and Figure 8.4, 16QAM requires much more severe jitter performance than BPSK. Moreover, if the symbol timing synchronization circuit operates very well in the case of 16QAM, it will also operate well in the case of QPSK or BPSK. Therefore, we will discuss performance in the following sections in terms of 16QAM only.

8.1.5 Roll-off Factor Dependency

The number of accumulated symbols (N_a) versus rms jitter greatly depends on the roll-off factor (α) of the transmitter and receiver filters because a smaller roll-off factor produces more ISI. Figure 8.5 shows the normalized rms jitter versus N_a with a parameter of α at (a) $E_b/N_0 = 2$ dB and (b) $E_b/N_0 = 10$ dB, where 16QAM is employed. We can find from these figures that smaller α requires more N_a to obtain a certain value of rms jitter, because smaller α causes larger ISI at the nonoptimum sampling point, which results in degrading discrimination performance of the sampling timing. For example, $\alpha = 0.2$ requires $N_a = 50$, which is twice as large as that for $\alpha = 0.5$.

8.1.6 Robustness to Offset Frequency

Offset frequency compensation is indispensable for the receiver in land mobile communications because carrier frequency is changing due to temperature variation.

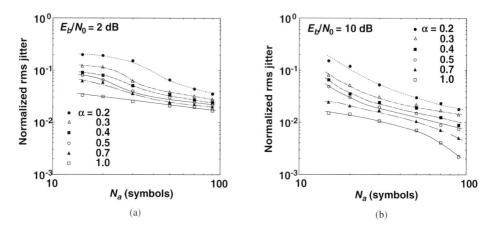

Fig. 8.5 Normalized rms jitter versus N_a with a parameter of α at (a) E_b/N_0 = 2 dB and (b) E_b/N_0 = 10 dB, where 16QAM is employed.

Offset frequency can be compensated for at either IF band or baseband. From the view of implementation, however, baseband signal processing is preferable as we will discuss later. For this purpose, symbol timing synchronization has to be taken before the offset frequency compensation stages, which means that the symbol timing synchronization should be robust to offset frequency.

Figure 8.6 shows normalized rms jitter versus offset frequency performances of 16QAM using MAM and WDM under AWGN conditions. Because the log-likelihood function for symbol timing does not include a phase component, jitter performances of MAM and WDM are not degraded by the offset frequency.

8.1.7 Robustness to DC Offset

DC offset can also be compensated for at either the IF band or baseband. However, for the same reason as that for the offset frequency compensation, baseband signal processing is preferable [8-7; 8-8; 8-9; 8-10]. Therefore, symbol timing synchronization should also be robust to DC offset.

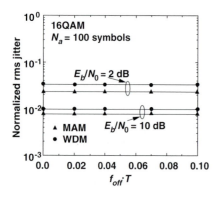

Fig. 8.6 Normalized rms jitter versus offset frequency performances of ZCM, MAM, and WDM under AWGN conditions.

Figure 8.7 shows normalized rms timing jitter versus DC offset under AWGN conditions for 16QAM, where $N_a = 100$, $\alpha = 0.5$, and the DC offset ratio is defined as

$$DC_{off} = \frac{\text{amplitude of DC} - \text{offset}}{\text{maximum amplitude of the transmitted symbols}} \qquad (8.26)$$

Because the average peak position or the average zero-slope position during a symbol is not affected by the DC offset value, the jitter performance of the MAM and WDM are not degraded by the DC offset.

8.1.8 BER Performance Degradation Due to Symbol Timing Synchronization in Flat Rayleigh Fading Environments

Figure 8.8 shows BER performance of 16QAM for nondiversity reception and two-branch selection diversity reception in flat Rayleigh fading environments, where $N_a = 100$ for any symbol timing synchronization scheme, and pilot symbol-aided fading compensation (discussed in chapter 4) is employed. When ZCM is employed, we can observe huge degradation due to large symbol timing jitter in both nondiversity and diversity reception cases. On the other hand, we can obtain very good BER performance when we employ MAM or WDM schemes.

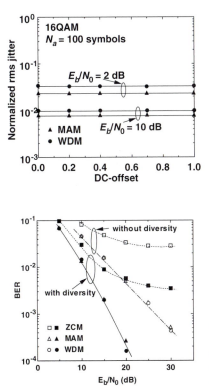

Fig. 8.7 Normalized rms timing jitter versus DC offset for 16QAM under AWGN conditions, where $N_a = 100$ and $\alpha = 0.5$.

Fig. 8.8 BER performance of 16QAM for nondiversity reception and two-branch selection diversity reception in flat Rayleigh fading environments, where $N_a = 100$ for any symbol timing synchronization scheme, and pilot symbol-aided fading compensation is employed.

8.1.9 Symbol Timing Synchronization in Frequency-Selective Fading Environments

In flat Rayleigh fading environments, the optimum sampling point is time-invariant as shown in equation (8.15). On the other hand, in frequency-selective fading environments, the maximum eye opening point is time-variant due to the instantaneous variation of the power ratio of the direct path to ISI caused by the delayed paths [8-11; 8-12]. Moreover, when we apply a diversity reception, which is one of the very effective and simple anti-frequency-selective fading techniques when delay spread is less than $0.1T_s$, the optimum sampling point of each branch could be different because the instantaneous complex delay profile of each branch is independent from the others. This suggests that it is a better strategy to prepare a symbol timing synchronization circuit for each branch.

Figure 8.9 shows the configuration of the diversity reception based on this idea [8-3]. In the receiver, symbol timing synchronization for each diversity branch is independently taken, and fading for each branch is compensated for using the independently regenerated symbol timing for each branch. After that, selection diversity is carried out, and the transmitted data sequence is regenerated.

Figure 8.10 shows the irreducible BER versus normalized delay spread performance for nondiversity reception and diversity reception cases in frequency-selective fading environments, where the modulation scheme is 16QAM and $f_d T_s = 3.125 \times 10^{-4}$ [8-3]. This figure also includes the performance for conventional symbol timing synchronization in which the optimum sampling timing is obtained by the average excess delay time of the delay profile [8-11; 8-12]. In this case, the sampling timing for each branch is the same.

First let's discuss the performance for nondiversity reception. As shown in Figure 8.10, MAM and WDM, in comparison to ZCM, are very effective for improving delay spread immunity because the half-symbol delayed timing from zero crossing is not always the maximum eye opening timing in frequency-selective fading environ-

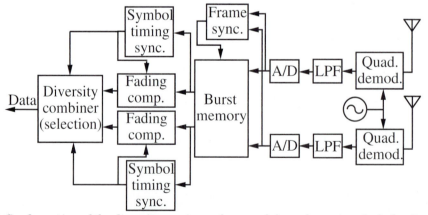

Fig. 8.9 Configuration of the diversity receiver, where each branch receiver includes its own symbol timing synchronization circuit.

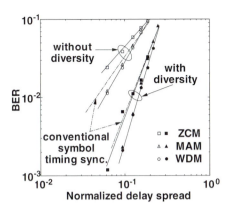

Fig. 8.10 Irreducible BER versus normalized delay spread performance for nondiversity reception and diversity reception cases in frequency-selective fading environments, where modulation scheme is 16QAM and $f_d T_s = 3.125 \times 10^{-4}$.

ments. Furthermore, the performances of MAM and WDM are slightly better than the performance of the conventional scheme in the case of nondiversity reception. This result means that the BER performance can be improved when the instantaneous maximum eye opening point is tracked as the symbol timing.

When we employ a diversity reception, superiority of MAM and WDM to ZCM becomes more remarkable. We can find that, although the performance of ZCM is worse than that for the conventional scheme, MAM and WDM show better performance than the conventional one. When we compare the performance at BER = 10^{-2}, the delay spread immunity of MAM and WDM is improved by about 30% over that of the conventional scheme.

8.2 CHIP TIMING SYNCHRONIZATION FOR CDMA SYSTEMS

In the case of CDMA, chip timing synchronization is more difficult than symbol timing synchronization for narrowband TDMA systems because ISI caused by frequency-selective fading completely closes the eye pattern of each chip.

One method for chip timing synchronization is to detect the average local maximum point of the delay profile [8-13]. As discussed in chapters 4 and 5, the delay profile of the channel can be obtained by taking correlation between the received baseband signal of the pilot channel and the spreading code. Figure 8.11 shows the configuration of the chip timing synchronization circuit using delay profile and its illustrative operation. Let's assume that the sampling timing for the A/D converter is defined as

$$t_{k,l,m} = kT_f + lT_c + \frac{m}{M}T_c \qquad (8.27)$$

where T_c is a chip duration; sampling frequency for A/D conversion is $f_{AD} (= Mf_{chip} = M/T_c)$; $T_f (= N_p T_c)$ is the accumulation interval for correlation; $k = 0, 1, 2, ..., N_{ap} - 1$ (N_{ap} is the number of accumulated delay profiles); $l = 0, 1, 2, ..., N_p - 1$ (N_p is the number of chips in one cycle of the spreading code); and $m = 0, 1, 2, ..., M-1$. As shown in Figure

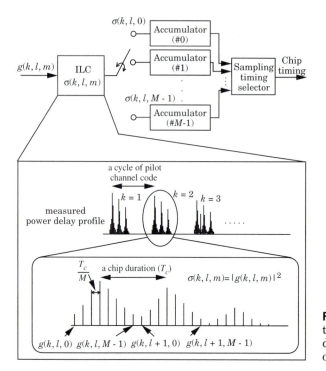

Fig. 8.11 Configuration of the chip timing synchronization circuit using delay profile and its illustrative operation.

8.11, the local peak of this delay profile can be obtained every T_c second, and this timing corresponds exactly to the chip timing. For example, in Figure 8.11, $m = 3$ corresponds to this timing. Therefore, when we obtain the output of the matched filter at $t = t_{k,l,m}$ ($g(t_{k,l,m}) = g(k, l, m)$), calculate its instantaneous likelihood value by

$$\sigma(k, \ l, \ m) = \left| g(k, \ l, \ m) \right|^2 \tag{8.28}$$

and accumulate it to the M-cyclic accumulator, the following log-likelihood function will be obtained in the m-th accumulator as

$$\Lambda(m) = \sum_{k=0}^{N_{ap}} \sum_{l=0}^{N_p-1} \sigma(k, \ l, \ m) \tag{8.29}$$

As a result, the optimum chip timing can be obtained by searching a cyclic accumulator that has the maximum accumulated value.

Figure 8.12 shows BER performance of a DS/CDMA system using a suppressed pilot channel-aided sounding technique discussed in chapter 4, section 4.3, and the chip timing synchronization scheme discussed here under flat Rayleigh fading and frequency-selective fading conditions, where $N_{ap} = 16$, $N_p = 256$, $f_d = 40$ Hz, and the two-ray Rayleigh model is assumed as the frequency-selective fading channel. We can confirm from this figure that the chip timing synchronization scheme works very well even under flat Rayleigh and frequency-selective fading conditions.

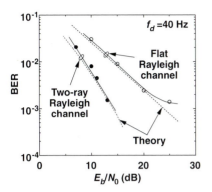

Fig. 8.12 BER performance of a DS/CDMA system with the chip timing synchronization scheme under flat Rayleigh fading and frequency-selective fading conditions, where f_d = 40 Hz and the two-ray Rayleigh model is assumed as the frequency-selective fading channel.

8.3 OFFSET FREQUENCY COMPENSATION

Offset frequency is an inevitable cause of performance degradation in wireless communication systems because the carrier frequency is gradually changing due to temperature drift.

If an offset frequency exists between the carrier frequency and the frequency of the local oscillator at the receiver (f_{off}), the power spectrum density of the channel can be observed by the receiver as [8-14]

$$S(f) = \frac{b_0}{\pi f_d \sqrt{1 - \left(\frac{\left(f - f_{off}\right)^2}{f_d}\right)}} \tag{8.30}$$

This means that the channel variation, including the effect of f_{off}, is determined by f_{off} + f_d. If various fading compensation techniques can also compensate for the phase variation due to offset frequency, we do not have to prepare an AFC circuit, thereby simplifying the receiver circuit configurations. However, this is not usually the case. Therefore, we need AFC to reduce the offset frequency to a smaller value that can be compensated for by the fading compensation techniques.

First, let's discuss the compensable range of the offset frequency by the fading compensation techniques. Figure 8.13 shows BER versus f_{off} + f_d performance for 16QAM at E_b/N_0 = 40 dB where a pilot symbol-aided fading compensation technique is employed [8-14]. This figure shows that fading compensation techniques can compensate for f_{off} if f_d + f_{off} is smaller than the acceptable maximum channel variation speed for these techniques. For example, when the acceptable highest-frequency component of the channel variation is 150 Hz and the maximum value of f_d is 80 Hz, we can accept f_{off} of less than 70 Hz.

In practical systems, the regulated frequency stability of the local oscillator is about 3×10^{-6}. This means that, if the carrier frequency is about 1 GHz, for example, the carrier frequency could have an uncertainty of up to 3 kHz. Therefore, some AFC function is necessary in the receiver.

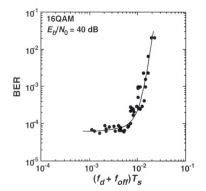

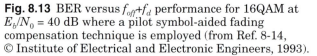

Fig. 8.13 BER versus $f_{off}+f_d$ performance for 16QAM at $E_b/N_0 = 40$ dB where a pilot symbol-aided fading compensation technique is employed (from Ref. 8-14, © Institute of Electrical and Electronic Engineers, 1993).

8.3.1 Phase Rotation Detection Method Using a Unique Word

One of the simplest schemes for AFC is to detect the average phase transition between consecutive unique word symbols [8-15; 8-16]. Assume the phase transition between $(k-1)$-th and k-th symbols in the transmitted unique word is θ_k, and the corresponding phase transition of the received signal is ϕ_k. The phase transition error is then given by

$$d\phi_k = \phi_k - \theta_k = 2\pi f_{off} T_s + \phi_{noise} \tag{8.31}$$

Practically, ϕ_k is equivalently obtained this way.

Let's assume that the received baseband signal of k-th unique word symbol is expressed as $u_k = u_{I,k} + ju_{Q,k}$. ϕ_k and $d\phi_k$ are then given by

$$\begin{aligned} \exp(j\phi_k) &= u_k u_{k-1}^* \\ &= (u_{I,k}u_{I,k-1} + u_{Q,k}u_{Q,k-1}) + j(u_{Q,k}u_{I,k-1} - u_{I,k}u_{Q,k-1}) \\ &= U_k + jV_k \end{aligned} \tag{8.32}$$

$$\begin{aligned} \exp(jd\phi_k) &= \exp(j\phi_k)\exp(-j\theta_k) \\ &= (U_k + jV_k)(\cos\theta_k - j\sin\theta_k) \\ &= (U_k\cos\theta_k + V_k\sin\theta_k) + j(V_k\cos\theta_k - U_k\sin\theta_k) \\ &= X_k + jY_k \end{aligned} \tag{8.33}$$

The f_{off} is then given by

$$f_{off} = \frac{1}{2\pi T_s}\tan^{-1}\left(\sum_{k=0}^{N_{uw}-1} Y_k \Big/ \sum_{k=0}^{N_{uw}-1} X_k\right) \tag{8.34}$$

where N_{uw} is the number of unique word symbols.

In this method, phase transition between two consecutive symbols has to be $-\pi/2$, 0, or $\pi/2$ because we cannot detect the direction of the phase rotation in the case of $-\pi$ or π phase transitions. Moreover, the maximum detectable f_{off} is limited to $|f_{off}| < 1/(4T_s)$ because $\tan^{-1}\theta$ is continuous in the range of $-\pi/2 < \tan^{-1}\theta < \pi/2$. For example, when a system applies a symbol rate of 20 ksymbol/s, the detectable f_{off} is limited to 5 kHz.

8.3.2 Delay Profile Rotation Detection Method using Pilot Channel in the CDMA System

In the CDMA system, a pilot channel is multiplexed in the downlink to measure the delay profile for the downlink at each terminal. Therefore, if a local oscillator in each terminal can take synchronization to the carrier frequency of the downlink, each terminal can generate its uplink carrier frequency synchronized to the BS, thereby canceling out the offset frequency between the BS and terminal in both downlink and uplink.

Because the phase of the pilot channel is not modulated over the whole range, we can arbitrarily select the time range (T_m) for measuring the average phase regardless of the symbol rate of the traffic channel [8-17; 8-18].

Let's assume that the chip timing and PN synchronization timing have already been taken. When we multiply the received baseband signal by the synchronized PN sequence and integrate it over the range of $(t - T_m/2, t + T_m/2)$ as discussed in chapter 5, we can obtain the received phase $\phi(t)$. Because the pilot channel carries an unmodulated signal, the phase difference $d\phi(t) = \phi(t) - \phi(t - T_m)$ represents phase rotation during a time interval of T_m. Therefore, when we average $d\phi(t)$ over a certain range, we can obtain f_{off} the same way we did it for the phase rotation detection method.

8.3.3 AFC for Wideband TDMA Systems

In the conventional receiver for narrowband wireless communication systems, a PLL-based AFC circuit such as the fourth power loop is usually employed [8-5]. However, when this type of AFC is applied to wideband wireless communication systems, it cannot normally operate under severe frequency-selective fading conditions because the fourth power circuit sometimes outputs nothing due to delayed waves [8-19]. We can remove this effect by inserting a matched filter whose impulse response is determined by the delay profile before the fourth power loop [8-19].

REFERENCES

8-1. Proakis, J. G., *Digital communications* (2d Edition), McGraw-Hill, New York, 1989.

8-2. Feher, K., ed., *Digital communications, satellite/earth station engineering*, Prentice Hall, New Jersey, 1981.

8-3. Sampei, S. and Feher, K., "Improvement of delay spread immunity by using symbol timing synchronization based on maximum likelihood estimation for 16QAM/TDMA diversity receivers," Electron. Letters, Vol. 29, No. 22, pp. 1917–18, October 1993.

8-4. Suzuki, T., Takatori, H., Ogawa, M. and Tomooka, K., "Line equalizer for a digital subscriber loop employing switched capacitor technology," IEEE Trans. Commun., Vol. COM-30, No. 9, pp. 2074–82, September 1982.

8-5. Gardner, F. M., *Phaselock technique*, John Wiley & Sons, Inc., New York, 1979.

8-6. Gardner, F. M., "A BPSK/QPSK timing-error detection for sampled receivers," IEEE Trans. Commun., Vol. COM-34, No. 5, pp. 423–29, May 1986.

8-7. Cavers, J. K., "Adaptive compensation for imbalance and offset losses in direct conversion receivers," 41th IEEE Veh. Tech. Conf. (St. Louis, Missouri), pp. 457–61, May 1991.

8-8. Kage, K., Sasaki, Y., Ichikawa, M. and Sato, T., "The feasibility study of the Nyquist baseband filtered 4-level FM for digital mobile communications," 35th IEEE Veh. Tech. Conf. (Boulder, Colorado), pp. 200–4, May 1985.

8-9. Sampei, S. and Feher, K., "Adaptive DC-offset compensation algorithm for burst mode operated direct conversion receivers," 42d IEEE Veh. Tech. Conf. (Denver, Colorado), pp. 93–96, May 1992.

8-10. Dien, W. S., Dang, N. and Feher, K., "Performance improvement method for DECT and other non-coherent GMSK systems," 42d IEEE Veh. Tech. Conf. (Denver, Colorado), pp. 97–100, May 1992.

8-11. Chuang, J. C. I., "The effects of multipath delay spread on timing recovery," IEEE ICC'86, pp. 3.1.1–3.1.5, June 1986.

8-12. Adachi, F. and Ohno K., "BER performance of QDPSK with post-detection diversity reception in mobile radio channel," IEEE Trans. Veh. Technol., Vol. VT-40, No. 1, pp. 237–49, February 1991.

8-13. Abeta, S., Sampei, S. and Morinaga, N., "DS/CDMA coherent detection system with a suppressed pilot channel," GLOBECOM'94 (San Francisco, California), pp. 1622–26, November 1994.

8-14. Sampei, S. and Sunaga, T., "Rayleigh fading compensation for QAM in digital land mobile communications," IEEE Trans. Veh. Technol., Vol. 42, No. 2, pp. 137–47, May 1993.

8-15. Kinoshita, N. et al., "Field experiments on 16QAM/TDMA and trellis coded 16QAM/TDMA systems for digital land mobile radio communications," IEICE Trans. Commun., Vol. E-77-B, No. 7, pp. 911–20, July 1994.

8-16. Yoshino, H. and Ueda, T., "Frequency offset compensation for adaptive equalizer," 1990 Autumn Natl. Conv. Rec. IEICE, P.2-285, March 1990.

8-17. Jeong, J. W., Sampei, S. and Morinaga, N., "Large Doppler frequency compensation techniques for DS/CDMA LEO mobile satellite communication systems," 2d Asia-Pacific Conf. on Commun. (Osaka, Japan), pp. 474–78, June 1995.

8-18. Ogura, K. and Serizawa, M., "Carrier and frame synchronization for TDMA digital mobile communications," ISITA'90 (Hawaii), pp. 801–4, November 1990.

8-19. Okanoue, K., Nagata, Y. and Furuya, Y., "A new MLSE receiver with carrier frequency offset compensator for TDMA mobile radio," Trans. IEICE (B-II), Vol. J73-B-II, No. 11, pp. 736–44, November 1990.

CHAPTER 9

System Performance Evaluation for the Development of Air Interfaces

When we specify air interfaces (physical layer) of a digital wireless system, we have to select the best combination of items such as

☞ Modulation/access scheme
☞ Symbol rate at the air interface
☞ Shape of the transmitter and receiver filters
☞ Frame format

While making our selections, we must consider, among other things, the

☞ Bit rate of terminals to be serviced
☞ Quality of services
☞ Ease of implementation
☞ Zone coverage
☞ Electric magnetic compatibility (EMC) with other systems
☞ Propagation path characteristics, especially delay spread statistics in the service area
☞ System capacity
☞ Protocol of higher layer signaling
☞ Ease of system extension in the future

For this radio interface and specification process, however, it is absolutely impossible to implement an experimental hardware system that can evaluate all the possible system parameter combinations. Fortunately, we have very powerful computer resources these days, and we can very accurately simulate hardware of a digital wireless system. Consequently, the mechanics of efficiently combining computer simulation; hardware development, including laboratory tests; and field trials are very important for the development of future digital wireless personal communication systems.

In this chapter, we will talk about how to conduct computer simulation, laboratory tests, and field trials, and how to combine these three evaluation techniques in an efficient manner.

9.1 COMPUTER SIMULATION TECHNIQUES

9.1.1 Advantage of Computer Simulation

These days, because we have powerful and low-cost computers such as workstations, we can simulate and evaluate almost all parts of the hardware on both the circuit level and the system level. Moreover, digital signal processing (DSP) technologies have been actively applied not only to the voice codec or FEC, but also to the modulator/demodulator and synchronization circuit in order to make small, maintenance-free, low-cost terminals (discussed in chapters 5 through 8). Consequently, we can develop and evaluate a new addition to the hardware, including its optimization, in a short time without actually making the hardware.

Although computer simulation is a very powerful tool in the development of digital wireless systems, we must be very careful with simulation. Most important, the simulation model must be very close to the real system—it should include all the parts that could degrade performance. This means that those who want to simulate a system should know exactly what the actual hardware problems are.

When performance of the developed hardware is degraded from the simulated result, we have to make a great effort to reduce the degradation or to find out reasons for such degradation. Furthermore, simulated results should agree with the theoretical performance if every cause of performance degradation is perfectly removed.

Figure 9.1 shows a configuration of a transmitter and receiver for an FDMA system. We usually achieve a complex baseband signal generator using DSP as shown in this figure [9-1]. In this system, there are many points that distort the signal. Some of them are

- Distortion and out-of-band radiation due to digital to analog (D/A) conversion in the complex baseband signal generator
- Imperfection of quadrature modulator due to carrier leakage, phase error of the quadrature carrier, and imbalance between in-phase and quadrature components
- Nonlinearity of the RF amplifier in the transmitter
- Imperfection of the AFC and AGC operations
- Timing jitter
- Imperfection of the demodulator operation

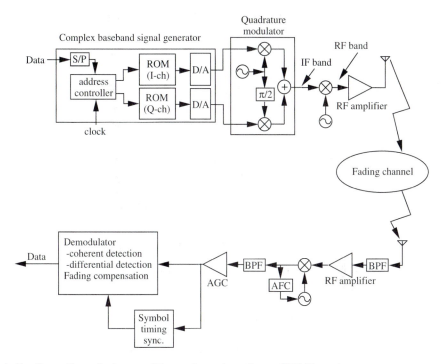

Fig. 9.1 Configuration of a transmitter and receiver for an FDMA system.

Moreover, propagation path characteristics cause severe performance degradation [9-2], [9-3]. For example, when the propagation path characteristics are subject to
flat Rayleigh fading, fast phase variation (random FM) causes irreducible error,
although the received signal level is very high as discussed in chapter 4. Thus, we also
have to simulate the propagation path characteristics.

On the other hand, some of the hardware-oriented problems—such as the optimization of the level diagram, interference between analog and digital circuits, and
so on—cannot be solved only by computer simulation. So we have to efficiently combine computer simulation, laboratory experiments, and field trials when developing a
system.

Figure 9.2 shows a flowchart of system development by computer simulation,
hardware development, and field trial. Computer simulation is very suitable for preliminary performance evaluation and optimization of some parameters. It can also be
used to investigate requirements of the hardware; for example, to what extent are the
linearity and power efficiency of the transmitter RF amplifier required. These preliminary investigations should be carried out simultaneously on both system performance and hardware development. After that, a test system will be developed.

In the test system, some hardware-oriented problems—such as interfaces between
each circuit and level diagrams—are investigated. Then, the performance of the developed system is compared with the computer simulation results. If the performance of

the test system is inferior to the simulation results, we will modify the hardware to reduce the degradation or to find out the reasons for it. Some of the results we get from doing this will be fed back to the simulation program to improve the accuracy of the sim-

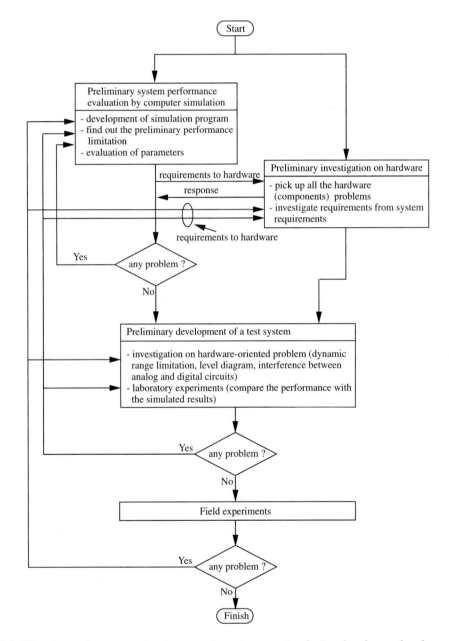

Fig. 9.2 Flowchart of a system development by computer simulation, hardware development, and field trial.

ulation. After that, the following field trial will verify usefulness of the test system under real propagation conditions. Of course, some field trial results will be fed back to computer simulation and hardware development.

9.1.2 Simulation Model

Figure 9.3 shows a simulation model of the FDMA system shown in Figure 9.1, where the simulation model is expressed by using equivalent low-pass systems. When signal bandwidth (B) is much smaller than its carrier frequency ($B \ll f_c$, f_c is carrier frequency), we can express a system using equivalent low-pass systems in which everything, such as band-pass filters, RF amplifiers, and so on, is expressed by the low-pass-equivalent characteristics [9-4]. Actually, performance of the equivalent low-pass system is very close to that of the real hardware because the only difference between band-pass and equivalent low-pass systems is whether or not frequency conversions are included; degradation caused by the frequency conversion is indeed very small if input signal levels and the local oscillator signals are appropriately adjusted.

The transmitted baseband signal can be expressed as

$$a(t) = \sum_{k=-\infty}^{\infty} a_k \delta(t - kT_s) \tag{9.1}$$

where a_k means k-th transmitted symbol, T_s is a symbol duration, and $\delta(t)$ means the delta function. This signal is then band-limited by the low-pass filter (LPF). When the impulse response of a transmitter is expressed as $h_T(t)$, the output of LPF is given by

$$
\begin{aligned}
s(t) &= s_I(t) + j \cdot s_Q(t) \\
&= a(t) \otimes h_T(t) \\
&= \sum_{k=-\infty}^{\infty} a_k h(t - kT_s)
\end{aligned} \tag{9.2}
$$

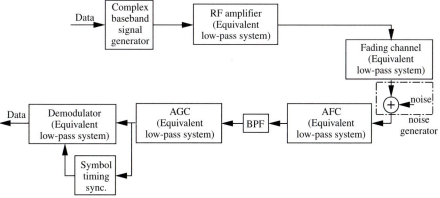

Fig. 9.3 Computer simulation model for FDMA system shown in Figure 9.1.

where $\otimes$ means convolutional integral. When we apply DSP techniques as shown in Figure 9.1, $s(t)$ can be generated by look-up ROM tables [9-1]. Especially, when the performance is highly dependent on filter characteristics as in the case of a small roll-off factor, the ROM look-up table method is very effective.

At the RF transmitter amplifier, the amplitude and phase of the signal are non-linearly distorted. To simulate such nonlinearity, the following quadrature model can be used [9-5].

$$A = |s(t)| \tag{9.3}$$

$$u(t) = u_I(t) + j \cdot u_Q(t) \tag{9.4}$$

$$u_I(A) = \sum_{i=0}^{\infty} f_i A^i \tag{9.5}$$

$$u_Q(A) = \sum_{i=0}^{\infty} g_i A^i \tag{9.6}$$

As a result, the output of this amplifier is given by

$$x(t) = u(A)s(t) \tag{9.7}$$

When the amplifier is completely linear and produces no phase distortion, the following relationships will be satisfied.

$$u_I(A) = 1.0 \tag{9.8}$$

$$u_Q(A) = 0.0 \tag{9.9}$$

On the other hand, when the amplifier has nonlinearity, we have to determine f_i and g_i by identifying the real amplifier input-output characteristics.

When we simulate the multipath radio channel, we usually use the N-ray model expressed as

$$y(t) = \sum_{i=1}^{N} c_i(t)x(t - \tau_i) \tag{9.10}$$

$$c_i(t) = c_{Ii}(t) + j \cdot c_{Qi}(t) \tag{9.11}$$

where τ_i is the delay time relative to the direct wave and $c_i(t)$ is the complex envelope variation of the i-th component [9-6]. The variation of $c_i(t)$ is subject to the complex Gaussian random process, and its bandwidth is limited to $-f_d$ to f_d, where f_d is the maximum Doppler frequency [9-7].

There are two methods of generating a complex band-limited Gaussian random signal—the multitone method [9.7] and the PN-method [9-8; 9-9]. The multitone method is suitable for computer simulation. On the contrary, the PN-method is suitable for hardware implementation of the fading simulator.

When we simulate the frequency-selective fading channel, we have to determine average power of $c_i(t)$, τ_i, and f_d. When, τ_i is negligible (small) compared with T_s, the channel can be regarded as a flat Rayleigh fading channel, and the received signal can be expressed as

$$y(t) = c(t)x(t) \tag{9.12}$$

$$c(t) = c_I(t) + j \cdot c_Q(t) \tag{9.13}$$

At the receiver, although the RF amplifier and BPF are employed in the actual hardware, these parts are removed in Figure 9.3 because degradation caused by these parts are negligible except that noise is generated in the RF amplifier. Therefore, the noise generator, expressed as an equivalent low-pass system, is inserted in Figure 9.3.

AFC, BPF, and AGC are also expressed by equivalent low-pass systems. AFC works to reduce the offset frequency (f_{off}) between the carrier and the local oscillator, and AGC controls the envelope of the received signal. When the effect of frequency offset exists, the received signal can be modified as

$$y(t) = c(t)x(t)\exp(j2\pi f_{off}t) \tag{9.14}$$

At the demodulator, $c(t)$ and the offset frequency are compensated for and the estimated data sequence $a'(t)$ is obtained. Demodulator and symbol timing synchronization circuits are crucial to receiver operation. Thus, development of these parts should be the main task for the development of digital wireless systems. At the same time, we have to check the spectrum at any point by using FFT to evaluate intrasystem and intersystem ACI and to determine the optimum combination of filters, symbol rate, and frame format. In the case of TDMA systems, development of slot synchronization will also be required.

Figure 9.4 shows simulated results of BER performance of a 16QAM/TDMA system in a flat Rayleigh fading environment, where the symbol rate is 16 ksymbol/s [9-9; 9-10]. Theoretical performance is also shown in this figure as a reference. The figure shows that the simulated results are degraded by 2 dB from the theoretical results. At this stage, we have to find out whether this degradation is reasonable (see Figure 9.2). In the case of Figure 9.4, theoretical investigation showed that this degradation is due to fading compensation, and it is quite reasonable [9-11]. With such consideration, we can find that the simulation software and the developed algorithm work well.

When we compare the transmitted data sequence $a(t)$ and regenerated sequence $a'(t)$ in this simulation, we can evaluate the BER performance and the bit error sequence, along with its statistics, such as p.d.f. of burst length, p.d.f. of error-free length, and so on. Analysis of the bit error sequence is very important for the development of error control technologies such as FEC, because the error sequence is bursty

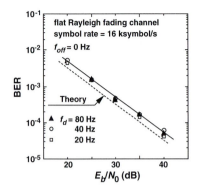

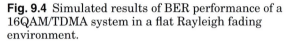

Fig. 9.4 Simulated results of BER performance of a 16QAM/TDMA system in a flat Rayleigh fading environment.

rather than random when the signal is transmitted via a Rayleigh fading channel, and the performance of FEC greatly depends on the burstiness of the error sequence [9-12]. Moreover, it is not always necessary to simulate all blocks in the hardware when we evaluate the performance of FEC or voice codec because its performance depends solely on the error sequence.

Figure 9.5(a) shows a digital radio channel model that expresses a relationship between sequences of transmitted data, regenerated data, and error. When we evaluate the performance of a voice codec with FEC and interleaver, its transmission model can be expressed as Figure 9.5(b). Voice bits $d(n)$ are first encoded and interleaved. Let's express the encoded data as $d_1(n)$ and the interleaving process as $\Phi(\cdot)$. After the interleaved data are transmitted via the channel as shown in Figure 9.5(a), the regenerated data sequence after demodulation is given by

$$d_2(n) = \Phi(d_1(n)) \oplus e(n) \tag{9.15}$$

When $d_2(n)$ is deinterleaved, we can obtain the deinterleaved sequence as

$$d_3(n) = \Phi^{-1}(d_2(n)) = d_1(n) + \Phi^{-1}(e(n)) \tag{9.16}$$

At this stage, burst error $e(n)$ is randomized by the deinterleaving. After FEC decoding, we can obtain a sequence

$$d_4(n) = d(n) + e'(n) \tag{9.17}$$

We can optimize the parameters of FEC and the interleaver by measuring the statistics of $e'(n)$. Thus, if we obtain an error sequence for some specific conditions, we can evaluate the quality of the voice codec as shown in Figure 9.5(c).

Figure 9.6 shows a configuration of a voice codec evaluation system for PDC systems [9-13]. First, a sample of a voice is transmitted from a personal computer (PC) to a voice coder, and the coded voice bits are returned to the PC. In the PC, a bit error is added using a bit error sequence file. This part corresponds to the digital radio channel model shown in Figure 9.5(a). Then the voice bits including error are fed to the voice decoder, and the decoded data are returned to the PC.

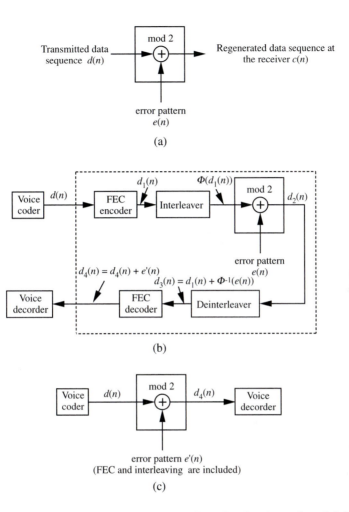

Fig. 9.5 Digital voice data transmission models: (a) digital radio channel model that expresses a relationship between the sequences of the transmitted data, received data, and error; (b) data transmission model for a voice codec including FEC and interleaver; (c) digital radio channel model including FEC and interleaver.

9.2 LABORATORY EXPERIMENTS

9.2.1 Tests for Transmitter and Receiver

In the air interface specifications of digital wireless and personal communication systems, although there are some parameters that can be determined by computer simulation, there are many parameters that should be determined based on laboratory experiments considering device quality, ease of adjustment, and cost. Based on

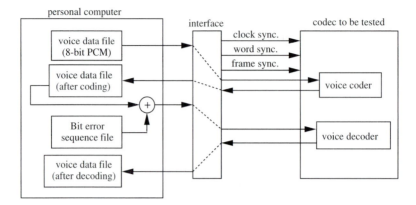

Fig. 9.6 Configuration of a voice codec evaluation system.

these extensive evaluations, some of the parameters related to radio signal as listed below are specified by the standardization institutes, such as the TIA committee in the United States, ETSI in Europe, and RCR (ARIB) in Japan.

For a Transmitter

☞ Tolerance of carrier frequency

☞ Tolerance of transmitted power

☞ Spurious emission

☞ 99% bandwidth

☞ Radiated signal power to the adjacent and next-adjacent channels

☞ Unwanted radiation

For a Receiver

☞ Spurious response

☞ Adjacent channel selectivity

☞ Intermodulation performance

☞ Unwanted radiation

9.2.2 BER Performance Evaluation Tests

Figure 9.7 shows a configuration of a laboratory experiment system for BER measurement. Binary sequence is generated by a pseudorandom binary sequence (PRBS) generator. Propagation path characteristics are generated by a fading simulator [9.8; 9.9]. First we have to measure the power of the desired signal (P_d) and the power of the interference signal (P_i) using an average power measurement instru-

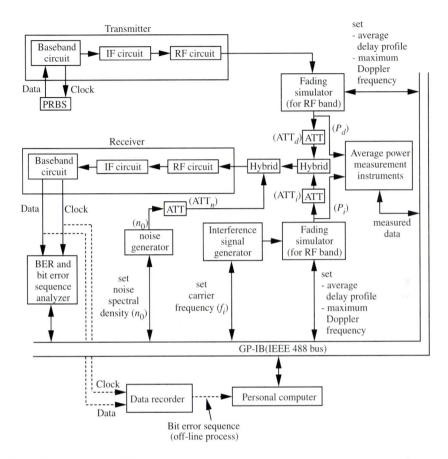

Fig. 9.7 Configuration of a laboratory experiment system.

ment, such as a power meter. When the signal level of the desired or interfered signal is dynamically changing due to fading, the average period for the power measurement should be sufficiently long.

Using the measured values of P_d and P_i and attenuators ATT_d and ATT_i, we can set the (C/I) as

$$\frac{C}{I}(\text{dB}) = P_d(\text{dBm}) - P_i(\text{dBm}) - \text{ATT}_d(\text{dB}) + \text{ATT}_i(\text{dB}) \qquad (9.18)$$

Moreover, we have to set E_b/N_0 by adjusting ATT_d and the attenuator for N_0 (ATT_n) as

$$\frac{E_b}{N_0}(\text{dB}) = P_d(\text{dBm}) - 10\log_{10} R_d(\text{bit/s}) \qquad (9.19)$$
$$- N_0(\text{dBm/Hz}) - \text{ATT}_d(\text{dB}) + \text{ATT}_n(\text{dB})$$

where R_d is a bit rate of the desired signal. When the noise generator does not have a function to set N_0, we can obtain N_0 by the measurements of the noise generator output power (P_n) and equivalent noise bandwidth (B_n) as $N_0 = P_n / B_n$. When we evaluate the receiver sensitivity, the relationship between the received signal level and E_b / N_0 is given by

$$\frac{E_b}{N_0}(\text{dB}) = P_d(\text{dBm}) - 10\log_{10} R_d(\text{bit/s}) - \text{NF(dB)}$$

$$- (10\log_{10} kT + 30)(\text{dBm/Hz}) - \text{ATT}_d(\text{dB}) + \text{ATT}_n(\text{dB})$$

(9.20)

where $k = 1.38 \times 10^{-23}$ Joules per degree Kelvin (J/K), T is temperature in Kelvin, 30 is a conversion factor from decibel(s) referred to 1 W (dBW) to dBm, and NF is a noise figure in dB.

As discussed in chapter 6, not only BER but also bit error sequence is very important to assess FEC and voice codec. When we compare the original data sequence and the regenerated data sequence, we can obtain the bit error sequence. We can also calculate the average BER by counting the number of errors per unit time.

Figure 9.7 also shows a configuration of an automatic BER and bit error sequence measurement system using a general purpose bus interface (GP-IB) (IEEE 488 bus). Because most of the measurement instruments have a GP-IB interface, we can set every parameter and collect any data using a PC.

If we do not have instruments with a GP-IB interface, we can manually set every parameter for those instruments. If we do not have an appropriate BER and bit error sequence analyzer, we can measure BER and bit error sequence by recording the waveform of regenerated data and associated clock in the data recorder, A/D converting these waveforms, and calculating the bit error sequence and BER with a computer using the digitized data. This procedure will be detailed next.

Figure 9.8 shows laboratory experimental results of BER performance of a 16QAM/TDMA system, where NF = 11.5 dB is used [9.14]. Theoretical BER performance is also shown in this figure. This degradation is almost the same as that of the simulated results except for the degradation in the high received signal level. This

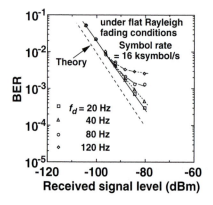

Fig. 9.8 Laboratory experimental results of BER performance for a 16QAM/TDMA system.

degradation is caused by imperfection of the analog circuit, such as DC offset, and by the lack of dynamic range of AGC and the A/D converter [9.14].

9.3 FIELD EXPERIMENTS

9.3.1 Path Loss Measurement

When we want to introduce a digital wireless communication system, we have to specify transmitted power, antenna gains for a BS and terminals, and so on, taking into consideration the service area and outage probability. Therefore, we usually estimate the coverage and outage probability using some empirical formula, such as Hata's equation [9-15], with some typical parameters, such as the path loss decay factor of $\alpha_p = 3.5$ and the standard deviation of log-normal distribution of $\sigma = 6$–8 dB. However, actual propagation path condition is more complicated. Thus, we need to measure the actual path loss of the service area.

Figure 9.9 shows a configuration of the path loss measurement system. Received signal level, time, and distance pulse that is generated when a mobile moves a fixed length—for example, 10 cm—are recorded in the data recorder [9-16]. At the same time, the landscape of the measured course is recorded by a video recorder to get landscape information of the test course. After the measurement, the recorded data are A/D converted. During the A/D conversion process, received signal level and time are converted to digital data by triggering distance pulse. With this process, we can obtain the received signal level variation with respect to distance. When distance pulse is generated with a very short period, we can obtain both large-scale and small-scale variation of the path loss.

Conversion from the received signal level to path loss is given by

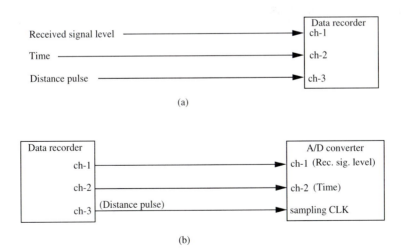

Fig. 9.9 Configuration of a path loss measurement system: (a) data recording process; (b) A/D conversion process.

$$L_p(\text{dB}) = P_T(\text{dBm}) + \left(G_T - L_{ft}\right)(\text{dB})$$
$$+ \left(G_R + L_{fr} - L_{BPF}\right)(\text{dB}) - P_R(\text{dBm}) \tag{9.21}$$

where

P_T = transmitted signal power
G_T = transmitter antenna gain
G_R = receiver antenna gain
L_{ft} = transmitter feeder loss
L_{fr} = receiver feeder loss
L_{BPF} = receiver filter loss
P_R = received signal level

Another important issue is how to obtain the received signal level with a large dynamic range. To do this, we usually use a logarithmic amplifier. In this case, however, we need a calibration of the input-output characteristics. Figure 9.10 shows a calibration procedure for the logarithmic amplifier. As a standard signal, we can use a frequency-calibrated frequency synthesizer. The procedure is

1. Set the frequency of the synthesizer.
2. Connect the output of the synthesizer to the input terminal of the signal level measurement circuit.

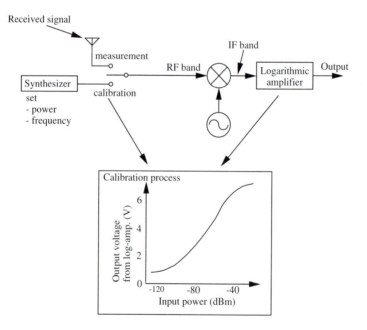

Fig. 9.10 Power calibration procedure for a logarithmic amplifier.

3. Set the output power level of the synthesizer and obtain the output voltage of the logarithmic amplifier.

4. Change the output level of the synthesizer by a fixed step, for example 5 dB, and repeat processes 3 and 4.

5. Interpolate the obtained data.

Then, we can obtain a function that shows the relationship between the output voltage and the input signal level as shown in Figure 9.10.

As for the logarithmic amplifier, a large dynamic range is required. The spectrum analyzer is a preferred logarithmic amplifier. When we set the frequency span of 0 Hz and we set the resolution bandwidth and video bandwidth to the appropriate values, it works as a logarithmic amplifier, having a very large dynamic range—100 dB or more.

9.3.2 Delay Profile Measurement

Delay profile measurement is very important in determining the maximum symbol rate. Figure 9.11 shows a delay profile measurement system using a spread spectrum. In the transmitter, a carrier with a frequency of 775 MHz is BPSK-modulated using a 10-stage PN signal with its chip rate of 5 MHz [9-17]. Resolution of the delay time of the delay profile is 0.2 μsec because chip rate is 5 Mchip/s.

At the receiver, the received signal is down-converted to the IF band, and correlation between the received signal and the PN sequence is taken for 1 frame (1,023 bits) to obtain the complex delay profile. When the phase of the PN sequence is scanned, we can obtain the delay profile.

Because the PN length is 1,023 bits and the chip rate is 5 Mchip/s, it takes 204.6 μsec to obtain the complex envelope of a point in a delay profile. Moreover, we have to scan the delay time to obtain a delay profile, which is carried out to the change phase of the PN sequence by 1/2 chip every 204.6 μsec. In this case, PN phase shift by 1/2 chip corresponds to scanning the delay time every 0.1 μsec because 1 chip duration is

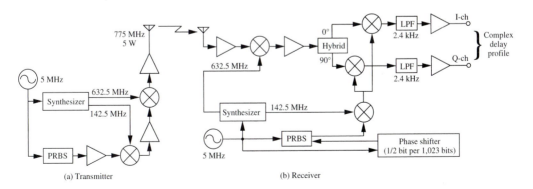

Fig. 9.11 Delay profile measurement system using a direct sequence spread spectrum: (a) transmitter; (b) receiver.

0.2 µsec. Thus, when we want to scan the delay time from 0 to k microseconds, it takes $(k/0.1) \times 204.6$ µs = $2.046k$ milliseconds to draw one delay profile.

On the other hand, when we want to analyze the spectrum of the delay profile in more detail, scan time to draw one delay profile should be twice or more than the maximum Doppler frequency according to the Nyquist theorem. For example, when we determine the maximum delay time for a delay profile of 5 µs, we can cope with the maximum Doppler frequency of up to 48.88 Hz.

Figure 9.12 shows a data acquisition and processing scheme for delay profile measurements. Instantaneous delay profiles of in-phase (I-ch) and quadrature phase (Q-ch) are recorded in the data recorder with other associated signals, such as time, distance pulse, and trigger pulse which indicate the start point of the delay profile. The landscape of the measured course is also recorded as mentioned in the previous section.

In the A/D conversion process, all data is A/D converted. We can arbitrarily determine the frequency of the sampling clock. When the frequency of sampling clock is equal to 4.89 kHz, we can obtain a discrete delay profile with an interval of 0.1 µs. Of course, we can directly A/D convert all data and store the data in the computer without a data recorder.

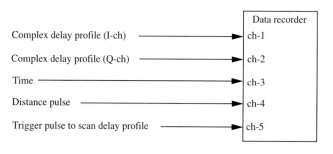

(a)

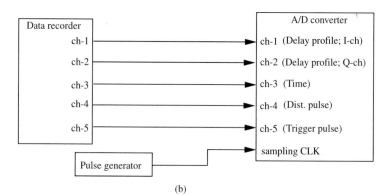

(b)

Fig. 9.12 Data acquisition and processing scheme for delay profile measurements: (a) data recording process; (b) A/D conversion process.

9.3.3 BER Performance Measurement

Figure 9.13 shows a bit error sequence measurement system for field experiments using a data recorder. When the bandwidth of the regenerated data sequence is wider than the bandwidth of the data recorder, we cannot record such a waveform. Figure 9.13(a) shows how to record the data sequence with a bit rate of 64 kbit/s in the data recorder when its bandwidth is 20 kHz. First, we convert 64 kbit/s serial sequence to 4-bit parallel data. After the S/P conversion, each sequence has a bit rate of 16 kbit/s. Next, 4-bit parallel data is divided into two groups, and each group is fed to a D/A converter to produce a 4-level analog signal. Then, these two sequences are fed to the data recorder.

Figure 9.13(b) shows the A/D conversion process of the recorded data. Because the recorded data sequence (ch-1 and ch-2 of the data recorder) as well as the sampling clock (ch-6) are distorted due to band limitation, we have to reshape the sam-

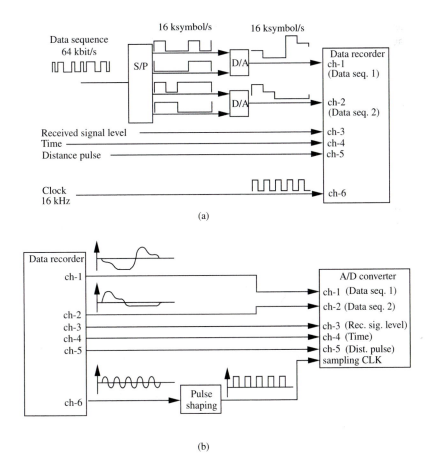

Fig. 9.13 Bit error sequence measurement system: (a) data recording process; (b) A/D conversion process.

pling clock. Then the received signal level, time, and distance pulse are A/D converted simultaneously.

Next we have to determine the averaging period to obtain average vehicular speed, average received signal level, and BER. We usually use the averaging period of 1–5 seconds. For example, when we calculate vehicular speed, the received signal level, and BER averaged over 1 second, we can obtain a table like the one shown in Table 9.1, where $e_i(n)$ means the error sequence for time i.

Figure 9.14 shows a photograph of a 16QAM/TDMA experimental system for field trial [9-14]. We will show you an example of the field experiment using this system. Figure 9.15 shows an example of the picture recorded in the VCR during the experiment. Along the course, time, average received signal levels (branch #1 and branch #2 for the diversity reception), distance from the start point, and a signal state diagram of the demodulated 16QAM signal are recorded. Figure 9.16 shows variations of average vehicular speed, average received signal level, and BER of a 16QAM/TDMA system as an example of the field measurement data, where the average time was selected as 1 second. When we compare these data and the recorded picture, we can check the relationship between the received signal level variation, the BER variation, and the manmade obstacles along the course.

Although maximum Doppler frequency and average E_b/N_0 are constant during BER measurements in the case of computer simulation or laboratory experiments, these two parameters are randomly changed with vehicular movement in the case of field experiments as shown in Figure 9.16. Thus, we have to classify the received signal level and the vehicular speed to compare the BER performance of the laboratory experiments and computer simulation.

For this purpose, we have to quantize their levels. In this case, when the quantization step is too small, the number of data to be averaged in each quantized level becomes small. On the contrary, when quantization level is too large, every condition is shrunk in one quantized level, and classification will lose sense.

Figure 9.17 shows median value of the BER versus an average received signal level with a parameter of the vehicular speed [9-14]. In this figure, the quantization step of the received signal level and the vehicular speed are 2 dB and 10 km/h, respectively. For example, the received signal of –109 dBm means –109 ± 1 dBm. When we

Table 9.1 An example of the processed data of field experiments.

Time (seconds)	Speed (km/h)	Received signal level (dBm)	BER	error sequence
1	0.0	–105.1	3.0e–2	$e_1(n)$
2	11.3	–106.2	4.0e–2	$e_2(n)$
3	25.0	–103.2	1.0e–2	$e_3(n)$
4	34.3	–101.3	3.5e–3	$e_4(n)$
.	.	.	.	.
.	.	.	.	.

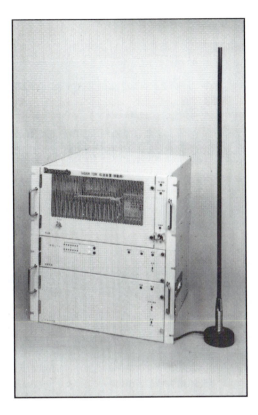

Fig. 9.14 Photograph of a 16QAM/TDMA experimental system for field trial (Courtesy of Matsushita Communication Industry Co. Ltd.)

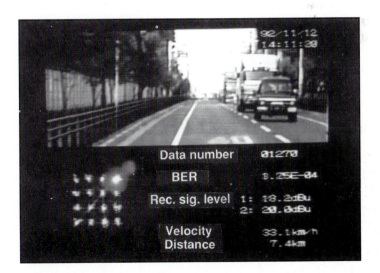

Fig. 9.15 Picture around the test course recorded by a VCR (Courtesy of Matsushita Communication Industry Co. Ltd.).

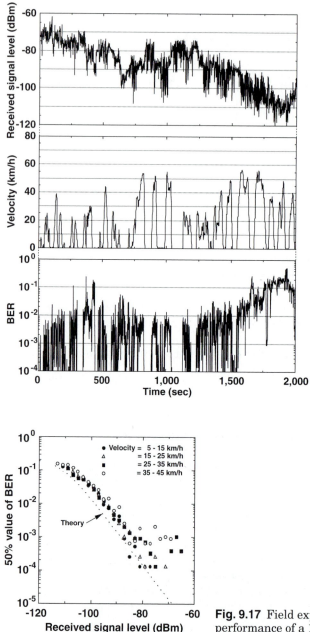

Fig. 9.16 Variation of the average received signal level, average vehicular velocity, and BER of a 16QAM/TDMA system.

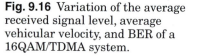

Fig. 9.17 Field experimental results of BER performance of a 16QAM/TDMA system.

compare Figure 9.8 and Figure 9.17, we may find that these performances are almost the same, and, with these results, we can conclude that the preliminary test of a digital wireless personal system is finished. Of course, we have to investigate more on the issues of quality and reliability improvements, IC developments, and so on.

REFERENCES

9-1. Okumura, Y., Ohmori, E., Kawano, T. and Fukada, K., "Field strength and its variability in VHF and UHF land mobile radio service," Rev. Elec. Commun. Lab., Vol. 16, pp. 825–71, September–October 1968.

9-2. Lee. C. Y., *Mobile communications design fundamentals* (2d Edition), John Wiley & Sons, Inc., New York, 1993.

9-3. Schwalts, M., Bennett, W. R. and Stein, S., *Communication systems and techniques*, McGraw-Hill, New York, 1966.

9-4. Jager, F. D. and Dekker, C. B., "Tamed frequency modulation, a novel method to achieve spectrum economy in digital transmission," IEEE Trans. Commun., Vol. COM-26, No. 5, pp. 534–42, May 1978.

9-5. Bhargava, V. K., Haccoun, D., Matyas, R. and Nuspl, P., *Digital communications by satellite*, John Wiley & Sons, Inc., New York, 1981.

9-6. COST 207 Management Committee, "Information technologies and sciences—digital land mobile radio communications," Commission of the European Communities, 1989.

9-7. Jakes, W. C., *Microwave mobile communications*, John Wiley & Sons, Inc., New York, 1974.

9-8. Hirade, K., Abe, F. and Adachi, F., "Fading simulator for land mobile radio communications," Trans. IECE of Japan, Vol. 58-B, No. 9, pp. 449–56, September 1975.

9-9. Kamio, Y., Sampei, S., Sasaoka, H. and Yokoyama, M., "A new type fading simulator with DSP," Trans. IEICE, Vol. E.70, No. 4, pp. 379–82, April 1987.

9-10. Kamio, Y. and Sampei, S., "Performance of a trellis-coded 16QAM/TDMA system for land mobile communications," IEEE Trans. Veh. Technol., Vol. 43, No. 3, pp. 528–36, August 1994.

9-11. Sampei, S. and Sunaga, T., "Rayleigh fading compensation for QAM in land mobile radio communications," IEEE Trans. Veh. Technol. Vol. 42, No. 2, pp. 137–47, May 1993.

9-12. Sampei, S. and Kamio, Y., "Performance of FEC with interleaving in digital land mobile communications," Trans. IECE of Japan, Vol. E68, No. 10, pp. 651–52, October 1985.

9-13. Miki, T. and Suda, H., "Speech CODEC evaluation system for Japanese digital cellular system," Technical Report of IEICE, RCS90-8, May 1990.

9-14. Kinoshita, N., Hiramatsu, K., Inogai, K., Honma, K., Sampei, S., Moriyama, E. and Sasaoka, H., "Field experiments on 16QAM/TDMA for land mobile communication systems," 2nd PIMPC (Yokohama, Japan), pp. 79–83, September 1993.

9-15. Hata, M., "Empirical formula for propagation loss in land mobile radio service," IEEE Trans. Veh. Technol., Vol. VT-29, No. 3, pp. 317–25, August 1980.

9-16. Moriyama, E., Iwama, T. and Saruwatari, T., "Experimental investigation of 1.5 GHz, 2.3 GHz and 2.6 GHz band land mobile radio propagation in urban and rural area," 39th IEEE Veh. Technol. Conf. (Orlando, Florida), pp. 311–15, May 1989.

9-17. Ohgane, T., Sampei, S., Kamio, Y., Sasaoka, H. and Mizuno, M., "UHF urban characteristics in wideband mobile radio communications," Trans. IEICE (B-II), Vol. J72-B-II, No. 2, pp. 63–71, February 1989.

CHAPTER 10

Radio Link Design

Spectral efficiency of a cellular system can be expressed both in terms of bit/s/Hz and in terms of the cell reuse factor. To improve the spectral efficiency, we usually apply a sector cell configuration. Furthermore, spectrum overlapping can be tolerated in cellular systems if adjacent channels are allocated not to the same cell but to geographically separated cells. This channel allocation is called *interleave channel allocation*. Another important factor for the radio link design is the coverage area because it is determined by the transmitter power of the terminal; thereby battery lifetime and terminal size are determined.

Therefore, we will discuss how to design coverage area and system capacity in wireless communication systems in this chapter.

10.1 COVERAGE AREA ESTIMATION

10.1.1 Level Diagram

When we evaluate the coverage area of a wireless communication system, we usually use the level diagram as shown in Figure 10.1. First of all, we have to calculate the minimum received signal level that satisfies a certain BER, which is called *receiver sensitivity* P_{th} (dBm). In the case of voice transmission, BER = 10^{-2} is usually assumed.

It is also important to know under which conditions the BER is evaluated—AWGN conditions or fading conditions. It depends on the holding time of a radio chan-

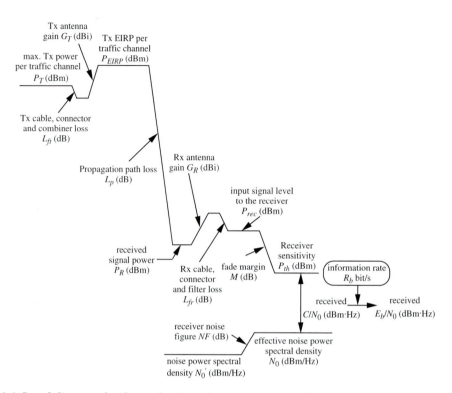

Fig. 10.1 Level diagram for the evaluation of coverage area estimation.

nel. When the holding time is relatively long compared to the fade duration as in the case of voice transmission, we have to evaluate the BER under fading conditions. In this case, the fade margin is obtained by the statistics of the large-scale signal variation, such as the log-normal fading.

On the other hand, when the holding time is relatively short as in the case of data transmission, we have to evaluate the BER under static conditions. In this case, fade margin is obtained by the combined statistics of both large-scale and small-scale signal variations. In practical wireless communication systems, however, the main service is considered to be voice communications. Therefore, we will assume only voice communication systems in our discussion.

At the receiver, dominant noise is generated at the front-end RF amplifier. At this stage, the power spectrum of the thermal noise is given by

$$N_0' = 10 \log_{10} kT \quad (\text{dBm} / \text{Hz}) \tag{10.1}$$

where $k = 1.38 \times 10^{-20}$ (mW/Hz/K) is Boltzmann's constant and T is temperature in degrees Kelvin (K). When $T = 300$ K, N_0' is about −173.8 dBm/Hz. Therefore, the noise power spectral density generated at the receiver RF amplifier is given by

$$N_0 \ (\mathrm{dBm} \,/\, \mathrm{Hz}) = 10 \log_{10} kT \ (\mathrm{dBm} \,/\, \mathrm{Hz}) + NF \ (\mathrm{dB}) \qquad (10.2)$$

where NF is the noise figure of the RF amplifier. Using equation (10.2), we can obtain the relationship between the minimum received signal level (P_{th}) and the required C/N_0 $[(C/N_0)_{req}]$ or required E_b/N_0 $[(E_b/N_0)_{req}]$ as

$$
\begin{aligned}
P_{th} \ (\mathrm{dBm}) &= \left(C \,/\, N_0 \right)_{req} \ (\mathrm{dB} \cdot \mathrm{Hz}) + N_0 \ (\mathrm{dBm/Hz}) \\
&= \left(E_b \,/\, N_0 \right)_{req} \ (\mathrm{dB}) \ + 10 \log_{10} R_b \ (\mathrm{Hz}) \\
& \qquad + N_0 \ (\mathrm{dBm/Hz})
\end{aligned}
\qquad (10.3)
$$

Let's define that P_T (dBm) is the transmitter power averaged over the assigned burst duration for a single traffic channel. Because there is a power loss due to the cable, connector, or combiner between the transmitter RF amplifier and the antenna, the transmitted power is reduced by a certain amount. Let's define the total power loss due to these factors as L_{ft} (dB). When we define the transmitter antenna gain as G_T dB relative to an isotropic radiator (dBi), the effective isotropically radiated power (EIRP) is given by

$$P_{EIRP} \ (\mathrm{dBm}) = P_T \ (\mathrm{dBm}) - L_{ft} \ (\mathrm{dB}) + G_T \ (\mathrm{dBi}) \qquad (10.4)$$

On the other hand, the received signal level is given by

$$P_{rec} \ (\mathrm{dBm}) = P_{EIRP} \ (\mathrm{dBm}) - L_p \ (\mathrm{dB}) + G_R \ (\mathrm{dBi}) - L_{fr} \ (\mathrm{dB}) \qquad (10.5)$$

where L_p (dB) is the propagation path loss due to distance, G_R (dBi) is the receiver antenna gain, and L_{fr} (dB) is power loss due to cable and connectors in the receiver. Moreover, we have to prepare a fade margin M (dB) to reduce outage probability due to shadowing. As a result, the maximum acceptable path loss L_{pmax} (dB) is given by

$$
\begin{aligned}
L_{pmax} \ (\mathrm{dB}) &= (P_T - P_{th}) \ (\mathrm{dB}) \\
& \quad + (G_T + G_R) \ (\mathrm{dB}) - (L_{ft} + L_{fr} + M) \ (\mathrm{dB})
\end{aligned}
\qquad (10.6)
$$

Using this result, we can estimate the coverage area of a system. In this process, the key factors are the relationship between the coverage area and path loss as well as how to obtain fade margin.

10.1.2 Relationship between the Propagation Path Loss and Coverage Area

Propagation path loss due to distance depends on the system operational environments, such as the macrocell outdoor system, microcellular system, and indoor system. Therefore, we have to create an appropriate path loss model for each operational environment as discussed in chapter 2, and then we have to evaluate the coverage area of a system using the created model. In this section, we will discuss how to obtain a relationship between the propagation path loss and coverage area in the case of an outdoor system as an example.

In the case of an outdoor system with a relatively large zone radius, Hata's equation is a good approximation of the propagation path loss due to distance. Figure 10.2 shows the relationship between the propagation path loss and distance obtained by Hata's equation for large cities, where the antenna height of the BS and that of the MS are assumed to be 100 m and 1.5 m, respectively. Using this chart, we can obtain a distance that gives the path loss L_{pmax} with 50% probability.

10.1.3 Fade Margin

The p.d.f. of the large-scale signal variation is approximated by the log-normal distribution given by

$$p(\Gamma) = \frac{1}{\sqrt{2\pi}\sigma\Gamma} \exp\left\{ -\frac{1}{2\sigma^2}\left(\ln^2 \frac{\Gamma}{\Gamma_m} \right) \right\} \tag{10.7}$$

where

Γ = short-term average received signal level (mW)
Γ_m = long-term average received signal level (mW)
σ_0 = standard deviation of log-normal fading in dB
σ = $\sigma_0 \ln 10/10$

Now, let's assume that the distance between the BS and a terminal is r, and the zone radius is R. When the long-term average signal level at the zone fringe ($r = R$) is defined as $\Gamma_m(R)$, and the receiver sensitivity is Γ_{th}, the outage probability at $r = R$ is given by

$$F_f(R) = \int_0^{\Gamma_{th}} p(\Gamma)d\Gamma = \frac{1}{2}\text{erfc}\left(\frac{\ln\left(\Gamma_m(R)/\Gamma_{th}\right)}{\sqrt{2}\sigma} \right) \tag{10.8}$$

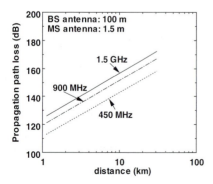

Fig. 10.2 Propagation path loss versus distance obtained by Hata's equation for large cities, where the antenna height of the BS and that of the MS are assumed to be 100 m and 1.5 m, respectively.

On the other hand, when a terminal is located at an arbitrary distance other than $r = R$, its received power is given by

$$\Gamma_m(r) = \Gamma_m(R)\left(\frac{R}{r}\right)^{\alpha_p} \qquad (10.9)$$

where α_p is a path loss decay factor, and the outage probability at distance r is given by

$$F_f(r) = \frac{1}{2}\mathrm{erfc}\left(\frac{\ln\left\{(\Gamma_m(R)/\Gamma_{th})(R/r)^{\alpha_p}\right\}}{\sqrt{2}\sigma}\right) \qquad (10.10)$$

Because the area outage probability is given by integrating $F_f(r)$ from $r = 0$ to $r = R$, the area outage probability [10-1] is given by

$$
\begin{aligned}
F_a(R) &= \frac{1}{\pi R^2}\int_0^R F_f(r)2\pi r\,dr \\
&= \frac{1}{\pi R^2}\int_0^R \frac{1}{2}\mathrm{erfc}\left[\frac{\ln(\Gamma_m(R)/\Gamma_{th}) + \alpha_p\ln(R/r)}{\sqrt{2}\sigma}\right]2\pi r\,dr \qquad (10.11) \\
&= \frac{1}{2}\mathrm{erfc}(X_0) - \frac{1}{2}e^{(2X_0Y_0+Y_0^2)}\mathrm{erfc}(X_0 + Y_0),
\end{aligned}
$$

where

$$X_0 = \frac{\ln\left[\Gamma_m(R)/\Gamma_{th}\right]}{\sqrt{2}\sigma} \qquad (10.12)$$

$$Y_0 = \frac{\sqrt{2}\sigma}{\alpha_p} \qquad (10.13)$$

Figure 10.3 shows fade margin (M) versus α_p performance with a parameter of σ_0 (a) for the coverage area ratio of 90% and (b) for the coverage area ratio of 99%. When we compare these two figures, we find that a higher coverage area ratio requires a higher fade margin, especially when σ_0 is large. For example, when $\alpha_p = 3.5$ and $\sigma_0 = 6$ dB, although M is about 4 dB in the case of a 90% coverage area ratio, a 6-dB higher margin is required for 99% coverage area ratio. When σ_0 is 14 dB and $\alpha_p = 3.5$, the fade margin difference between the coverage area ratio of 90% and that of 99% becomes 15 dB. In the commercial systems, however, we cannot prepare for a huge fade margin from a practical and economical point of view. Therefore, we usually employ the coverage area ratio of 90% (which corresponds to the outage probability of 10%). When we need much higher coverage area ratio, we also have to consider some other techniques, such as dynamic channel assignment (DCA) or the adaptive modulation scheme (these will be discussed in chapter 17).

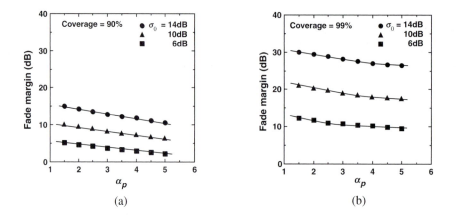

Fig. 10.3 Fade margin versus α_p performance with a parameter of σ_0, (a) for the coverage area ratio of 90% and (b) for the coverage area ratio of 99%.

10.1.4 Examples of the Coverage Area Estimation

As an example of the coverage area estimation, let's obtain the coverage area for the uplink of the PDC system with parameters of the required outage probability and the standard deviation of shadowing.

Table 10.1 summarizes the obtained coverage area as well as the parameters used for this evaluation. Because this table assumes the uplink of the PDC system (see chapter 11), we will assume that the transmitter power is 25 dBm (300 mW; Class III), which is a specified value of the terminal, and the receiver sensitivity of −106 dBm, which is the specified value for a diversity receiver (see chapter 11). As for the propagation path loss decay factor, $\alpha_p = 3.5$ is used.

The assumed model is classified into two groups—the outage probability of 10% (model 1 through model 3) and that of 1% (model 4 though model 6). Each group is further divided into three by the standard deviation of the log-normal fading (6, 10, and 14 dB).

We can find from this table that the coverage area becomes very narrow when the required outage probability is small or the standard deviation of shadowing is large because a larger fade margin is required.

10.2 BASICS OF THE SYSTEM CAPACITY ESTIMATION

In cellular systems, we have to maximize system capacity using the assigned system bandwidth. Therefore, we usually evaluate the spectral efficiency defined as the number of traffic/unit frequency/unit area. Let's assume that all of the channel is used only for the voice transmission services. In this case, the spectral efficiency [10-2] is given by

Table 10.1 Coverage area estimation for the uplink of the PDC system with parameters of the required outage probability and the standard deviation of shadowing, where $\alpha_p = 3.5$.

Factors		Model 1	Model 2	Model 3	Model 4	Model 5	Model 6
Tx	P_T	25 dBm	25 dBm	25 dBm	25 dBm	25 dBm	25 dBm
	L_{ft}	1 dB	1 dB	1 dB	1 dB	1 dB	1 dB
	G_T	1 dB	1 dB	1 dB	1 dB	1 dB	1 dB
	EIRP	25 dBm	25 dBm	25 dBm	25 dBm	25 dBm	25 dBm
Rx	L_{ft}	1 dB	1 dB	1 dB	1 dB	1 dB	1 dB
	G_R	10 dB	10 dB	10 dB	10 dB	10 dB	10 dB
	P_{th}	−106 dBm	−106 dBm	−106 dBm	−106 dBm	−106dBm	−106 dBm
Zone	outage prob.	10%	10%	10%	1%	1%	1%
	σ_0	6 dB	10 dB	14 dB	6 dB	10 dB	14 dB
	M	3.3 dB	7.6 dB	12.3 dB	10.5 dB	19.0 dB	27.5 dB
	L_{pmax}	136.7 dB	132.4 dB	127.7 dB	129.5 dB	121.0 dB	112.5 dB
	coverage	3.6 km	2.6 km	1.9 km	2.0 km	1.2 km	0.6 km

$$\eta_T = \frac{n_{zone} \cdot a_c}{AW} \quad (\text{Erl} / \text{Hz} / \text{m}^2) \tag{10.14}$$

where

n_{zone} (ch/zone)	= number of usable voice channels in each cell
a_c (erl/ch)	= traffic per channel
A (m²)	= area for each cell
W (Hz)	= assigned bandwidth to the system

When we define that f_{ch} is the channel spacing and L is the number of cells in each cell cluster, n_{zone} can be expressed as

$$n_{zone} = \frac{W}{f_{ch} \cdot L} \tag{10.15}$$

Using this equation, we can modify equation (10.14) as

$$\eta_T = \eta_s \eta_f \eta_t = \frac{1}{LA} \frac{1}{f_{ch}} a_c \tag{10.16}$$

where η_s is the spectral efficiency with respect to space given by

$$\eta_s = \frac{1}{LA} \quad (1/\text{m}^2) \tag{10.17}$$

η_f is spectral efficiency with respect to frequency given by

$$\eta_f = \frac{1}{f_{ch}} \ (\mathrm{ch}\,/\,\mathrm{Hz}) \tag{10.18}$$

and η_t is the spectral efficiency with respect to time given by

$$\eta_t = a_c \ (\mathrm{Erl}\,/\,\mathrm{ch}) \tag{10.19}$$

A modified definition of the capacity is Erl/Hz/cell or Erl/MHz/cell, in which coverage area is normalized. In this case, η_s is defined as $1/L$ (1/cell). This definition is more frequently used as a criterion for system capacity.

In this discussion, we have assumed voice transmission services. In the case of nonvoice transmission, bit/s/Hz/cell is frequently employed instead of Erl/Hz/cell. In this case, spectral efficiency in terms of bit/s/Hz/cell is defined as follows:

$$\eta_T = \eta_s \cdot \eta_f = \frac{1}{L} \frac{\log_2 M \cdot R_s}{f_{ch}} \ (\mathrm{bit}\,/\,\mathrm{s}\,/\,\mathrm{Hz}\,/\,\mathrm{cell}) \tag{10.20}$$

$$\eta_f = \frac{\log_2 M \cdot R_s}{f_{ch}} \ (\mathrm{bit}\,/\,\mathrm{s}\,/\,\mathrm{Hz}) \tag{10.21}$$

$$\eta_s = \frac{1}{L} \ (1\,/\,\mathrm{cell}) \tag{10.22}$$

When we define η_s as $1/(LA)$ instead of $1/L$, we can obtain the spectral efficiency in terms of bit/s/Hz/m^2. Therefore, we have several criteria of the spectral efficiency. Among them, because bit/s/Hz/cell is considered to be more general than the other definitions, and because we can easily convert bit/s/Hz/cell to Erl/Hz/cell, we will use bit/s/Hz/cell as spectral efficiency in our discussion.

Among the three factors of spectral efficiency, η_t is determined by the channel assignment process and is independent from the other two factors. On the other hand, η_s and η_f are determined by the modulation and access schemes, and they are dependent on each other. For example, when we employ a modulation with a higher modulation level, although we can increase η_f, we will degrade η_s because robustness to CCI is degraded with an increasing modulation level due to its shorter minimum signal distance. Therefore, we have to maximize $\eta_f \eta_s$ when we optimize parameters of the modulation and access schemes.

10.3 SPECTRAL EFFICIENCY ESTIMATION FOR FDMA AND TDMA SYSTEMS

10.3.1 Radio Resource Assignment to Each Cell

To simplify our discussion, let's assume that the assigned system bandwidth is W_{sys} (Hz), and the access scheme is FDMA. When each cell cluster consists of L cells, the number of usable channels per cell is given by

$$N_1 = \frac{W_{sys}}{f_{ch} \cdot L} \tag{10.23}$$

For example, when W_{sys} = 10 MHz (frequency band of f_L to f_L + 10 MHz), f_{ch} = 10 kHz, and $L = 3$, N_1 becomes 333 channels. Figure 10.4 shows an example of cell configuration for $L = 3$, where an omnidirectional antenna is assumed. When the center frequency for k-th channel is given by $f_k = f_L + (k - 0.5)f_{ch}$, assigned channels for each cell are given as shown in Table 10.2.

As shown in this table, adjacent channels are assigned, not to the same cell, but to different cells. For example, cell-1 and cell-3 include adjacent channels for cell-2. Therefore, cell-1 and cell-3 are called ACI cells for cell-2.

There are two strategies to determine f_{ch}. The first one is to suppress the ACI as much as possible. As a result, η_s is determined by only the CCI. Because f_{ch} is selected so neighboring channels do not overlap, the channel spacing is given by

$$f_{ch} \geq R_s(1+\alpha) + 2\delta f_c \tag{10.24}$$

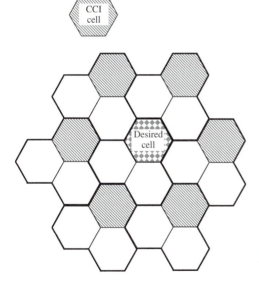

Fig. 10.4 An example of cell configuration for $L = 3$, where omnidirectional antenna is assumed.

Table 10.2 Assigned channels for each cell.

Cell	Assigned Channels (k)
Cell-1	1, 4, 7, 10, 13, 16, 19, 22, 25, 28, 31, 34, 37, 40, ...
Cell-2	2, 5, 8, 11, 14, 17, 20, 23, 26, 29, 32, 35, 38, 41, ...
Cell-3	3, 6, 9, 12, 15, 18, 21, 24, 27, 30, 33, 36, 39, 42, ...

where δ is the carrier frequency instability and f_c is the carrier frequency. When we ignore the effect of carrier frequency stability, η_f is given by

$$\eta_f = \frac{\log_2 M \cdot R_s}{f_{ch}} = \frac{\log_2 M}{1+\alpha} \text{ (bit/s/Hz)} \tag{10.25}$$

The second strategy for the definition of f_{ch} is to accept a certain amount of ACI, and L is determined by both ACI and CCI. This channel allocation is called *interleave channel allocation*. Because we can reduce f_{ch} by using this concept, we can achieve higher spectral efficiency than we can get in a non-interleaved system. Thus, the interleaving channel allocation is employed in most cellular systems.

10.3.2 Selection of the Number of Cells per Cell Cluster

When we select a modulation and access scheme as well as the associated fading compensation techniques, we can obtain its BER performance. Because the channel conditions for wireless communication systems are flat Rayleigh or frequency-selective fading, the BER has to be evaluated by the average BER under these fading conditions. On the other hand, when we select the supported services, we can define the required BER in the physical layer (BER_{th}). For example, BER_{th} of 10^{-2} to 5×10^{-2} is required for voice transmission.

To simplify discussion, let's consider only voice transmission with its BER_{th} of 10^{-2}. When we determine the BER_{th}, we can obtain the threshold C/I_c (Λ_m) that satisfies $\text{BER} = \text{BER}_{th}$.

Next, we will define the area outage probability given by [10-3]

$$P_0 = \text{Prob}(C / I_c \le \Lambda_m) \tag{10.26}$$

As for the area outage probability (P_0), $P_0 = 10\%$ is a typical value for existing cellular systems. On the other hand, when we discuss future personal communication systems, we also employ 5% and 1%. When we define the outage probability, we will search for L that satisfies the required outage probability.

First let's discuss the process for obtaining cumulative distribution of the C/I_c in the downlink. In this case, the desired signal is coming from the desired BS, and CCI signals are coming from the BSs of the CCI cells. As a result, the received signal level of the desired signal (C) and that coming from i-th CCI-cell BS (I_i) are given by

$$C = A \cdot r_0^{-\alpha_p} \cdot 10^{\frac{G_0(\theta_0)}{10}} \cdot 10^{\frac{\xi_0}{10}} \tag{10.27}$$

$$I_i = A \cdot r_i^{-\alpha_p} \cdot 10^{\frac{G_i(\theta_i)}{10}} \cdot 10^{\frac{\xi_i}{10}} \tag{10.28}$$

where

A	= constant
r_0	= distance from the desired BS to the terminal

r_i = distance from i-th interfering BS to the terminal
α_p = path loss decay factor with respect to distance
$G_0(\theta_0)$ = antenna directivity of the BS in the desired cell
$G_i(\theta_i)$ = antenna directivity of the BS in the i-th CCI cell
ξ_0 = long-term fading variation of the desired signal subject to a log-normal probability density function with its standard deviation of σ_0.
ξ_i = long-term fading variation of i-th CCI-cell BS with a log-normal probability density function with its standard deviation of σ_i

Cumulative distribution of C/I_c can be obtained by following these steps [10-4].

1. Randomly select a terminal location in the desired cell, where its p.d.f. is subject to uniform distribution.
2. Calculate r_0 and r_i.
3. ξ_0 and ξ_i for all CCI cells are randomly generated by a log-normal distributed random generator.
4. Calculate C/I_c at location x using values obtained by steps 2 and 3 as well as equations (10.27) and (10.28) as

$$C / I_c = C / \sum_i I_i \qquad (10.29)$$

5. Repeat steps 3 and 4 N_{log} times to simulate log-normal fading at each terminal location, where N_{log} is a sufficiently large number.
6. Go back to step 1 to change the terminal location, and repeat steps 1 through 5 N_{loc} times, where N_{loc} is a sufficiently large number.
7. Calculate the cumulative distribution of C/I_c obtained by steps 1 through 6.

To determine L, we have to obtain the cumulative distribution of C/I_c for each L using these steps. At the same time, we have to obtain the receiver sensitivity against CCI which is defined as the minimum C/I_c (C/I_{cth}) that satisfies BER $\leq$ BER$_{th}$ (for example, BER$_{th}$ = 10^{-2}), where BER is the average BER in the fading environments.

In the case of uplink, CCI signals are coming from terminals in the CCI cells. Therefore, the received signal of the desired signal (C) and the CCI signal from a terminal in i-th CCI cell (I_i) are given by

$$C = A \cdot r_0^{-\alpha_p} \cdot 10^{\frac{G_0(\theta_0)}{10}} \cdot 10^{\frac{\xi_0}{10}} \qquad (10.30)$$

$$I_i = A \cdot r_i^{-\alpha_p} \cdot 10^{\frac{G_0(\theta_i)}{10}} \cdot 10^{\frac{\xi_i}{10}} \qquad (10.31)$$

where r_i is the distance between the CCI interfering terminal in i-th CCI cell and the BS of the desired cell. Using these two equations, the cumulative distribution of C/I_c in the uplink can be obtained by following these steps.

1. A terminal location in the desired cell is randomly selected, where its p.d.f. is subject to uniform distribution.

2. In the same manner, terminal locations for all the CCI cells are randomly selected.

3. Calculate r_0 and r_i.

4. ξ_0 and ξ_i for all CCI cells are randomly generated by a log-normal distributed random generator.

5. Calculate C/I_c using values obtained by steps 2 and 3 as well as equations (10.30) and (10.31).

6. Repeat steps 3 and 4 N_{log} times to simulate log-normal fading at each terminal location, where N_{log} is a sufficiently large number.

7. Go back to step 1 to change the terminal location in the desired cell as well as the CCI cells, and repeat steps 1 through 5 N_{loc} times, where N_{loc} is a sufficiently large number.

8. Calculate the cumulative distribution of C/I_c obtained by steps 1 through 6.

Although we have assumed only CCI in our discussion, this process can also be extended for interleaving channel allocation in which there are ACI. The only difference between the ACI and CCI is that the ACI can be attenuated by the receiver filter whereas CCI cannot be attenuated [10-3; 10-4]. Thus, when we introduce a receiver filter attenuation factor of i-th interference signal (F_i), we can express the received ACI or CCI signal in both the downlink and the uplink as

$$I_i = A \cdot r_i^{-\alpha_p} \cdot 10^{\frac{-F_i}{10}} \cdot 10^{\frac{G_i(\theta_i)}{10}} \cdot 10^{\frac{\xi_i}{10}} \text{ (downlink)} \tag{10.32}$$

$$I_i = A \cdot r_i^{-\alpha_p} \cdot 10^{\frac{-F_i}{10}} \cdot 10^{\frac{G_0(\theta_i)}{10}} \cdot 10^{\frac{\xi_i}{10}} \text{ (uplink)} \tag{10.33}$$

where $F_i = 0$ dB when the signal is CCI and where $F_i > 0$ dB when it is ACI.

10.4 SPECTRAL EFFICIENCY FOR OMNIZONE SYSTEMS

When omnidirectional antennas are employed in the BS, $G_0(\theta)$ and $G_i(\theta)$ have constant values regardless of the angle in both downlink and uplink cases. Figure 10.5 shows cumulative distribution of C/I_c for omnizone configuration. When the acceptable outage probability is defined as 10%, this outage probability is obtained by C/I_c of 9.0 dB for $L = 7$, 10.5 dB for $L = 9$, 14.0 dB for $L = 12$, 16.2 dB for $L = 16$, and 18.8 dB for $L = 21$. In the case of QPSK with coherent detection and antenna selection diversity, C/I_c for BER $= 10^{-2}$ is theoretically 9 dB. From a practical point of view, however, we have to consider about 3 dB of degradation due to differential encoding and carrier phase jitter caused by fading. Thus, let's consider $C/I_c = 12$dB for BER $= 10^{-2}$. In this case, $L = 12$ is required. When we employ $\alpha = 0.5$ as the roll-off factor for transmitter and receiver filters and channel spacing is selected as $f_{ch} = (1 + \alpha)R_s$, spectral efficiency in terms of bit/s/Hz/cell is given by 1/9.

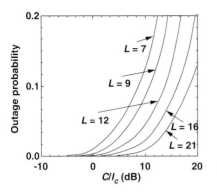

Fig. 10.5 Cumulative distribution of C/I_c for omnizone configuration.

10.5 SPECTRAL EFFICIENCY FOR SECTOR CELL SYSTEMS

10.5.1 Antenna Beam Pattern for Sector Cell

Figure 10.6 shows an antenna directive pattern of each cell site in the case of three sector cell systems. A dot in this figure shows a cell site.

The antenna directivity and its half power beam width are very important parameters in reducing interference from other cell sites. An example of the antenna directivity model is shown in Figure 10.7 in which its half power beam width is intro-

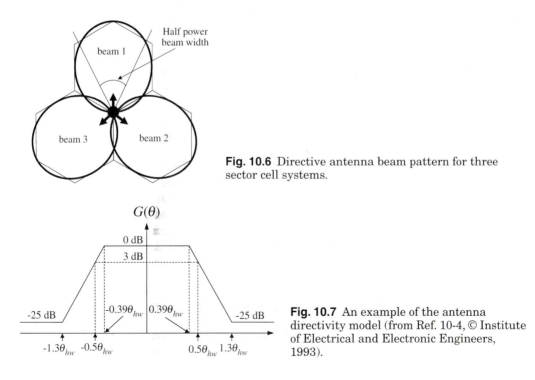

Fig. 10.6 Directive antenna beam pattern for three sector cell systems.

Fig. 10.7 An example of the antenna directivity model (from Ref. 10-4, © Institute of Electrical and Electronic Engineers, 1993).

duced as a parameter [10-3]. Although the actual pattern should be taken into account to evaluate an actual system, this pattern is very useful in evaluating a rough estimation of the system capacity, especially for evaluating with parameters of modulation and access schemes.

10.5.2 Characteristics of Interference from Neighboring Sectors

When an interleaving channel allocation is applied, one of the most important issues is the allocation of the adjacent channels. Figure 10.8 shows 90% value of the power ratio of the desired signal to the interference signal from each sector on the assumption that the filter attenuation is 0 dB [10-4]. A sector with no number indicates that the power ratio is more than 20 dB. From this figure, we can see that strong interference comes from sectors with different antenna beam directivity despite their geographical separation. Therefore, we can conclude that the adjacent channels should be allocated to sectors with the same antenna directivity as that of the desired sector.

Figure 10.9 shows an example of channel allocation for sector cells to satisfy this requirement. A full spectrum assigned to a system is divided into three parts to ensure

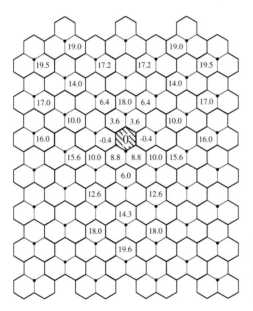

Fig. 10.8 Ninety-percent value of the power ratio of desired signal to the interference signal from each sector on the assumption that the filter attenuation is 0 dB (from Ref. 10-4, © Institute of Electrical and Electronic Engineers, 1993).

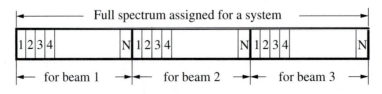

Fig. 10.9 An example of channel allocation for sector cells.

that no interference comes from sectors with different antenna directivity. Furthermore, the adjacent channel sectors should be allocated so as to reduce ACI as much as possible, considering the carrier to interference performance shown in Figure 10.5.

10.5.3 Sector Cell Layout

As for the sector cell configurations, we can simply design a sector cell by dividing an omnizone. Figure 10.10 shows a seven-site three-sector cell configuration. Because one omnizone is divided into three parts in this case, the acceptable number of cell sites is limited to 3, 4, 7, 9, 11, 13, 17, ... as in the case of omnizone configuration.

Another sector cell layout is the irregular parallel beam cell layout [10-5]. Figure 10.11 shows its cell layout for (a) $L = 4$, (b) $L = 5$, and (c) $L = 7$. As shown in this figure, this scheme reduces cell reuse distance in the direction orthogonal to the main beam direction, and it increases the distance in the main beam direction because CCI coming from the direction orthogonal to the main beam is smaller than that from the main beam direction owing to the antenna directivity. Moreover, this scheme can achieve a sector cell layout with an arbitrary value of L—for example $L = 2$, 5, and 6—which is impossible with conventional cell layout schemes. In Figure 10.11, the ACI sectors are also shown, where ACI sectors are allocated to minimize the received signal level of the ACI at the desired sector using the results shown in Figure 10.5.

10.5.4 Probability Distribution of Carrier to Interference Power Ratio in CCI and ACI Conditions

10.5.4.1 Effect of Half Power Beam Width The interference signal power depends on the half power beam width of the BS antenna in the case of a sector cell

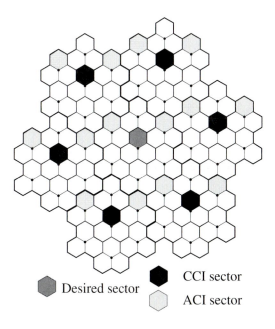

Desired sector
CCI sector
ACI sector

Fig. 10.10 Seven-site 3-sector cell configuration.

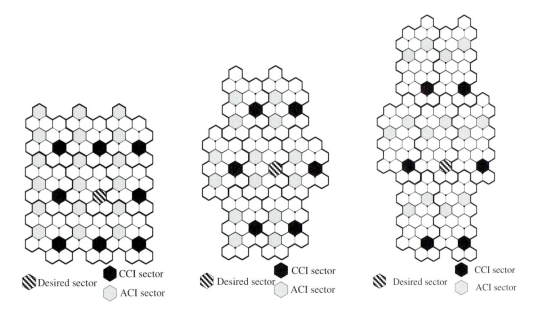

Desired sector   ■ CCI sector   ○ ACI sector
Desired sector   ■ CCI sector   ○ ACI sector
Desired sector   ■ CCI sector   ○ ACI sector

Fig. 10.11 Irregular parallel beam sector cell layout for (a) $L = 4$, (b) $L = 5$, and (c) $L = 7$.

layout. Figure 10.12 shows the 90% value of C/I_c versus the half power beam width (θ_{hw}), where no ACI is assumed and the antenna directivity shown in Figure 10.7 is employed. The results show that the half power beam width of approximately 70 degrees gives the highest 90% value of C/I_c. Therefore, $\theta_{hw} = 70$ degrees is considered to be optimum when we employ the antenna directivity given by Figure 10.7. When we employ an antenna with different antenna directivity, we have to obtain the optimum θ_{hw} for the directivity.

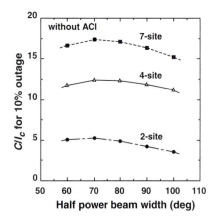

Fig. 10.12 Ninety-percent value of C/I_c versus half power beam width (θ_{hw}), where no ACI is assumed and the antenna directivity shown in Figure 10.7 is employed (from Ref. 10-4, © Institute of Electrical and Electronic Engineers, 1993).

10.5.4.2 Cumulative Distribution of C/I_c   Figure 10.13 shows the cumulative distribution of C/I_c in the downlink for the 3-sector irregular parallel beam cell layout with a parameter of L, where this result is obtained by using the steps outlined in Section 10.3. As shown in this figure, C/I_c for a 10% outage probability is 5.4 dB, 9.0 dB, 11.5 dB, 13.8 dB, and 17.6 dB for $L = 2$, 3, 4, 5, and 7, respectively. When we compare these results with those of the omnidirectional antenna, we find that the C/I_c for 10% outage probability is increased by about 8–9 dB when we employ the 3-sector irregular parallel beam cell layout.

10.5.4.3 Filter Attenuation Effect of the ACI   As discussed in Section 10.3, we can treat the effect of ACI just by introducing a filter attenuation factor (F_i) in the received signal level of the ACI. Therefore, even in the CCI and ACI conditions, we can obtain cumulative distribution performance the same way we did it in the case of only CCI conditions [10-3; 10-4].

Figure 10.14 shows $C/(I_c + I_A)$ for a 10% outage versus F_i at different values of L. $C/(I_c + I_A)$ for a 10% outage increases with filter attenuation and is almost constant when the filter attenuation is larger than 25 dB. This means that the ACI can be ignored if the filter attenuation is larger than 25 dB. When the filter attenuation is lower than 25 dB, $C/(I_c + I_A)$ for a 10% outage probability decreases with decreasing F_i.

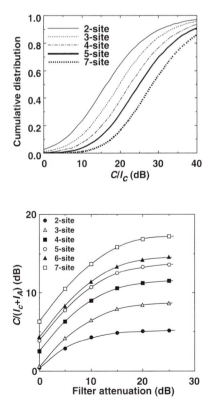

Fig. 10.13 Cumulative distribution of C/I_c in the downlink for the 3-sector irregular parallel beam cell layout with a parameter of L.

Fig. 10.14 $C/(I_c + I_A)$ for a 10% outage versus F_i at different values of L (from Ref. 10-4, © Institute of Electrical and Electronic Engineers, 1993).

In practice, F_i is defined as the power ratio of the desired signal after filtering (P_D) to that of the ACI after filtering (P_I) as

$$F_i = \frac{P_D}{P_I} \qquad\qquad (10.34)$$

Figure 10.15(a) shows the outline of this concept. However, this measurement method is not practical because the frequency response of the receiver filter of each terminal is slightly different from the others. Therefore, we usually employ another definition for F_i as shown in Figure 10.15(b), in which a brick-wall BPF with its center frequency of f_c (f_c is the carrier frequency of the desired signal) and its bandwidth of f_w are assumed as the receiver filter. In this case, P_D and P_I are defined as the integrated power of the desired signal and of the ACI over the frequency range of this brick-wall BPF. In the case of IS-54, $f_w = f_{ch}$ is selected, where f_{ch} is the channel spacing.

Figure 10.16, where QPSK is employed as the modulation scheme and channel spacing is normalized by the symbol rate ($R_s = 1/T_s$), shows F_i versus channel spacing with a parameter of the roll-off factor (α) for the specified receiver filter case and (b) for the brick-wall BPF case. When we compare these two figures, we find that the performance difference between the actual filter and the brick-wall filter is only a few dB. Therefore, it is practical to employ a brick-wall BPF for the definition of F_i.

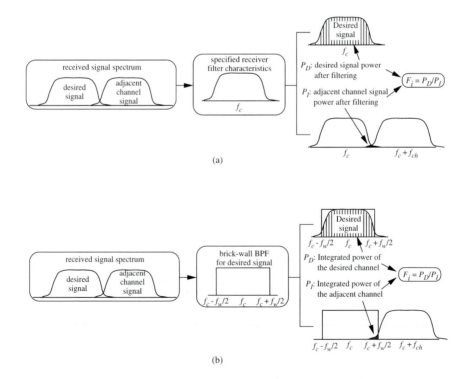

Fig. 10.15 Filter attenuation concept (a) for the actual receiver filter and (b) brick-wall filter case.

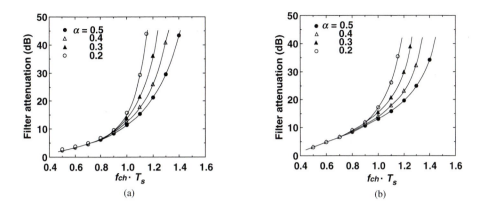

Fig. 10.16 F_i versus channel spacing with a parameter of the roll-off factor (a) for the specified receiver filter case and (b) for the brick-wall BPF case, where QPSK is employed as the modulation scheme and channel spacing is normalized by the symbol rate ($R_s = 1/T_s$).

10.5.4.4 Example of the Spectral Efficiency Evaluation   Let's evaluate spectral efficiency using the results in Figures 10.14 and 10.16. First, we will assume the following parameters:

☞ Modulation/demodulation scheme: QPSK with coherent detection

☞ Transmitter/receiver filters: root Nyquist filter with its roll-off factor of $\alpha = 0.5$

☞ Acceptable outage probability: 10%

☞ Anti-fading technique: antenna selection diversity

In the case of QPSK with coherent detection and antenna selection diversity, $C/(I_c + I_A)$ for BER = 10^{-2} is theoretically 9 dB. From a practical point of view, however, we have to consider about 3 dB of degradation due to differential encoding and carrier phase jitter caused by fading. Thus, we will consider that $C/(I_c + I_A)$ = 12 dB for BER = 10^{-2} in the following analysis.

From Figure 10.14, we can obtain $C/(I_c + I_A)$ = 12 dB using five or more cell sites if we select an appropriate filter attenuation. Moreover, when we employ $L \geq 5$, larger F_i is required with decreasing L. For example, when we define the required $C/(I_c + I_A)$ as $C/(I_c + I_A)$ = 12 dB, the required filter attenuation is 14 dB for five-site, 12 dB for six-site, or 7.5 dB for seven-site cells, respectively. On the other hand, we can find from Figure 10.16(b) that F_i = 14 dB, 12 dB, and 7.5 dB are achieved when $f_{ch}.T_s$ = 1.05, 0.96, and 0.76, respectively. When we employ QPSK, spectral efficiency in terms of frequency is given by $\eta_f = 2R_s/f_{ch} = 2/(f_{ch} \cdot T_s)$ as we have discussed earlier. On the other hand, spectral efficiency in terms of space is given by $\eta_s = 1/L$. Using these relationships, we can obtain spectral efficiency for each L as shown in Table 10.3 [10-4].

This table shows that total spectral efficiency η_T for L = 5, 6, and 7 are 0.38, 0.35, and 0.36, respectively. These results also show that, although L = 5 and f_c = 1.08R_s give the highest η_T, L = 6 with f_c = 0.96R_s and L = 7 with f_c = 0.70R_s also give accept-

Table 10.3 Evaluation of spectral effficiency.

L	5	6	7
F_i (dB)	14.0	12.0	7.5
η_f	1.90	2.08	2.63
η_s	0.2	0.17	0.14
$\eta_T = \eta_f \eta_s$	0.38	0.35	0.36

able spectral efficiency, which is only 10% degradation from the performance for $L = 5$ and $f_c = 1.08R_s$. Consequently, it is possible to improve spectral efficiency either by reducing L or by reducing channel spacing. When we compare these numbers with those for omnizone systems, we find that the numbers in Table 10.3 are approximately 3.5 times higher than that for omnizone systems.

REFERENCES

10-1. Feher, K., ed., *Advanced digital communications*, Prentice Hall, Englewood Cliffs, New Jersey, 1987.

10-2. Nagata, Y. and Akaiwa, Y. "Analysis for spectral efficiency in single cell trunked and cellular mobile radio," Trans. IEEE Veh. Tech., Vol. VT-35, No. 3, pp. 100–13, August 1987.

10-3. Nakajima, N. and Nakano, K., "Spectrum efficiency of cell layout," Technical report of IEICE, RCS89-18, October 1989.

10-4. Sampei, S., Leung, P. S. K. and Feher, K., "High capacity cell configuration strategy in ACI and CCI conditions," 43rd IEEE Veh. Technol. Conf., pp. 185–88, May 1993.

10-5. Kanai, T., "Channel assignment for sector cell layout," Trans. IEICE Vol. J73-B-II, No. 11, pp. 595–601, November 1990.

Personal Digital Cellular (PDC) Telecommunication System

The PDC telecommunication system [11-1] (previously called the Japanese digital cellular (JDC) system), was standardized in 1991 by the RCR as a second-generation cellular system; it is the first standardized digital wireless system in Japan. Initially, although the PDC system was developed mainly to increase system capacity for cellular voice communication systems, it has since developed into a more personalized system that supports both voice and nonvoice data.

In this chapter, we will outline this system, discussing mainly the specifications of the physical layer, including what kind of wireless technologies are integrated in the system.

11.1 HISTORIC OVERVIEW OF PDC DEVELOPMENT

The PDC system began to develop in 1989. On April 1989, MPT organized a study group on digital cellular telecommunication systems to look at technological aspects of the digital cellular system. The group's report was submitted to MPT in March 1990. Following this report, MPT consulted with the Telecommunication Technology Council (TTC) about the technological requirements for digital cellular systems, a report being issued in June 1990. Based on this report, MPT, in consultation with the Radio Regulation Council (RRC), promulgated the related ministerial ordinances.

Just after the TTC issued the final report, RCR (presently called ARIB) organized a standardization committee for the Digital Cellular Telecommunication System; its standard (RCR STD-27) was issued on April 1991. The first commercial operation started in April 1993 for the full-rate system. In April 1994, two new operators—Tu-Ka Cellular and Digital Phone groups—also started their services. As a result, the number of subscribers reached 10 million as of March 1996, about 8% of the population in Japan.

11.2 SYSTEM OVERVIEW

Because the PDC system was intended to support comprehensive telecommunication services including voice and data communications through a unique and general-purpose radio interface, certain functions were considered to be necessary for the PDC system.

11.2.1 Higher Capacity, Higher Quality, and Better Privacy

The PDC system was designed to offer system capacity, communications quality, and privacy surpassing those of conventional analog cellular systems because limited capacity, poor quality, and low privacy were recognized as serious disadvantages of the analog cellular system in comparison with the wired telephone network. Of the three, we can easily offer better privacy by introducing digital transmission technology. However, it is not so easy to offer high capacity because system capacity of the Japanese analog cellular system that employs 6.25 kHz interleave channel allocation is already five times higher than the U.S. AMPS system.

To achieve a high-quality, high-capacity system, the PDC system employs π/4-QPSK and low-bit-rate voice codec. In the first stage, full-rate codec with its bit rate of 11.2 kbit/s was introduced to achieve similar capacity and better voice quality than the existing analog systems because much lower voice codec technology was premature for the application of digital cellular systems from the viewpoint of voice quality as of 1993. To further improve system capacity, the RCR STD-27 was modified in 1995 to introduce half-rate codec with its bit rate of 5.6 kbit/s and its quality equivalent to that of full-rate codec.

11.2.2 Interconnectivity with Other Communication Networks

The PDC system has no backward compatibility with the existing analog cellular system because

☞ Spectrum for digital cellular systems was prepared in 800-MHz and 1.5-GHz bands.

☞ Backward compatibility was not required in the Japanese cellular market because terminals for analog cellular systems were supplied only by the lease system.

Of course, the PDC system offers interconnectivity with the existing analog cellular system.

Moreover, it also offers interconnectivity with the ISDN and packet data networks to integrate voice and nonvoice transmission services. To achieve such interconnectivity, the open system interconnect (OSI) reference model is employed in the signaling structure of the PDC system.

11.2.3 Roaming Capability

Roaming capability is the most important market-oriented requirement for the PDC system because it is not supported in Japanese analog cellular systems—for example, between the NTT and Japanese TACS systems. Therefore, all the digital cellular operators in Japan employ the PDC standard (RCR STD-27) as of 1996.

With these functions, the PDC system supports the following services:

☞ Voice (full-rate and half-rate codec)

☞ Supplementary services (call waiting, voice mail, three-party call, call forwarding, and so on)

☞ Nonvoice data (up to 9.6 kbit/s)

☞ Packet-switched wireless data (PDC-P)

Figure 11.1 shows the network configuration of the PDC system. This network consists of

☞ Mobile gateway switching centers (MGCs)—in charge of interconnection with the other networks and call routing

☞ Mobile communication control centers (MCCs)—in charge of BS control including handover and terminal location registration

☞ BSs

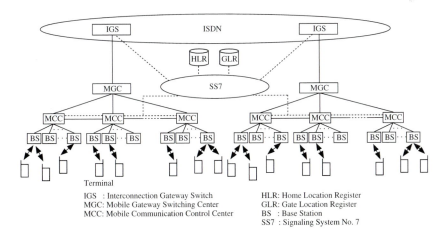

IGS : Interconnection Gateway Switch	HLR: Home Location Register
MGC: Mobile Gateway Switching Center	GLR: Gate Location Register
MCC: Mobile Communication Control Center	BS : Base Station
	SS7 : Signaling System No. 7

Fig. 11.1 Network configuration of the PDC system.

☞ Home location register (HLR)—includes terminal equipment numbers as well as their current locations

☞ Gate location register (GLR)—a temporal register for a roamed call.

Control signal transfer between this equipment are carried out using Signaling System No. 7 (SS7) protocol.

Figure 11.2 shows photos of (a) terminals, (b) a terminal connected with a portable digital facsimile machine, (c) BS equipment, and (d) BS antennas for the PDC system.

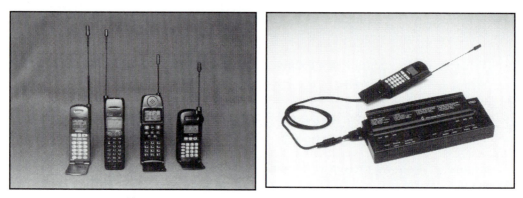

(a)   (b)

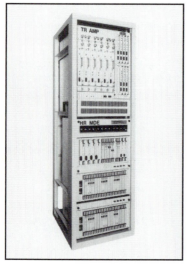

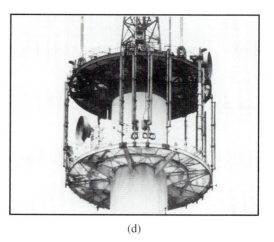

(d)

(c)

Fig. 11.2 Photos of the PDC system operated by NTT Mobile Communications Network Inc.: (a) terminals, (b) a terminal connected to a digital facsimile machine, (c) BS equipment, (d) BS antenna (Courtesy of NTT Mobile Communications Network Inc.).

11.3 SPECIFICATIONS OF THE RADIO INTERFACE

Table 11.1 shows specifications of the PDC system in comparison with the other digital cellular systems in the world—the GSM [11-2], the IS-54 [11-3], and the IS-95 [11-4]. Although IS-54 and IS-95 are dual-mode systems in which analog and digital systems employ the same frequency band, the Japanese PDC system uses a frequency band that is different from those of the analog systems.

$\pi/4$-QPSK is employed as a modulation scheme to achieve high system capacity as well as to mitigate a nonlinear effect at the transmitter amplifier. A root Nyquist filter with a roll-off factor (α) of 0.5 is employed as the transmitter and receiver filter.

Table 11.1 Specifications of the PDC system and its comparison with other digital cellular systems.

Specifications	PDC	GSM	IS-54	IS-95
Frequency band (downlink)	810–826 MHz 1429–1453 MHz	935–960 MHz	869–894 MHz	869–894 MHz
Frequency band (uplink)	940–956 MHz 1477–1501 MHz	890–915 MHz	824–849 MHz	824–849 MHz
Symbol rate	21 ksymbol/s (42 kbit/s)	271 ksymbol/s (271 kbit/s)	24.3 ksymbol/s (48.6 kbit/s)	1.288 kchips/s (chip rate)
Frequency band separation	130 MHz 48 MHz	45 MHz	45 MHz	45 MHz
Carrier spacing	50 kHz (25 kHz interleave)	400 kHz (200 kHz interleave)	60 kHz (30 kHz interleave)	1.25 MHz
Modulation	$\pi/4$-QPSK ($\alpha = 0.5$)	GMSK (BT=0.3) B: Bandwidth of premodulation filter T: a bit duration	$\pi/4$-QPSK ($\alpha = 0.35$)	- BPSK (source signal) - QPSK (spread signal for downlink) - OQPSK (spread signal for uplink)
Access	TDMA	TDMA	TDMA	CDMA
Channel/ carrier	3 (full rate) 6 (half rate)	8 (full rate) 16 (half rate)	3 (full rate) 6 (half rate)	N/A
Voice codec	VSELP - 11.2 kbit/s (full) source (6.7 k) FEC (4.5 k) - 5.6 kbit/s (half) (PSI-CELP) source (3.45 k) FEC (2.15 k)	RPE-LTP - 22.8 kbit/s (full) source (13 k) FEC (9.8 k) -11.4 kbit/s (half) (tbd)	VSELP - 13.0 kbit/s (full) source (7.95 k) FEC (5.05 k) - 6.5 kbit/s (half) (tbd)	QCELP (variable-rate codec) 9.6 kbit/s (max.)
Maximum delay time	10 µs	16 µs	50 µs	not specified

TDMA is employed as an access scheme. Initially, 3-channel multiplexing was employed for full-rate voice transmission using a 11.2 kbit/s voice codec. At present, a 6-channel multiplexed system using a 5.6 kbit/s [11-5] is also specified.

The propagation path characteristics are very peculiar in Japan. Because population density is very high and there are many houses or buildings along the roads in most of the area, multipath fading with a very long delayed wave is less probable. The extensive delay profile measurements show that delay spread is up to 5 µs in most cases [11-6]. When we employ a space diversity technique, which is a relatively simple anti-multipath fading technique, we can achieve BER $< 10^{-2}$ for the delay spread of $0.1T_s$ (T_s is a symbol duration) [11-7]. Therefore, the symbol rate of 21 ksymbol/s is selected so as not to apply the adaptive equalizer as a mandatory condition.

11.4 REQUIREMENTS FOR THE TRANSMITTER AND RECEIVER

In addition to the specifications shown in Table 11.1, RCR STD-27 also specifies requirements for the transmitter and receiver. Table 11.2 shows requirements for a PDC transmitter. In Table 11.2, spurious transmission, leakage power during carrier OFF period, and ACI power are defined as follows:

☞ **Spurious transmission.** The ratio of the average power of spurious emissions at each frequency fed to the feeder to the average power of the signal at the carrier frequency.

☞ **Leakage power during carrier OFF period.** The ratio of the average power radiated in the assigned bandwidth when no signal is transmitted to the transmitted average power during the assigned slot.

☞ **ACI power.** The power radiated within a bandwidth of ±10.5 kHz, of which the center frequency is separated by Δf kHz from the carrier frequency, where the carrier is modulated by the test signal sequence at the same bit rate as that of the specified one.

Table 11.3, on the other hand, shows requirements for PDC receivers. Included in this table, receiver sensitivity, spurious sensitivity, adjacent channel selectivity, carrier to co-channel interference power ratio (CIR) sensitivity, and intermodulation performance are defined as follows:

☞ **Receiver sensitivity.** Defined by the median value of the received signal level that yields a BER of 1×10^{-2} and 3×10^{-2} when a traffic channel (TCH) signal modulated by 2,556 bits of the 9-stage PN sequence is received.

☞ **Spurious sensitivity.** Defined by the power ratio of the desired signal to the interference signal that yields a BER of 1×10^{-2} in the TCH portion of the desired signal, where the desired signal level is set to 3 dB higher than the specified receiver sensitivity, the interference signal is unmodulated, and its carrier frequency is separated by 100 kHz from the desired signal.

☞ **Adjacent channel selectivity.** Defined by the power ratio of the desired signal level which is set to 3 dB higher than the specified receiver sensitivity to

Table 11.2 Requirements for PDC transmitter.

Parameters	Specifications
Carrier frequency instability ($\delta f/f_c$)	BS: $\delta f/f_c \leq 5 \times 10^{-8}$
	terminal: $\delta f/f_c \leq 3 \times 10^{-6}$ (800-MHz band) $\delta f/f_c \leq 2 \times 10^{-6}$ (1.5-GHz band) (tracking error $\leq 3 \times 10^{-7}$)
Spurious transmission	BS: −60 dB or less relative to the transmitted power, or 2.5 μW or less
	terminal: −60 dB or less relative to the transmitted power, or 0.25 μW or less.
Leakage power during carrier OFF period	BS: −60 dB or less compared with the power level during the carrier being turned ON, or 0.25 μW or less
	terminal: −60 dBm or less
99% bandwidth	32 kHz
Transmitter power for the terminal	Class I: 3.0 W Class II: 2.0 W Class III: 0.8 W Class IV: 0.3 W
Transmitter power accuracy	within +20% and −50% of the specified value
ACI power	Δf = 50 kHz: −45 dB or less Δf = 100 kHz: −60 dB or less
Cabinet radiation	25 μW or less

the interference signal level that yields a BER of 1×10^{-2} in the TCH part of the desired signal, where the interference signal is modulated by a 15-stage PN sequence, and its carrier frequency is separated by Δf kHz from the desired signal.

☞ **CIR sensitivity.** Defined by the power ratio of the desired signal to the interference signal that yields a TCH BER of 1×10^{-2} or 3×10^{-2} under Rayleigh fading conditions at f_d = 40 Hz, where the desired signal level is set to 3 dB higher than the specified receiver sensitivity, and the interference signal is modulated by a 15-stage PN sequence.

☞ **Intermodulation performance.** Defined by the power ratio of the desired signal level to the level of either of the two interference signals radiated to the TCH that yields a BER of 1×10^{-2}, where the desired signal level is set to 3 dB higher level than the specified receiver sensitivity, the interference signal is modulated by a 15-stage PN sequence, and its carrier frequency is separated by 100 kHz and 200 kHz from the desired signal.

Table 11.3 Requirements for PDC receivers.

Parameters	Specifications

Receiver sensitivity	under static conditions:
	    4 dBµ for BER = 10^{-2} and 2 dBµ for BER = 3×10^{-2}
	    (4 dBµ is called the specified receiver sensitivity)
	under Rayleigh fading conditions @ f_d = 40 Hz:
	    without diversity—
	        14 dBµ for BER = 10^{-2} and 8 dBµ for BER = 3×10^{-2}
	    with diversity—
	        7 dBµ for BER = 10^{-2} and 4 dBµ for BER = 3×10^{-2}

Spurious sensitivity	≥57 dB
Adjacent channel selectivity	≥1 dB for 25 kHz off ≥42 dB for 50 kHz off ≥57 dB for 100 kHz off

CIR sensitivity	under static conditions:
	    13 dB for BER = 10^{-2} and 11 dB for BER = 3×10^{-2}
	under Rayleigh fading condition @ f_d = 40 Hz:
	    without diversity—
	        22 dB for BER = 10^{-2} and 17 dB for BER = 3×10^{-2}
	    with diversity—
	        16 dB for BER = 10^{-2} and 13 dB for BER = 3×10^{-2}

Inter-modulation performance	≥57 dB

11.5 LOGICAL CHANNEL CONFIGURATION

Figure 11.3 shows a logical channel structure for the PDC system. The logical channels can be classified into two groups—the TCH, which carries the user's voice and data, and the control channel (CCH), which primarily carries system management messages.

The CCH can further be classified into the common access channel (CAC), which is primarily intended to carry signaling information for access management, and the

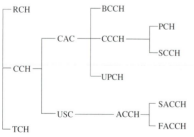

Fig. 11.3 Logical channel structure for the PDC system.

user-specific channel (USC), dedicated to carrying control signals specific to each user during a call. These CCHs are subject to the layered signaling protocols. The BCCH, SCCH, and PCH are multiplexed by using a superframe structure.

The radio-channel housekeeping channel (RCH) is an unlayered channel because real-time response is required to this channel.

BCCH. A point-to-multipoint unidirectional control signals from the network to the terminals. It carries, among others, the

☞ Operator identification number

☞ Restriction information

☞ Physical structure of the control channel (frequency, slot number, and so on) for each BS

☞ Maximum transmission power for the terminals

☞ Perch channel information

PCH. A point-to-multipoint unidirectional control channel from the network to the terminals. It is used for seeking a terminal when an incoming call has been placed on it.

SCCH. A point-to-point bidirectional control channel used for call establishment requirement and responses to inquiries. This channel transmits signaling information other than the paging message between the network and the terminal. Random access is applied for the uplink.

User Packet Channel (UPCH). A point-to-multipoint bidirectional channel used to transmit user packet data. Random access is applied for the uplink.

Associated Control Channel (ACCH). A point-to-point bidirectional channel to transmit signaling data during communication (after a point-to-point communication link has been established) associated with the TCH. The ACCH is classified into slow ACCH (SACCH) and fast ACCH (FACCH). The SACCH has a slow data transmission rate, and it is always allocated in the same slot as a TCH. On the other hand, FACCH has high data transmission rate that temporarily steals a part of the TCH.

RCH. A point-to-point, nonlayered signaling channel that transmits the received signal conditions (received signal level, BER), power control information, and time alignment information.

11.6 TDMA FRAME FORMAT

The PDC system employs 3-channel (for full rate) or 6-channel (for half rate) TDMA. Figure 11.4 shows frame formats for (a) full-rate transmission and (b) half-rate transmission. T_m and R_m ($m = 1, 2, ..., 6$) represent transmission and reception time slots for m-th user. Each slot length is 6.67 ms, and the frame length is 20 ms for full-rate transmission and 40 ms for half-rate transmission. To create 1 ms of the level monitoring period for the antenna selection diversity (LM) at the terminal, the transmission timing for the BS is delayed by 7.67 ms from the reception timing.

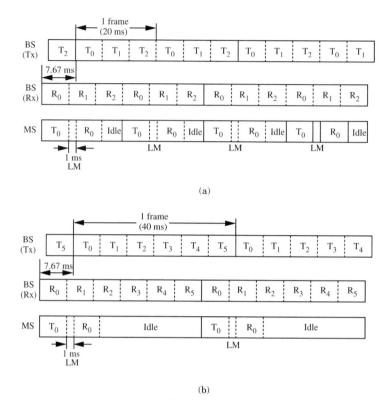

Fig. 11.4 Frame formats for (a) full-rate transmission and (b) half-rate transmission.

Figure 11.5 shows slot formats for (a) synchronization slot, (b) TCH, FACCH, SACCH, and RCH, and (c) CAC (BCCH, SCCH, and PCH). Every number represents the number of bits for each part. In any case, included in each slot are ramp-up symbols to suppress out-of-band radiation; a preamble to give initial data for the differential decoding; a synchronization word (SW) for the slot synchronization; and color code to identify the BS. For the uplink, guard time is also included to prevent slot collision due to a long delayed wave.

11.6.1 Format for Synchronization Slot

Synchronization slots are transmitted for the initial acquisition of the TCH or for resynchronization after handover or radio resource reassignment. Therefore, this slot includes a longer preamble and SW. In this slot, #1, #2, and #3 include color code, time alignment (TA) information, and a superframe synchronization counter (SSC) that represents relative slot position from the beginning of the superframe. B is the slot identification bit that represents whether it is for the first synchronization slot ($B = 0$)

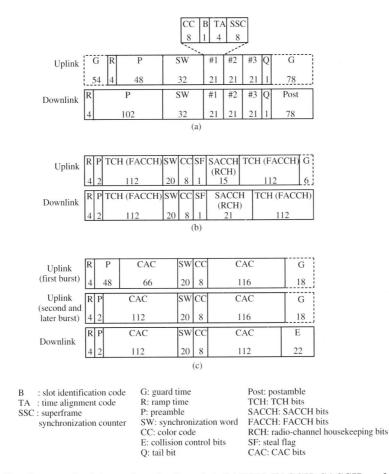

Fig. 11.5 Slot formats for (a) synchronization slot, (b) TCH, FACCH, SACCH, and RCH, and (c) CAC (BCCH, SCCH, and PCH).

or the second synchronization slot ($B = 1$). Again, #1, #2, and #3 include the same data, and they are detected by two-out-of-three majority voting.

11.6.2 Slot Format for TCH, RCH, and USC

In the downlink, each slot includes 4 bits of a ramp bit for windowing each slot signal, 224 bits of TCH, 20 bits of an SW, 8 bits of a color code, 1 bit of a steal flag (SF), and a 21-bit SACCH or RCH. SACCH and RCH are time division multiplexed on a superframe format. When an urgent or large amount of control signal message arises, such data are transmitted via the FACCH by stealing the TCH. The SF indicates whether the TCH is stolen by the FACCH.

In the uplink, the SACCH bits are reduced to 15 bits to prepare a 6-bit guard space.

11.6.3 Slot Format for CAC

CACs—BCCH, PCH, SCCH, and UPCH—are transmitted using a slot format shown in Figure 11.5(c). Among them, BCCH and PCH are transmitted only in the downlink; the others are transmitted both in the downlink and uplink.

The SCCH and UPCH in the uplink are operated in the random access mode called idle-signal casting multiple access (ICMA). In such an access scheme, long guard space is necessary to prevent slot signal collision because TA control, which is applied to the TCH slots, is not applicable. Therefore, 18 bits are assigned to the guard space. Moreover, in the random access scheme, symbol timing should be synchronized as fast as possible from the viewpoint of spectral efficiency. Therefore, the first slot for the uplink CAC has a longer preamble (48 bits).

11.7 SUPERFRAME FORMAT FOR BCCH, PCH, AND SCCH

In the downlink, BCCH, PCH, and SCCH are time division multiplexed on a super-frame format shown in Figure 11.6. In this superframe, full-rate transmission is applied in each frame. One superframe consists of 36 frames, and it corresponds to 720 ms. BCCH slots are located in the beginning of each superframe. PCH slots in a superframe are divided into several clusters, and each terminal is permitted to have access to only one of the groups determined by their own terminal numbers. A super-frame structure is determined by

☞ The number of BCCH slots in a superframe (A_B)
☞ The number of SCCH slots after each BCCH cluster (A_{S1})
☞ The number of PCH slots in each PCH cluster (A_P)
☞ The number of SCCH slots after each PCH cluster (A_{S2})
☞ The number of PCH clusters in a superframe (N_P)
☞ The number of frames in a superframe (A_w)

These parameters are broadcast to all the terminals via the BCCH. In the uplink, all the slots in a superframe are used for the SCCH.

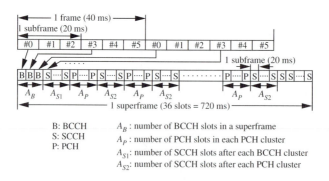

Fig. 11.6 Superframe format for CAC in the downlink.

11.8 SUPERFRAME STRUCTURE FOR TCH, ACCH, AND RCH

TCH and ACCH or RCH are time division multiplexed in each TDMA slot, and SACCH and RCH are time division multiplexed on a superframe as shown in Figure 11.7 for (a) full-rate transmission and (b) half-rate transmission. In any case, one superframe length is 720 ms.

The TCH is transmitted every 20 ms in the case of full rate and every 40 ms in the case of half rate. The SACCH and RCH are multiplexed on a superframe with a frame length of 720 ms. Therefore, one superframe consists of 36 slots for full rate and 18 slots for half-rate transmissions.

RCH is transmitted using 2 consecutive slots every 360 ms in the case of full-rate transmission and 2 consecutive slots every 720 ms in the case of half-rate transmission. On the other hand, SACCH is transmitted using 16 consecutive slots every 360

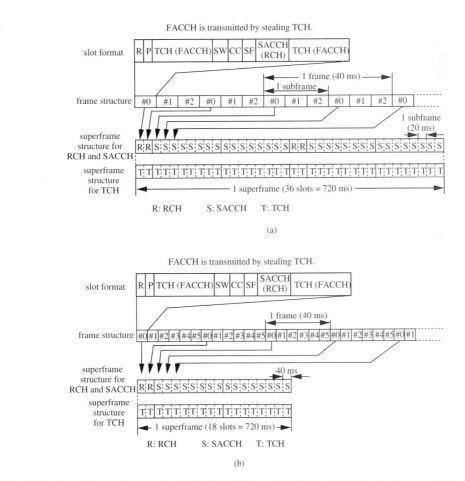

Fig. 11.7 Superframe format for TCH, SACCH, and RCH for (a) full-rate transmission and (b) half-rate transmission.

ms in the case of full-rate transmission and 16 consecutive slots every 720 ms in the case of half-rate transmission.

11.9 CHANNEL CODING FOR CAC

Because higher reliability is required for control signal transmission in comparison with voice channel transmission, these messages are protected by CRC as well as FEC.

Figure 11.8 (a) shows CAC data format for the first slot in the uplink. In each message unit, 8 bits of the message configuration information word (W), 104 bits of information, and 2 bits of dummy bit called *filler* are multiplexed, where W includes information on whether the slot is the first, last, or else slot, as well as how many slots are left to be transmitted. The multiplexed bits are encoded using ITV-T 16-bit CRC with its generator polynomial of $G(x) = x^{16} + x^{12} + x^5 + 1$. The CRC-encoded sequence is then divided into 13 blocks, each of which contains 10 bits of CRC-encoded data and is encoded by 1-bit shortened (14, 10) BCH code with its generator polynomial of $G(x) = x^4 + x + 1$. Of the BCH-encoded data sequence, 182 bits are interleaved using (13, 14) low-column conversion-type interleaver, divided into 66- and 116-bit groups, and mapped onto the physical channel slot.

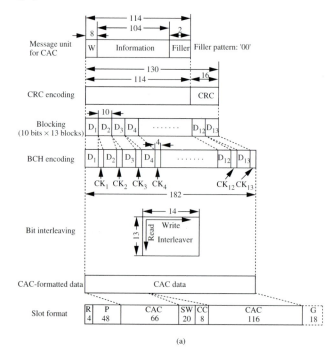

(a)

Fig. 11.8 Slot format and its encoding process for (a) the first CAC slot in the uplink, (b) the second or later CAC slots in the uplink, and (c) the CAC slot in the downlink.

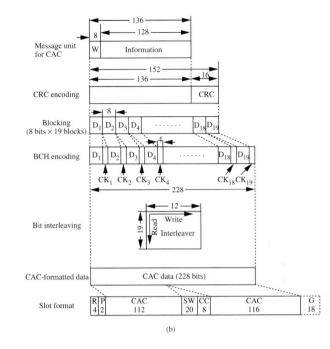

(b)

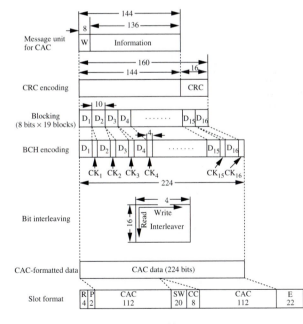

(c)

Fig. 11.8 Continued.

Figure 11.8(b) shows the slot format for the second or later CAC slots in the uplink. Each message unit consists of 128 bits of information and 8 bits of W. The encoding process is almost the same as that in Figure 11.8(a) except that the FEC is 3-bit shortened $(12, 8)$ BCH with its generator polynomial of $G(x) = x^4 + x + 1$, and the interleaver size is $(19, 12)$.

Figure 11.8(c) shows the slot format of the CAC in the downlink. Each message unit consists of 136 bits of information and 8 bits of W. The difference between Figure 11.8(a) and (c) is that the interleave is $(16, 14)$ instead of $(13, 14)$.

Figure 11.9 shows BCCH, SCCH, and PCH data mapping onto the physical channel with a superframe structure (a) in the downlink and (b) in the uplink, where the message length is assumed to be 1,000 bits. In the downlink, 136 bits of signaling information is transmitted in the first to the seventh superframes, and 48 bits are transmitted in the eighth (last) superframe. In the uplink, the first superframe transmits only 104 bits, and the latter superframe transmits 128 bits of signaling information.

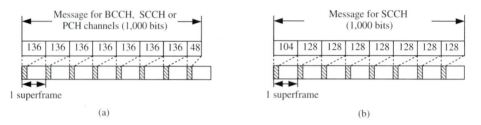

Fig. 11.9 BCCH, SCCH, and PCH data mapping onto the physical channel with a superframe structure (a) in the downlink and (b) in the uplink, where the message length is assumed to be 1,000 bits.

11.10 CHANNEL CODING FOR USC

Figure 11.10 shows encoding format for RCH (a) in the uplink and (b) in the downlink. In the uplink, 14 bits of source bits are encoded using an 8-bit CRC with its generator polynomial of $G(x) = x^8 + x^7 + x^4 + x^3 + x + 1$. The CRC encoded 22-bit data sequence is then divided into two blocks; each of them are encoded by $(15, 11)$ BCH code with its generator polynomial of $G(x) = x^4 + x + 1$, interleaved by the $(2, 15)$ interleaver, and mapped onto the RCH bit location in the assigned USC slot. In the case of the down-link, 22 bits of the source data are CRC encoded by using an 8-bit CRC with the same generator polynomial as that for the uplink to produce 30 bits of encoded sequence. The sequence is then divided into three blocks, each of which is encoded by 1-bit short-ened $(14, 10)$ BCH code with its generator polynomial of $G(x) = x^4 + x + 1$. After CRC- and BCH-encoded data are interleaved by the $(2, 14)$ interleaver, they are mapped onto the RCH bit location in two consecutive frames.

Figure 11.11 shows encoding format for SACCH (a) in the uplink and (b) in the downlink. In the uplink, 64 bits of source bits associated with 8 bits of W are encoded by the 16-bit CRC with its generator polynomial of $G(x) = x^{16} + x^{12} + x^5 + 1$. The CRC-encoded 88-bit data sequence is then divided into 8 blocks, each of them encoded by (15, 11) BCH code with its generator polynomial of $G(x) = x^4 + x + 1$, interleaved by the

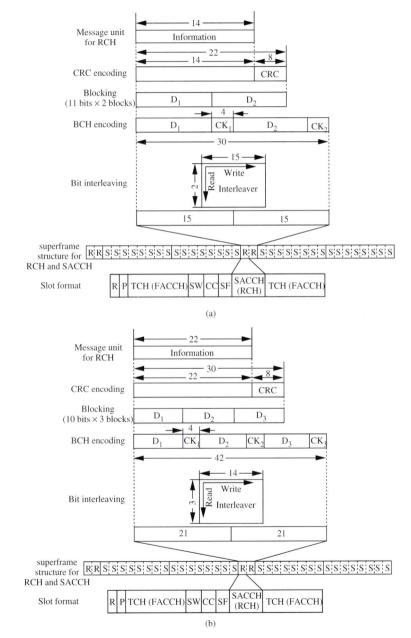

Fig. 11.10 Encoding format for RCH (a) in the uplink and (b) in the downlink.

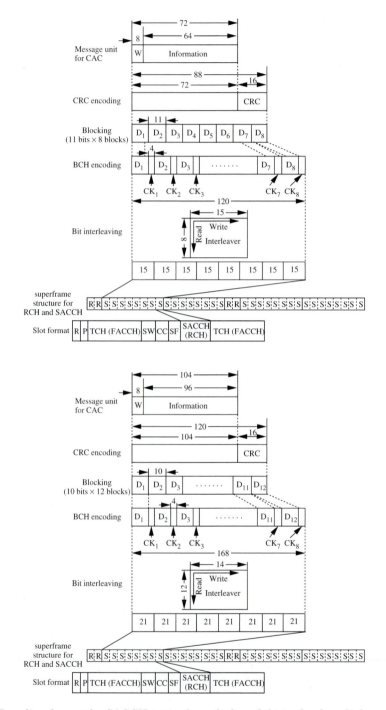

Fig. 11.11 Encoding format for SACCH (a) in the uplink and (b) in the downlink.

(8, 15) interleaver, and mapped onto the SACCH bit location in the assigned slot. In the case of the downlink, 96 bits of the source data with 8 bits of W are CRC encoded by using the 16-bit CRC to produce 120 bits of encoded sequence. The sequence is then divided into 12 blocks, each of which is encoded by 1-bit shortened (14, 10) BCH code with its generator polynomial of $G(x) = x^4 + x + 1$. After the CRC- and BCH-encoded data are interleaved by the (12, 14) interleaver, they are mapped onto the SACCH bit location in the assigned slots.

Figure 11.12 shows encoding format for FACCH. Because the number of bits for FACCH in each slot is the same in both the uplink and downlink, the same encoding scheme is employed for both the uplink and downlink. A message unit that consists of 8 bits of W and 88 bits of FACCH message is CRC encoded by the ITU-T 16-bit CRC with its generator polynomial of $G(x) = x^{16} + x^{12} + x^5 + 1$, and the encoded data sequence is divided into 28 blocks of 4-bit data. The blocked 4-bit data are then encoded by the 1-bit extended (8, 4) BCH code with its generator polynomial of $G(x) = x^3 + x + 1$. After the CRC- and BCH-encoded data sequence is interleaved by the (28, 8) interleaver, the former 112 bits are mapped onto the former part of the FACCH section in the present frame, and the latter 112 bits are mapped onto the latter part of the next frame.

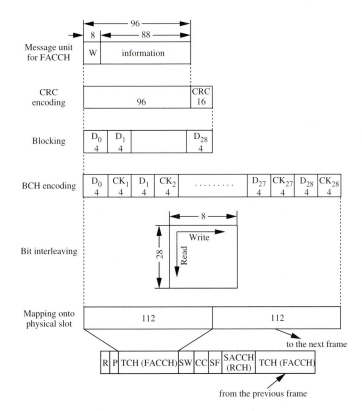

Fig. 11.12 Encoding format for FACCH.

11.11 TRANSMISSION QUALITY MONITORING

In PDC systems, each MS monitors the received signal level and the BER averaged over one superframe period. Each terminal also monitors the received signal level of the adjacent cells during idle time slots. Mobile-assisted handover in the PDC is carried out using these measured values in this way.

The received signal level and BER of each terminal is transmitted to the BS every superframe via the RCH. On the other hand, the received signal levels of the adjacent cells requested from the BS are transmitted to the BS via the SACCH. Using these data, the BS assesses the quality of the channel as well as the quality of the neighboring cells to be handed over. When the BS can find an appropriate cell to be handed over and that there are unused channels in the new cell, it sends handover permission to the terminal.

Using the received signal level information transmitted via the uplink RCH, the BS also sends a power control signal to each terminal.

11.12 INITIAL ACQUISITION FOR BCCH

In PDC systems, when the terminal power is turned on, each terminal first has to take synchronization with the BCCH before going to the standby mode (PCH waiting mode) because the BCCH broadcasts parameters for the CAC superframe structure. For this purpose, the PDC system prepares perch channels whose carrier frequency and slot are preliminarily determined as commonly used BCCH channels. In each cell, at least one of the perch channels is used as one of the BCCHs.

Figure 11.13 shows a flow diagram for perch channel acquisition and the CAC acquisition process. When the terminal power is turned on, the terminal begins to measure the received signal level (L_1) of all the perch channels. In this process, the measured value is memorized when the level is higher than a threshold level (L_{th}). Then a table in descending order of the measured power level is created.

When the power measurement and listing are finished, the receiver selects the channel on top of the list, takes synchronization of the superframe, frame, slot, and symbol timing; checks the color code; and evaluates the channel condition by using a CRC check. When these checks are not passed, the receiver selects the next channel in the list and goes through the same process. When none of the channels pass the checks, the receiver goes back to the perch channel signal level measurement stage.

On the other hand, when all the checks are passed, the receiver measures the received signal level of the BCCH slot (L_2). At the same time, the BCCH signal is detected and evaluated as to whether the BCCH is correctly detected. When L_2 is higher than a threshold level (L_{th2}) and BCCH is correctly detected, the received signal is switched from the perch channel to BCCH, and the receiver goes to the standby mode (PCH waiting mode).

When in the standby mode, the receiver intermittently receives the BCCH and the assigned PCH slots in the superframe. At the same time, the receiver measures the received signal level of the adjacent cells. The measured data are used for the handover process.

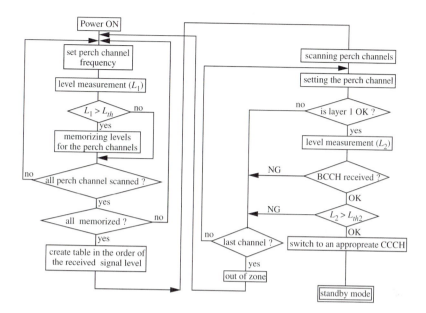

Fig. 11.13 Flow diagram of the perch channel acquisition and the CAC acquisition process.

11.13 RANDOM ACCESS PROTOCOL FOR SCCH AND UPCH

When a terminal receives a paging message from the BS initiating a call, the terminal and BS exchange necessary messages via SCCH. In the uplink SCCH, a random access protocol called *idle-signal casting multiple access with partial echo* (ICMA-PE) is employed [11-8]. In this protocol, the BS always broadcasts the following collision control signal bits to indicate random access channel occupancy.

☞ Channel idle/busy indicator (I/B = 111/000, 3-bit data)

☞ Correctly received/not received (R/N = 111/000, 3-bit data)

☞ 16 bits of the CRC check bits as an identification of the received signal, which is a PE (16 bits)

These bits are mapped onto the 22-bit collision control bits (E in Figure 11.5[c]) in the CAC slot for the downlink.

Figure 11.14 shows a flow diagram of the ICMA-PE protocols (a) for the BS and (b) for the terminal.

The BS demodulates the uplink slot signal and checks its CRC to evaluate whether the message is correctly received. When the message is not correctly received, the collision control bits are set to I/B = I, R/N = N, and PE = all zero. On the other hand, when the message is correctly received, 16 bits of the received CRC check bits are written in the PE, and R/N is set as R. When the signal is not the last slot, I/B is set as B because the channel will be used in the next frame. On the other hand, when

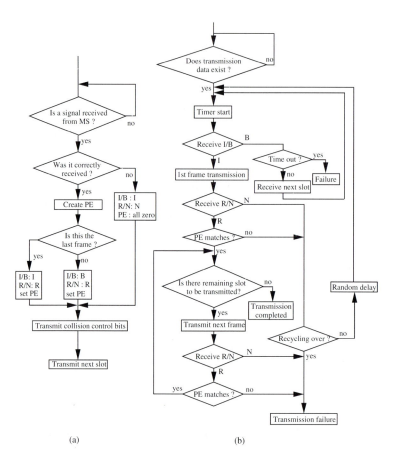

(a)   (b)

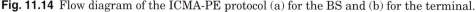

Fig. 11.14 Flow diagram of the ICMA-PE protocol (a) for the BS and (b) for the terminal.

it is the last slot, I/B is set as I. Then, I/B, R/N, and PE are transmitted with the SCCH message in the downlink.

When a terminal has a message to be transmitted at the terminal, the terminal checks whether the I/B signal in the downlink is idle or not where the channel is detected to be idle only when I/B is exactly 000. When I/B is detected to be busy, I/B detection is continued until the slot becomes idle or the trial time is expired. On the other hand, when I/B is detected to be idle, the terminal immediately begins message transmission.

When the terminal sends the first message in the uplink, the terminal checks R/N and PE of the collision control bits (E), where R/N is detected by a two-out-of-three majority voting and PE is detected with 1-bit error tolerance. When R/N = R and PE is detected to be a part of the transmitted message, the terminal continues transmission in the same manner. On the other hand, when the returned R/N is N or the returned echo includes errors, the message will be retransmitted after a randomly determined time period (between 0 and 0.5 second).

Because ICMA-PE is free from the hidden terminal problem owing to the I/B detection and it can effectively employ the capture effect by using a PE technique, it can achieve very high throughput performance.

11.14 LAYERED SIGNALING STRUCTURE

A three-layered signaling structure based on the OSI reference model is applied to the PDC system to allow independent development of the radio subsystem and signaling protocols. The layered structure is shown in Figure 11.15.

In layer 1, modulation, access, channel coding, time alignment, vox, and so on are determined. Layer 2 is in charge of transferring a data sequence in error-free condition. Layer 3 is in charge of establishing a communication path between the BS and a terminal. This layer includes three functions—radio transmission management (RT) in charge of radio resource management; mobility management (MM) in charge of terminal authentication and terminal location registration; and call control (CC) that controls call connection and disconnection.

11.14.1 Layer 2 Format

Layer 2, which defines the link access procedure for the digital mobile channel (LAPDM) is in charge of transferring a layer 3 message between the BS and a terminal in a predetermined order.

LAPDM is a protocol that operates at the data link layer of the OSI reference model, and it employs the following principles and functions:

☞ Consultative Committee for International Telegraph and Telephone (CCITT) Recommendation X.200 and X.210

☞ X.25 packet mode terminal interface

☞ ISO3309 and ISO4335 which determine high-level data link control (HDLC) protocol

LAPDM employs a frame structure independent from that of the physical layer. Figure 11.16 shows a frame format for layer 2 and layer 3 messages along with its mapping onto the physical slot. The layer 2 portion consists of the address field that includes the terminal ID and service access point ID, and the control field that

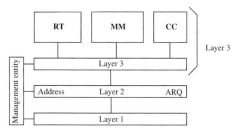

Fig. 11.15 The layered structure of signaling.

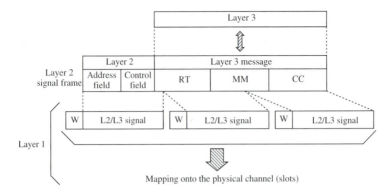

Fig. 11.16 Frame format for layer 2 and layer 3 messages and its mapping onto the physical slot.

includes messages and information necessary for data transfer control. On the other hand, the layer 3 message following the layer 2 portion includes signaling messages.

In the LAPDM, a layer 2 and 3 message having a bit length longer than the data section length of the physical slot is divided into several units, each of which is mapped onto physical slots; errors in each unit are corrected by retransmission control.

11.14.2 Layer 3 Format

Layer 3 is divided into three modules—RT, MM, and CC. RT manages radio resources including channel assignment, radio link maintenance, and handover. MM is in charge of authentication and location registration of the subscribers. CC is in charge of call control. Basic procedure for the CC is based on the ISDN user-network layer 3 interface (ITU-T Rec. I.451), and some other functions for supplementary services are included in the CC. In the PDC system, we need a very quick response for the call setup and the other call control services such as call waiting in order to save spectrum resources, although the average bit rate of the control signal is very low (about several hundred bit/s) compared to ISDN services (16 kbit/s). Therefore, RT, MM, and CC messages are multiplexed in the same signaling frame, thereby improving transmission and spectrum efficiency. Because there are so many layer 3 messages, we will just summarize the types of messages for RT, MM, and CC modules. Figure 11.17 shows a summary of the types of messages for each module. This figure also shows via which logical channels each message is transmitted. For example, when channel condition for a terminal is so degraded that handover to another cell is necessary, call control messages of the RT, such as the radio condition report, are exchanged between the terminal and network via SACCH.

11.15 EXAMPLES OF THE CONTROL SEQUENCE

In the preceding section, we discussed logical and physical channels in layer 1 and RT, MM, and CC management in layer 3. In this section, we will discuss how the layer 3

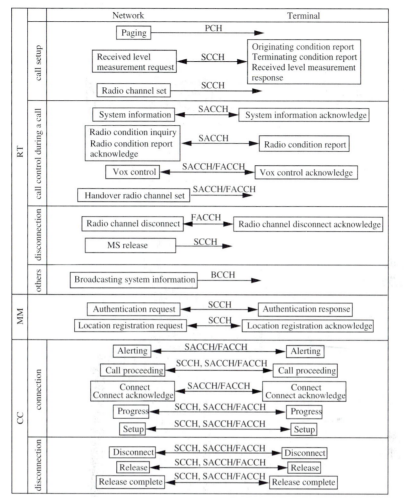

Fig. 11.17 Summary of the types of messages for each module.

messages are employed in the PDC system. For our discussion here, protocols for call origination, call reception, disconnection, and location registration will be shown as examples of layer 3 messages.

11.15.1 Call Origination and Reception Processes

Figure 11.18 shows a call origination process of the PDC system. When a terminal initiates a call, it sends a call request to the network via the SCCH. When the message is correctly received, authentication (MM) and channel condition (RT) check is started. When the process is successfully finished, a synchronization slot and radio resource assignment message is sent to the terminal. When the terminal receives these mes-

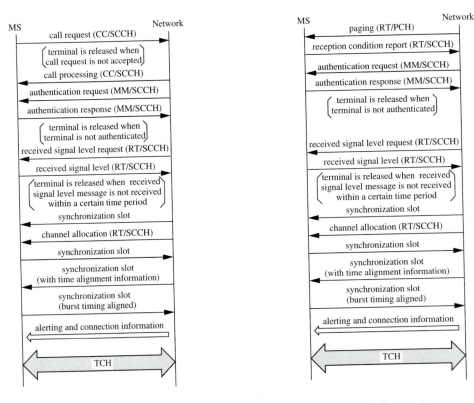

Fig. 11.18 Call origination process of the PDC system.

Fig. 11.19 Call reception process of the PDC system.

sages, it sends back a synchronization slot with the assigned radio resource information to the BS. When the BS receives this signal, it measures the slot timing and channel quality and sends a synchronization slot with TA information to the terminal. When the terminal receives this signal, it detects the TA information and sends back a slot-timing-aligned synchronization slot to the BS. When the BS confirms that the transmitted slot is correctly controlled, the communication phase using TCH is started.

Figure 11.19 shows a call reception process of the PDC system. When a terminal receives a paging signal, the terminal sends a terminal condition report that includes terminal ID, received signal level, and the received signal levels of the perch channels. After authentication, received signal measurement, radio resource assignment, and slot synchronization including time alignment, the terminal can start communication.

11.15.2 Disconnection

Figure 11.20 shows a disconnection process for the terminal. When a terminal wants to terminate a call, it sends a disconnect message (CC) to the BS. When the BS receives this message, it sends a release message (CC) to indicate that the BS has

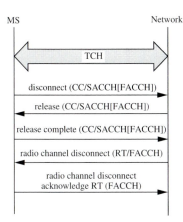
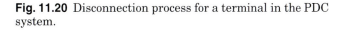

Fig. 11.20 Disconnection process for a terminal in the PDC system.

released the used TCH and the ID number of the called party. When this message is received by the terminal, it sends back the release complete message (CC) to indicate that the terminal has also released it. After that, the BS sends a radio channel disconnect message (RT) to the PS to indicate that the BS has released the radio channel. When the terminal receives this message, it also sends back the radio channel disconnect complete message (RT) to the BS to indicate that the terminal has also released the radio channel.

11.15.3 Location Registration

Figure 11.21 shows the location registration updating process. When a terminal wants to update a terminal location, it sends a location registration message to the BS. When the BS receives this message, it sends back an authentication request to the terminal. According to this request, the terminal sends the necessary information to the BS. When the authentication process is successfully finished, the BS sends a location registration acknowledge message to the terminal to indicate that the registration is finished.

When the location registration is carried out during the standby mode, these messages are exchanged via the SCCH. On the other hand, when it is carried out during a call, these messages are exchanged via the SACCH or FACCH.

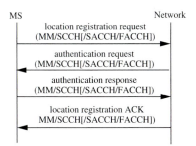
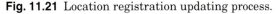

Fig. 11.21 Location registration updating process.

REFERENCES

11-1. RCR, "Personal digital cellular telecommunication system," RCR STD-27, April 1991.

11-2. GSM, "Physical layer on the radio-path: GSM system," GSM Recommendation, Vol. G, GSM Standard Committee, July 1988.

11-3. EIA/TIA, "Cellular system: Dual-mode mobile station—base station compatibility standard," IS-54, project 2215, Washington, D.C., December 1989.

11-4. TIA/EIA, "Mobile station—base station compatibility standard for dual mode wideband spread spectrum cellular system," TIA/EIA IS-95, 1993.

11-5. Satoh, T., Hoto, Y. and Murase, A., "TDMA half-rate digital cellular system based on PDC standard in Japan," 45th IEEE Veh. Tech. Conf. (Chicago, Illinois), pp. 301–5, July 1995.

11-6. Tanaka, T., Akeyama, A. and Kozono, S., "Urban multipath propagation delay characteristics in mobile communications," IEICE Trans. Vol. J73-B-II, No. 11, pp. 772–78, November 1990.

11-7. Sampei, S., Komaki, S. and Morinaga, N., "Adaptive modulation/TDMA scheme for personal multi-media communication systems," GLOBECOM'94 (San Francisco, California), pp. 989–93, November 1994.

11-8. Umeda, N. and Onoe, S., "Idle-signal casting multiple access with partial echo (ICMA-PE) for mobile packet communications," Electronics and Communications in Japan, Part 1 (Communications), Vol. 77, No. 4, pp. 92–102, April 1994.

Personal Handyphone System (PHS)

The personal handyphone system (PHS) is a digital microcellular system that is intended to support personal multimedia communication systems. This system was standardized by the RCR (ARIB at present) in 1993, and its service was launched in July 1995 in Tokyo and Sapporo and in October 1995 in other areas including Osaka and Kyoto.

In this chapter, we will outline the PHS, with special emphasis on its physical layer characteristics.

12.1 HISTORIC OVERVIEW OF PHS DEVELOPMENT

The standardization process of the PHS, including preliminary investigation of PHS technology, began in 1989. In January 1989, MPT organized a study group on the Next Generation of Portable Telephone Systems. In this study group, the technological aspects of the next-generation portable phone were summarized, and its report was submitted to MPT in February 1990. Following this report, MPT consulted the TTC about the technological requirements for PHS, and its interim and final reports were issued in June 1992 and April 1993. Based on these results, MPT, in consultation with the RRC, promulgated the PHS-related ministerial ordinances in October 1993. Along with the TTC discussion, RCR started standardization of the PHS system in June 1990 and issued its standard RCR STD-28 in December 1993 [12-1].

Although these are the technological and regulative studies on PHS, commercialization of PHS was also studied. In October 1992, MPT organized a study group in which the commercialization aspect of PHS was discussed, and its interim and final study reports were issued in June 1993 and October 1993.

Based on the report of the TTC and the interim study report by the study group on PHS, a study group on field trials was organized in July 1993, and extensive field trials from the viewpoint of technology and commercialization were conducted until the end of 1994.

In June 1994, based on the field trial results, MPT issued the basic licensing guideline for PHS with a goal of achieving, as quickly as possible, cheap and good-quality nationwide PHS services under fair, competitive, and good regionally balanced conditions. A summary follows (from MPT press release of June 24, 1994):

☞ For the time being, 12 MHz from the 1.9-GHz band that is exclusively allocated for PHS public use will be allocated for three PHS operators at maximum in each regional block. Further frequency allocation will be considered after the start of PHS business considering the use of frequency resources and demand trends.

☞ PHS operators are requested to make an effort to provide services in the area where more than 50% of the population of the licensed regional block resides within 5 years after they start services.

☞ Public telecommunications operators are requested to interconnect PHS operators to their networks with fair conditions to promote sound development of PHS business in competitive circumstances.

Furthermore, MPT also issued additional licensing guidelines that relate to concrete business development of PHS in November 1994. A summary follows (from MPT press release of November 1, 1994):

☞ PHS is expected to significantly promote personal and multimedia communication as new types of communication media for ease. To develop PHS, operators are expected to provide services at cheaper and diversified prices and with wider service areas including industrial parks, exhibition sites, and so on.

☞ Public telecommunication operators are requested to make an effort to consult with PHS operators when they make a plan to prepare facilities for the interconnection of PHS networks and to ensure a fair interconnection as well as to settle reasonable interconnection charges and technical conditions through the establishment of interconnection agreements with PHS operators.

☞ The numbering system for PHS operators should be on the pattern 050 (service identification number)-XX (operator identification number)-YYYYY (subscriber number).

☞ PHS operators are requested to make an effort to realize as early as possible interregional roaming that enables nationwide utilization of PHS services.

☞ It is desirable that related organizations should take necessary measures to enable the use of public facilities and buildings such as traffic signals, electricity supply poles, and public telephone booth and stations on reasonable terms and fair conditions. Each PHS operator is requested to develop base stations that every operator can share.

Based on these guidelines, 28 operators were licensed as PHS service operators from MPT during January and June 1995, and PHS services began in July 1995.

12.2 SYSTEM OVERVIEW

PHS is a system that makes it possible to use the handset of a digital cordless phone outside the home or office, including in stations, commercial areas, and any other public spaces. In this system, a BS is called a cell station (CS) and a terminal is called a personal station (PS). The zone radius of PHS is restricted to up to 100–500 m, which is called a *microcellular system*, to achieve high system capacity, to reduce transmission power for saving battery life, and to reduce cost of the CS.

Figure 12.1 shows a system concept of the PHS. Because PHS is basically a digital cordless phone system, a PHS terminal can be used as a handset of a cordless phone in the business office or home as in the case of an analog cordless phone. This operation is called a *private mode*. In this case, a parent phone at home or a PBX is connected to the public-switched telephone network (PSTN). When we are outside the home or business office, we can access the public CS using this terminal. This operation is called a *public mode*. When we want to use a PHS terminal in a public mode, the terminal has to have another 10-digit subscriber number. In this case, each CS is connected to the ISDN. Moreover, the PHS has a mode that enables direct communication between terminals, which is called the *transceiver mode*. This mode is available when the terminals are located close together. Each user can select one of these three modes, taking into consideration the operational environments.

Figure 12.2 shows (a) PHS terminals, (b) a parent phone for a digital cordless phone (private-mode operation), (c) a CS installed on top of a building, and (d) antenna elements installed on the ceiling of an underground town in downtown Tokyo.

Features of the PHS system at present are

☞ High-quality voice communication and enhanced security

☞ High capacity owing to microcellular and DCA technologies

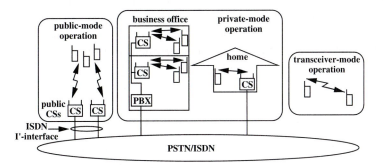

Fig. 12.1 System concept of the PHS.

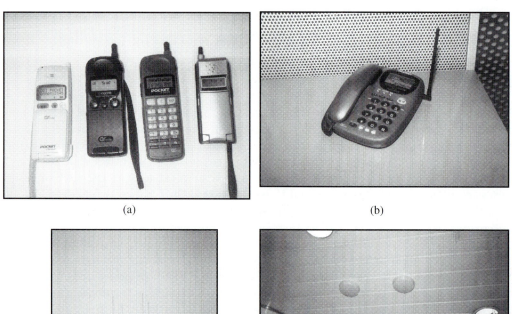

(a)   (b)

(d)

(c)

Fig. 12.2 The PHS system operated by DDI Pocket Telephone, Inc. (a) PHS terminals, (b) a parent phone for a digital cordless phone (private-mode operation), (c) a CS installed on top of a building, and (d) antenna elements for 2-branch space diversity installed on ceiling of an underground town in downtown Tokyo (courtesy of DDI, Inc.).

☞ Long continuous call operating time (5 hours) and standby time (17 days) due to small cell size, intermittent reception during standby mode, and some other battery-saving technologies

☞ Multimedia service capability using its bearer service function with its bit rate of up to 64 kbit/s

☞ Simplicity of spectrum management because of the DCA

☞ Terminal commonality between cordless phone, public microcellular phone and transceiver

Table 12.1 shows a comparison between PHS and PDC systems from the services point of view. Because PHS is a microcellular system, its coverage is much narrower than that of the PDC system, thereby reducing the cost of CSs as well as the transmission power of both CSs and PSs. On the other hand, in the PHS, we can employ high-quality and high-bit-rate voice codec because we can achieve very high system capacity owing to the small cell radius.

Because a simple algorithm is employed for the handoff and its capability is limited only when the CSs are connected to the same local switch, we cannot use PHS in a vehicle or fast-moving public transportation. The PHS employs a DCA technique. As a result, each PHS operator co-uses the same frequency band for the TCH.

Table 12.1 Comparison between PHS and PDC from the services point of view.

	PHS	PDC
BS (CS) coverage	100–300 m (radius)	1.5–5.0 km (radius)
Population coverage	low (around 50%)	high (>90%)
Relative number of BS (CS)	40–120	1
Relative cost per BS (CS)	0.001–0.025	1
Voice codec	32 kbit/s adaptive differential pulse code modulation (ADPCM)	11.2 kbit/s vector sum excited linear prediction (VSELP)
Terminal mode	- private mode - public mode - transceiver mode	public mode
Tx power for BS (CS)	≤20 mW for standard type ≤500 mW for high-power type	not specified
Tx power for MS (PS)	≤10 mW	≤3.0 W
Handover	slow handover	fast handover
Channel assignment	DCA	fixed channel assignment (FCA)

12.3 NETWORK CONFIGURATION

PHSs greatly depend on the networking facilities of the PSTN and ISDN of the currently operated companies, such as NTT. At present, there are two types of network configurations—NTT network-dependent type and NTT network-connection type.

Figure 12.3 shows the network configuration of the NTT network-dependent-type system. In this system, facilities of the PHS operator are only the radio subsystems between CSs and PSs, monitoring systems for CSs, and customer database. Each public CS is connected to the digital local switch (LS) by the ISDN I'-interface, where I'-interface is I-interface plus some mobility management functions, and one

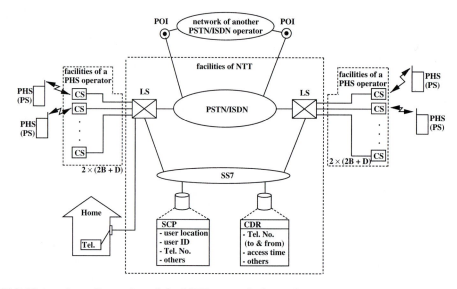

Fig. 12.3 Network configuration of the NTT network-dependent-type system.

physical channel is transferred using one B-channel. Because each CS supports four channels, two ISDN I'-interface lines are used to connect a CS and the digital LS. Although the PHS operation company has a customer database, it employs most of the customer information database in the NTT network, such as the mobility management information, authentication information and subscriber numbers stored in the service control point (SCP), and call detail record (CDR) of the subscribers, such as the telephone number of the calling and called parties and access time. Such information is transferred via SS7.

When a call is originated from a telephone in the PSTN to a PS, an LS that accommodates the call-initiated telephone accesses the SCP to find out the LS near the PS and then transmits a call setup request signal to the LS. When the LS receives a call setup request signal, it sends a paging signal to all the CSs connected to the LS.

Figure 12.4 shows the network configuration of the NTT network-connection-type system. The difference between NTT network-dependent- and NTT network-connection-type systems is that the NTT network-connection-type system employs networking facilities of the regional PSTN/ISDN operation company (other than NTT). In this type of system, the network of the regional operator is connected with the NTT PSTN/ISDN network or another new common carrier (NCC) network at the points of interfaces (POIs).

12.4 SPECTRUM ALLOCATION FOR PHSs

For the PHSs, 77 carriers with a bandwidth of 300 kHz are prepared in the frequency range of 1895–1918.1 MHz (23.1 MHz of the system bandwidth). Among them, usage of each carrier is determined as shown in Table 12.2.

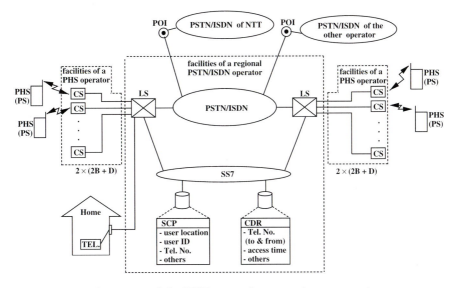

Fig. 12.4 Network configuration of the NTT network-connection-type system.

Carriers 1–37 (37 carriers) are used for private-mode operation. Among them, carriers 1–10 are also used for the transceiver-mode operation. Carriers 38–77 (40 carriers) are used for public-mode operation. Among these, the high-power CSs can use only carriers 38–53 for the TCHs, and low-power CSs can use carriers 38–69 for the TCHs, where high power means the average transmission power for the CS of more than 20 milliwatts (mW). As for the CCHs for the public mode, one dedicated carrier is assigned to each PHS operator.

Although carriers for TCHs are divided into two groups—for the private mode and public mode—the carrier (and slot) for each call is selected from an appropriate group by the CS using a DCA technique based on carrier sensing.

Table 12.2 Frequency allocation for PHSs.

Carrier No.	Frequency (MHz)	Usage
1	1895.150	TCHs for both transceiver mode and private mode
—	—	
10	1897.850	
11	1898.150	TCHs for private mode
12	1898.450	CCH for private mode
13	1898.750	TCHs for private mode
—	—	
17	1899.950	

Table 12.2 Frequency allocation for PHSs. (Continued)

Carrier No.	Frequency (MHz)	Usage
18	1900.250	CCH for private mode
19	1900.550	TCHs for private mode
—	—	
—	—	
37	1905.950	
38	1906.250	TCHs for public mode (for both low-power and high-power CSs)
—		
—		
53	1910.750	
54	1911.050	TCHs for public mode (for only low-power CSs)
—		
—		
69	1915.550	
70	1915.850	guard channel
71	1916.150	CCH (public mode) (reserved)
72	1916.450	guard channel
73	1916.750	CCH (public mode) (operator #1)
74	1917.050	guard channel
75	1917.350	CCH (public mode) (operator #2)
76	1917.650	guard channel
77	1917.950	CCH (public mode) (operator #3)

12.5 SPECIFICATION OF THE RADIO INTERFACE

Table 12.3 shows specifications of the PHS and compares PHS with the other digital cordless telephone systems, such as the DECT [12-2] and the CT-2 [12-3]. Although CT-2 is operated in an 800-MHz band, PHS and DECT systems are both operated in a 1.9-GHz band which overlaps with the FPLMTS band identified by WARC-92 [12-4]. $\pi/4$-QPSK is employed as a modulation scheme for the PHS to achieve high system capacity as well as to mitigate the nonlinear effect at the transmitter amplifier. A root Nyquist filter with a roll-off factor (α) of 0.5 is employed as the transmitter and receiver filter.

Four-channel multiplexed TDMA/TDD is employed as an access scheme. The voice codec for the PHS is a 32-kbit/s ADPCM because its quality is equivalent to a 64-kbit/s PCM, processing delay is very low in comparison with the low-rate voice codec, and the bit rate of 32 kbit/s is very attractive for nonvoice services. A TDMA/TDD frame length of 5 ms is employed to achieve short buffering delay time.

Because PHS is operated in microcellular systems with a zone radius of 100–500 m, the maximum delay spread is much smaller than that of digital cellular systems such as PDC. The extensive measurement of the propagation path characteristics shows that the maximum delay spread under such microcellular system conditions is up to 250 nanoseconds (ns) [12-5]. As we discussed in chapter 7, we can employ trans-

Table 12.3 Specifications of the PHS and its comparison with the other digital cordless systems.

Specifications	PHS	DECT	CT-2
Frequency (MHz) (bandwidth)	1895.00–1918.05 (23.1 MHz)	1880.00–1990.00 (110 MHz)	864–868 (4 MHz)
Symbol rate (bit rate)	192 ksymbol/s (384 kbit/s)	1,152 ksymbol/s (1,152 kbit/s)	72 ksymbol/s (72 kbit/s)
Carrier spacing	300 kHz	1,728 kHz	100 kHz
Access	4-ch TDMA	12-ch TDMA	FDMA
Duplex	TDD	TDD	TDD
Frame length	5 ms	10 ms	2 ms
Modulation	differentially encoded $\pi/4$-QPSK ($\alpha = 0.5$)	GMSK (BT = 0.5)	GMSK (BT = 0.3)
Voice codec	32-kbit/s ADPCM	32-kbit/s ADPCM	32-kbit/s ADPCM
Number of usable carriers for each operator	77	10	40
Channel/carrier	4	12	1
Channel/system	308	120	40

mitter diversity in the PHS because it employs TDD. As a result, when we define the acceptable irreducible BER due to delay spread of BER = 10^{-3}, we can accept the delay spread of less than 0.08 times a symbol duration, which corresponds to the symbol rate of around 300 ksymbol/s in the case of a delay spread of 250 ns. With this under consideration, the PHS employs the symbol rate of 192 ksymbol/s at the air interface.

12.6 REQUIREMENTS FOR TRANSMITTER AND RECEIVER

In addition to the specifications shown in Table 12.3, RCR STD-28 also specifies requirements for the transmitter and receiver. Table 12.4 shows requirements for a PHS transmitter. Parameters in Table 12.4 are the same as those for the PDC system shown in chapter 11, and the definitions of those parameters are the same except for ACI power, which is defined as the power radiated within a bandwidth of ±96 kHz, of which the center frequency is separated by Δf kHz from the carrier frequency, where the carrier is modulated by the test signal sequence at the same bit rate as that of the specified one.

On the other hand, Table 12.5 shows the requirements for PHS receivers. The definition of each item is the same as that for the PDC system except for spurious sensitivity and carrier sense level threshold. These definitions are as follows:

☞ **Spurious sensitivity.** Defined by the power ratio of the desired signal to the interference signal that yields a BER of 1×10^{-2} in the TCH portion of the desired signal, where the desired signal level is set to 3 dB higher than the specified receiver sensitivity, the interference signal is unmodulated, and its carrier frequency is set to a spurious frequency.

☞ **Carrier sense level threshold.** Because DCA is employed for the channel assignment algorithm in the PHS, the channel occupancy is detected by the measurement of the average power received signal level (carrier sense level). Two carrier sense thresholds called *first threshold* and *second threshold* are prepared in the PHS.

In the carrier sense process, each station searches a channel with the measured signal level of less than or equal to the first threshold. When no channel can be detected, the same carrier sensing is carried out again using the second threshold level.

12.7 LOGICAL CHANNEL STRUCTURE

Figure 12.5 shows the logical channel structure for a PHS. This structure is almost the same as the structure for a PDC system except that the UPCH is divided into the user-specific control channel (USCCH) and the user-specific packet channel (USPCH). The USCCH is multiplexed with the BCCH, PCH, and SCCH. On the other hand, the USPCH is multiplexed with the TCH, SACCH, and FACCH. The usage and specifications of the USCCH and USPCH are not defined. Therefore, we can optionally use these logical channels.

Table 12.4 Requirements for PHS transmitter.

Parameters	Specifications
Carrier frequency instability ($\delta f / f_c$)	$\delta f / f_c \leq 3 \times 10^{-6}$
Spurious transmission	≤ 250 nW (inside the PHS band) ≤ 2.5 μW (outside the PHS band)
Leakage power during carrier OFF period	≤ 80 nW
99% bandwidth	288 kHz
Transmitter power	≤ 500 mW (for public CS) ≤ 10 mW (for the other CS and PS)
Transmitter power accuracy	within +20% and −50% of the specified value
ACI power	≤ 800 nW @ $\Delta f = 600$ kHz ≤ 250 nW @ $\Delta f = 900$ kHz
Cabinet radiation	≤ 2.5 μW

Table 12.5 Requirements for the PHS receiver.

Parameters	Specifications
Receiver sensitivity	16 dBμ for BER = 10^{-2} (specified receiver sensitivity) 25 dBμ for BER = 10^{-5} (for public mode station)
Spurious sensitivity	≥47 dB
Adjacent channel selectivity	≥50 dB @ 600 kHz frequency separation
Carrier sense level threshold	26 dBμ (first level) 40 dBμ (second level)
Unwanted radiation during standby mode	≤4 nW
Intermodulation performance	≥47 dB

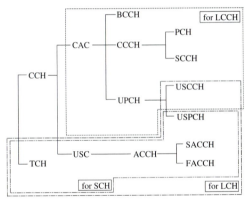

Fig. 12.5 Logical channel structure for a PHS.

12.8 BASIC CONCEPT OF THE RADIO LINK ACQUISITION PROCESS

In wired systems, the protocol to establish a communication link is divided into the call connection phase and the communication phase. However, because the PHS is a wireless system and its terminal should be small, the PHS must

☞ Support various types of services using CCHs whose capacity and quality are inferior to that of wired services

☞ Reduce the amount of software installed on the PS as well as to maximize protocol commonality between PHS and wired systems

☞ Make it easy to accommodate local protocol specific for PHSs

Therefore, the call setup process for PHS is divided into the link channel (LCH) acquisition phase and the service channel (SCH) acquisition phase.

In the LCH acquisition phase, only the radio link between a CS and a PS is handshaken using a common access control channel. Because this phase is specific for the PHS and its protocol should be as simple as possible, this protocol takes no layered structure.

On the other hand, in the SCH acquisition phase, the layered structure of the OSI reference model (layers 1–3) is employed.

An outline of each phase is as follows:

☞ **LCH acquisition phase.** This phase establishes an LCH that transmits information for the call setup process. During this phase, a CAC is used to transmit necessary information for LCH establishment, and a CS assigns radio resources (carrier frequency and slot position) to be used for the SCH acquisition phase. In this phase, BCCH, PCH, SCCH, and USCCH are used. These channels are called *link control channels* (LCCHs).

☞ **SCH acquisition phase.** This phase establishes an SCH for user information transmission using the established LCH. During this phase, CC, RT, and MM are processed via the USC, such as SACCH and FACCH. In this phase, USCCH, USPCH, SACCH, FACCH, and TCH are used. Moreover, a synchronization burst is also used in this phase. These logical channels including the synchronization burst are called the *link channel*.

☞ **Communication phase.** In this phase, user information is transmitted via the TCH. Logical control channels such as USPCH, SACCH, and FACCH are also transmitted during this phase to keep a certain radio transmission quality as well as to manage terminal mobility during each call. These CCHs and TCHs as well as the synchronization burst and Vox signal are called the *service channel*.

12.9 TDMA FRAME FORMAT

The PHS employs a 4-channel TDMA/TDD scheme. Figure 12.6 shows the frame format where T_m and R_m (m = 1, 2, 3, and 4) represent the transmission and reception time slots for the m-th user. Each slot length is 0.625 ms and the frame length is 5 ms. The former 4 slots are used for the downlink, and the latter 4 slots are used for the uplink.

In the PHS, a slot to transmit a logical CCH is called a *physical control slot*, and a slot to transmit a TCH associated with some other USC is called a *physical commu-*

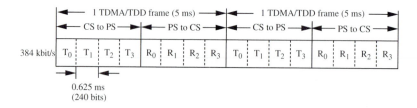

Fig. 12.6 Frame format; T_m and R_m (m = 1, 2, 3, and 4) represent the transmission and reception time slots for the m-th user.

nication slot. Figure 12.7 shows logical channel mapping onto the physical channels. Because the physical control slot is operated in the random access mode and the physical communication slot is operated in the circuit switching mode, each physical channel has a different slot format.

Figure 12.8(a) shows slot formats for SCCH, BCCH, and PCH that are mapped onto the physical control slot. Every number represents the number of bits for each part. In the downlink, these logical channels are multiplexed on a superframe format as in the case of the PDC system. Because this superframe format is an option for the operator, this structure is informed by the BCCH.

Figure 12.8(b) shows slot formats for USCCH. Two types of USCCH formats are defined in the PHS specification for future application.

Figure 12.8(c) shows slot formats for a synchronization slot. When the LCH is established, a synchronization slot is transmitted to take slot synchronization as the first process for the SCH acquisition process.

Figure 12.8(d) shows slot formats for TCH, FACCH, SACCH, and the Vox signal. TCH and SACCH are multiplexed onto a slot. Usually, the user-specific control signal is transmitted via the SACCH. When very quick control signal transmission is needed, a TCH slot is temporarily stolen to quickly transmit the USC signal. This channel is called the FACCH. Moreover, the PHS can optionally employ Vox to lengthen battery life.

Figure 12.8(e) shows slot formats for USPCH. Two types of USPCH formats are defined in the PHS specification for future application.

In any case, data other than the ramp symbol (R), start symbol, preamble, and SW are protected by the ITU-T 16-bit CRC code with its generator polynomial of $G(x) = x^{16} + x^{12} + x^5 + 1$.

12.10 LAYERED STRUCTURE OF THE LCH AND SCH ACQUISITION PHASES

In the PHS, control signals for the RT, MM, and CC are transmitted via the logical control channels using a protocol based on the OSI reference model.

		LCCH	LCH	SCH
Physical Control Slots	Uplink	SCCH USCCH (option)	USCCH (option)	
	Downlink	BCCH PCH SCCH USCCH (option)	USCCH (option)	
Physical Communication Slots			TCH FACCH SACCH USPCH (option) Sync. slot	TCH FACCH SACCH USPCH (option) Sync. slot Vox (option)

Fig. 12.7 Logical channel mapping onto the physical channels.

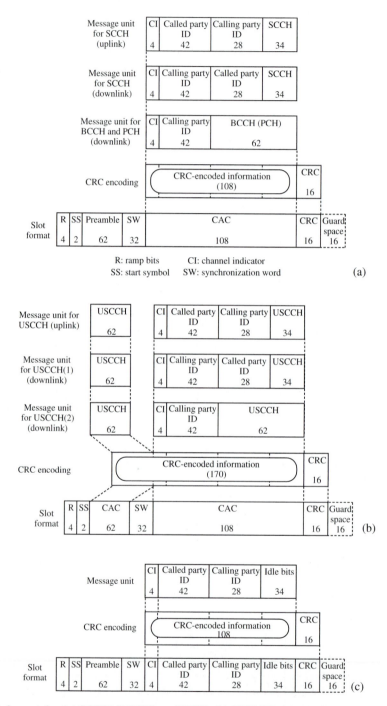

Fig. 12.8 Slot format for (a) SCCH, BCCH, and PCH; (b) USCCH; (c) synchronization slot; (d) TCH, FACCH, SACCH, and VOX; (e) USPCH.

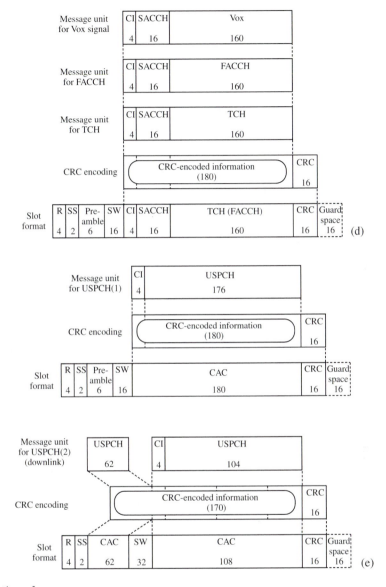

Fig. 12.8 Continued.

Figure 12.9 shows the layered structure of the LCH acquisition phase. In this phase, layer 2 and layer 3 are merged to shorten the LCH acquisition time as well as to save radio spectrum. Therefore, the control signal transmission protocol for the LCH acquisition phase obeys the following principles:

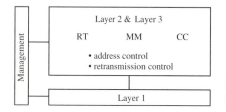

Fig. 12.9 Layered structure of the LCH acquisition phase.

☞ Every physical slot transmits an RT, MM, or CC message using 34 bits of the SCCH (see Figure 12.8(b)) because the layer 2 frame format is not employed.

☞ Error is detected in layer 1 rather than in layer 2.

☞ Retrial for LCH acquisition is limited to three tries per call.

Figure 12.10 shows the layered structure of the SCH acquisition phase, in which the control signal transmission protocol has three layers. The link access procedure for digital cordless (LAPDC) is employed as the layer 2 data transmission protocol via the SACCH and FACCH. Layer 2 is in charge of address control and message retransmission control to guarantee the error-free condition of the transmitted message. Error detection is not included in this process because error detection is carried out by the CRC check in the layer 1.

Figure 12.11 shows the relationship of the frame formats between the SACCH physical slot, the layer 2 frame, and the layer 3 frame. One layer 2 frame consists of the 14-bit layer 2 area and the 16-bit layer 3 area. In the layer 2 area, various control signals for message transmission are included. On the other hand, the layer 3 signal is framed every 16 bits (2 octets), and a 2-octet message is mapped onto the 16-bit layer 3 area in each layer 2 frame. Then, a layer 2 frame is mapped onto two consecutive SACCH physical frames as shown in Figure 12.11. Retransmission of the layer 2 frames is controlled frame by frame.

Figure 12.12 shows the relationship of the frame format between the FACCH physical slot, the layer 2 frame, and the layer 3 frame. In this case, the layer 3 signal is framed every 136 bits (17 octets), and a 17-octet message is mapped onto the layer 3 area in the layer 2 frame, where the layer 2 frame consists of 24 bits of the layer 2 area and 136 bits of the layer 3 area. Then, one layer 2 frame is mapped onto the 160 bits of the FACCH physical slot. Retransmission of the layer 2 frames is controlled frame by frame as in the case of SACCH.

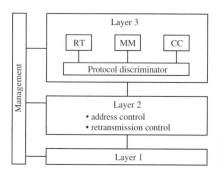

Fig. 12.10 Layered structure of the SCH acquisition phase.

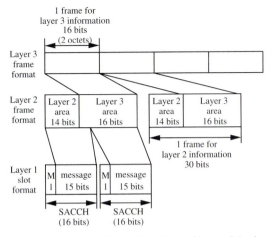

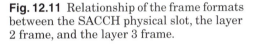

M: message unit assemble and deassemble control signal

Fig. 12.11 Relationship of the frame formats between the SACCH physical slot, the layer 2 frame, and the layer 3 frame.

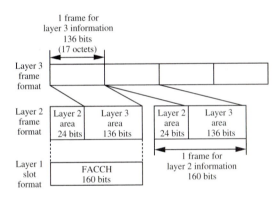

Fig. 12.12 Relationship of the frame format between the FACCH physical slot, the layer 2 frame, and the layer 3 frame.

When a layer 3 message is received, its type (RT, MM, or CC) is discriminated by the protocol discriminator included in each message.

12.11 LAYER 3 FUNCTIONS

In the PHS, management functions for RT, MM, and CC are specified as the protocols. Each protocol includes certain functions.

12.11.1 Radio Transmission Management

RT management includes the following functions:

☞ Radio resource assignment and reassignment for the TCH

☞ Radio channel disconnect

☞ Radio channel condition measurement
☞ Radio channel condition information exchange between the CS and the PS
☞ Transmitter power control
☞ Encryption
☞ Vox control

Of these functions, the radio resource reassignment is the most important function of the PHS because the PHS employs a DCA technique. For this purpose, CS and PS regularly measure the received signal level and exchange the averages of this information, which is used to decide whether or not the carrier frequency and slot position for the TCH should be changed.

Furthermore, this protocol includes various radio transmission options, such as the transmitter power control and Vox control.

12.11.2 Mobility Management

MM includes the following functions:

☞ Authentication request/response
☞ Location registration request/acknowledge/reject
☞ Location registration area report

These functions are split between authentication of the PS and its location registration. For the location registration, each PS regularly transmits its PS-ID number and the received signal level information to the CS.

12.11.3 Call Control

CC management includes the following functions:

☞ Call processing
☞ Call setup
☞ Connection/disconnection

These control signals are transmitted during the SCH acquisition phase via the SACCH and FACCH physical channels.

12.12 EXAMPLES OF THE CONTROL SEQUENCE

We have completed our discussion of logical and physical channels in layer 1 and RT, MM, and CC management in layer 3. In this section, we will discuss how these layer 1 channels and layer 3 management functions are related, as well as how they are practically employed in the PHS. We will show some examples of how these management functions are applied.

12.12.1 Call Origination and Reception Processes

Figure 12.13 shows the call origination process from a PS. In the LCH acquisition process, each PS regularly receives the BCCH in order to know various system parameters of the CS nearby and to obtain the superframe structure of the CAC. When a PS initiates a call (off hook), the PS first sends an LCH acquisition request to the CS using an uplink SCCH slot in the slotted ALOHA random access mode. When the CS receives this message, it assigns a carrier frequency to the PS considering the interference signal level of all the available channels. At this point, the signaling protocol phase changes from the LCH acquisition phase to the SCH acquisition phase.

In the SCH acquisition phase, the PS and CS mutually transmit synchronization slots to identify the ID numbers of the calling and called parties as well as to take

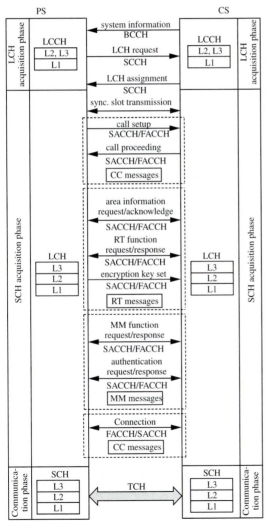

Fig. 12.13 Call origination process from a PS.

frame, slot, and symbol timing synchronization. After that, the PS and CS execute call setup request/processing (an RT function), area information request/acknowledge (RT), and optional functions request/response which includes Vox and encryption (RT) to setup radio transmission function to be used in the communication phase. After the optional functions for the authentication process is setup using messages of function request for the authentication/response (MM), the authentication is processed using messages of the authentication request/response. When the PS is not authenticated, the LCH is disconnected, and the radio channel is released. On the other hand, when the PS is authenticated, the CS sends call processing status information, and the signaling phase is switched to the communication phase.

Figure 12.14 shows a call reception process for a PS. Because call origination and call reception processes are both basically radio link establishment processes, the two protocols shown in Figure 12.13 and Figure 12.14 are almost the same.

12.12.2 Disconnection Process

Figure 12.15 shows a disconnection process for a PS. When a call is terminated, the PS sends a disconnect message (CC) to the CS. When the CS receives this message, it sends a release message (CC) indicating that the CS has released the used TCH and the ID number of the called party. When the PS receives this message, it sends back the release complete message (CC) to indicate that the PS has also released them. After that, the CS sends a radio channel disconnect message (RT) to the PS to indicate that the CS has released the radio channel. When the PS receives this message, it also sends back the radio channel disconnect complete message to the CS to indicate that the PS has also released the radio channel.

12.12.3 Location Registration

Figure 12.16 shows the location registration updating process. The LCH acquisition phase is exactly the same as it is for the call setup process. When the LCH is established, the PS sends a location registration request (MM) to the CS. When some message exchange—such as the area information request/acknowledgment and encryption key set—is necessary, these processes are first executed. Then the CS sends back an authentication request. According to this request, the PS sends the necessary information to the CS. When this authentication process is successfully finished, the CS sends location information of the PS to the network, and it sends a location registration acknowledge to the PS when this update is finished.

12.12.4 Channel Switching

In the PHS, channel quality during a call is measured by the frame error rate (FER) which is detected by counting the number of error frames using a 16-bit CRC. When the FER is increasing and it is beyond a certain value, the PS sends a TCH reassign request to the CS. When the CS receives this request and it investigates its capability and finds that channel switching is possible, the channel switching process is started. Figure 12.17 illustrates the process.

12.12.5 Handover

When a terminal is located near the fringe area, another CS connected to the same local switch is expected to give better FER performance. In such a case, the terminal is handed over to another CS. Figure 12.18 illustrates the handover process of the PHS. When the received signal level and/or FER performance are getting degraded, the PS sends a TCH reassign request to the CS. When the new CS is found to be capable of accepting another call, a TCH reassign message is transmitted to the

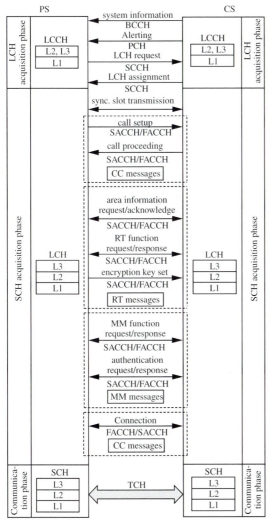

Fig. 12.14 Call reception process for a PS.

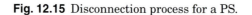

Fig. 12.15 Disconnection process for a PS.

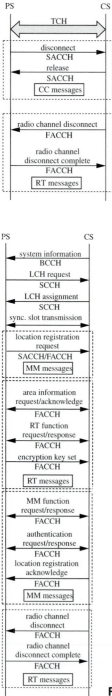

Fig. 12.16 Location registration update process.

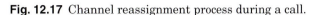

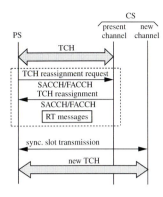

Fig. 12.17 Channel reassignment process during a call.

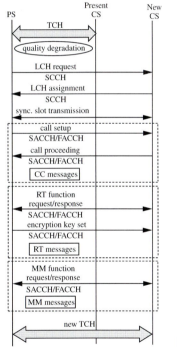

Fig. 12.18 Handover process.

PS. Then the PS and the new CS mutually transmit a synchronization slot to establish a new communication link between the PS and the new CS. This process is the same as the call origination process.

12.13 TRANSCEIVER FUNCTION OF THE PHS TERMINAL

When we want to establish a communication link between two PSs in the uncovered area of the public CSs, we can establish a link using a transceiver mode of the PS. In

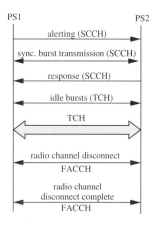

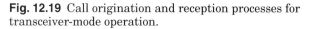

Fig. 12.19 Call origination and reception processes for transceiver-mode operation.

this mode, the call-originated PS selects one of the unused slots of carrier 1–10 as shown in Table 12.2, and one of the available carriers and slots is used to send a call origination message to the other PS. In this mode, the same carrier and slot are used during the LCH and SCH acquisition phases and the communication phase.

Figure 12.19 shows the call origination and reception processes for transceiver-mode operation. When PS1 originates a call, it sends a call origination message to the destination PS (PS2). When PS2 receives this message, a synchronization slot is exchanged between the PSs to take synchronization as well as to process authentication. When authentication is successfully completed, a communication link is established after idle TCH bursts are exchanged to take synchronization.

When one of the terminals sends a radio channel disconnect message and the other one sends back a radio channel disconnect complete message, the communication link is disconnected.

REFERENCES

12-1. RCR, "Personal handy phone system," RCR STD-28, December 1993.

12-2. DECT, "Digital European cordless system—common interface specifications," Code RES-3(89), DECT, 1989.

12-3. CT-2, "Second generation cordless telephone (CT-2), Common Air Interface Specifications," Dept. of Trade and Industry, London, May 1989.

12-4. ITU-R, "Requirements for the radio interface(s) for future public land mobile telecommunication systems (FPLMTS)," Recommendation ITU-R M.1034, 1994.

12-5. Telecommunication Technology Council, "Interim report for personal handy phone system," June 1992.

Digital Multichannel Access (MCA) System

At present, in addition to public radio communication systems, PMR systems are digitalized in Japan to support various types of information. Among many PMR systems, the digital MCA system, similar to the shared mobile radio (SMR) in the United States, is the most successfully introduced digital private mobile system in Japan.

In this chapter we will discuss the system overview, the physical layer interface, and specifications of digital MCA systems.

13.1 HISTORIC OVERVIEW OF DIGITAL MCA SYSTEM DEVELOPMENT

The MCA system is a PMR system for Japanese business use with the following features:

☞ To achieve a cost-effective private radio system, a large-zone system with a radius of 20–30 km was introduced.

☞ To achieve a high-capacity system, first-come-first-served queuing, group communication, connection time limitation and a half-duplex scheme were introduced.

A first generation MCA system (analog system) was introduced by the Mobile Radio Center (MRC) in 1982 and by Japan Shared Mobile Radio (JSMR) in 1987. The number of subscribers increased rapidly, with an annual growth rate of more than 50%, reaching about 460,000 subscribers by March 1991. As a result, spectrum shortage became a severe problem by this time. At the same time, demand for high-bit-rate data transmission service also increased.

Thus, MPT organized a study group for digitalization of land mobile communication systems, which discussed a digital MCA system as well as the other private radio systems; its report was submitted to MPT in March 1991. Following this report, MPT consulted the TTC about the technological requirements for a digital MCA system; TTC's final report was submitted to MPT in June 1992. Based on this report, MPT, in consultation with the RRC, promulgated the digital MCA-related ministerial ordinances in March 1993.

Along with the radio regulatory process, RCR organized a standardization committee to make a standard for the digital MCA system in May 1991, which also conducted field experiments to evaluate key technologies for digital MCA. In November 1992, the committee issued the digital MCA standard (RCR STD-32) [13-1], and the first commercial operation began in March 1994.

13.2 SYSTEM OVERVIEW

Figure 13.1 shows a concept of the MCA system. An MCA system is a large-zone, private radio system that consists of a BS, operation stations, and terminals in each zone. Each user group consists of an operation station and terminals.

The BS covers a radius of approximately 20–30 km, and terminals in a user group communicate with each other via the BS. In this system, communication is operated mainly in half-duplex mode.

Table 13.1 shows a summary of the basic functions of the MCA system. The group communication mode is one of the most important features of the MCA system. In this mode, when a terminal in a user group (subgroup) want to initiate a call, it sends a call request message to the BS via the CCCH, and the BS assigns a TCH to the user group (subgroup). The assigned TCH is shared by all the terminals in the user group (subgroup) during the call. For example, when the operation station of user

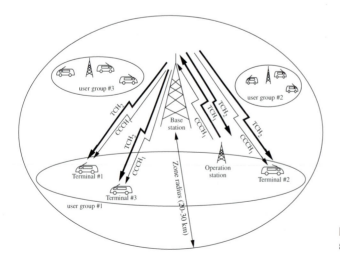

Fig. 13.1 Concept of the MCA system.

Table 13.1 Summary of the basic functions of the MCA system.

Categories		Functions
Service mode	voice	Voice transmission is supported in the half-duplex (push-to-talk) mode.
	nonvoice	Fax, picture, and data transmission services are supported.
Communication mode	group communication	Each terminal can transmit information to the terminals including the operation station of the same user group.
	point-to-point communication	Each terminal can transmit information to a specific terminal or the operation station of the same user group.
Service zone configuration	single-zone system	Each terminal or the operation station can communicate with each other when they are located in the same zone.
	multizone system	Each terminal or the operation station can communicate with each other even though they are located in different zones.

group #1 originates a call, it sends a connection request to the BS via the CCCH. According to this request, the BS assigns TCHs for the uplink (TCH1) and downlink (TCH2). When the operator has finished using TCH1, another user in the user group can use TCH1. For such group-mode communication, each terminal has its own terminal ID as well as its user group ID or user subgroup ID.

Each user group can be divided into several subuser groups. When there are only two users in a subuser group, the communication mode is equivalent to the point-to-point communication mode. Moreover, interzone connection, although it was not supported in analog MCA systems, is introduced in digital MCA systems.

Figure 13.2 shows photos of (a) an MS, (b) an operation room for a central site, (c) BS equipment, and (d) BS antennas for a digital MCA system.

Table 13.2 shows a summary of the additional (optional) functions of digital MCA services. Because digital MCA employs a 6-channel TDMA/FDD scheme as we will discuss later, it can support simultaneous transmission of voice and nonvoice data; high-bit-rate data transmission of up to 64 kbit/s; full-duplex voice transmission; or subgroup communication by allocating multiple slots to each user group, if such multi-slot mode operation does not severely affect the other user groups operating in the basic service mode. Furthermore, some special services, such as encryption and priority control, are also supported.

13.3 SPECIFICATIONS OF THE RADIO INTERFACE

Table 13.3 shows specifications of the digital MCA system. This system is operated in a 1.5-GHz band, which is a newly allocated band for digital MCA systems. It employs 16QAM because introduction of a higher modulation level is the only way to enhance

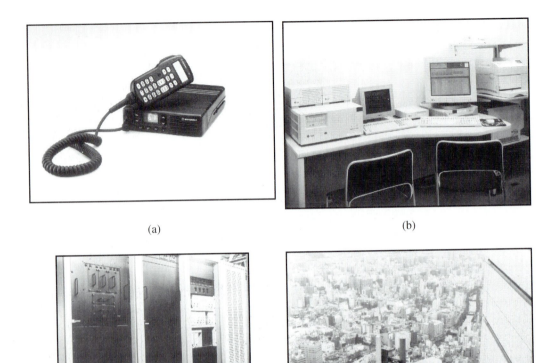

(a)   (b)

(c)   (d)

Fig. 13.2 Photos of the digital MCA system operated by JSMR: (a) an MS, (b) operation center for central site, (c) BS equipment, and (d) BS antennas (courtesy of Motorola, Inc.).

the system capacity of large-zone land mobile radio systems. In this case, however, 16QAM is less robust to delay spread due to its shorter minimum signal distance. To improve delay spread immunity, the digital MCA system introduced multicarrier transmission with four subchannels, which is called M16QAM [13-2]. Figure 13.3 shows the channel structure of M16QAM. Each subchannel is band-limited by a root Nyquist filter with a roll-off factor of 0.2, and the subchannel spacing is 4.5 kHz.

A 6-channel TDMA with its frame length of 90 ms is employed. When a call request originates from a user group (subgroup), the BS assigns a slot in a carrier to

Table 13.2 Summary of additional functions of digital MCA systems.

Categories		Functions
Multislot assignment	simultaneous transmission of voice and nonvoice data	multislot assignment to simultaneously transmit different types of information
	high-bit-rate data transmission	multislot assignment for high-bit-rate data transmission
	full-duplex	multislot assignment to support full-duplex transmission
	subgroup communication mode	a communication mode to support two or more group communications in a user group
Special communication mode set	priority control	a priority assign mode to emergency communications
Special services	encryption	to support encryption mode
Others	PSTN/PSDN connection	to support PSTN/PSDN connection at the BS or operation stations

Table 13.3 Specifications of the digital MCA system.

Parameters	Specifications
Frequency band	800 MHz (could be used in the future) 1501–1513 MHz (uplink), 1453–1465 MHz (downlink)
Frequency band separation	55 MHz (800-MHz band) 48 MHz (1.5-GHz band)
Modulation	M16QAM (Multicarrier 16QAM) number of multicarriers is four roll-off factor of Tx and Rx filters is 0.2
Access	TDMA (uplink)/TDM (downlink)
Duplex	half-duplex for terminals and operation station full-duplex using FDD for the BS
Channel/carrier	6 (full rate)
Slot length	15 ms
Frame length	90 ms
Carrier spacing	25 kHz
Voice codec	not specified; maximum bit rate including FEC should be less than 7.467 kbit/s
Other features	- high delay spread immunity (up to 10 μs) - pilot symbol-aided flat Rayleigh fading - AGC preamble embedded in the preamble to achieve high-speed AGC in uplink

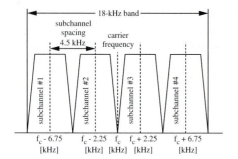

Fig. 13.3 Channel structure of M16QAM.

the group (subgroup). A half-duplex scheme is employed as the duplex scheme for terminals and the operation station. On the other hand, full duplex using FDD is employed at the BS.

13.4 REQUIREMENTS FOR THE TRANSMITTER AND RECEIVER

Table 13.4 shows the requirements for digital MCA transmitters. Definitions of the parameters are the same as those given in chapter 11 for the PDC system.

Table 13.4 Requirements for digital MCA transmitters.

Parameters	Specifications
Carrier frequency instability ($\delta f/f_c$)	BS: $\delta f/f_c \leq \pm 0.1 \times 10^{-6}$ terminal: $\delta f/f_c \leq \pm 0.15 \times 10^{-6}$ (When an AFC is employed, $\delta f/f_c$ at a free-running mode has to be smaller than $\pm 2.0 \times 10^{-6}$)
Spurious transmission	BS: 60 dB lower than the transmitted signal level, or lower than 2.5 μW terminal: 60 dB lower than the transmitted signal level, or lower than 0.25 μW
99% bandwidth	$\leq$20 kHz
Transmitter power	BS: $\leq$ 40W terminal: $\leq$2W
Transmitter power accuracy	within +20% and −50% of the specified value
ACI power	BS: $\leq$−55 dB terminal: $\leq$−50 dB as well as smaller than 6.3 μW @ 25 kHz channel separation
Leakage power during carrier OFF period	to the same band: $\leq$32 nW during idle time to the other frequency band: $\leq$4 nW
Cabinet radiation	$\leq$2.5 μW

Table 13.5 shows the requirements for a digital MCA receiver. The definition of receiver sensitivity is the same as that for the PDC system given in chapter 11. On the other hand, definitions for spurious sensitivity, adjacent channel selectivity, and inter-modulation characteristics are as follows:

☞ **Spurious sensitivity.** The transmitted signal level of the desired signal is set at 3 dB higher than the nominal receiver sensitivity (9 dBμ), and an undesired signal modulated by a 15-stage PN code is located 50 kHz apart from the carrier frequency of the desired signal. Spurious response is defined as the power ratio of the undesired signal to the desired signal that results in BER = 10^{-2}.

☞ **Adjacent channel selectivity.** The transmitted signal level of the desired signal is set at 3 dB higher than the nominal receiver sensitivity (9 dBμ), and an undesired signal modulated by a 15-stage PN code is located 25 kHz apart from the carrier frequency of the desired signal. Adjacent channel selectivity is defined as the power ratio of the undesired signal to the desired signal that results in BER = 10^{-2}.

☞ **Intermodulation performance.** The transmitted signal level of the desired signal is set at 3 dB higher than the nominal receiver sensitivity (9 dBμ), and two types of undesired signals are prepared. The first undesired signal is an unmodulated signal located 50 kHz apart from the desired signal, and the other one is a signal modulated by a 15-stage PN code located 100 kHz apart from the desired signal. Intermodulation characteristics are defined as the power ratio of the undesired signal to the desired signal that results in BER = 10^{-2}.

Table 13.5 Requirements for digital MCA receivers.

Parameters	Specifications
Rreceiver sensitivity	BS 9 dBm for BER = 10^{-2} in static conditions (nominal receiver sensitivity) 7.0 dBμ for BER = 3×10^{-2} in Rayleigh fading conditions terminal 9 dBμ for BER = 10^{-2} in static conditions 13.0 dBμ for BER = 3×10^{-2} in Rayleigh fading conditions (f_d = 40 Hz for 800 MHz band, and f_d = 70 Hz for 1.5-GHz band)
Spurious sensitivity	≥53 dB
Adjacent channel selectivity	≥42 dB @ Δf = ±25 kHz
Intermodulation performance	≥53 dB

13.5 LOGICAL CHANNEL CONFIGURATION

Figure 13.4 shows a logical channel structure of the digital MCA system. Logical channels are classified into the CAC, USC, and radio control channel (RCCH). Most

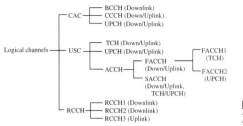

Fig. 13.4 Logical channel structure of the digital MCA system.

notably, the RCCH is a specific logical channel in digital MCA systems prepared to achieve quick and simple radio channel control. The following are the specifications of these logical channels for digital MCA systems.

BCCH. This is a downlink channel that broadcasts the following information from the BS to all the terminals and operation stations.

☞ System code
☞ Physical channel structure (TDMA frame structure, superframe structure of the downlink CAC, supported services, restricted conditions)
☞ Supported services and restrictions
☞ Connection time limitations

CCCH. This channel carries the following control signals between terminals and the BS.

☞ Connection request (uplink)
☞ PSTN connection/disconnection request (uplink)
☞ Interzone connection request (uplink)
☞ Channel reservation/assignment acknowledge/release

UPCH. A point-to-multipoint bidirectional, end-to-end channel used to transmit user packet data between the BS and terminals.

TCH. An end-to-end bidirectional channel that carriers various user information streams.

FACCH. A point-to-point high-throughput signaling channel that temporarily steals the TCH. Two types of FACCH (FACCH1 and FACCH2) are defined. FACCH1 temporarily steals a TCH slot, whereas FACCH2 steals a UPCH. FACCH1 and FACCH2 are used mainly to transmit disconnection messages.

SACCH. This channel indicates a type of the logical channel (TCH/UPCH, FACCH1/FACCH2, RCCH3, or filler) carried in the slot. Selection of TCH/UPCH and FACCH1/FACCH2 are informed by the RCCH1.

RCCH. This channel carries the physical channel control signal. There are three types of RCCH (RCCH1, RCCH2, and RCCH3), which have different functions.

RCCH1. Associated with the downlink slot of both the CAC and USC, it transmits the attribute of the slot as well as the operation mode information. When RCCH1 is associated with the CAC, it carries the following information:

☞ Whether the slot carries CAC or USC

☞ Transmission mode (normal transmission or retransmission)

☞ CAC mapping information

☞ Access conditions of subslot 1 and subslot 2 (channel idle/busy, CCCH/UPCH, received signal ACK/NACK)

☞ User information

On the other hand, when RCCH1 is associated with the USC, it carries this information.

☞ Whether the slot carries CAC or USC

☞ Transmission mode of the paired channels (TCH or UPCH, with or without time alignment and transmission power control, whether the slot is stealed)

☞ Information for TA and TPC

☞ Repeater function at the BS (without FEC decode, with FEC decode)

☞ User information

RCCH2. Associated with BCCH, CCCH, or UPCH in the downlink, it carries the following information:

☞ BCCH/CCCH/UPCH indication

☞ The number of the remaining CCCH frames

☞ LCN (this is a logical channel number temporarily assigned to each user group by the BS during a call origination process)

☞ Standby request signal for simultaneous call

RCCH3. Transmits transmission mode switching request as well as the requests for acquisition and release of the single user mode. This channel is transmitted in the uplink by temporarily stealing a TCH slot.

13.6 MODULATION SCHEME

In the digital MCA system, a 4-multicarrier M16QAM is employed. Figure 13.5 shows the modulator configuration of M16QAM systems. The transmitted data are first con-

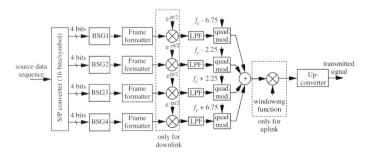

Fig. 13.5 Modulator configuration of M16QAM.

verted to 16-bit parallel data. These parallel data are further blocked into 4-bit data, and each 4-bit data block is fed to the baseband signal generator (BSG1–BSG4). These signals are then fed to the frame formatter for each subchannel, in which pilot symbols, synchronization word, and the other symbols are inserted. Because the phase of each subchannel at the beginning of each burst is specified to be equal to compensate for fading using pilot symbols in both its own subchannel and the other subchannels, the output signal of each BSG at the BS is phase-shifted by $\pi/2$ for BSG1 and BSG3, and $-\pi/2$ for BSG2 and BSG4 when the modulators are operated in a continuous mode. On the other hand, such phase shifters are not necessary when the terminal is operated in the burst mode. These signals are then band-limited by a root Nyquist filter with its roll-off factor of 0.2, followed by the quadrature modulator. After that, they are multiplexed in the frequency domain, windowed to suppress out-of-band radiation in the case of burst-mode transmission, up-converted to the assigned carrier frequency, and transmitted.

13.7 FRAME AND SLOT FORMAT

13.7.1 Frame Format

Figure 13.6 shows a frame format for a digital MCA system. For QAM which includes information not only in the phase but also in the amplitude, we have to control the received signal level to a certain value using an AGC before demodulation, especially in the uplink. For this purpose, a certain time duration of a preamble signal is necessary to preset the AGC. However, a longer AGC preamble is not preferable because it could degrade the frame efficiency of the uplink signal. Therefore, the digital MCA system prepares 1.25 ms of AGC preamble in the uplink signal, and it employs a relatively longer frame length and slot length compared to those of the PDC and PHS systems.

One frame consists of 6 time slots, each slot length is 15 ms, and a frame length is 90 ms. Because the symbol rate of each subchannel is 4 ksymbol/s (16 kbit/s), one

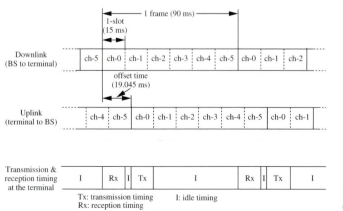

Fig. 13.6 Frame format for digital MCA system.

slot can transmits 960 bits in total. Transmission timing of the uplink is delayed by 19.045 ms from that of the downlink time slot. This delay time is called the *transmission and reception timing offset*. As a result, transmission and reception timing for terminals are as shown in Figure 13.6.

13.7.2 Subslot Format for the CAC in the Uplink

In the digital MCA system, the number of transmitted bits in the downlink physical slot is larger than that in the uplink physical slot because the uplink slot requires more redundant bits, such as guard time and AGC preamble. On the other hand, the transmitted information bits of the bidirectional logical channels should be symmetrical with each other. Therefore, the downlink slots transmit RCCH1 (and RCCH2 in the case of the CAC) associated with such bidirectional logical channels using extra bits.

As we have discussed, a slot length of 15 ms is employed in this system. However, such a long slot length could degrade throughput of the CAC operated in the random access mode. Therefore, a subslot format with a subslot length of 7.5 ms is employed in the CAC.

Figure 13.7 shows the subslot format for the CAC in the uplink. At the front of each subslot is located the AGC preamble with its duration of 1.25 ms (5 symbols) to control the received signal level to a proper level. To quickly control the received signal level during such a short time period, the waveform of the AGC preamble is defined as shown in Figure 13.8. In this figure, 0 dB means the average signal level of

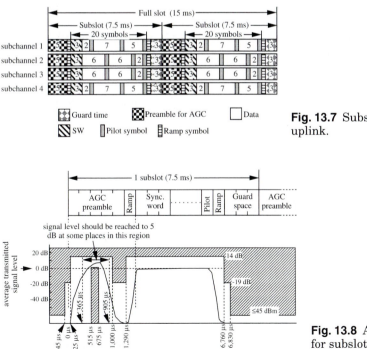

Fig. 13.7 Subslot format for CAC in the uplink.

Fig. 13.8 AGC preamble waveform for subslot format in the uplink.

the transmitted signal. The waveform of the AGC preamble has to be located inside the shaded area. During 365 μs < t < 905 μs, the transmitted signal level has to be reached at 5 dB higher level than the average level. Moreover, the radiated power level during no signal transmission, which is called the *carrier OFF level*, is specified as 32 nW (–45 dBm). Therefore, the signal level has to go below this level during guard time.

Ramp symbols located after the AGC preamble and at the end of the slot are used to prevent spectrum spreading due to rapid waveform transition. The SW is used to take slot timing synchronization at the receiver.

When we define the symbol pattern of the SW and pilot symbols, we have to consider the following factors:

☞ The word can easily have been discriminated.

☞ Misdetection probability is as low as possible.

☞ Peak to average power ratio at the symbol timing should be as low as possible.

Suppression of the peak to average power ratio is an especially important factor among these factors because power efficiency is degraded when we increase this power ratio. The SW is selected considering these factors. Figure 13.9 shows the symbol phase of each subchannel in the (a) first synchronization symbol, (b) second synchronization symbol, and (c) third synchronization symbol. The amplitude of the synchronization word is equal to the maximum amplitude of the 16QAM constellation for each subchannel.

Pilot symbols are used to compensate for fading using a pilot symbol-aided technique described in chapter 4. To prevent spectral efficiency reduction due to pilot symbol insertion, a known pilot symbol is inserted as shown in Figure 13.7. Moreover, pilot symbol insertion timing of subchannels 2 and 3 is staggered by 3- or 4-symbol duration from those of the subchannels 1 and 4 in order to prevent an increase in the peak to average power ratio. Another advantage of this pilot symbol allocation is that we can compensate for fading 3- or 4-symbol duration if we employ pilot symbols of the other subchannels. Furthermore, the phase of each pilot symbol is also assigned to

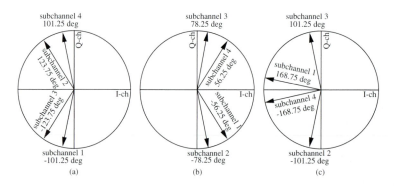

Fig. 13.9 Phase of SW for (a) first symbol, (b) second symbol, and (c) third symbol.

suppress the peak to average power ratio. Figure 13.10 shows the phase of each pilot symbol for the subslot format in the uplink.

At the end of each subslot, 0.75 ms of guard time is prepared to buffer the slot timing jitter due to the difference in propagation path distance. The guard time of 0.75 ms is determined by consideration of the zone radius of up to 100 km. In this format, because 14 symbols are included in each subchannel, 224 bits of data can be transmitted by this subslot.

On the other hand, the slot format for the CAC in the downlink is the same as that for the USC in the downlink (this will be discussed in a later section).

13.7.3 Full-Slot Format for the USC in the Uplink

In the case of USC transmission, a full-slot format with a slot length of 15 ms is used. Figure 13.11 shows a full-slot format for USC in the uplink. In this format, 43 symbols are included in each subchannel. Therefore, 688 bits of data can be transmitted by this slot. One big difference between the subslot and full-slot formats in the uplink is that no guard time is prepared in the full-slot format because transmission

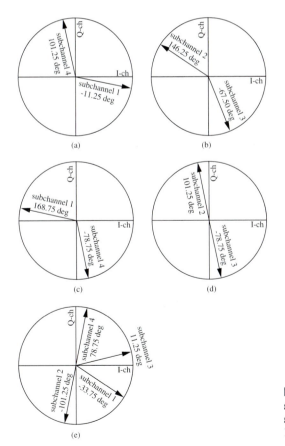

Fig. 13.10 Phase of each pilot symbol for subslot format in the uplink for (a) 6th symbol, (b) 10th symbol, (c) 14th symbol, (d) 17th symbol, and (e) 20th symbol.

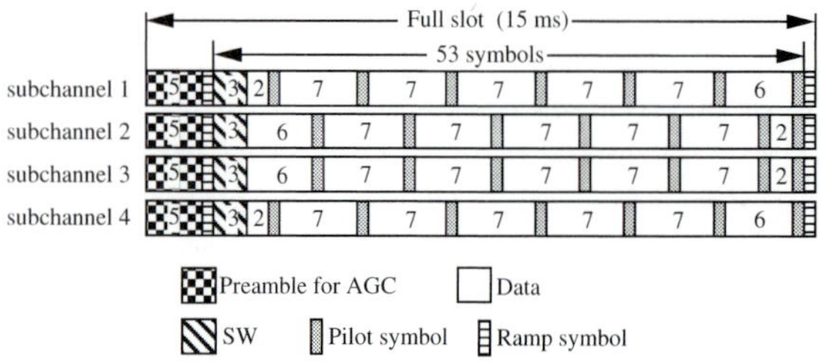

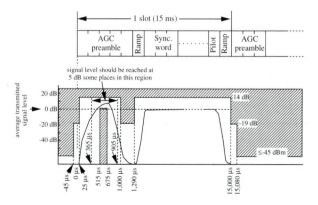

Fig. 13.11 Full-slot format for USC in the uplink.

timing of the full-slot uplink signal is precisely controlled by a time alignment control before USC is established. Such control is done by the BS via the RCCH1.

The AGC preamble is also included in this slot format, where the waveform of the AGC preamble is defined as shown in Figure 13.12.

The SW for this slot is the same as that for the subslot format. On the other hand, a different pattern of the pilot symbol is employed. Figure 13.13 shows the phase of pilot symbols.

13.7.4 Slot Format in the Downlink

In the downlink, the same slot format is employed for both CAC and USC. Figure 13.14 shows the slot format for the downlink. Because the downlink signal is transmitted in the continuous mode, there is no AGC preamble nor guard time. The synchronization symbol pattern is the same as that for the full-slot and subslot format of the uplink channels. On the other hand, the pilot symbol phases shown in Figure 13.15 are employed.

13.8 LOGICAL CHANNEL MAPPING ON THE PHYSICAL SLOT

In the downlink slot, 800 bits of the data section are included. On the other hand, 688 bits of data are included in the uplink full slot, and 244 bits are included in the uplink

Fig. 13.12 AGC preamble waveform for full-slot format in the uplink.

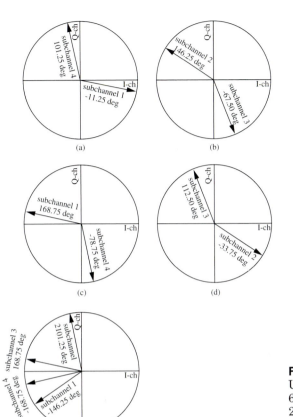

Fig. 13.13 Phase of pilot symbols for USC in the uplink full-slot format for (a) 6th, 22nd, and 38th symbols; (b) 10th, 26th, and 42nd symbols; (c) 14th, 30th, and 46th symbols; (d) 18th, 34th, and 50th symbols; and (e) 53rd symbol.

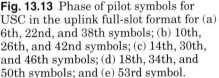

Fig. 13.14 Slot format for the downlink.

subslot. The logical channels are mapped onto these data sections. Figure 13.16 shows the relationship between logical channels and physical slots or subslots.

In the downlink CAC, BCCH, CCCH, and UPCH are multiplexed on a super-frame format, where each frame takes a full-slot format. In the uplink, however, CCCH and UPCH are transmitted in the subslot format. Because CAC has to be symmetrical in both the downlink and in the uplink, two CACs are multiplexed onto each slot in the downlink, where suffix -F indicates a subslot located in front and suffix -B indicates a subslot located in back. The downlink CAC slot also includes 112 bits of the RCCH1, and each CAC is accompanied by the RCCH2.

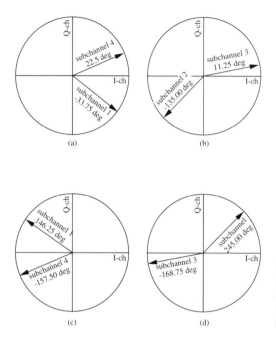

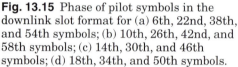

Fig. 13.15 Phase of pilot symbols in the downlink slot format for (a) 6th, 22nd, 38th, and 54th symbols; (b) 10th, 26th, 42nd, and 58th symbols; (c) 14th, 30th, and 46th symbols; (d) 18th, 34th, and 50th symbols.

In the case of the USC, a 112-bit RCCH1, 16-bit SACCH, and 672-bit TCH are multiplexed on an 800-bit data section of each downlink slot. The FACCH1 or filler can be temporarily transmitted instead of a TCH slot during each call. On the other hand, only a 16-bit SACCH and a 672-bit TCH or FACCH1 are multiplexed on the uplink USC slot.

RCCH3 can also be mapped using the same slot format. The RCCH3 slot is transmitted prior to the TCH transmission because it is used for the time alignment and power control of the terminal by the BS.

The UPCH is also defined as an optional user-specific logical channel. This channel is transmitted by the slotted ALOHA random access mode in the uplink using a subslot format. As in the case of CAC transmission, two sets of a 16-bit SACCH and a 208-bit UPCH are mapped onto a slot in the downlink.

13.9 CHANNEL-CODING SCHEME FOR LOGICAL CHANNELS

In the digital MCA system, convolutional encoding and CRC are employed for each logical slot to improve the received signal quality of the USC as well as to improve throughput of the CAC.

13.9.1 Channel Coding for the CAC

The CAC in the downlink is composed of an RCCH1 and two groups of BCCH, CCCH, or UPCH associated with an RCCH2.

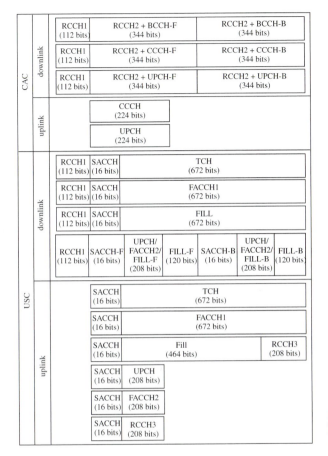

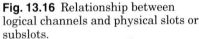

Fig. 13.16 Relationship between logical channels and physical slots or subslots.

Figure 13.17 shows the channel-encoding process for the RCCH1. Of the RCCH1 message, 46 bits are CRC-encoded by the 5-bit CRC with its generator polynomial of $G(x) = x^5 + x^4 + 1$ to create 51 bits of the CRC-encoded message. After 5 bits of 0s are attached at the end of the message, the data are convolutionally encoded by a rate-1/2 convolutional encoder with its generator polynomials of $G_1(x) = x^5 + x^3 + x + 1$ and

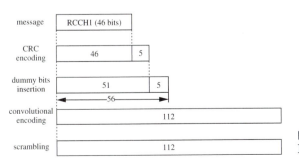

Fig. 13.17 Channel-encoding process for RCCH1.

$G_2(x) = x^5 + x^4 + x^3 + x^2 + 1$. After this coded signal is scrambled, 112 bits of the coded RCCH1 message is mapped onto the location of the RCCH1 section in the downlink CAC physical slot.

Figure 13.18 shows the channel-encoding process for the BCCH, CCCH, or UPCH associated with the RCCH2. After 26 bits of the RCCH2 and 104 bits of the BCCH, CCCH or UPCH messages are multiplexed to create 130 bits of the composite message, it is 17-bit CRC-encoded with a generator polynomial of $G(x) = x^{17} + x^{16} + x^{15} + x^{13} + x^{12} + x^8 + x^7 + x^5 + x^2 + 1$. After 0 is inserted every 7 bits and 5 bits of 0 are attached at the end of the sequence, the sequence is convolutionally encoded by a rate-1/2 convolutional encoder with its generator polynomials of $G_1(x) = x^5 + x^3 + x + 1$ and $G_2(x) = x^5 + x^4 + x^3 + x^2 + 1$ and scrambled to create 344 bits of the coded sequence. After this channel-encoding process is carried out for two CACs, 344 bits of RCCH2 + BCCH/CCCH/UPCH-F and RCCH2 + BCCH/CCCH/UPCH-B are interleaved and mapped onto the 688-bit data section in the CAC slot.

On the other hand, in the uplink CAC, 90 bits of data of a CCCH or UPCH are encoded and mapped onto 224 bits of the data section in each slot. Figure 13.19 shows the channel-encoding process for the CCCH or UPCH in the uplink. After 90 bits of the CCCH or UPCH message is 17-bit CRC encoded with a generator polynomial of $G(x) = x^{17} + x^{16} + x^{15} + x^{13} + x^{12} + x^8 + x^7 + x^5 + x^2 + 1$, and 5 bits of 0 are added at the end of the sequence, the CCCH or UPCH message is encoded by the same convolutional encoder as that for RCCH1, followed by a scrambler and interleaver. The 224 bits of encoded data sequence are then mapped onto the subslot format for the uplink CAC.

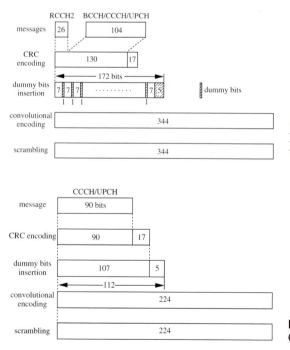

Fig. 13.18 Channel encoding process for BCCH, CCCH, or UPCH associated with the RCCH2.

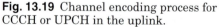

Fig. 13.19 Channel encoding process for CCCH or UPCH in the uplink.

A block diagram of channel encoding for CAC, logical channel mapping on a physical slot, and superframe format are shown in Figure 13.20.

13.9.2 Channel Coding for the USC

There are two types of USCs. One is the USC operated in a circuit-switching mode, such as the TCH; the other is the channel operated in a packet-switching mode, such as the UPCH.

13.9.2.1 For Logical Channels Operated in the Circuit-Switching Mode   In the downlink USC operated in the circuit-switching mode, although channel coding for the TCH, FILL (filler pattern), and UPCH are not specified, channel coding for SACCH, FACCH1, and FACCH2 are specified. Figure 13.21 shows the channel-encoding process for the SACCH. First, a bit inversion pattern of the 2 bits of the SACCH message is added at the end of the SACCH message, and 4 bits of 0 are added. This 8-bit

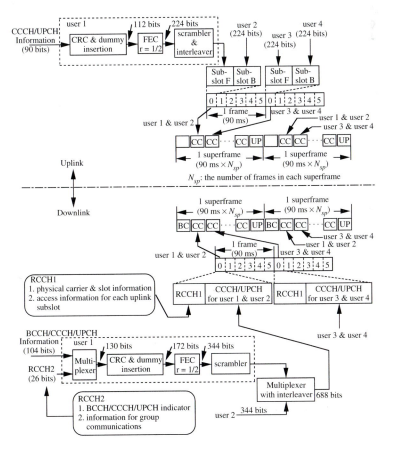

Fig. 13.20 Block diagram of the channel encoding for CAC, logical channel mapping on a physical slot, and superframe format.

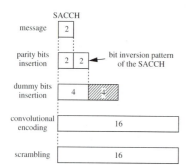

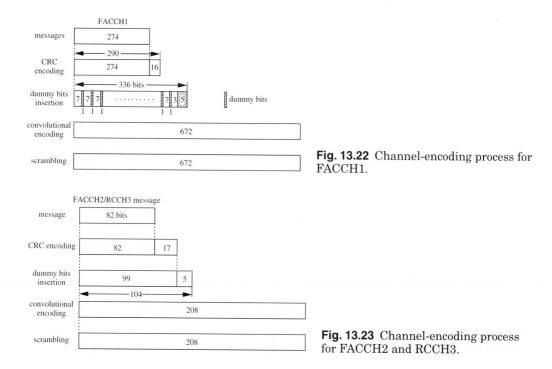

Fig. 13.21 Channel-encoding process for SACCH.

sequence is then convolutionally encoded by a rate-1/2 convolutional encoder with its generator polynomials of $G_1(x) = x^4 + x^3 + 1$ and $G_2(x) = x^4 + x^2 + x + 1$. After this message is scrambled, a 16-bit SACCH-coded message is created.

Figure 13.22 shows the channel encoding process for the FACCH1. After 274 bits of an FACCH message is 16-bit CRC encoded, 0 is inserted every 7 bits, 5 bits of 0s are added at the end of the message, and then the sequence is convolutionally encoded with generator polynomials of $G_1(x) = x^5 + x^3 + x + 1$ and $G_2(x) = x^5 + x^4 + x^3 + x^2 + 1$. The encoded sequence is then scrambled and 672 bits of a coded FACCH1 message is created.

In the uplink, channel encoding for RCCH3 is specified. Figure 13.23 shows the channel-encoding process for RCCH3. After 82 bits of an RCCH3 message is 17-bit

Fig. 13.22 Channel-encoding process for FACCH1.

Fig. 13.23 Channel-encoding process for FACCH2 and RCCH3.

CRC encoded and 5 bits of 0s are added at the end of the message, the sequence is convolutionally encoded with generator polynomials of $G_1(x) = x^5 + x^3 + x + 1$ and $G_2(x) = x^5 + x^4 + x^3 + x^2 + 1$. The encoded sequence is then scrambled and 672 bits of a coded RCCH3 message is created.

Figure 13.24 shows a block diagram of each encoding, the multiplexing process of the logical channels, and the logical channel mapping on a physical slot for the USC operated in the circuit-switching mode in both the downlink and the uplink. As shown in Figure 13.24, after 672 bits of TCH, FACCH, or RCCH3 are multiplexed with a 16-bit SACCH and the multiplexed sequence is interleaved, the 688-bit sequence is mapped onto a physical slot.

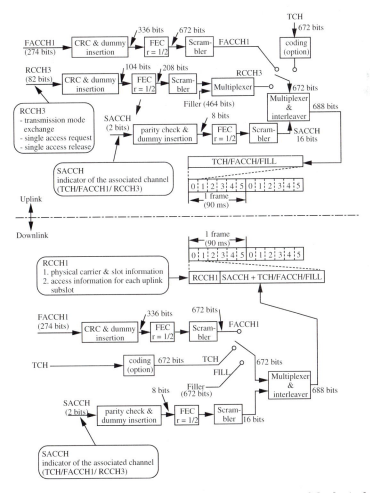

Fig. 13.24 Block diagram of each encoding, the multiplexing process of the logical channels, and the logical channel mapping on a physical slot for the USC operated in the circuit-switching mode in both downlink and uplink.

13.9.2.2 For Logical Channels Operated in the Packet-Switching Mode In the packet-switching mode, because the subslot format is employed to improve throughput performance, a message length is shorter than the length operated in the circuit-switching mode.

In the downlink, SACCH is multiplexed with a UPCH or FACCH2. The channel-coding scheme for the UPCH is not specified, and the SACCH is the same as the SACCH for the circuit-switching mode. On the other hand, FACCH2 is a specific logical channel for this mode, whose encoding process is the same as the encoding process for the RCCH3 for the circuit-switching mode shown in Figure 13.23.

In the uplink, RCCH3 is also mapped onto a subslot whose channel coding scheme is the same as that for RCCH3 applied in the circuit-switching mode.

Figure 13.25 shows a block diagram of each encoding, the multiplexing process of the logical channels, and the logical channel mapping on a physical slot for the USC operated in the packet-switching mode in both the downlink and the uplink. As shown in this figure, after the 16-bit SACCH is multiplexed with the 208 bits of UPCH, FACCH1, or RCCH3 and the multiplexed sequence is interleaved, the interleaved sequence is mapped onto a physical slot.

13.10 LAYERED SIGNALING STRUCTURE

The signaling structure of the digital MCA employs a three-layer structure—a basic interface layer that defines the physical structure of the channel; a transmission control layer that controls random access and individual access controls; and a call control layer that processes call setup and release. However, this layered structure is quite different from those of the PDC or PHS systems because

- Response time for the signaling should be fast.
- Signaling protocol should be matched to the push-to-talk operation.
- It should support both circuit-switched and packet-switched modes.
- Algorithm should be simple.

Of these layers, the transmission control layer for digital MCA is especially unique.

In the PDC or PHS systems, the frame structures of layers 2 and 3 are independent from the frame structure of the physical slot. Conversely, the frame structure of the transmission control layer and call control layer in digital MCA is synchronized to the physical slot or subslot. Moreover, this system has special logical channels called *radio control channels* (RCCH1, RCCH2, and RCCH3) that are dedicated to the transmission control. Because any radio control message is transmitted by each RCCH slot or subslot with error detection, we can achieve very quick response time for the radio transmission control using simple protocols.

Furthermore, this scheme is suitable for packet transmission because each slot can simultaneously transmit data and the media access control message. Moreover,

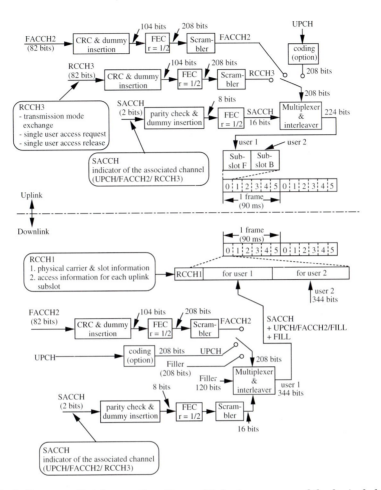

Fig. 13.25 Block diagram of each encoding, the multiplexing process of the logical channels, and the logical channel mapping on a physical slot for the USC operated in the packet-switching mode in both downlink and uplink.

because we do not have to prepare handover or channel reassignment functions owing to MCA's large-zone and single-zone features, the number of the message types is very small compared to the PDC and PHS.

13.11 ACCESS CONTROL IN THE CAC

There are two access modes for the CAC—random access mode and individual access mode. Although the CAC is basically operated in the random access mode, it is occasionally used in the individual access mode to transmit supplemental signals.

13.11.1 Transmission Control Protocol for Random Access Mode

In the downlink CAC slot, the following information is indicated in the RCCH1 of each subslot.

☞ Access mode indicator (random access or individual access mode) for the slot in the next frame

☞ Logical slot indicator (CCCH or UPCH) for the slot in the next frame

☞ Received uplink slot condition (OK or NG) for the slot in the previous frame

Initially, the BS sets the access mode indicator as the random access mode to accommodate call requests from terminals. When a terminal confirms that both or one of the subslots are in the random access mode, it can transmit call requests to the BS at the subslot of the next frame. When the BS receives a message from a terminal, it transmits OK (when there is no error in the received signal) or NG (when the message includes errors) via the RCCH1 of the next downlink slot. When the terminal detects NG, it goes to the retransmission mode.

13.11.2 Transmission Control Protocol for Individual Access Mode

When the BS receives an individual access request from a terminal, it changes the access mode indicator to the individual access mode, and it also indicates the terminal ID at the RCCH1 of the corresponding downlink subslot. When the terminal confirms that the subslot is in the individual access mode and that its own terminal ID is indicated in the RCCH1 of the corresponding downlink slot, it can start message transmission.

13.12 Access Control in the USC

In the USC, each terminal transmits information to the BS via the TCH or UPCH, and the signal is repeated to all the terminals of the same group or subgroup in the downlink. In the case of TCH transmission, there are two modes—user access mode, in which all the terminals in the group can access the USC slot, and individual access mode, in which only the accepted terminals can access the USC slot. Whether the slot is in the user access or individual access mode is indicated in RCCH1 in the corresponding downlink slot.

When RCCH1 indicates that the corresponding uplink slot is in the user access mode and a terminal wants to use the channel, the terminal sends a call request message, including its terminal ID and the associated information, to the BS via the RCCH3. When the BS receives this message, it changes the access mode indicator in the RCCH1. At the same time, the BS detects the received slot timing and the received signal level. According to these measured data, the amount of time alignment and the transmission power level increment/decrement are decided, and they are transmitted to the terminal via the RCCH1 in the downlink. After the terminal confirms that the received uplink message condition is OK, the uplink mode is changed to the individual access mode, and after the terminal ID is indicated in the RCCH1, the terminal starts TCH transmission. When the received uplink message condition is NG, the same process is repeated again.

13.13 CALL CONTROL LAYER

The call control layer for the digital MCA system corresponds to layer 3 (network layer) of the OSI reference model, and it is divided into three phases.

☞ Communication link acquisition phase

☞ Communication phase

☞ Communication link release phase

13.13.1 Communication Link Acquisition Phase

The communication link acquisition phase's function is to initialize the terminal conditions to be the idle state and to allocate radio resources to each terminal. To initialize terminal conditions to be the idle state, the BS regularly broadcasts various information, such as the physical structure of the CAC. When a terminal enters the service area of a BS or when power of the terminal is turned on, it receives all the messages of the BCCH. The terminal can transmit a call request to the BS only when all the BCCH messages are received and the terminal is in the idle mode.

When a terminal wants to initiate a call, it has to send a call request to the BS. When the BS receives the call request, it assigns radio resources to the terminal when there are available channels. When there are no available channels, however, the call request is put into a queue until a channel becomes available for the call request, and the BS informs the terminal that the call request is booked. When the queue is full, the call request is rejected.

13.13.2 Communication Phase

Call control during the communication phase is reserved as an option for the operator. When such control is defined by the operator, the message is transmitted via the FACCH.

13.13.3 Communication Link Release Phase

There are four cases in which the communication link release phase is used in digital MCA systems.

☞ When the accepted talk time is expired

☞ When the BS receives no message from the terminal after a radio channel is assigned to it

☞ When a terminal finds that the channel condition in the downlink is too poor a quality

☞ When a terminal sends a call termination message to the BS

When a terminal wants to terminate a call, it sends the channel release request to the BS. When the BS receives this request or the assigned time is expired, the BS sends the channel release command to the terminal and the radio channel is released.

13.14 EXAMPLES OF SIGNALING PROCESSES

As examples of the signaling process for the digital MCA system, we will discuss the call origination process and the call termination process.

Figure 13.26 shows the call origination process. When an MS wants to originate a call, it has to check CCCH occupancy. When the MS finds a usable CCCH, it sends a call request to the BS via the uplink CCCH. When the message is correctly received by the BS and the queue is not full, the call request is queued and a reservation message is sent to the MS. When a usable channel is found, the channel is assigned to the terminal as well as to all the terminals of the user group.

After that, when a user pushes the talk request button, it can obtain the right to use the uplink TCH. In this case, the access mode of the channel is changed to the

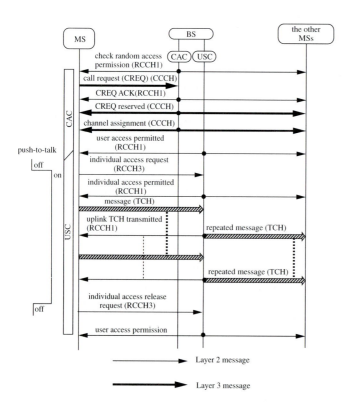

Fig. 13.26 Call origination process.

individual access mode. This TCH message is transmitted to all the users in the user group via the downlink TCH. The user can talk while he or she pushes the button. When the talk request button is released, the channel status is changed to the user access permission mode again.

Figure 13.27 shows the communication link release process. When a terminal (operator station, in most cases) sends a disconnection request to the BS via the FACCH1 and the message is correctly received by the BS, the BS sends a disconnection message to all the users in the user group and the radio channel is released.

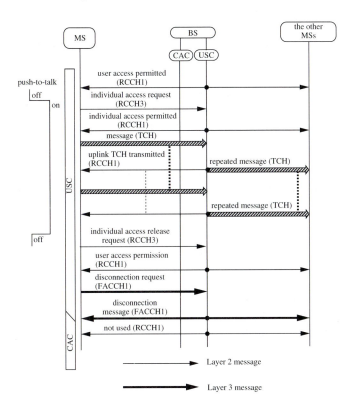

Fig. 13.27 Communication link release process.

REFERENCES

13-1. RCR, "Digital MCA system," RCR STD-32, November 1992.

13-2. Birchler, M. A. and Jasper, S. C., "A 64 kbps digital land mobile radio system employing M-16QAM," 1992 IEEE International Conf. on Select. Topics in Wireless Commun. (Vancouver, B.C.), pp. 158–62, June 1992.

CHAPTER 14

Digital Public PMR System

The digital public private mobile radio (PMR) system is another standardized digital land mobile communication system in Japan. The most distinct feature of this system is that three radio interfaces are standardized because the application field of the public PMR system is widely diverged because of how different the public service is that will be supported by the system.

This chapter will discuss system overview, the physical layer interface, and specifications of the digital public PMR system.

14.1 SYSTEM OVERVIEW

14.1.1 Outline of System Configurations

In Japan, there are many companies and organizations that use the public PMR band allocated around 400 MHz. Because spectrum shortage is the most serious problem even in this system, its digitalization was started in 1992, and RCR issued a standard for digital public PMR systems (RCR STD-39) in December 1993 [14-1].

Because each public PMR system has a different system configuration, different system size, and different services, we have to expect various types of system configurations in developing the specification process of digital public PMR systems.

Figure 14.1 is one example of the basic system configuration for public PMR systems. Because most of the PMR systems do not have too many terminals, a large-zone system as shown in Figure 14.1 is preferable to a small-zone cellular configuration

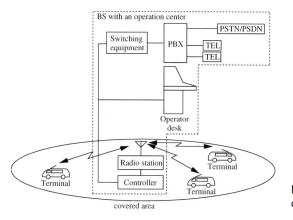

Fig. 14.1 An example of the basic system configuration for public PMR systems.

from an economical point of view. In this system, the operator desk connected to the BS along with the PBX supervises all the terminals. Terminals can also be connected to the terminals in the PSTN or the public-switched data network (PSDN). Although a large-zone configuration is economical, its serious drawback is that its outage probability is large due to shadowing and irregular terrain. Especially, when a tall building cluster, such as a cluster of apartment buildings, exists in the service area, such a cluster could produce relatively large shadowed areas.

When the dominant outage area is restricted to a limited area, putting a remote zone in such an outage area is a good solution to reduce outage probability with a relatively low cost. Figure 14.2 shows a remote-zone system configuration as an example of such systems. In this case, however, we have to carefully select the location of the remote-zone site.

When a system has to cover a large area with a relatively large number of terminals, we can select multizone systems as shown in Figure 14.3. If the area is extremely large, we can also use multicontrol station systems as shown in Figure 14.4.

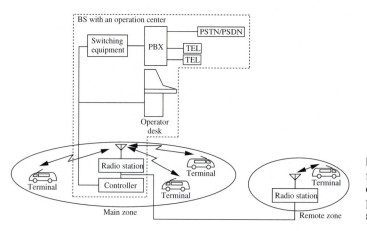
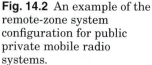

Fig. 14.2 An example of the remote-zone system configuration for public private mobile radio systems.

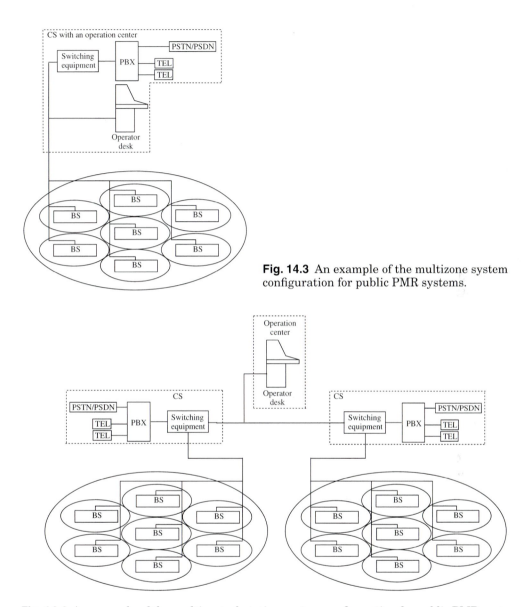

Fig. 14.3 An example of the multizone system configuration for public PMR systems.

Fig. 14.4 An example of the multicontrol station system configuration for public PMR systems.

To flexibly support such a variety of system configurations, the standard for the digital public PMR systems (RCR STD-39) specifies three types of modulation schemes—M16QAM, 16QAM, and π/4-QPSK [14-1]. Features of these schemes will be detailed later in this chapter.

14.1.2 Supported Services

In digital public PMR systems, various communication modes are supported to cope with a variety of system configurations as well as service requirements. Table 14.1 summarizes the communication modes and the expected services of digital public PMR systems.

Table 14.1 Summary of the communication modes and expected services supported by digital public PMR systems.

Categories	Functions	Features
Communication mode	point-to-point communication	In this mode, a radio link is established between two arbitrary terminals or between an operation center and a terminal.
	group communication	In this mode, a message from a terminal or an operator is transmitted to all of the terminals or the operator in a group.
	broadcasting mode	One-way communication from the operation center to all the terminals.
	controlled communication	Emergency message to all the terminals from the operation center.
Expected services	Voice, data, picture, fax, automatic vehicle monitoring (AVM).	
Special communication mode	simultaneous transmission of voice and nonvoice data	Multislot assignment to simultaneously transmit nonvoice data during voice communications.
	high-bit-rate data communication	Multislot assignment for higher-bit-rate data communication.
	full duplex	Multislot assignment to support full-duplex communication between terminals.
	encryption	Optional function for special encryption.

14.2 SPECIFICATIONS OF THE RADIO INTERFACES

Table 14.2 shows the radio interface specifications for the digital public PMR systems. Three types of systems are standardized in RCR STD-39.

Specifications of the M16QAM system is almost the same as the specifications for the digital MCA system except that the layer 3 messages are slightly different due to a difference in the services supported. One big advantage of the M16QAM is that it can achieve high system capacity for large-zone single-cell systems because improving spectral efficiency in terms of frequency is the only way to achieve high system capacity in such systems. Moreover, it is robust to a large delay spread due to a longer symbol duration than π/4-QPSK or 16QAM.

Table 14.2 Radio interface specifications for the digital public PMR system.

Parameters	Specifications		
Modulation	M16QAM (M = 4) (α = 0.2)	π/4-QPSK (α = 0.5)	16QAM (α = 0.5)
Access	TDM (downlink), TDMA (uplink)		
Frequency band	400 MHz		
Frequency band separation	>18 MHz		
Carrier spacing	25 kHz		
Channel/carrier	6 (full slot) 12 (subslot)	4	6 (full slot) 12 (subslot)
Symbol rate	4 ksymbol/s per each subcarrier	16 ksymbol/s	16 ksymbol/s
99% bandwidth	20 kHz	21 kHz	21 kHz
Slot length	15 ms (full slot) 7.5 ms (subslot)	10 ms	15 ms (full slot) 7.5 ms (subslot)
Frame length	90 ms	40 ms	90 ms
Voice codec	not specified bit rate ≤ 7.5 kbit/s (including FEC)	not specified bit rate ≤ 6.4 kbit/s (including FEC)	not specified bit rate ≤ 7.5 kbit/s (including FEC)
Layers 2 & 3 structures	same as digital MCA	same as PDC	not specified
Other features	- high capacity in large-zone systems - high delay spread immunity - pilot symbol-aided fading compensation	- high capacity in cellular systems - high power efficiency - no fading compensation required	- high capacity in large-zone systems - pilot symbol-aided fading compensation

π/4-QPSK is a modulation scheme applied to the PDC and PHS systems. However, for digital public PMR systems, various parameters of the physical layer, such as the symbol rate, the number of multiplexed channels in a carrier, and slot length are different from those of PDC and PHS. On the other hand, structures of layer 2 and layer 3 of this system are almost the same as those of the PDC systems except that layer 3 messages are different due to the difference of the applied systems. Because π/4-QPSK is more robust to CCI than M16QAM or 16QAM because of its higher receiver sensitivity, this modulation scheme is suitable if a public PMR system employs a cellular configuration with a zone radius of around 1 km.

16QAM is a single carrier scheme. Therefore, it is also suitable for large-zone single-cell systems as in the case of M16QAM. Although its delay spread immunity is worse than that of M16QAM because of its shorter symbol duration, its power efficiency is higher than that of M16QAM because its peak/average power ratio is lower. As for the higher layer specification of 16QAM systems, it is not specified in the RCR STD-39.

Because specifications for M16QAM are the same as specifications for the digital MCA system, we will discuss specifications for π/4-QPSK and 16QAM here.

14.3 π/4-QPSK Systems

14.3.1 Frame and Slot Format

Figure 14.5 shows the frame format of the π/4-QPSK system. Its frame length is 40 ms and it consists of 4 slots with a slot length of 10 ms. Transmission timing of the uplink is delayed by 20 ms from that of the downlink. As a result, each terminal has 10 ms of idle time after each reception and transmission time slot.

In digital public PMR systems, three types of slot format are employed.

☞ Synchronization slot

☞ CAC

☞ USC

Figure 14.6 shows the slot format for a synchronization slot. This slot is transmitted before the TCH is established to take frame, slot, and symbol timing synchronization. This slot includes TA to adjust uplink burst timing, SSC to detect superframe timing, terminal ID (ID) to identify terminals for group communications, and color code to identify the cell cluster. The slot identification bit (B) indicates whether a time alignment and superframe counter are used in the terminal. In general, a synchronization slot is transmitted twice from both the terminal and the BS, and only the second synchronization slot employs these functions. Therefore, B = 0 indicates the first synchronization slot, and B = 1 indicates the second synchroniza-

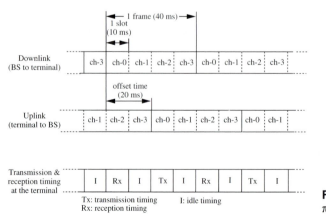

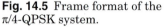

Fig. 14.5 Frame format of the π/4-QPSK system.

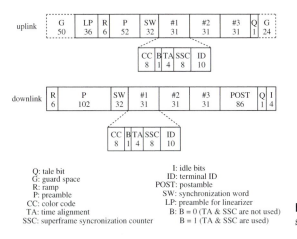

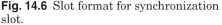

Q: tale bit
G: guard space
R: ramp
P: preamble
CC: color code
TA: time alignment
SSC: superframe syncronization counter

I: idle bits
ID: terminal ID
POST: postamble
SW: synchronization word
LP: preamble for linearizer
B: B = 0 (TA & SSC are not used)
B = 1 (TA & SSC are used)

Fig. 14.6 Slot format for synchronization slot.

tion slot. In each synchronization slot, color code, B, TA, SSC, and ID are transmitted three times, and these data are detected by a two-out-of-three majority vote.

Figure 14.7 shows the slot format for the CAC. Because the CACs in the uplink are operated in the random access mode, a long guard space (24 symbols = 0.75 ms) is prepared to prevent slot collision. Moreover, a longer preamble is prepared in the first unit to improve initial acquisition performance of slot and symbol timing synchronization. Because a different system might be located in the 25-kHz separated adjacent channel in public PMR systems, we have to sufficiently suppress out-of-band radiation by introducing a linearizer. For this purpose, 36 bits of preamble for a linearizer (LP) is also allocated in the beginning of each slot. Figure 14.8 shows specification of the uplink CAC slot, including waveform specification of the LP.

Figure 14.9 shows the slot format for the USC. An FACCH message is transmitted by stealing a TCH slot. SACCH or RCH is time division multiplexed on a superframe format.

14.3.2 Superframe Format

In this system, a superframe with its frame length of 720 ms is employed to multiplex BCCH, PCH, and SCCH in the CAC downlink. It is also used to multiplex

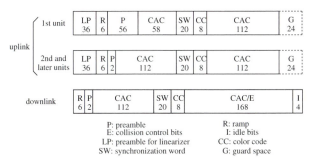

P: preamble
E: collision control bits
LP: preamble for linearizer
SW: synchronization word

R: ramp
I: idle bits
CC: color code
G: guard space

Fig. 14.7 Slot format for CAC.

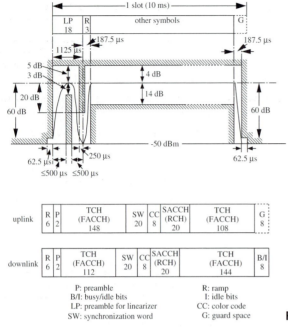

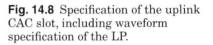

Fig. 14.8 Specification of the uplink CAC slot, including waveform specification of the LP.

uplink	R 6	P 2	TCH (FACCH) 148	SW 20	CC 8	SACCH (RCH) 20	TCH (FACCH) 108	G 8

downlink	R 6	P 2	TCH (FACCH) 112	SW 20	CC 8	SACCH (RCH) 20	TCH (FACCH) 144	B/I 8

P: preamble   R: ramp
B/I: busy/idle bits   I: idle bits
LP: preamble for linearizer   CC: color code
SW: synchronization word   G: guard space

Fig. 14.9 Slot format for USC.

SACCH and RCH in both the downlink and uplink of the USC. Figure 14.10 shows the superframe format of this system. Superframe configuration of the BCCH, PCH, and SCCH is broadcast to all the terminals via the BCCH. Figure 14.11 shows RCH and SACCH multiplexed on a superframe.

14.3.3 Channel-Coding Scheme for Each Logical Channel

14.3.3.1 For the CAC In π/4-QPSK systems, the CAC frame structure for layer 2 and layer 3 is independent from that of the physical layer. Therefore, layer 2 and 3 messages are mapped onto several physical slots.

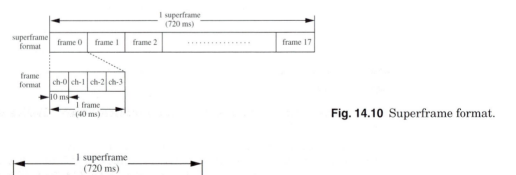

Fig. 14.10 Superframe format.

Fig. 14.11 RCH and SACCH multiplexing on a superframe format.

For Downlink Channels. Figure 14.12 shows the channel encoding and mapping process for CACs in the downlink. First, a message is divided every 93 bits, and a message assemble/disassemble information word (W) and collision control bits (E) are attached to this message unit to produce 119 bits of a blocked message. This blocked message unit is then encoded by the 16-bit CRC with its generator polynomial of $G(x)$ = $x^{16} + x^{12} + x^5 + 1$. After 5 bits of dummy bits are attached at the end of the sequence, the sequence is convolutionally encoded by a rate-1/2 convolutional encoder with generator polynomials of $G_1(x) = x^5 + x^3 + x + 1$ and $G_2(x) = x^5 + x^4 + x^3 + x^2 + 1$. After the encoded signal is interleaved by using a (20, 14) low-column conversion-type interleaver, the data are mapped onto the CAC data sections in each slot.

For Uplink Channels. Figure 14.13 shows the channel encoding and mapping process for CACs in the uplink. Because the first uplink CAC slot prepares a longer preamble, data area for the first slot and data for the second or later slots have different lengths. Therefore, two types of channel encoding schemes are specified for the uplink CACs.

Each CAC message is divided into a 56-bit block for the first unit and 83-bit blocks for the second or later slots. In the first unit, after W is attached in front of the first message unit, it is encoded by the 16-bit CRC with its generator polynomial of $G(x) = x^{16} + x^{12} + x^5 + 1$. After 5 bits of dummy bits are attached at the end of the sequence, the sequence is convolutionally encoded by a rate-1/2 convolutional encoder with generator polynomials of $G_1(x) = x^5 + x^3 + x + 1$ and $G_2(x) = x^5 + x^4 + x^3 + x^2 + 1$. After the encoded signal is interleaved by using a (17, 10) low-column conversion-type interleaver, the data are mapped onto the CAC data sections in the first slot. On the other hand, for the second or later units, 83 bits of the CAC message unit associated with W is CRC encoded, convolutionally encoded, interleaved, and mapped onto the CAC data sections in each slot, where the CRC encoder and convo-

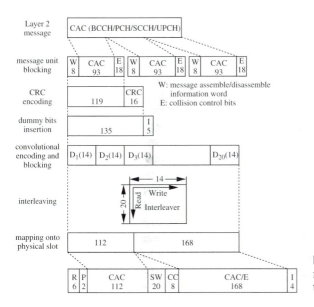

Fig. 14.12 Channel encoding and mapping onto physical slot for CACs in the downlink.

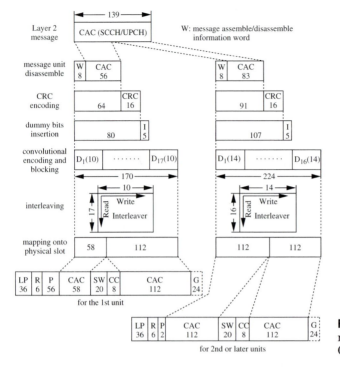

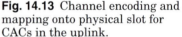

Fig. 14.13 Channel encoding and mapping onto physical slot for CACs in the uplink.

lutional encoder are the same as those used for the first unit and an interleaver size of (16, 14) is employed.

14.3.3.2 For FACCH   Figure 14.14 shows the channel encoding and mapping process for the FACCH. First, each FACCH message is divided into 99-bit units. After W is attached in front of the unit, it is encoded by the 16-bit CRC with its generator polynomial of $G(x) = x^{16} + x^{12} + x^5 + 1$. After 5 bits of dummy bits are attached at the end of the sequence, the sequence is convolutionally encoded by a rate-1/2 convolutional encoder with generator polynomials of $G_1(x) = x^5 + x^3 + x + 1$ and $G_2(x) = x^5 + x^4 + x^3 + x^2 + 1$. After the encoded signal is interleaved by using a (16, 16) low-column conversion-type interleaver, 256 bits of the interleaved data are divided into two blocks, each of which contains 128 bits, where the former 128 bits are mapped onto the same frame and the latter section is mapped onto the next frame.

14.3.3.3 For SACCH   Figure 14.15 shows the channel encoding and mapping process for SACCH. After 8 bits of W is attached in front of the 51-bit unit, it is encoded by the 16-bit CRC with its generator polynomial of $G(x) = x^{16} + x^{12} + x^5 + 1$. After 5 bits of dummy bits are attached at the end of the sequence, the sequence is convolutionally encoded by a rate-1/2 convolutional encoder with generator polynomials of $G_1(x) = x^5 + x^3 + x + 1$ and $G_2(x) = x^5 + x^4 + x^3 + x^2 + 1$. After the encoded signal is interleaved by using a (16, 10) low-column conversion-type interleaver, the sequence

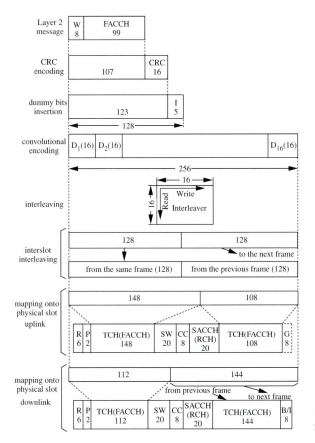

Fig. 14.14 Channel encoding and mapping onto physical slot for FACCH.

is blocked into eight blocks, each of which has 20 bits of data, and these blocks are mapped onto eight consecutive frames as shown in Figure 14.15.

14.3.3.4 For RCH   Figure 14.16 shows the RCH encoding and mapping process. Each RCH message has 14 bits, and the message is first encoded by the 6-bit CRC with its generator polynomial of $G(x) = x^6 + x + 1$ to produce a 20-bit CRC-encoded message. After this message is blocked into 4-bit data blocks, each block is encoded by the following generator matrix:

$$G = \begin{bmatrix} 1 & 0 & 0 & 0 & 1 & 0 & 1 & 1 \\ 0 & 1 & 0 & 0 & 1 & 1 & 1 & 0 \\ 0 & 0 & 1 & 0 & 1 & 1 & 0 & 1 \\ 0 & 0 & 0 & 1 & 0 & 1 & 1 & 1 \end{bmatrix} \tag{14.1}$$

This message is then interleaved by using a (5, 8) low-column conversion-type interleaver, divided into two 20-bit data blocks, and these two blocks are mapped onto two consecutive frames as shown in Figure 14.16.

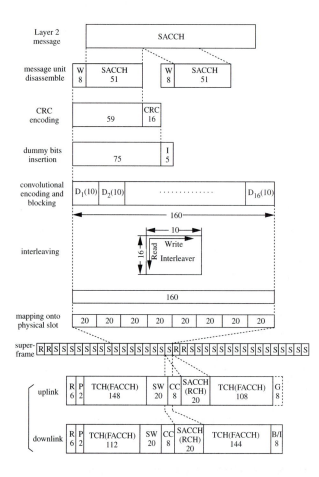

Fig. 14.15 Channel encoding and mapping onto physical slot for SACCH.

14.3.4 Requirements for the Transmitter and Receiver

Other than the specifications shown in Table 14.2, RCR STD-39 also specifies requirements for the transmitter and receiver. Table 14.3 shows requirements for a π/4-QPSK transmitter for public PMR radio systems. Definitions of the items in Table 14.3 are the same as the definitions for the PDC system except for the ACI power, which is defined as the power radiated within a bandwidth of ±9 kHz, of which the center frequency is separated by ±25 kHz from the carrier frequency, where the carrier is modulated by the test signal sequence at the same bit rate as that of the specified one.

On the other hand, Table 14.4 shows requirements for a π/4-QPSK receiver for public PMR systems.

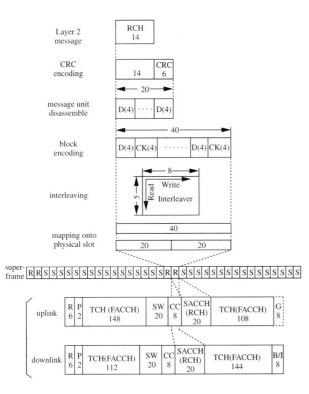

Fig. 14.16 Channel encoding and mapping onto physical slot for RCH.

Table 14.3 Requirements for the transmitter of digital public private mobile radio systems.

Parameters	Specifications
Carrier frequency instability ($\delta f/f_c$)	BS: $\delta f/f_c \leq 5 \times 10^{-7}$ terminal: $\delta f/f_c \leq 3 \times 10^{-6}$ (without AFC) $\delta f/f_c \leq 6 \times 10^{-7}$ (tracking error to AFC)
Spurious transmission	BS: −60 dB or less, relative to the transmitted power, or 2.5 μW or less terminal: −60 dB or less, relative to the transmitted power, or 0.25 μW or less
Leakage power during carrier OFF period	≤−50 dBm within the assigned band ≤4 nW outside the assigned channel in the system band −60 dB or less compared with the power level during the carrier being turned ON
99% bandwidth	21 kHz
Transmitter power for the terminal	≤40 W (BS) ≤5 W (terminal)

Table 14.3 Requirements for the transmitter of digital public private mobile radio systems. (Cont.)

Parameters	Specifications
Transmitter power accuracy	within +20% and –50% of the specified value
ACI power	$\leq$–55 dB @ Δf = 25 kHz (BS) $\leq$–55 dB @ Δf = 25 kHz, or –50 dB or less than the power controlled level (terminal)
Cabinet radiation	25 µW or less

Table 14.4 Requirements for the receiver of digital public private mobile radio systems.

Parameters	Specifications

Receiver sensitivity	under static condition
	   6 dBµ for BER = 10^{-2}
	     specified receiver sensitivity
	under Rayleigh fading condition @ f_d = 20 Hz
	   10 dBµ for BER = 3×10^{-2} (without diversity)

Spurious sensitivity	$\geq$53 dB for ±50 kHz separated unmodulated carrier
Adjacent channel selectivity	$\geq$42 dB for ±25 kHz off
Intermodulation performance	$\geq$53 dB for ±50 kHz and ±100 kHz separated unmodulated carrier

14.4 16QAM Systems

14.4.1 Frame and Slot Format

Figure 14.17 shows a frame format for a 16QAM system. Its frame length is 90 ms, and it consists of 6 slots with a slot length of 15 ms. Transmission timing of the uplink is delayed by 22.5 ms from that of the downlink. As a result, each terminal has 7.5 ms and 52.5 ms of idle time after each reception and transmission time slot, respectively.

Because signaling is not specified for the 16QAM public PMR radio systems, only full-slot and subslot formats are specified in this system. Figure 14.18 shows the slot format for the (a) full-slot format and (b) subslot format. In either format, data symbols (D in Figure 14.18) are 16QAM-modulated symbols by using the symbol mapping rule shown in Figure 14.19. On the other hand, SW and pilot symbols (P1 to P12) are QPSK-modulated symbols with an amplitude equal to the maximum amplitude of 16QAM symbols as shown in Figure 14.19. When we define the symbol pattern of the SW, however, we have to consider ease of synchronization, reduction of misdetection probability, and randomization of phase. Therefore, the SW pattern is specified as shown in Table 14.5. In this table, each slot in a frame employs a different SW pattern.

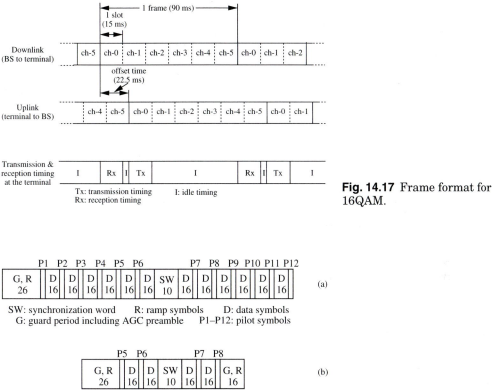

Fig. 14.17 Frame format for 16QAM.

Fig. 14.18 Slot format for 16QAM: (a) full-slot format and (b) subslot format.

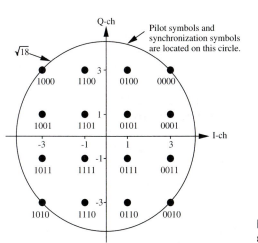

Fig. 14.19 Signal state diagram for the 16QAM system.

On the other hand, the phase patterns of the pilot symbols are defined as shown in Table 14.6.

14.4.2 Requirements for the Transmitter and Receiver

Other than the specifications shown in Table 14.2, RCR STD-39 also specifies requirements for the transmitter and receiver. Table 14.7 shows requirements for a

Table 14.5 Phase pattern of the SW for each slot.

	1st slot	2nd slot	3rd slot	4th slot	5th slot	6th slot
SW1	45°	45°	45°	45°	45°	45°
SW2	135°	45°	45°	−45°	45°	45°
SW3	135°	−45°	135°	−45°	135°	−45°
SW4	−135°	−135°	135°	−135°	−135°	−45°
SW5	135°	135°	−135°	−45°	−45°	−135°
SW6	135°	135°	−135°	−45°	−45°	−135°
SW7	−135°	−135°	135°	−135°	−135°	−45°
SW8	135°	−45°	135°	−45°	135°	−45°
SW9	135°	−45°	45°	−45°	45°	45°
SW10	45°	−45°	45°	45°	45°	45°

Table 14.6 Phase of pilot symbols.

Pilot symbols	Phase
P1	45°
P2	−135°
P3	135°
P4	−45°
P5	135°
P6	−135°
P7	−135°
P8	135°
P9	−45°
P10	135°
P11	−135°
P12	45°

Table 14.7 Requirements for the transmitter of digital public private mobile radio systems.

Parameters	Specifications
Carrier frequency jitter ($\delta f/f_c$)	BS: $\delta f/f_c \leq 5 \times 10^{-7}$ terminal: $\delta f/f_c \leq 3 \times 10^{-6}$ (without AFC) $\delta f/f_c \leq 6 \times 10^{-7}$ (tracking error to AFC)
Spurious transmission	BS: −60 dB or less, relative to the transmitted power, or 2.5 µW or less terminal: −60 dB or less, relative to the transmitted power, or 0.25 µW or less
Leakage power during carrier OFF period	$\leq$−50 dBm within the assigned band $\leq$4 nW outside the assigned channel in the system band −60 dB or less compared with the power level during the carrier being turned ON
99% bandwidth	20 kHz
Transmitter power for the terminal	$\leq$40 W (BS) $\leq$5 W (terminal)
Transmitter power accuracy	within +20% and −50% of the specified value
ACI power	$\leq$−55 dB @ Δf = 25 kHz (BS) $\leq$−55 dB @ Δf = 25 kHz, or −50 dB or less than the power-controlled level (terminal)
Cabinet radiation	25 µW or less

16QAM transmitter for public PMR systems. Definitions for items in Table 14.7 are the same as the definitions for the items in Table 14.3.

Table 14.8 shows requirements for a 16QAM receiver for public PMR systems. Definitions of these parameters are also the same as those for π/4-QPSK shown in Table 14.3.

Table 14.8 Requirements for 16QAM receiver of the digital public private mobile radio systems.

Parameters	Specifications
Receiver sensitivity	under static condition 9 dBµ for BER = 10^{-2} specified receiver sensitivity under Rayleigh fading condition @ f_d = 20 Hz 13 dBµ for BER = 3×10^{-2} (without diversity)
Spurious sensitivity	$\geq$53 dB for $\pm$50 kHz or more separated unmodulated carrier
Adjacent channel selectivity	$\geq$42 dB for $\pm$25 kHz off
Intermodulation performance	$\geq$53 dB for $\pm$50 kHz and $\pm$100 kHz separated unmodulated carrier

14.5 COMPARISON OF M16QAM, 16QAM, AND π/4-QPSK SYSTEMS

Because the application field for the public private mobile radio system is diverged, three modulation schemes are specified in the RCR STD-39. Table 14.9 summarizes the features of these three modulation schemes.

The advantage of π/4-QPSK is its wider coverage area compared to that of M16QAM and 16QAM. On the other hand, M16QAM and 16QAM are advantageous from the viewpoint of spectral efficiency for large-zone systems. When we compare spectral efficiency for small-zone systems, it is almost the same because a spectral efficiency improvement owing to a higher modulation level is offset by a spectral efficiency degradation due to a shorter minimum signal distance of 16QAM and M16QAM.

From the viewpoint of delay spread immunity, M16QAM has very high delay spread immunity. However, it requires higher linearity for the transmitter amplifier. With these results, we find we have to select the modulation scheme taking into consideration of required coverage, spectral efficiency, and delay spread statistics of the service area.

From the viewpoint of supported services, it is important to consider user information bit rate per slot and information bit rate per carrier. When these are considered, M16QAM and 16QAM are more effective for high-bit-rate data transmission.

We also have to bear in mind that short message transmission such as AVM is an important service for this system. Therefore, high throughput for short message transmission would be strongly required. When we employ M16QAM, we can transmit

Table 14.9 Comparison between M16QAM, 16QAM, and π/4-QPSK.

Parameters	M16QAM	16QAM	π/4-QPSK
Coverage area	narrower than π/4-QPSK		wider than M16QAM & 16QAM
Spectral efficiency for large-zone system	higher than π/4-QPSK		lower than M16QAM & 16QAM
Spectral efficiency for small-zone system	almost the same		
Delay spread immunity without diversity [14-2; 14-3; 14-4]	8.8 μs (BER = 1%) 15 μs (BER = 3%)	3.2 μs (BER = 1%) 6.0 μs (BER = 3%)	4.9 μs (BER = 1%) 9.0 μs (BER = 3%)
Required linearity for Tx amplifier	high	medium	low
Information bit rate per slot	7.64 kbit/s	8.53 kbit/s	6.4 kbit/s
Information bit rate per carrier	45.87 kbit/s	51.2 kbit/s	25.6 kbit/s
Packet transmission throughput	high	N/A	low

a short message by using only 1 time slot. On the other hand, when we employ $\pi/4$-QPSK, we need at least 4 slots to transmit a short message because it requires one CAC, two synchronization bursts, and a message slot. From this standpoint, M16QAM is more suitable for short message packet transmission.

REFERENCES

14-1. RCR, "Digital public private mobile radio systems," RCR STD-29, December 1993.

14-2. Birchler, M. A. and Jasper, S. C., "A 64kbps digital land mobile radio system employing M-16QAM," 1992 IEEE International Conf. on Select. Topics in Wireless Commun. (Vancouver, B.C.), pp. 158–62, June 1992.

14-3. Kinoshita, N. et al., "Field experiments on 16QAM/TDMA and trellis coded 16QAM/TDMA systems for digital land mobile radio communications," IEICE Trans. Commun., Vol. E77-B, No. 7, pp. 911–20, July 1994.

14-4. Adachi, F. and Ohno, K., "BER performance of QDPSK with post detection diversity reception in mobile radio channels," IEEE Trans. Veh. Technol., Vol. 40, No. 1, pp. 237–49, February 1991.

CHAPTER 15

System Design Strategy for TDMA-Based Systems

$\mathbf{I}$n this chapter we'll discuss how to design future-generation wireless communication systems using TDMA technology [15-1; 15-2; 15-3; 15-4; 15-5]. In the case of TDMA systems, there are so many parameters relating to the modulation schemes as well as to the TDMA frame format. Moreover, there are many constraint conditions, such as the propagation path characteristics; restrictions from the viewpoint of handy personal terminal implementation; spectrum allocation policy; and so on.

We will first discuss how these constraint conditions and the user service requirements are related, and then we'll discuss how to design TDMA-based radio interfaces in light of these constraint conditions.

15.1 BASIC DESIGN CONCEPT OF TDMA SYSTEMS

Figure 15.1 shows a flow diagram for the design of the modulation scheme and frame format for TDMA-based wireless communication systems. On the left side of this figure, we show conditions specific to wireless communication systems—service requirements, constraint conditions for the hardware implementation, propagation path conditions, and system requirements. Considering these conditions, we have to determine the frame format and modulation scheme for TDMA systems.

When we temporarily determine these parameters, we have to evaluate the coverage area and its spectral efficiency and whether these parameters can satisfy preliminary determined services and system requirements. When the parameters cannot

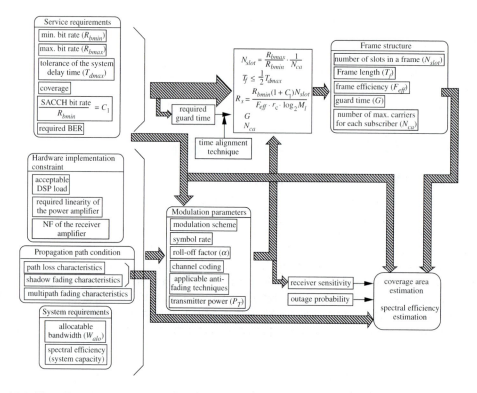

Fig. 15.1 Flow diagram for the design of the modulation scheme and frame format for a TDMA-based wireless communication systems.

satisfy these requirements, we will modify the modulation scheme and TDMA frame parameters and evaluate the coverage and spectral efficiency again. When we finalize the specification of these parameters, we can go to the hardware verification process of the system, including the field trial. Although this verification process is a very important step, especially for the system standardization process, we will focus only on the system design process that is shown in Figure 15.1 in our discussion.

15.2 CONDITIONS SPECIFIC TO WIRELESS COMMUNICATION SYSTEMS

For wireless communication system design, we have to consider certain conditions that are specific to wireless communication systems.

15.2.1 Service Requirements

All wireless systems have to support user requirements as much as possible. Although the main services of existing wireless communication systems are voice transmission and low-bit-rate data transmission with a bit rate of up to around 10 kbit/s even in digital cellular systems, multimedia services using variable-bit-rate transmission are expected to be in use in the future. For example, Study Group 8 of the ITU-R is now developing FPLMTSs that support voice services as well as the data transmission services with a bit rate of up to 2 Mbit/s.

One of the important factors specific to TDMA systems is the processing delay caused by TDMA framing. Therefore, a TDMA frame length has to be determined by the acceptable maximum processing delay time. Moreover, we have to consider the other service requirements, such as the coverage, the bit rate for the SACCH, and the required BER performance for each type of information.

15.2.2 Constraint Conditions for Hardware Implementation

Constraint conditions for hardware implementation are important factors in making small, light, and less expensive terminals. Particularly when we want to support high-bit-rate transmission, we have to care for the DSP load necessary for the anti-frequency-selective fading techniques. For example, when we employ a DFE, its load rapidly increases with the symbol rate as we discussed in chapter 5. When the DSP load is found to be too high to implement a handheld terminal, we have to consider some DSP load reduction techniques, such as the reduction of the symbol rate, application of a higher modulation level, increase of the maximum number of carriers to be assigned to a terminal, or modification of the fading compensation algorithm.

Another important constraint condition is the required linearity of the power amplifier because higher linearity decreases the battery-saving time due to its lower power efficiency unless an appropriate nonlinearity compensation technique is employed. On the other hand, a high spectral efficient modulation scheme such as QAM requires higher linearity. Therefore, the required linearity has to be determined by the trade-off between the battery lifetime and spectral efficiency.

The NF of the receiver amplifier is also an important constraint condition because a lower NF raises the terminal cost.

15.2.3 Propagation Path Conditions

As discussed in chapter 2, propagation path conditions are very important in implementing terminals as well as designing cell configuration. Among three components of the propagation path conditions, path loss due to distance and shadow fading determine the cell configuration and the coverage area estimation (discussed in chapter 10). On the other hand, multipath fading, which rapidly changes the received signal level, influences terminal design decisions, including the selection of anti-fading compensation techniques.

15.2.4 System Requirements

In wireless communication systems, the allocatable bandwidth (W_{alo}) is determined by the radio regulation policy of each country. Such regulation policy is determined taking into consideration of the number of operators that support similar services using the same frequency band, as well as the geographical locations and specifications of different services that use the same or adjacent channels. Moreover, we have to determine the required spectral efficiency (system capacity) from the potential demand for the system.

15.3 MODULATION/ACCESS SCHEME SELECTION

In the TDMA system, the following modulation parameters have to be determined by the requirements we've just discussed.

- ☞ Modulation scheme (modulation level is M_l)
- ☞ Symbol rate (R_s)
- ☞ Roll-off factor of the transmitter and receiver root Nyquist filter (α)
- ☞ Channel encoding scheme including interleaver
- ☞ Applicable anti-fading technique
- ☞ Transmitter power (P_T)

As discussed in chapters 4 and 5, we can select any modulation scheme as a result of the recent development of anti-fading techniques. However, we have to be careful when selecting anti-fading techniques. When we integrate various anti-frequency-selective fading techniques in a terminal, although we can improve its robustness to the fading, this will increase terminal size as well as shorten battery life. Needless to say, such a terminal would be very expensive. Therefore, we have to determine the anti-fading technique taking into consideration the complexity of the terminal and the value of the supported services.

The symbol rate is determined by various factors. First of all, to support high-bit-rate transmission, we have to consider the maximum delay time of the delayed wave (τ_{max}) to be compensated for. For example, in the case of DFE, its number of computations increases linearly with the square of $\tau_{max}R_s$ as we have discussed in chapter 5. Such an increase in the number of computations (DSP load) prevents making hand-held portable terminals.

Countermeasures to reduce the DSP load include [15-6; 15-7]

- ☞ Reduction of R_s by increasing the number of assigned carriers to a terminal (N_{ca})
- ☞ Reduction of R_s by increasing M_l
- ☞ Neglect of long-delayed waves
- ☞ Employment of a long-delayed wave-insensitive anti-frequency-selective fading technique

Among these countermeasures, although an increase in N_{ca} or M_l is effective in reducing the DSP load, it requires much higher linearity of the transmitter amplifier, which could reduce battery life. Therefore, the required linearity of the RF amplifier is also determined by the trade-off between battery life and DSP load in addition to the trade-off between battery life and spectral efficiency.

Neglecting the long-delayed wave is another option used to reduce DSP load because a longer delay time is less probable than shorter ones. If the probability of the long-delayed wave is smaller than the required outage probability, such neglect is not a problem at all. In other words, we have to determine the τ_{max} to be compensated for considering the delay spread statistics and the required outage probability.

Another effective means of coping with a long-delayed wave is to employ a delayed time-insensitive anti-frequency-selective fading technique. One such technique is the adaptive array antenna [15-8]. Of course, adaptive array antenna techniques have different types of disadvantages. For example, they require more RF devices, which have relatively large mass.

With these technical considerations, we can define the modulation and demodulation scheme, including anti-frequency-selective fading techniques.

Next we have to consider the selection of transmitter and receiver filters. When we employ a linear modulation such as π/4-QPSK, a root Nyquist filter is the optimum filter for the transmitter and receiver. In this case, selection of the roll-off factor (α) is very important. When we employ smaller α, we can suppress ACI. However, smaller α is less robust to delay spread [15-9]. Figure 15.2 shows delay spread versus α with a parameter of the irreducible BER for (a) QPSK and (b) 16QAM, where delay spread is normalized by a symbol duration and selective combining is employed as the diversity combining scheme. As shown in this figure, delay spread for a certain BER is reduced when α is reduced. In other words, smaller α is less tolerable against delay spread. Moreover, it is even more remarkable in the case of 16QAM. Therefore, α has to be determined by the trade-off between ACI protection performance and delay spread immunity. In the PDC and PHS systems discussed in chapters 11 and 12, $\alpha = 0.5$ is employed because of this.

Channel coding is also a key technology for improving receiver sensitivity and spectral efficiency. In a cellular system, we can improve spectral efficiency with respect to space because we can improve CCI immunity by using a channel coding. On

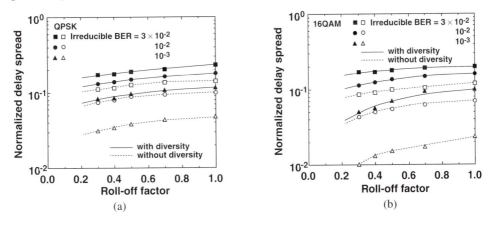

Fig. 15.2 Delay spread versus α with a parameter of the irreducible BER for (a) QPSK and (b) 16QAM, where delay spread is normalized by a symbol duration and selective combining is employed as the diversity combining scheme.

the other hand, channel coding degrades spectral efficiency with respect to frequency because it requires bandwidth expansion. Therefore, the optimum coding scheme is selected to maximize total spectral efficiency. Moreover, the coding scheme should be optimized at an appropriate BER depending on the type of transmitted information. For example, the coding scheme for the voice codec is optimized at BER = $10^{-3} - 10^{-2}$, whereas that for nonvoice data is optimized at BER = $10^{-5} - 10^{-3}$.

With these considerations, we can obtain the transmitted power (P_T) using the process discussed in chapter 10.

15.4 TDMA FRAME STRUCTURE DESIGN

In the TDMA system, we can support variable-bit-rate transmission by changing the number of the assigned slots. When such variable-bit-rate transmission is supported by using a single carrier, a minimum bit rate (R_{bmin}) is supported by assigning one slot in each TDMA frame to the user, whereas the maximum bit rate (R_{bmax}) is supported by assigning all the slots in a TDMA frame to one user. When we can assign N_{ca} carriers to a user, we can assign N_{ca} times more slots to one user. Therefore, the required number of slots in each frame is given by

$$N_{slot} = \frac{R_{bmax}}{R_{bmin}} \cdot \frac{1}{N_{ca}} \tag{15.1}$$

One TDMA frame length is determined by the tolerance of the processing delay time (T_{dmax}). When a frame length is expressed as T_f, T_{dmax} due to the TDMA frame is given by

$$T_f \le \frac{1}{2} T_{dmax} \tag{15.2}$$

Next let's assume that the bit rate for SACCH is C_1 times the minimum information bit rate, modulation level is M_l, coding rate of the channel encoder is r_c, and the acceptable TDMA frame efficiency is F_{eff}. The required symbol rate is then given by

$$R_s = \frac{R_{bmin}(1+C_1)N_{slot}}{F_{eff} \cdot r_c \cdot \log_2 M_l} \tag{15.3}$$

Another important factor for the TDMA frame is the guard time, which is prepared to prevent slot collision due to terminal location difference. When we prepare a guard time of G μs, we can cope with the zone radius R_0 of $0.5Gc$, where c is the velocity of light. Although preparing the guard time is a simple idea and is effective in preventing slot collision, it could degrade frame efficiency as well as spectral efficiency. Alignment for the uplink transmission timing controlled by the BS is an effective technique for reducing the guard time. This technique is included in the PDC and digital MCA systems. In these systems, the slot timing is measured by using the synchronization slot transmitted from each terminal after the TCH is assigned by the BS.

15.5 EXAMPLE OF A TDMA SYSTEM DESIGN

15.5.1 Modulation Parameter Design

When we select the modulation scheme for a TDMA system, we have to decide which anti-fading techniques to apply. Because future wireless communication systems are expected to support multimedia transmission using an air interface with high-bit-rate transmission capability, we will have to assume a frequency-selective fading channel as the propagation path condition. To cope with such conditions, let's assume that we will employ two-branch space diversity and BDDFE, both of which were discussed in chapter 5. As discussed in chapter 5, because the number of computations for the BDDFE with diversity rapidly increases with the number of taps, the maximum delay time to be compensated for has to be limited to several symbols in duration, say 5–6 symbols.

Table 15.1 shows the relationship between delay time of the delayed wave in terms of symbols and the corresponding delay time for each symbol rate. When we limit the compensatable delay time of the delayed wave to up to 6 symbols, the compensated maximum delay time of the delayed wave is given by 3.0 µs for 2 Msymbol/s, 6.0 µs for 1 Msymbol/s, 12.0 µs for 500 ksymbol/s, and 24 µs for 250 ksymbol/s. In the case of microcellular systems with its zone radius of 100–500 m, we can expect the maximum delay time of the delayed wave of less than 1 µs. Therefore, we can employ a symbol rate of 2 Msymbol/s. When the zone radius is around 1 km and the BS antenna height is not too high, we can employ 1 Msymbol/s in most cases because the maximum delay in such a case is expected to be less than several µs. On the other hand, in the case of a large-zone system with a zone radius of around 10 km, we have to reduce the symbol rate to 500 ksymbol/s or less. Because the symbol rate also has to be determined by the other parameters, we will discuss symbol rate selection later.

Table 15.1 Relationship between the compensated maximum delay time of the delayed wave in terms of symbols and the corresponding delay time.

Compensated Max. Delay Time of the Delayed Wave in Terms of Symbols	Compensated Max. Delay Time of the Delayed Wave (µs) Symbol Rate (symbol/s)			
	2 M	1 M	500 k	250 k
3 symbols	1.5 µs	3.0 µs	6.0 µs	12.0 µs
4 symbols	2.0 µs	4.0 µs	8.0 µs	16.0 µs
5 symbols	2.5 µs	5.0 µs	10.0 µs	20.0 µs
6 symbols	3.0 µs	6.0 µs	12.0 µs	24.0 µs
7 symbols	3.5 µs	7.0 µs	14.0 µs	28.0 µs
8 symbols	4.0 µs	8.0 µs	16.0 µs	32.0 µs
9 symbols	4.5 µs	9.0 µs	18.0 µs	36.0 µs
10 symbols	5.0 µs	10.0 µs	20.0 µs	40.0 µs

Roll-off factor α is also an important factor even when we employ the adaptive equalizer. In this section, we will assume $\alpha = 0.5$ because $\alpha = 0.5$ is frequently employed in practical digital wireless systems.

When we select the modulation scheme, we have to evaluate the relationship between the transmitter power and the zone radius as well as the spectral efficiency. Figure 15.3 shows the BER performance versus E_s/N_0 (C/I_c) for QPSK and 16QAM, where the maximum delay time of the delayed wave is assumed to be a 6-symbol duration and selective combining is employed as the diversity combining technique. From this figure, we can obtain C/I_c that satisfies BER = 10^{-2} and BER = 10^{-3} for QPSK and 16QAM systems as shown in Table 15.2.

First let's evaluate the relationship between transmitter power and zone radius. Because the uplink channel is more power limited than the downlink due to battery size of the terminal, we will evaluate this relationship for the uplink. In this evaluation, we will assume the following conditions:

☞ Path loss is given by Hata's equation.
☞ Carrier frequency is 900 MHz.
☞ Large-scale signal variation: log-normal fading ($\sigma_0 = 6$ dB).
☞ Antenna gain for the terminal: 1 dB.
☞ Antenna gain for the BS: 10 dB.
☞ Cable and connector loss at the terminal: 1 dB.
☞ Cable and connector loss at the BS: 1 dB.
☞ NF of the receiver amplifier at the BS: 3 dB.
☞ Outage probability: 10% and 1%.

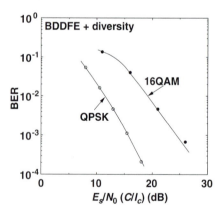

Fig. 15.3 BER performance versus E_s/N_0 (C/I_c) for QPSK and 16QAM, where the maximum delay time of the delayed wave is assumed to be a 6-symbol duration and selective combining is employed as the diversity combining technique.

Table 15.2 C/I_c that satisfies BER = 10^{-2} and BER = 10^{-3} for QPSK and 16QAM.

	BER = 10^{-2}	BER = 10^{-3}
QPSK	12.0 dB	15.6 dB
16QAM	19.0 dB	24.5 dB

First we can obtain the noise power spectral density as –170.8 dBm/Hz because NF = 3 dB. Next, let's obtain the receiver sensitivity for BER = 10^{-2} and BER = 10^{-3} for QPSK and 16QAM using Table 15.2. Table 15.3 shows the results for the symbol rate of 2 Msymbol/s, 1 Msymbol/s, 500 ksymbol/s, and 250 ksymbol/s. Using this table, we can obtain the relationship between the transmitter power and the zone radius. Figure 15.4 shows the relationship between the transmitter power and the zone radius for 2-Msymbol/s QPSK and 16QAM systems, where required BER = 10^{-2}, the antenna height of the BS is 100m and that of the MS is 1.5 m. Table 15.4 summarizes the required transmitter power for a typical zone radius and outage probability. In this table, we have assumed that the antenna height of the BS (h_B) and the carrier frequency (f_c) are 100 m and 900 MHz, respectively. When we employ different values, we can approximate the required transmitter power from Hata's equation as

$$P_T \ (\mathrm{dBm}) = P_{T0} \ (\mathrm{dBm}) - 13.82 \log_{10}(h_B / 100) \ (\mathrm{dB})$$
$$+ 26.16 \log_{10}(f_c / 900) \ (\mathrm{dB}) \tag{15.4}$$

Table 15.3 Receiver sensitivity with parameters of the required BER and symbol rate.

Required BER	Symbol rate	QPSK	16QAM
BER = 10^{-2}	2 Msymbol/s	–95.8 dBm	–88.8 dBm
	1 Msymbol/s	–98.8 dBm	–91.8 dBm
	500 ksymbol/s	–101.8 dBm	–94.8 dBm
	250 ksymbol/s	–104.8 dBm	–97.8 dBm
BER = 10^{-3}	2 Msymbol/s	–92.2 dBm	–83.3 dBm
	1 Msymbol/s	–95.2 dBm	–86.3 dBm
	500 ksymbol/s	–98.2 dBm	–89.3 dBm
	250 ksymbol/s	–101.2 dBm	–92.3 dBm

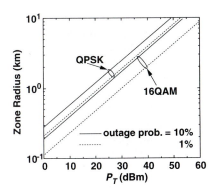

Fig. 15.4 Relationship between the transmitter power and zone radius for 2-Msymbol/s QPSK and 16QAM systems, where required BER = 10^{-2}, antenna height of the BS is 100 m and that of the MS is 1.5 m.

Table 15.4 Relationship between the transmitter power and zone radius for 2-Msymbol/s QPSK and 16QAM systems.

	Transmitter power			
	QPSK		16QAM	
Zone radius	outage = 10%	outage = 1%	outage = 10%	outage = 1%
10 km	48.8 dBm	52.5 dBm	54.0 dBm	62.0 dBm
1 km	17.2 dBm	21.3 dBm	22.5 dBm	30.0 dBm

where P_{T0} is the transmitter power for $h_B = 100$ m and $f_c = 900$ MHz. For example, when we employ $h_B = 30$ m and $f_c = 2$ GHz, we need 16.3 dB more power. In this case, we need 33.5 dBm for QPSK systems when the zone radius is 1 km and the outage probability is 10%.

As for the transmitter power for the terminals, the acceptable transmitter power is up to 1 W at most. From this viewpoint, the zone radius of 1 km is considered to be an appropriate value for the 900-MHz to 2-GHz system with its BS antenna height of 30–100 m. When we need a much larger coverage area, we have to employ a higher gain antenna, higher antenna height, and lower NF for the receiver amplifier of the BS, or we have to improve receiver sensitivity by applying FEC and so on.

Using Table 15.4 and equation (15.4), we will modify system parameters such as the antenna height to meet various constraint conditions. Table 15.5 shows the

Table 15.5 Designed system parameters for small-zone and large-zone systems.

Parameters	Small-zone systems (Zone Radius = 1 km)		Large-zone systems (Zone radius = 10 km)
BS antenna	height = 30 m gain = 10 dB	height = 30 m gain = 13 dB	height = 150 m gain = 16 dB
MS antenna	height = 1.5 m gain = 1 dB	height = 1.5 m gain = 1 dB	height = 1.5 m gain = 1 dB
Tx power (MS)	1 W (30 dBm)	1 W (30 dBm)	1 W (30 dBm)
Receiver NF for BS	4.5 dB	2.0 dB	2.0 dB
Modulation	QPSK	16QAM	QPSK
Symbol rate	2 Msymbol/s	1 Msymbol/s	250 ksymbol/s
Max. delay to be compensated for	3.0 μs (6-symbol duration)	6.0 μs (6-symbol duration)	24 μs (6-symbol duration)
Outage probability	1%	1%	10%

designed system parameters for small-zone systems with a zone radius of 1 km and for large-zone systems with a zone radius of 10 km after modification of system parameters, where the outage probability is defined as 1% for small-zone systems and 10% for large-zone systems. Because 16QAM is applicable in the case of small-zone systems, both QPSK and 16QAM cases are shown in this table.

15.5.2 Frame Format Design

Let's design a system that supports a variable bit rate from 8 kbit/s to 2.048 Mbit/s. When $N_{ca} = 1$, N_{slot} becomes 256 slots. When a plural number of carriers are accepted, N_{slot} becomes 128 for $N_{ca} = 2$ and 64 for $N_{ca} = 4$. When we employ QPSK ($M_l = 4$), a coding rate of $r_c = 3/4$, frame efficiency of $F_{eff} = 0.75$, and the symbol rate of SACCH of $C_1 = 0.1R_s$, R_s is given by 2 Msymbol/s for $N_{ca} = 1$, 1 Msymbol/s for $N_{ca} = 2$, and 500 ksymbol/s for $N_{ca} = 4$. When we employ 16QAM, R_s is given by 1 Msymbol/s for $N_{ca} = 1$, 500 ksymbol/s for $N_{ca} = 2$, and 250 ksymbol/s for $N_{ca} = 4$. With these assumptions, we can design the frame format.

When we design a frame format, the most important factor is the TDMA frame length and the guard time (G). Table 15.6 shows the slot length, the required guard time, and the guard time ratio in each slot (G/T_{slot}) with a parameter of the zone radius. When the zone radius becomes larger or T_f becomes small, the G/T_{slot} is increased. From a practical viewpoint, G/T_{slot} has to be less than 10% when we define the frame efficiency as 70–80%.

When we assume the zone radius of less than 1 km, we can employ the symbol rate of 2 Msymbol/s and T_f of 20–80 ms for both QPSK and 16QAM cases. However, when we assume the zone radius of 10 km, which could sometimes be introduced in suburban and mountainous areas, Table 15.6 suggests that the symbol rate has to be limited to 500 ksymbol/s for QPSK systems or 250 ksymbol/s for 16QAM systems to satisfy the G/T_{slot} of less than 10%.

Table 15.7 shows the applicable frame length and the corresponding G/T_{slot} for small-zone and large-zone systems, where system parameters except for frame format are the same as those shown in Table 15.5.

One way to achieve small G/T_{slot} for a higher symbol rate in the case of $R = 10$ km is to apply a time alignment technique as in the case of PDC systems and digital MCA systems [15-10]. In this case, time alignment is achieved by transmitting a synchronization slot in the uplink just after the radio resource is assigned to each terminal. At the BS, the transmission timing of the synchronization slot is detected, and its timing control signal is transmitted from the BS to the terminal. For example, when time alignment can compensate for a transmission delay time of around 40 μs in a QPSK-system with a frame length of $T_f = 80$ ms and a symbol rate of 2 Msymbol/s, we can reduce G to be less than 30 μs which satisfies $G/T_{slot} < 10\%$.

In our discussion here, we have considered two types of systems because it is very difficult to obtain optimum system parameters that satisfy requirements for both large-zone and small-zone systems. Adaptive modulation techniques provide a solution for such difficulties. We will discuss them in chapter 17.

Table 15.6 Slot length, required guard time , and guard time ratio in each slot with a parameter of zone radius.

factors		QPSK			16QAM		
N_{ca}		1	2	4	1	2	4
R_s (Msymbol/s)		2	1	0.5	1	0.5	0.25
N_{slot}		256	128	64	256	128	64
T_f = 20 ms	T_{slot} (µs)	78.125	156.25	312.5	78.125	156.25	312.5
R = 100 m	G (µs)	0.667	0.667	0.667	0.667	0.667	0.667
	G/T_{slot} (%)	0.854	0.427	0.213	0.854	0.427	0.123
R = 1 km	G (µs)	6.667	6.667	6.667	6.667	6.667	6.667
	G/T_{slot} (%)	8.54	4.27	2.13	8.54	4.27	2.13
R = 10 km	G (µs)	66.67	66.67	66.67	66.67	66.67	66.67
	G/T_{slot} (%)	85.4	42.7	21.3	85.4	42.7	21.3
T_f = 40 ms	T_{slot} (µs)	156.25	312.5	625	156.25	312.5	625
R = 100 m	G (µs)	0.667	0.667	0.667	0.667	0.667	0.667
	G/T_{slot} (%)	0.427	0.213	0.107	0.427	0.213	0.107
R = 1 km	G (µs)	6.667	6.667	6.667	6.67	6.67	6.67
	G/T_{slot} (%)	4.27	2.13	1.07	4.27	2.13	1.07
R = 10 km	G (µs)	66.67	66.67	66.67	66.7	66.7	66.7
	G/T_{slot} (%)	42.7	21.3	10.7	42.7	21.3	10.7
T_f = 80 ms	T_{slot} (µs)	312.5	625	1250	312.5	625	1250
R = 100 m	G (µs)	0.667	0.667	0.667	0.667	0.667	0.667
	G/T_{slot} (%)	0.213	0.107	0.0534	0.213	0.107	0.0534
R = 1 km	G (µs)	6.667	6.667	0.667	0.667	6.667	6.667
	G/T_{slot} (%)	2.13	1.07	0.534	2.13	1.07	0.534
R = 10 km	G (µs)	66.67	66.67	66.67	66.67	66.67	66.67
	G/T_{slot} (%)	21.3	10.7	5.34	21.3	10.7	5.34

Table 15.7 Applicable frame length and the corresponding G/T_{slot} for small-zone and large-zone systems

Parameters	Small-zone systems (Zone radius = 1 km)		Large-zone systems (Zone radius = 10 km)
Modulation	QPSK	16QAM	QPSK
Symbol rate	2 Msymbol/s	1 Msymbol/s	500 ksymbol/s
applicable T_f	$T_f \geq 20$ ms	$T_f \geq 20$ ms	$T_f \geq 40$ ms
G/T_{slot}	8.54%	8.54%	10.7%

REFERENCES

15-1. G.S.M., "Physical layer on the radio-path: G.S.M. system," G.S.M. Recommendation 05.02, Vol. G, G.S.M. Standard Committee, July 1988.

15-2. DECT, "Digital European cordless system—common interface specifications," Code RES-3(89), DECT, 1989.

15-3. EIA/TIA, "Cellular system: Dual-mode mobile station—base station compatibility standard," IS-54, project 2215, Washington, D.C., December 1989.

15-4. RCR, "Personal digital cellular telecommunication system," RCR STD-27, April 1991.

15-5. RCR, "Personal handyphone system," RCR STD-28, December 1993.

15-6. Sampei, S., "Development of Japanese adaptive equalizing technology toward high bit rate data transmission in land mobile communications," IEICE Trans., Vol. E74, No. 6, pp. 1512–21, June 1991.

15-7. Nagayasu, T., Sampei, S. and Kamio, Y., "Complexity reduction and performance improvement of a decision feedback equalizer for 16QAM in land mobile communications," IEEE Trans. Veh. Technol., Vol. 44, No. 3, pp. 570–78, August 1995.

15-8. Ohgane, T., Shimura, T., Matsuzawa, N. and Sasaoka, H., "An implementation of a CMA adaptive array for high speed GMSK transmission in mobile communications," IEEE Trans. Veh. Technol., Vol. VT-42, pp. 282–88, August 1993.

15-9. Chang, C.I., "Effect of time delay spread on portable radio communications," *IEEE Journal of Selected Areas in Communications*, Vol. SAC-5, No. 5, pp. 879–88, June 1987.

15-10. RCR, "Digital MCA system," RCR STD-32, December 1992.

CHAPTER 16

System Design Strategy for CDMA-Based Systems

$\mathbf{I}$n this chapter, we will discuss how to design future wireless communication systems using CDMA technologies. Although the constraint conditions for the CDMA system design are almost the same as those for TDMA systems, the basic approach for CDMA system design is quite different from that for TDMA systems. We will discuss how to cope with the constraint conditions and how to satisfy user requirements.

16.1 BASIC DESIGN CONCEPT OF CDMA SYSTEMS

Figure 16.1 is a flow diagram for designing modulation and spread spectrum parameters for DS/CDMA systems. As in the case of the flow diagram for TDMA systems in chapter 15, we have to satisfy requirements shown on the left side of Figure 16.1. When we design a DS/CDMA system, we can classify the parameters for the physical layer into the source data modulation stage and the direct sequence/spread spectrum (DS/SS) stage. Parameters for the source data modulation stage are determined mainly by the service requirements. On the other hand, parameters for the DS/SS stage are determined mainly by hardware implementation constraints, propagation path conditions, and system requirements. Of course, parameters for both the source data modulation stage and the DS/SS stage are mutually related; therefore, they have to be adjusted after all the parameters are temporarily determined by the verification processes for the coverage area estimation, the spectral efficiency estimation, and the hardware implementation.

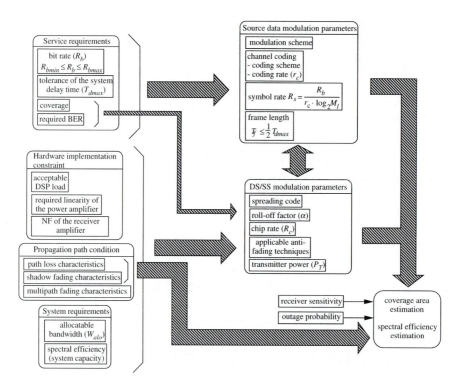

Fig. 16.1 Flow diagram for designing modulation and spread spectrum parameters for DS/CDMA systems.

16.2 PARAMETER DESIGN FOR SOURCE DATA MODULATION AND DS/SS STAGES

16.2.1 Parameter Design for Source Data Modulation Stage

In the source data modulation stage, various parameters (listed below) have to be determined by service requirements.

☞ Modulation scheme

☞ Channel-coding scheme including coding rate

☞ Symbol rate

☞ Frame length

First we have to decide on the source data modulation scheme. In the case of CDMA systems, the system's coverage area and capacity are determined only by the receiver sensitivity (acceptable minimum value of E_b/N_0), where E_b is the energy per information bit. From this viewpoint, BPSK and QPSK are considered to be appropriate modulation schemes for source data modulation because they require the smallest

E_b/N_0 for a certain BER among M-ary PSK and M-ary QAM, as discussed in chapter 3. Therefore, we will assume BPSK as the source data modulation scheme in our discussion.

Selection strategy for the channel-coding scheme for CDMA systems is also quite different from that for TDMA systems. For TDMA systems, a higher coding rate with higher coding gain is preferable to achieve high spectral efficiency as discussed in chapter 15. On the other hand, for CDMA, the bandwidth expansion due to channel coding is not a problem at all provided that

☞ The bandwidth after channel encoding is still much narrower than that after spectrum spreading.

☞ Channel-coding gain is larger than the reduction of the processing gain due to channel coding.

On the contrary, if we can improve receiver sensitivity by introducing convolutional coding with a low coding rate, we can increase system capacity because we can accept more CCI [16-1; 16-2]. Because fast power control is essential for the DS/CDMA to solve the near-far problem, we can define the receiver sensitivity from the BER performance under AWGN conditions. Figure 16.2 shows the BER performance of convolutionally encoded BPSK systems under static conditions, where constraint length is K = 7. In the case of r_c = 2/3 and r_c = 3/4, a puncture code using an r_c = 1/2 encoder is employed (refer to chapter 6). Table 16.1 shows the required E_b/N_0 with parameters of r_c and required BER (BER_{req}). As we can see from this table, a lower coding rate gives

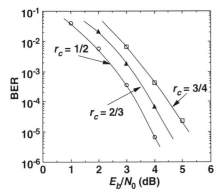

Fig. 16.2 BER performance of convolutionally encoded BPSK systems under static conditions, where constraint length is K = 7.

Table 16.1 Required E_b/N_0 with parameters of r_c and BER_{req}.

BER_{req}	r_c = 1/2	r_c = 2/3	r_c = 3/4
10^{-2}	1.7 dB	2.3 dB	2.8 dB
10^{-3}	2.6 dB	3.2 dB	3.7 dB
10^{-4}	3.4 dB	3.9 dB	4.5 dB
10^{-5}	3.9 dB	4.5 dB	5.2 dB

better receiver sensitivity. Moreover, we find that the BER of 10^{-5} is achieved when the transmitter power is increased by only 2.2 dB from the power for BER = 10^{-2} in the case of $r_c = 1/2$. This means that a lower coding rate is very effective in achieving a high-quality and high-capacity system without much increase in the transmitter power.

Next we have to decide the symbol rate for the source data modulation. Let's assume that the information bit rate is R_I (bit/s), the bit rate for SACCH is $C_1 \cdot R_I$ (bit/s), the coding rate for channel coding is r_c, and the modulation level of the source data is M_l ($M_l = 2$ in the case of BPSK). In this case, we can obtain symbol rate as

$$R_s = \frac{R_I(1+C_1)}{r_c \cdot \log_2 M_l} = \frac{R_b}{r_c \cdot \log_2 M_l} \tag{16.1}$$

where

$$R_b = R_I(1+C_1) \tag{16.2}$$

When R_I = 8 kbit/s, C_1 = 0.2, r_c = 1/2, and M_l = 2, R_b and R_s are given by 9.6 kbit/s and 19.2 ksymbol/s.

The frame length is determined by the following factors:

☞ Frame length of the voice codec (T_{finf})
☞ Update period of the closed-loop power control (T_{pwr})

In most of the voice codec, T_{finf} is 20 ms. On the other hand, T_{pwr} should be around 1 ms or less to compensate for the fast-varying-received signal level. Moreover, we can select a shorter frame length than that for TDMA systems because CDMA does not require any guard time. Therefore, frame length for CDMA systems is determined by the power control updating period, which is around 1 ms.

16.2.2 Parameter Design for DS/SS Stage

The most important parameter for the DS/SS stage is the selection of a spreading code, which is based on the following requirements:

☞ Cross-correlation between two arbitrary spreading codes should be as small as possible to improve signal discrimination performance of CDMA signals.
☞ The number of codes should be as high as possible to achieve high system capacity.
☞ Spreading code should have as much privacy as possible.

One way to achieve low cross-correlation is to employ a PN sequence with relatively a long period (it is usually called a *long code*) because cross-correlation for a long code is relatively low [16-3]. This method is especially effective when the desired signal and interference signals are not synchronized with each other—for example, in the uplink channels. In the IS-95 system, a 42-stage PN is employed in the uplink [16-4]. When the desired signal and CCI signals are perfectly synchronized with each other as in the case of downlink channels, we can achieve orthogonal spreading codes by concatenating a PN code with an orthogonal code, such as the Walsh code. For example, in the downlink of IS-95, 64 spreading codes are prepared by the concatenation of a 15-stage PN code and 64-symbol Walsh functions [16-4]. Another advantage of the long code is that it can prepare a large amount of code. Furthermore, when we employ a long code, it is also very effective in achieving high security.

On the other hand, one disadvantage of the long code is that it requires very a precise time base because it is almost impossible to take self-synchronization with a long code. In the case of IS-95, all the BSs are synchronized with each other, and each of them broadcasts timing information to the terminal via the synchronization channels in the downlink [16-4].

The modulation scheme for DS/SS is also an important parameter for DS/CDMA systems. When we select the modulation scheme, we have to consider the following:

☞ It is desirable that the phase of the DS/CDMA signal is uniformly distributed on the phase plane as much as possible to guarantee phase randomness of the CCI signal.

☞ A modulation scheme with smaller envelope fluctuation is preferable, especially for the terminals, for achieving high power efficiency, thereby saving the battery.

To satisfy the first requirement, a higher-level phase modulation is preferable. On the other hand, to satisfy the second requirement, OQPSK, FQPSK including SQAM, and any other constant envelope modulation schemes are preferable [16-5]. For these reasons, IS-95 employs OQPSK in the uplink.

The next parameter is the transmitter and receiver filters. When we select a linear modulation scheme for the DS/SS process, a root Nyquist filter with a small roll-off factor of 0.2, for example, is preferable because it can reduce the occupied bandwidth. Although smaller α could produce intersymbol interference due to the timing jitter, it can be compensated for when we employ a Rake receiver.

16.2.3 Selection of the Chip Rate

The maximum chip rate is limited by the allocatable bandwidth. Moreover, a higher chip rate also requires high-speed digital signal processing including high-speed A/D converters. However, a higher chip rate is preferable because the ability to resolve delayed paths is improved by increasing the chip rate [16-6]. A higher chip rate is also preferable because processing gain of at least 10 is needed to sufficiently resolve each multipath component at the Rake receiver. As a result, the chip rate is upper limited by the spectrum allocation policy and the restriction of the hardware implementation; it is lower limited by the bit rate of the source information and the required ability of the Rake receiver.

16.2.4 Variable-Bit-Rate Transmission Capability

When a variable-bit-rate transmission is supported in the DS/CDMA system, only the bit rate for the source information is changed while the chip rate is kept constant regardless of the source bit rate. In other words, the processing gain is changed according to the source bit rate. Of course, the controlled received signal level should be changed according to the supported bit rate and the required BER performance.

When the maximum bit rate is limited to lower than the maximum bit rate to be serviced, we can employ a multicode transmission in which two or more DS/CDMA TCHs are assigned to a terminal. In this case, however, higher linearity is required for the transmitter amplifier.

16.2.5 Selection of Anti-Frequency-Selective Fading Techniques

The Rake receiver is a very effective anti-frequency-selective fading technique for DS/CDMA wireless communication systems, as discussed in chapter 5. There are two types of the Rake receiver—the coherent detection type, such as the pilot channel-aided scheme [16-7], and the differential detection type, such as the DPSK/PDI receiver [16-8; 16-9]. In the downlink DS/CDMA channels, we can easily achieve coherent Rake diversity by using a pilot channel multiplexed with TCHs because a pilot channel can be shared by all the TCHs in the downlink.

In the uplink, such a pilot channel has to be individually associated with each TCH. To make such pilot signal-assisted coherent Rake diversity techniques feasible in the uplink, there are some proposals, such as the suppressed pilot channel scheme [16-7] and the pilot symbol-aided scheme [16-10]. Because these proposals can reduce the power ratio of the pilot signal power to the TCH power, they can achieve high-capacity coherent Rake diversity combining with small power redundancy.

Although the Rake receiver is very effective for improving receiver sensitivity, it is not helpful if there are no delayed paths. Such situation could occur when the assigned bandwidth is not wide enough. Several techniques have been proposed to solve this problem, such as space diversity [16-11], antenna array [16-12], time diversity [16-13], and symbol splitting [16-14].

16.2.6 Soft Handover Parameters

When we introduce a soft handover technique [16-15], it is very important to decide the signal margin that represents the threshold level difference between the add threshold and drop threshold [16-16] as shown in Figure 7.17. When we increase the signal margin (γ_m), the soft handover region is increased, and, as a result, transmission quality of a terminal at the fringe area can be improved owing to the macro-diversity effect as we discussed in chapter 4. However, a larger margin degrades the downlink capacity because a terminal occupies several radio channels to transmit only one voice channel.

Figure 16.3 shows the soft handover area ratio versus γ_m performance for a DS/CDMA system. As shown in this figure, the soft handover area ratio is 25–35% when γ_m = 4–6 dB, which suggests that the downlink capacity is reduced by about 15%. Therefore, γ_m is determined by the tradeoff between the uplink capacity improvement because of the macro-diversity effect and the downlink capacity reduction due to plural channel occupancy by a terminal in the soft handover area. In the IS-95 system, γ_m of around 5 dB is employed.

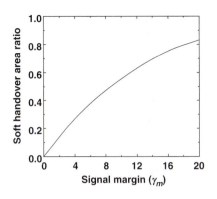

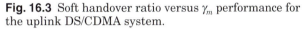

Fig. 16.3 Soft handover ratio versus γ_m performance for the uplink DS/CDMA system.

16.3 COVERAGE AREA ESTIMATION FOR DS/CDMA SYSTEMS

When we evaluate the coverage area for DS/CDMA systems, we have to take the effect of soft handover into account. Because the average path loss is considered to be almost the same in the downlink and uplink, we will discuss the estimation of the coverage area in the case of the downlink.

Even in CDMA systems, the maximum transmitter power is limited to a certain level. Therefore, when a terminal is located at the fringe of the covered area, we cannot compensate for the path loss, shadowing, or multipath fading by the power control techniques. As for multipath fading, the BER performance is degraded by 2–3 dB from the performance under AWGN conditions when we combine several paths using a Rake receiver, even if fast power control does not work at all. Therefore, we will assume that the required E_b/N_0 is a value 3 dB higher than the theoretical one.

Let's assume that the distance between the BS and a terminal is r, and the long-term average value of the received signal level at the zone fringe ($r = R$) is defined as $\Gamma_m(R)$, where R is the zone radius. With this assumption, the long-term average received signal level normalized by $\Gamma_m(R)$ at a distance r is given by

$$\Gamma_m(r)/\Gamma_m(R)\ [\text{dB}] = -10\alpha_p \log_{10}(r/R)\ [\text{dB}] \tag{16.3}$$

where α_p is the path loss decay factor. Moreover, let's assume that the large-scale signal variation is subject to log-normal fading with its standard deviation of σ_0 (dB). When we receive a desired signal from several BSs, the large-scale signal variation from each BS is independent from the others when the cause of the shadowing is located far away from the terminal. Conversely, when the cause of the shadowing is located near the terminal, the large-scale signal variation from each BS tends to be correlated with the other BSs. One shadowing model that accommodates such shadowing correlation was proposed by Viterbi et al. [16-17]. In the proposed model, the short-term signal variation from i-th BS (ζ_i) is given by

$$\zeta_i = a\xi + b\xi_i \tag{16.4}$$

$$a^2 + b^2 = 1 \tag{16.5}$$

where ξ is the large-scale signal variation that is common to all the BSs, ξ_i is the variation specific for i-th BS, and a^2/b^2 represents the power ratio of the correlated and uncorrelated components. Without loss of generality, we can assume that

$$E(\zeta_i) = E(\xi) = E(\xi_i) = 0 \text{ for all } i \tag{16.6}$$

$$E(\zeta_i^2) = E(\xi^2) = E(\xi_i^2) = \sigma_0^2 \text{ for all } i \tag{16.7}$$

$$E(\xi_i \xi_j) = 0 \text{ for all } i \text{ and } j, i \neq j \tag{16.8}$$

Let's assume that the BS 0 and BS 1 are connected to a terminal for soft handover and r_i ($i = 0, 1$) is the distance between the terminal and i-th BS. The average received signal level from i-th BS is then given by

$$\Gamma(r_i, \zeta_i) \text{ [dB]} = -10\alpha_p \log_{10}(r_i / R) + \zeta_i \text{ [dB]} \tag{16.9}$$

When we employ a soft handover technique, the received signal level for a terminal located at a distance of r_0 from BS 0 and at a distance of r_1 from BS 1 is given by

$$
\begin{aligned}
&\Gamma(r_0, r_1) / \Gamma_m(R) \\
&= \text{Max}\left[-10\alpha_p \log_{10}(r_0 / R) + \zeta_0, -10\alpha_p \log_{10}(r_1 / R) + \zeta_1\right]
\end{aligned}
\tag{16.10}
$$

Therefore, cumulative distribution of the received signal level for a terminal located at the distance from the BS 0 of r_0 and that from the base station 1 of r_1 is given by

$$
\begin{aligned}
&P_{r_0, r_1}(X) \\
&= \text{Pr}\left(\Gamma(r_0, r_1) / \Gamma_m(R) \le X\right) \\
&= \text{Pr}\left[\text{Max}\{-10\alpha_p \log(r_0 / R) + a\xi + b\xi_0, \right. \\
&\qquad\qquad \left. -10\alpha_p \log(r_1 / R) + a\xi + b\xi_1\} \le X\right] \\
&= \frac{1}{(2\pi\sigma_0)^{\frac{3}{2}}} \int_{-\infty}^{\infty} e^{-\xi^2/(2\sigma_0^2)} d\xi \cdot \int_{-\infty}^{\frac{X + 10\alpha_p \log_{10}(r_0 / R) - a\xi}{b}} e^{-\xi_0^2/(2\sigma_0^2)} d\xi_0 \cdot \int_{-\infty}^{\frac{X + 10\alpha_p \log_{10}(r_1 / R) - a\xi}{b}} e^{-\xi_1^2/(2\sigma_0^2)} d\xi_1 \\
&= \frac{1}{\sqrt{2\pi}} \int_{-\infty}^{\infty} e^{-x^2/2} \left[1 - \frac{1}{2}\text{erfc}\left(\frac{X + 10\alpha_p \log_{10}(r_0 / R) - a\sigma_0 x}{\sqrt{2}\sigma_0 b}\right)\right] \\
&\qquad \cdot \left[1 - \frac{1}{2}\text{erfc}\left(\frac{X + 10\alpha_p \log_{10}(r_1 / R) - a\sigma_0 x}{\sqrt{2}\sigma_0 b}\right)\right] dx
\end{aligned}
\tag{16.11}
$$

The cumulative distribution of the received signal level for an area is then obtained by integrating the probability given by equation (16.12) for all of the terminal locations (area A) as

$$P_{a0}(X) = \int_A P_{r_0, r_1}(X)dA \qquad (16.12)$$

In actual cellular systems, the soft handover is achieved by receiving up to the three nearest BSs. Therefore, when we evaluate the cumulative distribution of the received signal level in a cell, we have to consider seven BSs in total. In this case, equation (16.11) is modified as [16-17]

$$P_{r_0,r_1}(X)$$

$$= \frac{1}{\sqrt{2\pi}} \int_{-\infty}^{\infty} e^{-x^2/2} \prod_{i=0}^{6} \left[1 - \frac{1}{2} \mathrm{erfc}\left(\frac{X + 10\alpha_p \log_{10}(r_i / R) - a\sigma_0 x}{\sqrt{2}\sigma_0 b} \right) \right] dx \qquad (16.13)$$

Figure 16.4 shows the cumulative distribution of the normalized received signal level, where it is assumed that $a = 0$ and $b = 1$(uncorrelated case). As shown in this figure, when the handover is not employed, the normalized received signal level for the outage probability of 10% is −7.7 dB. This means that 7.7 dB of power margin is necessary to guarantee an outage probability of 10%. On the other hand, when we employ soft handover, the received signal level for the 10% outage probability becomes 3.0 dB. In this case, the required power margin is −3.0 dB. In the same manner, we can obtain the required power margin for the 1% outage probability as shown in Table 16.2.

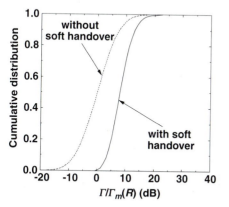

Fig. 16.4 Cumulative distribution of the normalized received signal level, where it is assumed that $a = 0$ and $b = 1$ (uncorrelated case).

Table 16.2 Required power margin for the outage probability of 10% and 1%.

Outage Probability	Without Soft Handover	With Soft Handover
10%	7.7 dB	−3.0 dB
1%	13.8 dB	−0.2 dB

Using these results, let's obtain the relationship between the transmitter power and zone radius for CDMA systems in the uplink. In this evaluation, we will assume the following conditions:

☞ Path loss: Hata's equation
☞ Carrier frequency: 900 MHz
☞ Large-scale signal variation: log-normal fading (σ_0 = 6 dB)
☞ Antenna gain for the terminal: 1 dB
☞ Antenna gain for the BS: 10 dB
☞ Cable and connector loss at the terminal: 1 dB
☞ Cable and connector loss at the BS: 1 dB
☞ NF of the receiver amplifier at the BS: 3 dB
☞ Outage probability: 10% and 1%

First we have to calculate the receiver sensitivity. As shown in Table 16.1, the required E_b/N_0 for BER = 10^{-2} and BER = 10^{-3} are given by 1.7 dB and 2.6 dB, respectively, under AWGN conditions. When we add a 3-dB margin, considering degradation due to imperfection of power control and some other devices, and completely disregard the constraint conditions for hardware implementation and limitation of the allocatable bandwidth, we can obtain the receiver sensitivity for BER = 10^{-2} and BER = 10^{-3} as follows.

Table 16.3 Receiver sensitivity with parameters of the required BER and bit rate.

Required BER	Bit Rate	Receiver Sensitivity
BER = 10^{-2}	2.048 Mbit/s	−103.0 dBm
	256 kbit/s	−112.0 dBm
	64 kbit/s	−118.0 dBm
	8 kbit/s	−127.1 dBm
BER = 10^{-3}	2.048 Mbit/s	−102.1 dBm
	256 kbit/s	−111.1 dBm
	64 kbit/s	−117.1 dBm
	8 kbit/s	−126.2 dBm

Figure 16.5 shows the relationship between transmitter power and zone radius for 2-Mbit/s transmission. As shown in this figure, the performance difference between BER = 10^{-2} and BER = 10^{-3} is very small due to convolutional encoding. Moreover, the performance difference between the outage performances of 10% and 1% is also small because of the macro-diversity effect of the soft handover. Using these results, we can obtain the required transmitter power for the zone radii of 1 km and 10 km as shown in Table 16.4, where the bit rate is 2 Mbit/s.

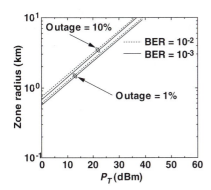

Fig. 16.5 Relationship between transmitter power and zone radius for 2-Mbit/s transmission.

Table 16.4 Relationship between transmitter power and zone radius for 2 Mbit/s CDMA systems.

Zone radius	Outage = 10%		Outage = 1%	
	BER = 10^{-2}	BER = 10^{-3}	BER = 10^{-2}	BER = 10^{-3}
10 km	36.0 dBm	36.9 dBm	38.8 dBm	39.7 dBm
1 km	4.2 dBm	5.1 dBm	7.0 dBm	7.9 dBm

When we compare the results shown in Table 15.4 for TDMA systems, we find that CDMA can achieve the same zone radius with lower transmitter power than TDMA systems. However, we have to bear in mind that this difference is caused not only by introducing CDMA instead of TDMA, but also by introducing some other techniques such as the low-coding-rate channel coding, fast power control, and soft handover techniques. The fast power control is a particularly important technique for improving receiver sensitivity because it equivalently changes dynamically varying fading channels to additive white Gaussian channels.

In Table 16.4, we have assumed that the antenna height of the BS (h_B) is 100 m, the carrier frequency (f_c) is 900 MHz, and the bit rate is (R_b) 2,000 kbit/s. When we employ different values, we can approximate the required transmitter power as

$$P_T \ (\text{dBm}) = P_{T0} \ (\text{dBm}) - 13.82 \log_{10}(h_B \ / \ 100)$$
$$+ 26.16 \log_{10}(f_c \ / \ 900) + 10 \log_{10}(R_b \ / \ 2000) \tag{16.14}$$

as in the case of TDMA system design. When we employ $h_B = 10$m, $f_c = 2$ GHz, and $R_b = 8$ kbit/s, we can reduce the transmitter power by 28.7 dB.

16.4 SYSTEM CAPACITY FOR CDMA SYSTEMS

16.4.1 Estimation of C/I_c Performance

When we evaluate system capacity for DS/CDMA systems, we have to estimate receiver sensitivity as well as cumulative distribution of the C/I_c. Because we have already discussed receiver sensitivity, we will mainly discuss how to estimate cumulative distribution of C/I_c here. Although CCI conditions in the downlink and uplink are different, we will discuss only the uplink channel capacity here because system capacity is limited mainly by the uplink channel [16-17; 16-18; 16-19].

When no power control is employed in DS/CDMA systems, the received signal power of the desired signal is given by

$$P_{Rd} = A \cdot P_T \cdot r_d^{-\alpha_p} \cdot 10^{\frac{\zeta_d}{10}} \cdot R_d \qquad (16.15)$$

and the received signal power of i-th CCI received at the connected BS is given by

$$P_{Ri} = A \cdot P_T \cdot r_i^{-\alpha_p} \cdot 10^{\frac{\zeta_i}{10}} \cdot R_i \qquad (16.16)$$

where A is a constant, P_T is the transmitted power; r_d is the distance between a subscriber and the receiving BS; ζ_d is the log-normal fading of the desired signal with its standard deviation of σ_0; and R_d and R_i represent Rayleigh fading. When the received power of this signal is perfectly controlled by a power control technique, the power-controlled received signal level is given by

$$P_{RCd} = A \cdot P_T \qquad (16.17)$$

Intracell interference signals are also controlled to the same received signal level at the BS.

However, in the case of intercell interference, because the transmitter power of the CCI is controlled by the connected BS, the transmitter power of a terminal in i-th BS is given by

$$P_{Ti} = P_T \cdot r_i^{\alpha_p} \cdot 10^{\frac{-\zeta_i}{10}} \cdot R_i^{-1} \qquad (16.18)$$

As a result, the intercell interference signal level is given by

$$\begin{aligned} P_{Ri} &= A \cdot P_{Ti} \cdot r_{di}^{-\alpha_p} \cdot 10^{\frac{\zeta_{di}}{10}} R_{di} \\ &= A \cdot P_T \cdot r_i^{\alpha_p} \cdot 10^{\frac{-\zeta_i}{10}} \cdot R_i^{-1} \cdot r_{di}^{-\alpha_p} \cdot 10^{\frac{\zeta_{di}}{10}} R_{di} \end{aligned} \qquad (16.19)$$

where R_{di} represents the envelope variation of i-th interference signal, and the cumulative distribution of C/I_c can be obtained as follows:

Terminal location setup process

Step 1. Determine the number of terminals in each cell (N_{ter}).

Step 2. Randomly select terminal locations for all the terminals in each cell, where p.d.f. is subject to uniform distribution.

Power control for terminals in the desired cell

Step 3. Calculate r_d, generate random variables of ζ_d and R_d, and obtain the power-controlled signal for all the terminals in the desired cell.

Power control for terminals in the interference cells

Step 4. Calculate r_i, generate random variables of ζ_i and R_i, and obtain the transmitter power P_{Ti}. Then calculate r_{di}, generate random variables of ζ_{di} and R_{di}, and obtain the intercell interference signal level for a terminal in i-th cell.

Step 5. Repeat Step 4 for all the terminals in i-th BS as well as the other cells.

C/I_c calculation

Step 6. Calculate the interference signal level by summing up all the interference signals, and obtain C/I_c.

Log-normal fading and Rayleigh fading simulation

Step 7. Repeat Step 3 through Step 6 N_{fade} times to simulate randomness due to log-normal fading and Rayleigh fading, where N_{fade} is a sufficiently large number.

Terminal location randomization

Step 8. Go back to Step 1 to change locations for all the terminals, and repeat Step 1 through Step 7 N_{loc} times to simulate terminal location randomness, where N_{loc} is a sufficiently large number.

Calculation of the cumulative distribution of C/I_c

Step 9. Calculate cumulative distribution of C/I_c.

Figure 16.6 shows the cumulative distribution of C/I_c both with and without soft handover, where $N_{ter} = 10$ and a 4-dB signal margin for handover is employed. We can find from this figure that the probability for small C/I_c is drastically reduced in the case of soft handover owing to the macro-diversity effect.

16.4.2 Estimation of System Capacity

One criterion for system capacity estimation is the outage probability defined as the probability that C/I_c is lower than a C/I_c value that satisfies the required BER (BER$_{req}$) as

$$P_{outage}(N_{ter}) = \Pr\left(C/I_c \leq (C/I_c)_{BER=BER_{req}} \right) \tag{16.20}$$

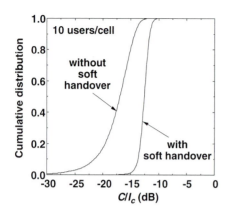

Fig. 16.6 Cumulative distribution of C/I_c both with and without soft handover, where $N_{ter} = 10$ and a 4-dB signal margin for handover is employed.

As we discussed previously, BER = 10^{-3} is achieved by E_b/N_0 = 2.6 dB. When we employ the processing gain of G_p = 21 dB (= 128), the required C/I_c is given by C/I_c = –18.4 dB. From Figure 16.6, we can find that the probability that $C/I_c \leq$ –18.4 dB is 30.7% for the nonsoft handover system and 0.07% for the soft handover system. In the same manner, we can obtain the cumulative distribution of C/I_c with a parameter of N_{ter} as shown in Figure 16.7. When we define the BER_{req} as 10^{-2}, 10^{-3}, 10^{-4}, and 10^{-5}; the coding rate for the convolutional encoder of 1/2; and G_p = 21 dB; using Table 16.1 we can obtain the required C/I_c for each BER_{req} as –19.3 dB, –18.4 dB, –17.6 dB, and –17.1 dB, respectively. Using these results and Figure 16.7, we can obtain the outage probability versus N_{ter} for a system using a soft handover as shown in Figure 16.8. From Figure 16.8, N_{ter} for the outage probability of 1% is given as shown in Table 16.5.

These numbers are obtained based on the assumption that each zone employs an omnidirectional antenna and voice activation is not employed. When we employ N_{sec}-sector cell layout and voice activation with its factor of v_{act}, we can improve system capacity by a factor of approximately $N_{sec} \times v_{act}$ [16-15].

In our discussion, we have assumed perfect operation of the power control. In practice, however, the power control does not operate perfectly due to the limitation of the tracking ability to channel variation and estimation error of the received signal level. Several papers reported that a power control error of 1–2 dB results in a capac-

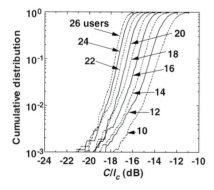

Fig. 16.7 Cumulative distribution of C/I_c with a parameter of N_{ter}, where soft handover is employed.

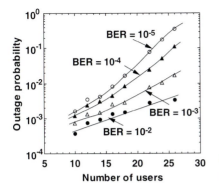

Fig. 16.8 Outage probability versus N_{ter} with a parameter of the required BER.

Table 16.5 N_{ter} versus the outage probability of 1% for each required BER.

Required BER	Number of Terminals for 1% Outage
10^{-2}	31
10^{-3}	23.5
10^{-4}	18.7
10^{-5}	16.8

ity reduction of about 20–50% [16-20; 16-21]. Therefore, we have to prepare a 20–50% margin when designing CDMA systems.

REFERENCES

16-1. Viterbi, A. J., "Very low rate convolutional codes for maximum theoretical performance of spread-spectrum multiple-access channels," *IEEE Journal on Selected Areas in Communication*, Vol. 8, No. 4, pp. 641–49, May 1990.

16-2. Monogioudis, P., Tafazolli, R. and Evans, B. G., "Multirate 3rd generation CDMA systems," IEEE ICC'93 (Geneva, Switzerland), pp. 151–55, May 1993.

16-3. Simon, M. K., Omura, J. K., Scholtz, R. A. and Levitt, B. K., *Spread spectrum communications*, Computer Science Press Inc., Rockville, Maryland, 1985.

16-4. TIA/EIA, "Mobile station-base station compatibility standard for dual mode wideband spread spectrum cellular system," TIA/EIA IS-95, 1993.

16-5. Wan Z. and Feher, K., "Improved efficiency CDMA by constant envelope SQAM," 42d IEEE Veh. Tech. Conf. (Denver, Colorado), pp. 51–54, May 1992.

16-6. Allpress, S. A., Beach, M. A. and McGeehan, J. P., "On the optimum DS-CDMA channel bandwidth for personal communication systems," 43d IEEE Veh. Tech. Conf. (Secaucus, New Jersey), pp. 436–39, May 1993.

16-7. Abeta, S., Sampei, S. and Morinaga, N., "DS/CDMA coherent detection system with a suppressed pilot channel," IEEE GLOBECOM'94 (San Francisco, California), pp. 1622–26, November 1994.

16-8. Sanada, Y., Kajiwara, A. and Nakagawa, M., "Adaptive RAKE receiver for mobile communications," IEICE Trans., Vol. E76-B, No. 8, pp. 1002–7, August 1993.

16-9. Higashi, A. and Matsumoto, T., "BER performance of adaptive RAKE diversity (ARD) in DPSK DS/CDMA mobile radio," IEICE Technical Report, SST92-16, June 1992.

16-10. Ohno, K., Sawahashi, M. and Adachi F., "Wideband coherent DS-CDMA," IEEE VTC'95 (Chicago, Illinois), pp. 779–83, August 1995.

16-11. Ishikawa, H. and Kobayashi, H., "A novel selection diversity method with decision feedback equalizer," IEICE Trans. Commun., Vol. E77-B, No. 5, pp. 566–72, May 1994.

16-12. Xia, H. H., Herrera, A. B., Kim, S. and Rico, F. S., "A CDMA-distributed antenna system for in-building personal communications services," *IEEE Journal on Selected Areas in Communication*, Vol. 14, No. 4, pp. 644–50, May 1996.

16-13. Kubota, S., Kato, S. and Feher, K., "A time diversity CDMA scheme employing orthogonal modulation for time varying channels," 43d IEEE Veh. Tech. Conf. (Secaucus, New Jersey), pp. 444–7, May 1993.

16-14. Meyer, M., "Improvement of DS-CDMA mobile communications systems by symbol splitting," 45th IEEE Veh. Tech. Conf. (Chicago, Illinois), pp. 689–693, July 1995.

16-15. Gilhousen, K. S., Jacobs, J. M., Padovani, R., Viterbi, A. J., Weaver, L. A. and Wheatly, C. E., III, "On the capacity of a cellular CDMA system," IEEE Trans. Veh. Tech., Vol. 40, No. 2, pp. 303–12, May 1991.

16-16. Hong, D. and Moon, Y., "Performance analysis of the soft handover technique for the CDMA cellular systems," 2d Asia-Pacific Conf. on Commun., pp. 336–40, June 1995.

16-17. Viterbi, A. J., Viterbi, A. M., Gilhousen, K. S. and Zehavi, E., "Soft handover extends CDMA cell coverage and increases reverse link capacity," *IEEE Journal on Selected Areas in Communication*, Vol. 12, No. 8, pp. 1281–88, October 1994.

16-18. Viterbi, A. M. and Viterbi, A. J., "Erlang capacity of a power controlled CDMA system," *IEEE Journal on Selected Areas in Communication*, Vol. 11, No. 6, pp. 892–99, August 1993.

16-19. Abeta, S., Sampei, S. and Morinaga, N., "Adaptive coding rate and processing gain control for cellular DS/CDMA systems," 4th IEEE ICUPC (Tokyo, Japan), pp. 241–45, November 1995.

16-20. Cameron, R. and Woerner, B. D., "An analysis of CDMA with imperfect power control," 42d IEEE Veh. Tech. Conf. (Colorado, Denver), pp. 977–80, May 1992.

16-21. Falciasecca, G., Gaiani, E., Missiroli, M., Muratore, F., Palestini,V. and Riva, G., "Influence of propagation parameters on cellular CDMA capacity and effects of imperfect power control," IEEE 2d Int. Symp. on Spread Spectrum Tech. and Appl. (Yokohama, Japan), November 1992.

CHAPTER 17

Flexible Radio Interface Design Strategy for Future Wireless Multimedia Communication Systems

$\mathbf{A}$s discussed in previous chapters, very extensive studies have been made on the development of digital wireless communication systems in the past 10 years. In the beginning of these studies, digital mobile communication technologies were developed mainly to increase system capacity for voice transmission. However, this now changing. During the past couple of years, demand for multimedia transmission systems rather than the dedicated voice transmission systems has rapidly increased even in wireless communication systems because of the rapid growth of multimedia communications via the Internet as well as the emerging personal digital assistance (PDA) terminals. As a result, digital wireless communication systems are now changing from voice communication systems to wireless multimedia communication systems.

To achieve such wireless multimedia communication services, however, conventional radio link design strategies as mentioned in chapters 15 and 16 are insufficient because they are not flexible against dynamically changing traffic, QoS requirements, and channel conditions peculiar to wireless multimedia communication systems. In other words, both the radio link and its network should be more flexible and intelligent to cope with such requirements.

In this chapter, we will first discuss how to offer flexibility of the radio link between the BS and terminals—the adaptive modulation techniques. This chapter also introduces the concept of *radio highway network* which offers a flexible network extension and an efficient utilization of the existing infrastructure of optical fiber networks and can flexibly cope with the modification of the radio link interface.

17.1 REQUIREMENTS FOR WIRELESS PERSONAL COMMUNICATION SYSTEMS

Figure 17.1 shows differences in system requirements using as an example large cities and small cities. In a large city like Tokyo or New York, achieving very high capacity using microcell or picocell technologies is the most important issue because traffic density is very high. In a small city, on the other hand, higher delay spread immunity is more important than higher capacity because the zone radius in a small city is relatively large and the traffic density is relatively low.

In conventional systems, because the modulation parameters are fixed at any time, they are selected to satisfy the regulated quality with a certain outage probability (P_{out}), say 10%. For example, when we specify the symbol rate for a TDMA system, we determine it considering the maximum delay spread in the service area. As a result, this design concept prevents high-bit-rate transmission capability in large cities, although the delay spread is very small and demand for such high-bit-rate service is especially high in such cities. Although Figure 17.1 takes into consideration only the voice traffic, if multimedia services are going to be introduced, traffic and the required QoS will also dynamically change with respect to time, space, and the type of information. In such cases, the conventional radio communication system design strategy, namely fixed parameter design strategy, is no longer effective. On the contrary, we have to introduce more intelligent functions, such as adaptive radio resource management and adaptive radio transmission techniques, and intelligent network management functions to radio communication systems.

17.2 CONCEPT OF INTELLIGENT RADIO COMMUNICATION TECHNIQUES

When channel conditions, traffic distribution, and the type of information are dynamically changing with respect to space and time, radio transmission subsystems should have more flexible and intelligent functions

☞ To mitigate dynamic variation of traffic

☞ To control QoS according to the type of information

City size	Large ←→	Small
Traffic density	High ←→	Low
Cell size	Micro/picocell ←→	Macrocell
Delay spread	Small ←→	Large
Required spectral efficiency	High ←→	Low
Requirement for high-bit-rate services	High ←→	Medium

Fig. 17.1 Differences in requirements for radio communication systems in large cities and small cities.

☞ To mitigate channel variation due to fading

☞ To increase system capacity

Fortunately, at present, system design engineers can apply most of the modulation techniques, such as GMSK, π/4-QPSK, and 16QAM, and access techniques, such as TDMA and CDMA, to wireless communication systems due to recent emerging wireless technologies as we have discussed in the previous chapters. Therefore, the key is how to effectively combine these techniques to construct intelligent radio resource management and radio transmission techniques to satisfy the aforementioned four requirements.

Figure 17.2 shows the classification and effects of the key techniques for intelligent radio communication systems [17-1]. These techniques are roughly classified into adaptive radio *resource management* and adaptive radio *transmission technologies*. Adaptive radio resource management technologies include DCA and slow adaptive modulation techniques. Adaptive radio transmission technologies include transmission power control, fast adaptive modulation, and adaptive zone configuration techniques. We will discuss some of these techniques in this chapter.

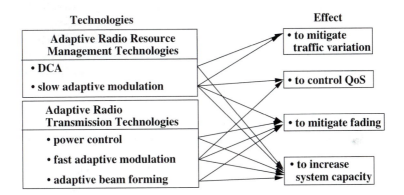

Fig. 17.2 Classification and effects of the key techniques for intelligent radio communication systems (from Ref. 17-1, © Institute of Electronics, Information and Communication Engineers, 1996).

17.3 TRANSMISSION POWER CONTROL TECHNIQUES

Transmission power control is a very basic technique to compensate for received signal loss due to fading. Therefore, a lot of studies have been done on its application to various radio communication systems. For two-way duplex high-frequency (HF) radio links in which fading is caused by ionospheric or tropospheric scatters, a power control technique was proposed by Hayes [17-2]. For satellite communication systems, a power control technique to compensate for the rain attenuation was investigated in the experiments of the communication satellite of Japan called *Sakura* using a 32/20-GHz band [17-3]. Because channel variation is very slow in HF radio links and satellite communication systems, and because they are power-limited systems rather than interference-limited systems, these transmitter power control techniques were shown to be very effective for improving BER performance.

When transmission power control techniques are applied to interference-limited systems like cellular systems, performance is determined not only by the receiver sensitivity improvement due to the constancy of the received signal level, but also by the increase in the CCI due to short-duration peaks in transmission power needed for compensation of deep fades. Ariyavisitakul investigated the trade-off of these two factors for TDMA systems [17-4], showing that, although the transmitter power control can reduce the required carrier-to-CCI power ratio (C/I_c) by 7–8 dB, the average C/I_c increases by about 4 dB, thereby resulting in an overall power control gain of 3–4 dB.

Although power control is an optional technique for the HF radio link, satellite communication systems, or TDMA cellular systems, it is essential for DS/CDMA cellular systems because of the near-far problem (discussed in chapter 7). Therefore, a lot of studies have been done on power control techniques, such as the effect of the imperfect power control [17-5], the C/I_c-based power control technique [17-6], and the TDD-type power control technique [17-7]. Even in CDMA systems, increase of the average C/I_c due to short-duration peaks in transmission power is still a problem as in the case of TDMA systems. As a solution to prevent short-duration peaks of the interference signal in the DS/CDMA system, Abeta et al. [17-8] have proposed a soft power control technique combined with the coding rate and processing gain control. Extensive research has been done on these power control techniques in an effort to increase CDMA system capacity.

17.4 Concept of Adaptive Modulation Systems

The most primitive adaptive modulation was first proposed by Cavers in 1972 as an alternative to the power control technique in Rayleigh-faded HF links [17-9]. In this technique, symbol rate is adaptively controlled according to the received signal level, and its performance is theoretically analyzed including the effect of feedback delay as well as the effect of noise in the feedback channel [17-9]. One of its important advantages is that it does not produce any short-duration peaks of the CCI because the system operates to keep the ratio of the received energy per bit to the noise spectral density (E_b/N_0) constant, not by changing the transmission power but by changing the bit rate. However, it has a disadvantage—such an adaptive bit rate requires very complicated hardware to generate various clock rates as well as to prepare transmitter and receiver filters matched to the possible bit rates.

On/off transmission, which is a special case of the symbol-rate-controlled system, was also proposed by Bello and Cowan. The results show that on/off transmission, although it is a simple algorithm, has a large gain [17-10].

Another possible modulation parameter for adaptive modulation is the modulation level. Webb and Hanzo proposed a modulation-level-controlled adaptive modulation using star-QAM [17-11]. The advantage of this system is that the hardware is simpler than that for the symbol-rate-controlled systems because we can employ almost the same modem configuration for any modulation level.

Although the aforementioned adaptive modulations have been developed as alternative schemes for power control techniques, adaptive modulation is also effective for controlling system capacity according to the traffic conditions. Komaki [17-12] proposed a modulation-level-controlled adaptive modulation using square-QAM to control capacity of microwave radio links [17-12; 17-13; 17-14]. In this system, a lower modulation level is selected when the traffic is not so heavy to keep the BER performance to as high a level as possible. On the other hand, when the traffic gets heavier, a higher modulation level is selected to reduce the blocking rate to as low as possible. Komaki's extensive research along with that of his associates has shown the capacity control scheme using adaptive modulation to be effective for improving throughput in fading environments.

As discussed before, when adaptive modulation is applied to compensate for fading in HF radio links or satellite systems, it has been confirmed to be very effective for improving transmission quality. However, when we want to apply it to land mobile communication systems, the concept of the modulation parameter adaptation becomes quite different because its channel variation speed is much faster than that for fixed services due to terminal mobility.

In land mobile communication systems, local mean value of the received signal level is dynamically changing due to the irregular configuration of the natural terrain as well as many irregularly arranged artificial structures. Moreover, the instantaneous value of the received signal varies quite rapidly. This makes it advantageous to divide adaptive modulation techniques into slow adaptive modulation and fast adaptive modulation. Figure 17.3 shows the concept of adaptive modulations for wireless communication systems. In slow adaptive modulation systems, carrier frequency, slot

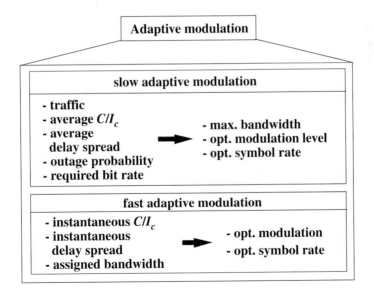

Fig. 17.3 Concept of adaptive modulations for wireless communication systems (from Ref. 17-1, © Institute of Electronics, Information and Communication Engineers, 1996).

(in the case of TDMA), bandwidth, and modulation parameters to be used during a call are assigned to each terminal in the radio link connection process taking into consideration traffic, average C/I_c, average delay spread, outage probability, and required bit rate. Therefore, the slow adaptive modulation is considered to be an adaptive radio resource management technique.

On the other hand, in fast adaptive modulation systems, both BS and each terminal dynamically change the modulation parameters according to the channel conditions within the assigned bandwidth. We will discuss both adaptive modulations for mobile communications.

17.5 SLOW ADAPTIVE MODULATION

17.5.1 Basic Concept of Slow Adaptive Modulation

Figure 17.4 shows the concept and the effect of slow adaptive modulation in the case of macrocell and microcell systems. In this figure, fixed-rate QPSK is considered to be a conventional system. In adaptive modulation systems, the modulation scheme is assumed to be selected from full-rate 16QAM, full-rate QPSK, 1/2-rate QPSK, and 1/4-rate QPSK as an example.

In the conventional system, when a terminal is located farther from the BS, its performance gets degraded. As a result, the coverage area is determined to satisfy a certain BER_{req} in a certain outage probability (P_0) in conventional systems. The shadowed area in Figure 17.4 shows such a performance-degraded area.

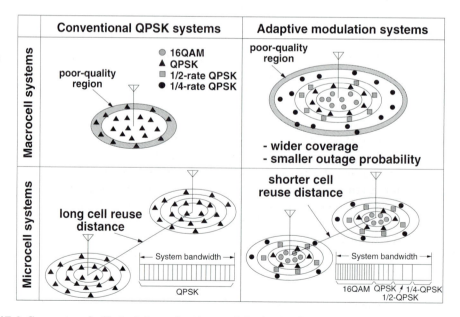

Fig. 17.4 Concept and effect of slow adaptive modulation in the case of macrocell and microcell systems.

On the other hand, when we apply the slow adaptive modulation scheme to macrocell systems, we employ 16QAM when a terminal is located near the BS, and we employ QPSK or lower-rate QPSK when a terminal is located at the fringe of the covered area. Because the receiver sensitivity for lower-rate QPSK is higher than it is for the full-rate QPSK, we can increase the coverage area or we can reduce outage probability in macrocell systems.

When we apply adaptive modulation to microcell systems, we can reduce the cell reuse distance owing to receiver sensitivity improvement. At the same time, we can increase the number of channels in the assigned bandwidth because terminals near the BS can employ 16QAM. As a result, when we employ slow adaptive modulation systems, we can expect higher system capacity than conventional systems have.

Moreover, adaptive modulation has one more big advantage—it is robust to delay spread because reduction of symbol rate is one of the most effective ways to improve delay spread immunity.

17.5.2 Frame Format for the Slow Adaptive Modulation Scheme

When we apply the adaptive modulation scheme to fixed-bit-rate services, such as voice transmission, we have to control the assigned bandwidth according to the employed modulation scheme. For example, when we employ 16QAM for voice transmission, we can reduce its bandwidth by half, compared to the case of QPSK. However, such bandwidth control is not practical because it requires the preparation of a lot of receiver filters matched to all the possible symbol rates.

TDMA is a very suitable access scheme for solving this problem because it can equivalently control the assigned bandwidth just by changing the number of assigned slots in each TDMA frame.

Figure 17.5 shows an example of the frame format for the slow adaptive modulation scheme [17-15]. One frame consists of 96 slots, and 1 frame length is 80 ms. Symbol rate for the full rate is 80 ksymbol/s, and the modulation level is selected from 4, 16, and 64. Consequently, a burst length is 833.3 μs (66.67 symbols).

In each slot, preamble and tail symbols consist of the BPSK-modulated symbols using symbols having the maximum amplitude in the first and third quadrant. In the preamble and tail symbols, the symbol rate is fixed at the full rate because we can measure the channel impulse response with higher delay time resolution. The preamble includes guard time as well as ramp-up and frame synchronization symbols. Frame synchronization symbols are also used to measure the channel impulse response (channel condition). Tail symbols are for ramp down. Preamble and tail symbols are also used as pilot symbols to compensate for fading [17-16].

On the other hand, in the information symbol section, a lower symbol rate is equivalently achieved by consecutively transmitting the identical multilevel QAM symbols at the full-rate symbol rate. For example, when 1/4 rate is employed, 4 identical symbols are consecutively transmitted. In this section, 48 symbols are included in the case of full-rate transmission, lowered to 24 and 12 symbols in the case of 1/2-rate and 1/4-rate transmissions, respectively.

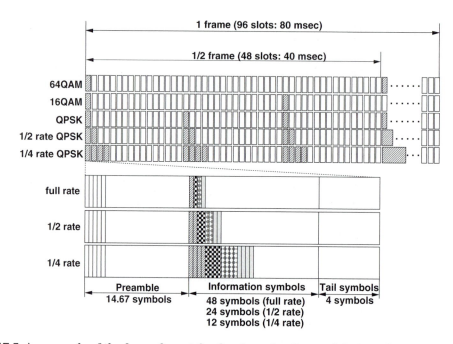

Fig. 17.5 An example of the frame format for the slow adaptive modulation scheme.

For symbol rate control, figuring out how to simplify transmitter and receiver filters is the most important problem because they have to be matched to each symbol rate. Let's assume that we will prepare only a pair of the transmitter and receiver filters consisting of a root Nyquist filter with its bandwidth of equal to the full rate. When we select the full rate as the information symbol rate, it is optimum. However, when we employ a lower rate, it reduces S/N after demodulation because the bandwidth is two or four times wider than the optimum bandwidth. This problem can be solved when we coherently accumulate the consecutively transmitted identical symbols. For example, when 1/4-rate transmission is employed, four consecutive identical symbols are accumulated. Such accumulation yields the maximum S/N in the AWGN channel because it is equivalent to the integrate-and-dump filter which is a matched filter for the rectangular pulse. As a result, we can always maximize S/N after demodulation for any symbol rate.

For voice transmission, two slots in each frame are used when 64QAM is assigned as a modulation level. In this case, voice data including FEC and SACCH are transmitted at the bit rate of 7.2 kbit/s. When a lower modulation level is assigned for voice transmission, more slots in each frame are used in order to keep the bit rate of 7.2 kbit/s; that is, three slots for 16QAM and six slots for QPSK.

When the received C/I_c is too low or the delay spread is too large, the symbol rate is lowered to 1/2 rate (40 ksymbol/s) or 1/4 rate (20 ksymbol/s).

When we employ such an adaptive modulation scheme, various modulations could be assigned to each slot. Figure 17.6 shows an example of the slot assignment

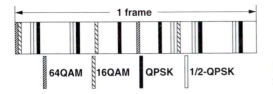

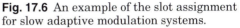

Fig. 17.6 An example of the slot assignment for slow adaptive modulation systems.

for slow adaptive modulation systems. In this example, four subscribers with different modulation schemes are allocated in the same carrier.

When a higher bit rate is required to transmit nonvoice data, more slots in each frame will be assigned according to the required bit rate. Consequently, the maximum bit rate of 345.6 kbit/s per carrier (96 slots × 3.6 kbit/s) is available using 64QAM.

17.5.3 C/I_c Estimation

In slow adaptive modulation systems, measurement of the average C/I_c is the most important issue because it determines system performance. To measure C/I_c, we usually embed a known symbol sequence, such as the PN sequence.

Let's assume that the received signal is given by

$$s_R(t) = \text{Re}\left[y(t)e^{j2\pi f_c t}\right] \tag{17.1}$$

$$y(t) = s(t) \otimes c(t) + n(t) \tag{17.2}$$

where $s(t)$ is the baseband signal waveform of the PN sequence, and delay spread is so small that the channel impulse response can be identified as the two-ray Rayleigh fading model given by

$$c(t) = c_0(t)\delta(t) + c_1(t)\delta(t - T_s) \tag{17.3}$$

When equation (17.3) is substituted into equation (17.2), $y(t)$ is given by

$$y(t) = c_0(t)s(t) + c_1(t)s(t - T_s) + n(t) \tag{17.4}$$

Now let's assume that f_d is so small that the delay profile is time invariant during each burst. In this case, we can assume that $c_0(t) = c_0$ and $c_1(t) = c_1$. To measure c_0 and c_1, a known sequence is embedded. Among many sequences suitable for the delay profile measurement, a modified PN sequence [17-17] is very useful for measuring the delay profile, especially when the length of the sequence is strictly limited.

In the modified PN sequence, a certain value of DC offset (D) is simply added to the PN sequence to suppress the side lobe of the autocorrelation function. To satisfy this requirement, D is given by

$$D = \frac{-1 \pm \sqrt{1 + N_p}}{N_p} \tag{17.5}$$

where N_P is the number of symbols in a PN frame.

Table 17.1 shows D for each N_p obtained by equation (17.5). As a result, the autocorrelation value for the time lag of lT_s (l-symbol duration) is given by

$$R(l) = \begin{cases} N_P + 1; & l = 0 \mod N_P \\ 0 & ; \; l \neq 0 \mod N_P \end{cases} \tag{17.6}$$

Table 17.1 Relationship between PN stage and DC offset.

PN stage	N_p	DC-offset (D)
3	7	0.261 or –0.547
4	15	0.200 or –0.333
5	31	0.150 or –0.215
6	63	0.111 or –0.143

Let's define the modified PN sequence as $(\beta_0, \beta_1, \beta_2, ..., \beta_{Np-1})$. When we measure the delay profile of the two-ray Rayleigh fading channel with its delay time of T_s, we have to add 1 symbol (β_{Np-1}) prior to the PN sequence to suppress the effect of the delayed wave being 0 when we measure $c_0(t)$. For the same reason, we have to add 1 symbol (β_0) after the PN sequence when we measure $c_1(t)$. As a result, we need $N_p + 2$ symbols as the delay profile estimation word for the two-ray Rayleigh fading channel with a delay time of up to 1 symbol duration. Thus, the embedded delay profile estimation word is given by

$$s_{dp}(t) = \beta_{N_p-1}\delta(t + T_s) + \sum_{k=0}^{N_p-1} \beta_k \delta(t - kT_s) + \beta_0 \delta(t - N_p T_s) \tag{17.7}$$

where $t = 0$ is defined as the timing for the first β_0 point.

At the receiver, we will take a correlation between the received signal $y(t)$ and the modified PN code given by

$$s_{PN}(t) = \sum_{k=0}^{N_p-1} \beta_k \delta(t - kT_s) \tag{17.8}$$

When the sampled value of $y(t)$ at $t = iT_s$ is expressed as $y(i)$ and $t = 0$ is defined as the first β_0 point in the delay profile estimation word, we can estimate c_0 and c_1 as follows:

$$\hat{c}_j = \frac{1}{N_p+1} \sum_{i=j}^{N_p-1+j} y(i)\beta_{i-j}^* = c_j + \Delta c_j \quad (j = 0, \; 1) \tag{17.9}$$

where

$$\Delta c_j = \frac{1}{N_p + 1} \sum_{i=j}^{N_p - 1 + j} n(i)\beta_{i-j}^* \qquad (17.10)$$

is the estimation error of the delay profile. Using equation (17.9), we can obtain the delay spread as follows:

$$\sigma = \frac{\sqrt{p_1 / p_0}}{1 + p_1 / p_0} T_s \qquad (17.11)$$

where $p_0 = |\hat{c}_0|^2$ and $p_1 = |\hat{c}_1|^2$.

Next, we need to obtain the noise component. When we take the convolution between the obtained delay profile

$$\hat{c} = \hat{c}_0 \delta(t) + \hat{c}_1 \delta(t - T_s) \qquad (17.12)$$

and the delay profile estimation word $s_{dp}(t)$, we can obtain a replica of the received signal given by

$$\hat{y}(i) = \hat{c}_0 s(i) + \hat{c}_1 s(i - 1) \qquad (17.13)$$

Consequently, we can obtain the noise component as

$$\begin{aligned} n'(i) &= y(i) - \hat{y}(i) \\ &= n(t) + \Delta c_0 s(i) + \Delta c_1 s(i - 1) \end{aligned} \qquad (17.14)$$

When the delay profile estimation error is negligible, $n'(i) = n(i)$ is satisfied. As a result, instantaneous S/N of the received signal can be obtained as

$$S/N = \left[\frac{1}{2} E\left[|s(i)|^2\right]\left(|\hat{c}_0|^2 + |\hat{c}_1|^2\right)\right] \bigg/ E\left[|n'(i)|^2\right] \qquad (17.15)$$

When we need only the average S/N information, we can obtain it simply by averaging this instantaneous S/N.

In the case of macrocell systems, we can control the modulation parameters according to the averaged S/N given by averaging the instantaneous S/N, given in equation (17.15), over a certain period. In the case of microcell systems, CCI rather than noise becomes dominant. In this case, $y(i) - \hat{y}(i)$ gives the CCI component. Therefore, C/I_c becomes identical to equation (17.14).

17.5.4 Zone Radius Enlargement Effect for the Macrocell Systems

In this section, we will evaluate the zone radius enlargement effect and the spectral efficiency in terms of bit/s/Hz. Moreover, we will discuss how to effectively employ symbol rate and modulation level control techniques for slow adaptive modulation systems.

First let's define spectral efficiency for macrocell systems. Because macrocell systems do not employ the idea of geographical frequency reuse, spectral efficiency is simply defined in terms of bit/s/Hz. When a $1/k$-rate M-ary modulation is employed in macrocell systems, spectral efficiency is given by

$$\eta_T = \frac{R_{smax}F_{eff}(\log_2 M - \log_2 k)}{f_{ch}}$$ (17.16)

where

R_{smax} = symbol rate for the full-rate transmission
F_{eff} = frame efficiency
f_{ch} = carrier spacing

In this case, the outage probability is defined as the probability that a required BER (BER_{req}) cannot be satisfied in the service area. When the received E_s/N_0 that satisfies $\text{BER} < \text{BER}_{req}$ is L_{req}, the outage probability is defined as

$$P_0 = \Pr(\Lambda < \Lambda_{req})$$ (17.17)

When we employ a $1/k$-rate QPSK ($k = 1, 2, ..., k_1$, k_1 is the number of the prepared symbol rate); full-rate 16QAM; and full-rate 64QAM as selectable modulation schemes for slow adaptive modulation systems, spectral efficiency is given by

$$\eta_T = \frac{R_{smax}F_{eff}}{f_{ch}}\left[\sum_{k=1}^{k_1} 2^{2-k}\cdot\Pr(1/2^{k-1}-\text{rate QPSK}) + \sum_{l=2}^{3} 2^l\cdot\Pr(2^{2l}-\text{QAM})\right](\text{bit}/\text{s}/\text{Hz})$$ (17.18)

Figure 17.7 shows computer-simulated results of the BER versus E_s/N_0 (C/I_c) performances of QPSK, 16QAM, and 64QAM with a pilot symbol-aided maximal ratio combining space diversity in Rayleigh fading environments. Gray coding with absolute phase-coherent detection is employed for any modulation. Because the required

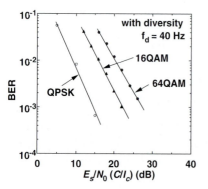

Fig. 17.7 Computer-simulated results of BER versus E_s/N_0 (C/I_c) performance of QPSK, 16QAM, and 64QAM with a pilot symbol-aided maximal ratio combining space diversity in Rayleigh fading environments, where Gray coding with absolute phase-coherent detection is employed for any modulation (from Ref. 17-15, © Institute of Electronics, Information and Communication Engineers, 1994).

BER for voice transmission is 10^{-2}, and because voice transmission may be a main service even for future personal multimedia communication systems, BER = 10^{-2} will be used to evaluate spectral efficiency in our discussion here. From Figure 17.7, the required E_s/N_0 (C/I_c) for BER = 10^{-2} is 9 dB for QPSK, 16 dB for 16QAM, and 20 dB for 64QAM. When we design the zone radius of the coverage area (R), we have to guarantee a threshold BER—for example, BER = 10^{-2}—with a certain outage probability (P_0).

First let's discuss the relationship between the outage probability and R. Figure 17.8 shows the cumulative distribution of $\Lambda = C/I_c$, where we assume a path loss decay factor of 3.5, a standard deviation of log-normal fading of 6 dB, and Λ normalized by the Λ_Q that gives BER = 10^{-2} in the case of QPSK of 9 dB.

Now, let's define that R_0 is the zone radius that gives $P_0 = 0.1$ for the fixed-rate QPSK and that the outage probability (P_0) with respect to Λ/Λ_Q is $P_0 = P(\Lambda/\Lambda_Q)$. When the zone radius (R) is increased, the path loss due to distance is increased by $35\log_{10}(R/R_0)$ (dB) at the fringe. Moreover, when we employ 16QAM or 64QAM, we need more power margin Λ_m. From Figure 17.7, we can find that $\Lambda_m = 0$ dB for QPSK, $\Lambda_m = 7$ dB for 16QAM, and $\Lambda_m = 11$ dB for 64QAM. On the other hand, when we employ a $1/k$-rate modulation, we can reduce the required received signal level by $10\log_{10}k$ (dB). As a result, when the zone radius is R, the required power margin for a $1/k$-rate modulation is given by

$$\text{Power margin} = 35\log_{10}(R/R_0) + \Lambda_m - 10\log_{10}k \ \ (\text{dB}) \tag{17.19}$$

This means that the outage probability for a $1/k$-rate modulation is given by

$$P_0 = P\big(35\log_{10}(R/R_0) + \Lambda_m - 10\log_{10}k\big) \tag{17.20}$$

Figure 17.9 shows the outage probability versus R performance for the full-rate $(k = 1)$ QPSK, 16QAM, and 64QAM. We can find from this figure that the zone radius for $P_0 = 10\%$ is reduced to $0.65R_0$ in the case of 16QAM and $0.49R_0$ in the case of 64QAM, although their spectral efficiency in terms of bit/s/Hz is increased to 4 bit/s/Hz and 6 bit/s/Hz, respectively.

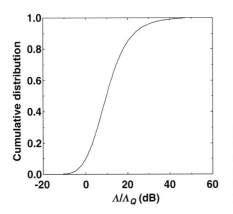

Fig. 17.8 Cumulative distribution of $\Lambda = , C/I_c$ where we assume a path loss decay factor of 3.5, a standard deviation of log normal fading of 6 dB, and Λ normalized by the Λ_Q of 9 dB that gives BER = 10^{-2} in the case of QPSK (from Ref. 17-21, © Institute of Electrical and Electronic Engineers, 1995).

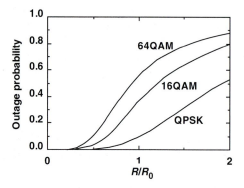

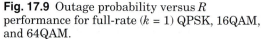

Fig. 17.9 Outage probability versus R performance for full-rate ($k = 1$) QPSK, 16QAM, and 64QAM.

Next we will discuss the zone radius enlargement effect of the slow adaptive modulation system. First we will consider modulation scheme candidates of only 1/4-rate QPSK, 1/2-rate QPSK, and full-rate QPSK. In this case, the modulation scheme is selected on the following basis:

☞ 1/4-rate QPSK: $\Lambda/\Lambda_Q \leq 35\log_{10}(R/R_0) - 3.0$
☞ 1/2-rate QPSK: $35\log_{10}(R/R_0) - 3.0 < \Lambda/\Lambda_Q \leq 35\log_{10}(R/R_0)$
☞ full-rate QPSK: $35\log_{10}(R/R_0) < \Lambda/\Lambda_Q$

Because spectral efficiency is 2 bit/s/Hz, 1 bit/s/Hz, and 0.5 bit/s/Hz for full-rate, 1/2-rate, and 1/4-rate QPSK, respectively, when the roll-off factor (α) is 0, spectral efficiency is given by

$$\eta_T = \frac{R_{smax}F_{eff}}{f_{ch}}\big[0.5\cdot\mathrm{Pr}(1/4-\mathrm{QPSK})+\mathrm{Pr}(1/2-\mathrm{QPSK})$$

$$+\,2\cdot\mathrm{Pr}(\mathrm{full-rate\ QPSK})\big] \tag{17.21}$$

where

$$\mathrm{Pr}(1/4-\mathrm{QPSK})$$
$$=\mathrm{Pr}\big(\Lambda/\Lambda_Q \leq 35\log_{10}(R/R_0)-3.0\big) \tag{17.22}$$

$$\mathrm{Pr}(1/2-\mathrm{QPSK})$$
$$=\mathrm{Pr}\big(35\log_{10}(R/R_0)-3.0 < \Lambda/\Lambda_Q \leq 35\log_{10}(R/R_0)\big) \tag{17.23}$$

$$\mathrm{Pr}(\mathrm{QPSK})=\mathrm{Pr}(35\log_{10}(R/R_0) < \Lambda/\Lambda_Q) \tag{17.24}$$

Here, P_0 is the probability that BER = 10^{-2} cannot be obtained even if 1/4-rate QPSK is employed. As a result, P_0 is given by

$$P_0 = \mathrm{Pr}\big(\Lambda/\Lambda_Q \leq 35\log_{10}(R/R_0)-6.0\big) \tag{17.25}$$

Figure 17.10 shows (a) spectral efficiency and (b) outage probability of the slow adaptive modulation with respect to the zone radius. As shown in this figure, we can reduce the outage probability when we employ a lower symbol rate, although spectral efficiency will be a little bit degraded. In other words, when we introduce a symbol-rate-controlled adaptive modulation, we can increase the zone radius without degrading so much spectral efficiency. For example, when we define the acceptable outage probability as $P_0 = 0.1$, we can increase the zone radius by 50% although spectral efficiency will be sacrificed by 25%.

So far we have discussed only the symbol-rate-controlled QPSK system. However, if we can effectively introduce the modulation level control to slow adaptive modulation systems, we can expect further spectral efficiency improvement.

When we introduce full-rate 16QAM and full-rate 64QAM to the 1/4-rate QPSK-QPSK system, we will select the modulation parameters based on the following:

☞ 1/4-rate QPSK: $\Lambda/\Lambda_Q \le 35\log_{10}(R/R_0) - 3.0$
☞ 1/2-rate QPSK: $35\log_{10}(R/R_0) - 3.0 < \Lambda/\Lambda_Q \le 35\log_{10}(R/R_0)$
☞ full-rate QPSK: $35\log_{10}(R/R_0) < \Lambda/\Lambda_Q \le 35\log_{10}(R/R_0) + 7.0$
☞ full-rate 16QAM: $35\log_{10}(R/R_0) + 7.0 < \Lambda/\Lambda_Q \le 35\log_{10}(R/R_0) + 11.0$
☞ full-rate 64QAM: $35\log_{10}(R/R_0) + 11.0 < \Lambda/\Lambda_Q$

In this case, spectral efficiency is given by equation (17.18), where $k_1 = 3$; the selection probability for 1/4-rate QPSK, 1/2-rate QPSK, and full-rate QPSK is given by equations (17.22) through (17.24); and selection probability for 16QAM and 64QAM is given by

$$\begin{aligned}&\Pr(16\text{QAM})\\&= \Pr(35\log_{10}(R/R_0) + 7.0 < \Lambda/\Lambda_Q \le 35\log_{10}(R/R_0) + 11.0)\end{aligned} \qquad (17.26)$$

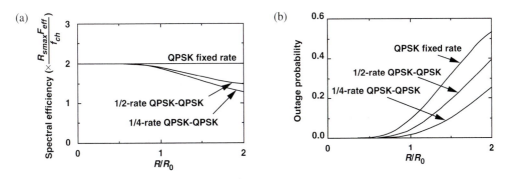

Note: 1/2-rate QPSK-QPSK means the selectable modulation schemes of 1/2-rate QPSK and full-rate QPSK, and 1/4-rate QPSK-QPSK means the selectable modulation schemes of 1/4-rate QPSK, 1/2-rate QPSK and full-rate QPSK.

Fig. 17.10 Slow adaptive modulation versus zone radius: (a) spectral efficiency and (b) outage probability, where the modulation parameter is selected from 1/4-rate QPSK, 1/2-rate QPSK, and full-rate QPSK.

$$\Pr(64QAM) = \Pr(35 \log_{10}(R/R_0) + 11.0 < \Lambda / \Lambda_Q) \qquad (17.27)$$

The outage probability of this system is exactly the same for the 1/4-QPSK-QPSK system given by equation (17.25) because it is determined only by the lowest spectral efficient modulation scheme among the selectable modulation schemes.

Figure 17.11 shows spectral efficiency versus zone radius for the modulation-level- and symbol-rate-controlled slow adaptive modulation scheme. When only the 1/4-rate QPSK, 1/2-rate QPSK, and full-rate QPSK are selectable, spectral efficiency is upper limited by $2.0 R_{smax} F_{eff}/f_{ch}$. When we add 16QAM to the selectable modulation schemes, the upper limit becomes $4.0 R_{smax} F_{eff}/f_{ch}$. It becomes $6.0 R_{smax} F_{eff}/f_{ch}$ when we further add 64QAM to the selectable modulation schemes. Furthermore, we can increase spectral efficiency at any R when we employ modulation schemes with higher spectral efficiency. These results show that we can achieve zone radius enlargement and an increase in spectral efficiency at the same time when we employ both a lower-rate QPSK and a high spectral efficient modulation scheme as the selectable modulation schemes for slow adaptive modulation systems.

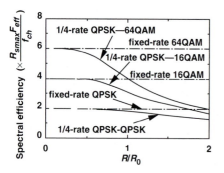

Fig. 17.11 Spectral efficiency versus zone radius for the modulation level and symbol-rate-controlled slow adaptive modulation scheme.

17.5.5 Definition of Spectral Efficiency of Slow Adaptive Modulation Systems in Microcell Systems

When the modulation level is fixed in microcell systems, we can express its spectral efficiency in terms of bit/s/Hz per number of cell sites in a cell cluster as

$$\eta_T = \frac{l \cdot R_s \cdot F_{eff}}{f_{ch} \cdot L} \qquad (17.28)$$

$$l = \log_2 M \qquad (17.29)$$

where

M = modulation level
R_s = symbol rate
f_{ch} = channel spacing
F_{eff} = frame efficiency
L = number of cell site in each cell cluster

In the case of slow adaptive modulation systems, spectral efficiency is given by

$$\eta_T = \sum_{k=1}^{k_1} \frac{1}{2^{k-2}} R_{smax} \frac{F_{eff}}{f_{ch} \cdot L} \Pr(\frac{1}{2^{k-1}} - \text{rate QPSK})$$

$$+ \sum_{l=2}^{l_{max}} \frac{(2l) \cdot R_{smax} \cdot F_{eff}}{f_{ch} \cdot L} \Pr(2^{2l} - \text{QAM})$$

(17.30)

$$l_{max} = \log_2 M_{max}$$

(17.31)

where R_{smax} is the symbol rate for full-rate transmission, $\Pr(\text{mod}_x)$ is a probability that a modulation (mod_x) is selected, k_1 is the number of symbol rates to be used, and M_{max} is the maximum modulation level. $k = 1$, $k = 2$, and $k = 3$ correspond to full-rate QPSK, 1/2-rate QPSK, and 1/4-rate QPSK, respectively, and $l = 2$ and $l = 3$ correspond to 16QAM and 64QAM, respectively. The channel spacing is determined under the constraint that the power ratio of the desired signal to the ACI (C/I_A) is much higher than C/I_c. When a roll-off factor of 0.5 is used for both the transmitter and receiver filters and f_{ch} is greater than $1.2R_{smax}$, this condition will be satisfied [17-18]. Thus, we will disregard the effect of C/I_A in our discussion because $f_{ch} = 1.2R_{smax}$ is a very typical number for digital cellular systems such as PDC and IS-54 systems.

As for the cell configuration, we will employ the irregular parallel beam three-sector cell layout [17-19] because it gives very high spectral efficiency. This scheme reduces L by reducing the co-channel reuse distance in the direction orthogonal to the main beam because CCI coming from this direction is very small due to antenna directivity. Moreover, it can take an arbitrary integer number for L. Here we will use a directive antenna with a beam halfwidth of 70 degrees [17-18].

Furthermore, we will assume the following for both the desired and interference signals.

☞ Large-scale signal variation is subject to path loss with respect to distance (attenuation factor of 3.5) and log-normal fading with a standard deviation of 6.0 dB.

☞ Small-scale signal variation is subject to Rayleigh fading and delay spread.

17.5.6 Spectral Efficiency in Flat Rayleigh Fading Environments

For the selection of L, we have to decide on a threshold C/I_c that satisfies a required BER. Figure 17.12 shows the cumulative distribution of C/I_c (probability that C/I_c is lower than abscissa) with a parameter of L for the irregular parallel beam three-sector cell layout. When the required C/I_c for BER = 10^{-2} is Λ_{req}, Λ is determined to satisfy that $\Pr(C/I_c < \Lambda_{req})$ is lower than a certain outage probability (P_0). Because E_s/N_0 is equivalent to C/I_c when we employ the root Nyquist filter as the transmitter and receiver filters, we can also obtain this Λ_{req} from Figure 17.7. For example, in the case of QPSK, Λ_{req} = 9 dB. At this C/I_c, we can find from Figure 17.12 that the outage probability is P_0 = 20.3% when we employ the cell configuration of $L = 2$. In the same manner, we can obtain P_0 = 10.9%, 5.8%, 3.4%, 2.5%, 1.1%, and 0.76% for $L = 3, 4, 5, 6,$

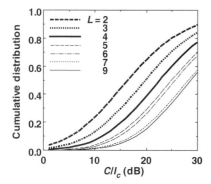

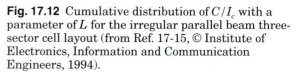

Fig. 17.12 Cumulative distribution of C/I_c with a parameter of L for the irregular parallel beam three-sector cell layout (from Ref. 17-15, © Institute of Electronics, Information and Communication Engineers, 1994).

7, and 9, respectively. These results mean that $L = 3$ is applicable when we accept $10.9\% \leq P_0 \leq 20.3\%$.

Figure 17.13 shows spectral efficiency versus P_0 performance for QPSK, 16QAM, and 64QAM fixed modulation parameter systems. The required L for each P_0 is also shown in this figure. We can find from this figure that the spectral efficiency improvement obtained by increasing the modulation level is very small because a larger modulation level requires larger L, thereby increasing cell reuse distance, although a larger modulation level can increase spectral efficiency with respect to frequency (η_f). This means that QPSK is the most appropriate modulation scheme when the modulation level is fixed.

Next let's consider spectral efficiency of adaptive modulation systems in which only the modulation level is controlled according to the average C/I_c. In this case, the outage probability is defined as the probability that C/I_c is lower than the required C/I_c for a modulation with the smallest modulation level—QPSK (= 9.0 dB) in this case. In this analysis, the terminal mobility is assumed to be very low. With this assumption, large-scale signal variation is assumed to be constant during each call.

Figure 17.14 shows spectral efficiency of the adaptive modulation with the available modulation level of 4 and 16 (QPSK—16QAM in Figure 17.14), and that with the available modulation level of 4, 16, and 64 (QPSK—64QAM in Figure 17.14). When the modulation level is selected from 4, 16, and 64, the modulation level is controlled on these criteria.

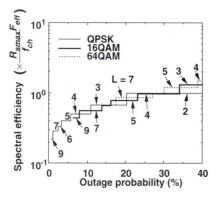

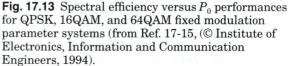

Fig. 17.13 Spectral efficiency versus P_0 performances for QPSK, 16QAM, and 64QAM fixed modulation parameter systems (from Ref. 17-15, (© Institute of Electronics, Information and Communication Engineers, 1994).

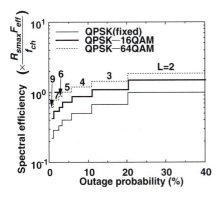

Fig. 17.14 Spectral efficiency of the modulation level controlled adaptive modulation systems (from Ref. 17-15, © Institute of Electronics, Information and Communication Engineers, 1994).

☞ $C/I_c < 16$ dB: QPSK

☞ $16 \leq C/I_c \leq 20.0$ dB: 16QAM

☞ 20.0 dB $< C/I_c$: 64QAM

When the maximum modulation level is limited to 16, 16QAM is selected whenever $C/I_c \geq 16.0$ dB.

Figure 17.14 shows that spectral efficiency is 1.8 times higher than that of conventional QPSK systems at $P_0 = 10\%$ even if the maximum modulation level of the adaptive modulation system is limited to 16. When the maximum modulation level is 64, its spectral efficiency becomes 2.4 times higher than that of conventional QPSK systems.

When we introduce symbol rate control for QPSK, we can further improve spectral efficiency. When we introduce $1/k$-rate QPSK, we can reduce the required C/I_c by $10\log_{10}(k)$ (dB). In this case, the outage probability is defined as the probability that the C/I_c is lower than the required C/I_c for QPSK with the lowest symbol rate.

Figure 17.15 shows the spectral efficiency of the modulation-level- and symbol-rate-controlled adaptive modulation systems. When we introduce 1/2-rate QPSK to the QPSK—64QAM adaptive modulation systems (1/2-rate QPSK—64QAM in Figure 17.15), spectral efficiency becomes 2.8 times higher than that of conventional QPSK systems at the outage probability of 10%. When we further introduce 1/4-rate QPSK (1/4-rate QPSK–64QAM in Figure 17.15), spectral efficiency becomes 3.5 times higher

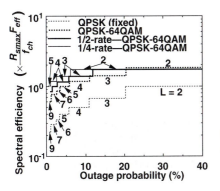

Fig. 17.15 Spectral efficiency of the modulation-level- and symbol-rate-controlled adaptive modulation systems (from Ref. 17-15, © Institute of Electronics, Information and Communication Engineers, 1994).

than that of conventional QPSK systems. Moreover, this figure shows that the improvement in spectral efficiency is more remarkable at the lower P_0. For example, spectral efficiency of the 1/4-rate QPSK—64QAM system is 4.5 times higher than that of conventional QPSK systems at $P_0 = 1\%$.

Although the adaptive modulation system shows very high spectral efficiency, it has a problem in that the maximum bit rate is lowered as the distance from the BS increases. Figure 17.16 shows the selection ratio of each modulation with respect to distance from the BS (d) normalized by the cell radius (R) in the case of 1/4-rate QPSK—64QAM adaptive modulation systems, where $L = 5$ is employed. In this case, $P_0 = 1\%$ is satisfied. The figure shows that, although the selection ratio of the lower-level modulation increases with d/R, the selection ratio of 64QAM is still 46% at the fringe area ($0.9 \leq d/R \leq 1.0$). In this area, P_0 was confirmed to be 1.2%. This result means that multimedia services with a bit rate of 7.2–345.6 kbit/s is possible with a probability of more than 46% even at the fringe area, provided that the delay spread is very small and the traffic density is not too high. Of course, BER = 10^{-2} is not a sufficient transmission quality for multimedia services. Thus we should further apply FEC and ARQ to improve transmission quality.

17.5.7 Spectral Efficiency in Frequency-Selective Fading Environments

When the propagation path is under frequency-selective fading conditions, we have to consider not only C/I_c but also delay spread for modulation parameter control of slow adaptive modulation systems.

Figure 17.17 shows the cumulative distribution of the delay spread (τ_{rms}) at Shinjyuku and Kofu in Japan [17-20]. Shinjyuku is typical of a city that is filled with high-rise buildings and has very high traffic density; Kofu is typical of a city surrounded by mountains with moderate traffic density.

Before we discuss spectral efficiency, we have to know the effect of the delay spread. Figure 17.18 shows computer-simulated results of the irreducible BER (BER caused by the delay spread under noise-free and slow-mobility conditions) versus

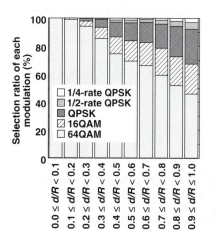

Fig. 17.16 Selection ratio of each modulation with respect to d/R in the case of 1/4-rate QPSK-64QAM adaptive modulation systems, where $L = 5$ ($P_0 = 1\%$) is employed (from Ref. 17-15, © Institute of Electronics, Information and Communication Engineers, 1994).

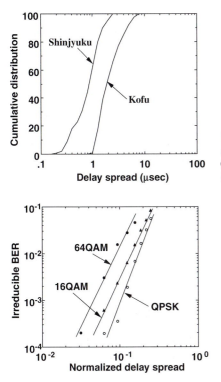

Fig. 17.17 Cmulative distribution of the delay spread at Shinjyuku and Kofu in Japan (from Ref. 17-15, © Institute of Electronics, Information and Communication Engineers, 1994).

Fig. 17.18 Computer-simulated results of the irreducible BER caused by the normalized delay spread versus delay spread performances of QPSK, 16QAM, and 64QAM (from Ref. 17-15, © Institute of Electronics, Information and Communication Engineers, 1994).

delay spread performances of QPSK, 16QAM, and 64QAM, where 2-branch space diversity is employed and the delay spread is normalized by a symbol duration. This figure shows that the normalized delay spread should be smaller than 0.18, 0.13, and 0.09 for QPSK, 16QAM, and 64QAM, respectively, to achieve irreducible BER of less than 10^{-2}.

When the modulation scheme is fixed to QPSK, the symbol rate is limited to 40 ksymbol/s to satisfy BER = 10^{-2} with the outage probability of 10% because the normalized delay spread for BER = 10^{-2} is 0.18 and the 90% value of the delay spread at Kofu is 4.5 µs. On the other hand, in the case of modulation-level- and symbol-rate-controlled adaptive modulation systems, when the delay spread is normalized by the symbol duration of the highest symbol rate ($T_{smax} = 1/R_{smax}$), τ_{rms}/T_{smax} for 64QAM, 16QAM, QPSK, 1/2-rate QPSK, and 1/4-rate QPSK are given by 0.09, 0.13, 0.18, 0.36, and 0.72, respectively.

For the selection of the modulation scheme in adaptive modulation systems, BER is determined by both the average C/I_c and the average τ_{rms}/T_{smax}. Figure 17.19 shows the decision criteria of adaptive modulation systems that considers both C/I_c and delay spread when the required BER is determined as BER = 10^{-2}.

To achieve BER = 10^{-2} using 64QAM, $C/I_c \geq 20$ dB and $\tau_{rms}/T_{smax} \leq 0.09$ should be satisfied. Thus, the selection probability of 64QAM is given by

$$\text{Pr(64QAM)} = \text{Pr}(C/I_c \geq 20 \text{ dB}) \cdot \text{Pr}(\tau_{rms}/T_{smax} \leq 0.09) \qquad (17.32)$$

In the case of 16QAM, BER = 10^{-2} is satisfied when $C/I_c \geq 16$ dB and $\tau_{rms}/T_{smax} \leq$ 0.13. However, this condition also includes the acceptable condition for 64QAM. Thus, Pr(16QAM) is given by

$$
\begin{aligned}
\text{Pr(16QAM)} \\
= \text{Pr}(C/I_c \geq 16 \ \text{dB}) \cdot \text{Pr}(\tau_{rms}/T_{smax} \leq 0.13) - \text{Pr(64QAM)} \\
= \text{Pr}(16 \ \text{dB} \leq C/I_c < 20 \ \text{dB}) \cdot \text{Pr}(\tau_{rms}/T_{smax} \leq 0.13) \\
+ \text{Pr}(C/I_c \geq 20 \ \text{dB}) \cdot \text{Pr}(0.09 < \tau_{rms}/T_{smax} \leq 0.13).
\end{aligned}
\tag{17.33}
$$

In the same manner, Pr(QPSK) and Pr(1/2-rate QPSK) are given by

$$
\begin{aligned}
\text{Pr(QPSK)} \\
= \text{Pr}(9 \ \text{dB} \leq C/I_c < 16 \ \text{dB}) \cdot \text{Pr}(\tau_{rms}/T_{smax} \leq 0.18) \\
+ \text{Pr}(C/I_c \geq 16 \ \text{dB}) \cdot \text{Pr}(0.13 < \tau_{rms}/T_{smax} \leq 0.18)
\end{aligned}
\tag{17.34}
$$

$$
\begin{aligned}
\text{Pr(1/2-rate QPSK)} \\
= \text{Pr}(6 \ \text{dB} \leq C/I_c < 9 \ \text{dB}) \cdot \text{Pr}(\tau_{rms}/T_{smax} \leq 0.36) \\
+ \text{Pr}(C/I_c \geq 9 \ \text{dB}) \cdot \text{Pr}(0.18 < \tau_{rms}/T_{smax} \leq 0.36)
\end{aligned}
\tag{17.35}
$$

When other modulation schemes cannot satisfy BER = 10^{-2}, 1/4-rate QPSK is selected. Thus, Pr(1/4-rate QPSK) is given by

$$
\begin{aligned}
\text{Pr(1/4-rate QPSK)} \\
= \text{Pr}(C/I_c < 6 \ \text{dB}) + \text{Pr}(C/I_c \geq 6 \ \text{dB}) \cdot \text{Pr}(0.36 > \tau_{rms}/T_{smax})
\end{aligned}
\tag{17.36}
$$

In this case, P_0 is the probability that BER = 10^{-2} cannot be obtained even if 1/4-rate QPSK is applied. Thus, it is given by

$$
P_0 = \text{Pr}(C/I_c < 3 \ \text{dB}) + \text{Pr}(C/I_c \geq 3 \ \text{dB}) \cdot \text{Pr}(\tau_{rms}/T_{smax} > 0.72)
\tag{17.37}
$$

In Figure 17.19, the shadowed area corresponds to the condition of the outage probability.

Figure 17.20 shows spectral efficiency versus R_{smax} for a 1/4-rate QPSK—64QAM system for P_0 = 1% and 10% in the Shinjyuku and Kofu areas. We can find from this figure that spectral efficiency is almost constant when $R_{smax} \leq 80$ ksymbol/s in the Shinjyuku area. On the other hand, spectral efficiency for $R_{smax} = 80$ ksymbol/s is

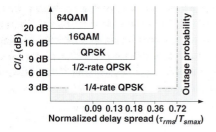

Fig. 17.19 Decision criteria of adaptive modulation systems that considers both C/I_c and delay spread when the required BER is determined as BER = 10^{-2} (from Ref. 17-15, © Institute of Electronics, Information and Communication Engineers, 1994).

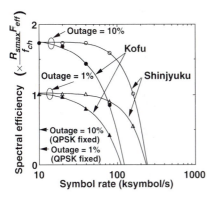

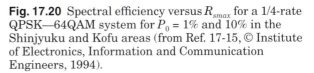

Fig. 17.20 Spectral efficiency versus R_{smax} for a 1/4-rate QPSK—64QAM system for $P_0 = 1\%$ and 10% in the Shinjyuku and Kofu areas (from Ref. 17-15, © Institute of Electronics, Information and Communication Engineers, 1994).

degraded by about 50% in the Kofu area when compared with the value in the Shinjyuku area due to a larger delay spread. However, the traffic density at Kofu is not as high, and its spectral efficiency is still higher than that of conventional QPSK systems. Thus we can apply $R_{smax} = 80$ ksymbol/s in both the Shinjyuku and Kofu areas.

17.6 SLOW ADAPTIVE MODULATION SCHEME WITH ADAPTIVE EQUALIZER FOR HIGHER-BIT-RATE TRANSMISSION

In the previous section, we discussed the slow adaptive modulation scheme to transmit up to several hundred kbit/s under moderate frequency-selective fading conditions ($\tau_{rms} < 5$ μs). Actually, this bit rate is very attractive for wireless personal communication systems because it is much higher than those of the existing digital cellular systems. However, in the near future a bit rate of up to 2 Mbit/s will be required in wireless personal communication systems. In the far future 10 Mbit/s will be required. Moreover, when we want to extend services to suburban or mountainous areas, we have to cope with much larger delay spread, perhaps 25 μs. We will discuss how we can combine anti-frequency-selective fading techniques with slow adaptive modulation systems to solve this problem. Among the many anti-frequency-selective fading techniques, we will focus on a combination of adaptive modulation, space diversity, and adaptive equalizing techniques in this section.

17.6.1 Basic Strategy for High-Bit-Rate Wireless Data Transmission

When we design wireless communication systems, the maximum delay time to be compensated for is the most important factor because it greatly depends on service area conditions. The PDC system operated in Japan can compensate for a maximum delay time of up to 10 μs. GSM can cope with up to 16 μs, and IS-54 can cope with 50 μs of maximum delay time of the delayed wave. Although the maximum delay spread is determined by regional conditions, when we want to develop global wireless systems that cover all areas, we have to cope with the maximum delay time of these areas. This means we have to cope with the maximum delay time of the delayed wave of up to 50 µs.

Coping with such a long delayed wave, however, the DFE requires too many taps. For example, when we want to achieve 2 Mbit/s transmission, we need 73 taps in the case of 500 ksymbol/s 16QAM/TDMA and 146 taps in the case of QPSK/TDMA as discussed in chapter 5. Even if the ability of the digital signal processor is rapidly being improved, these numbers are still far from the practical number.

Slow adaptive modulation with space diversity and adaptive equalizer is a solution for this problem [7-21]. In this system, we will assume that a 7-tap BDDFE combined with two-branch space diversity is employed to compensate for the maximum delay time of up to 3 full-rate symbol duration. When the delay time exceeds 3 full-rate symbol duration, the symbol rate is lowered to 1/2-rate, 1/4-rate, or 1/8-rate so as to satisfy that the maximum delay time is lower than 3 symbol duration. We will also assume that we can select the modulation scheme from QPSK and 16QAM to improve spectral efficiency.

17.6.2 Basic BER Performance of BDDFE with Space Diversity and Decision Criteria for Adaptive Modulation

Because we have already discussed the basic operation of the BDDFE with space diversity in chapter 5, we will use only a part of the results from chapter 5. Figure 17.21 shows computer-simulated BER performance versus E_s/N_0 (which is equal to C/I_c in the case of CCI-limited conditions) for QPSK/TDMA and 16QAM/TDMA systems with BDDFE and two-branch space diversity. In this simulation, we have employed the two-ray Rayleigh fading model, in which the average power ratio of the direct and delayed waves are the same and the delay time of the delayed wave is $3T_s$, and Gray coding with absolute phase-coherent detection. Figure 17.21 shows that the required C/I_c is 12 dB for full-rate QPSK and 19 dB for full-rate 16QAM. When we express the threshold C/I_c for full-rate QPSK as Λ_Q, we get the threshold C/I_c for each rate of QPSK/TDMA and 16QAM/TDMA with BDDFE and two-branch space diversity as shown in Table 17.2.

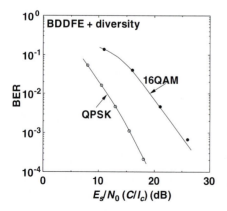

Fig. 17.21 Computer-simulated BER performance versus E_s/N_0 (C/I_c) for QPSK/TDMA and 16QAM/TDMA systems with BDDFE and two-branch selection combining space diversity (from Ref. 17-21, © Institute of Electrical and Electronic Engineers, 1995).

Table 17.2 C/I_c threshold for each rate of QPSK/TDMA and 16QAM/TDMA with BDDFE and two-branch space diversity.

Rate	QPSK	16QAM
full rate	Λ_Q	$\Lambda_Q + 7$
1/2-rate	$\Lambda_Q - 3$	$\Lambda_Q + 4$
1/4-rate	$\Lambda_Q - 6$	$\Lambda_Q + 1$
1/8-rate	$\Lambda_Q - 9$	$\Lambda_Q - 2$

At the same time, symbol rate is also controlled by the maximum delay time of the delayed wave (τ_{max}). When τ_{max} is shorter than $3/R_{smax}$, we will employ full-rate transmission. On the other hand, we will employ 1/2-rate, 1/4-rate, and 1/8-rate transmission in the case of $3/R_{smax} < \tau_{max} < 6/R_{smax}$, $6/R_{smax} < \tau_{max} < 12/R_{smax}$, and $12/R_{smax} < t_{max}$, respectively. The resulting decision criteria for modulation level and symbol rate is given by Figure 17.22.

For the evaluation of this adaptive modulation scheme, we will use the delay spread statistics shown in Figure 17.23. In this figure, delay spread in the Vancouver area [17-22] is also included because it is a good example of the mountainous terrain. The other two curves are the same as those shown in Figure 17.17.

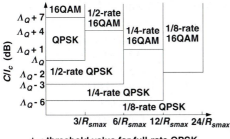

Λ_Q: threshold value for full-rate QPSK
R_{smax}: symbol rate for full rate

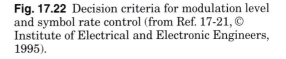

Fig. 17.22 Decision criteria for modulation level and symbol rate control (from Ref. 17-21, © Institute of Electrical and Electronic Engineers, 1995).

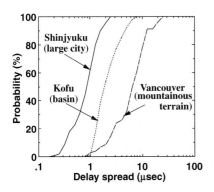

Fig. 17.23 Cumulative distribution of delay spread in the Shinjyuku, Kofu, and Vancouver areas (from Ref. 17-21, © Institute of Electrical and Electronic Engineers, 1995).

17.6.3 Spectral Efficiency and Average-Bit-Rate Performance

Figure 17.24 shows the average bit rate and spectral efficiency performances of the Shinjyuku, Kofu, and Vancouver areas for the cell reuse factor of (a) $L = 2$, (b) $L = 3$, and (c) $L = 4$, where delay spread statistics shown in Figure 17.23 are used.

In conventional QPSK systems, $L = 4$ is necessary for $P_0 = 0.1$. Therefore, spectral efficiency of the conventional QPSK system is given by $0.5(\times R_s F_{eff}/f_{ch})$. Figure 17.24 shows that, when we employ the slow adaptive modulation with BDDFE and space diversity, we can achieve higher spectral efficiency than that of conventional systems, and spectral efficiency becomes higher when L is decreased. However, the average bit rate is decreased when we employ a lower L.

When we compare the performance of the slow adaptive modulation system with $L = 4$ to that of conventional QPSK systems, more than 2-Mbit/s transmission is possi-

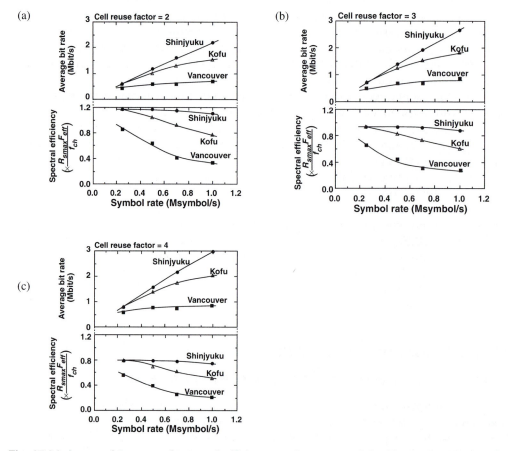

Fig. 17.24 Average bit rate and spectral efficiency performances of the Shinjyuku, Kofu, and Vancouver are as for the cell reuse factor of (a) $L = 2$, (b) $L = 3$, and (c) $L = 4$.

ble in the Shinjyuku area achieving about 50% higher spectral efficiency. In the Kofu area, 2-Mbit/s transmission is still possible achieving the same spectral efficiency as that of conventional QPSK systems. Even in the Vancouver area, 1-Mbit/s transmission is possible, although the spectral efficiency is degraded by 60% compared to that of conventional QPSK systems. However, this degradation is not a serious problem because subscriber density is not very high in such a mountainous area.

As discussed early in this chapter, it is preferable to achieve high spectral efficiency in large cities and a high delay spread immunity in small-to-medium cities and suburban and mountainous areas. From this viewpoint, slow adaptive modulation combined with BDDFE and space diversity is a very useful technique for satisfying such requirements with reasonable hardware complexity.

17.6.4 Other Applications of the Slow Adaptive Modulation

We have discussed outage probability, average bit rate, and spectral efficiency of slow adaptive modulation for voice transmission services. For wireless multimedia communication systems, however, we have to cope with various kinds of information having different traffic statistics and required QoS. Slow adaptive modulation can also cope with such requirements.

First of all, when we want to transmit various types of information having different requirements for QoS, we will control QoS by adjusting BER_{th} for the modulation parameter selection. For example, when we define BER_{th} for voice and facsimile as 10^{-2} and 10^{-4}, respectively, and prepare a two-modulation parameter selection chart, we can control the QoS for each service as well as maximize spectral efficiency.

The slow adaptive modulation system can also cope with temporally and spatially varying traffic by controlling the BER_{th}. In the conventional systems, when the whole channel is occupied, any call is blocked. In slow adaptive modulation systems, on the other hand, we can temporarily increase system capacity by increasing the required BER (BER_{th}) because higher BER_{th} increases the selection probability of a higher modulation level, although it will slightly degrade the transmission quality of each channel. In other words, we can introduce the concept of *graceful degradation* into TDMA systems using such BER_{th} control. As a result, the system capacity is softly limited by the traffic.

17.7 FAST ADAPTIVE MODULATION

17.7.1 Basic Concept of Fast Adaptive Modulation Systems

In fast adaptive modulation systems, the optimum modulation parameters are controlled slot by slot according to the instantaneous channel conditions [17-11; 17-23]. Therefore, how to precisely estimate the instantaneous channel conditions for the transmission slot is a very important factor for this system. Moreover, it is also important to accurately inform the modulation parameters of each slot which are controlled slot by slot. For this purpose, the fast adaptive modulation [17-23] introduces the following specific ideas:

☞ TDD is employed to expect the channel conditions of the transmission time slot by the delay profile sequence of the received signal, because the fading correlation between the TDD downlink and uplink is very high if the time interval between them is sufficiently short.

☞ The channel condition estimation word (CE) which consists of a modified PN code is embedded in each burst to measure the instantaneous delay profile variation of the received signal.

☞ The modulation parameter codeword (MC) which indicates the used modulation parameters in the slot is embedded in each slot.

Figure 17.25 shows the concept of fast adaptive modulation/TDMA/TDD systems. At the transmitter, the carrier is modulated using the selected modulation parameters in the modulation parameter controller.

At the receiver, modulation parameters are first estimated by decoding the modulation parameter codeword embedded in each slot, and the received signal is demodulated using the obtained modulation parameters as well as the pilot symbol-aided fading compensation technique. At the same time, an instantaneous delay profile for each reception slot is measured using a method discussed earlier in this chapter. Then the channel conditions for the next transmission time slot are estimated by using the measured delay profile sequence of the received signal, and the optimum modulation parameters for the next transmission time slot are selected.

There are two ways to handle modulation parameter selection [17-11]. One is to select parameters that give the highest bit rate with an acceptable BER_{th}. The other is to select the parameters to achieve a constant average bit rate while accepting a variable BER. We will employ the first one, aiming at increasing the maximum bit rate for data transmission even in severe frequency-selective fading environments.

Three modulation parameters—modulation level [17-11], symbol rate [17-9], and coding rate of the FEC [17-24]—have been proposed and their performances have

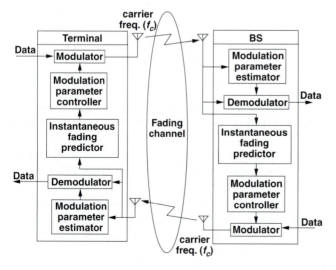

Fig. 17.25 Concept of fast adaptive modulation/TDMA/TDD systems.

been evaluated. Unfortunately, most of them are considered only in flat Rayleigh fading environments despite the fact that delay spread is an important factor for the design of digital wireless communication systems. Therefore, we will discuss the modulation-level-controlled adaptive modulation that can cope with frequency-selective fading.

17.7.2 Frame and Slot Format

Figure 17.26 shows an example of the frame and slot formats for the fast adaptive modulation/TDMA/TDD systems. Preamble and postamble include guard space, ramp-up symbols, ramp-down symbols, and frame SW. The CE and an MC are embedded in the midamble in each burst. The last symbol in the preamble, the first symbol in the postamble, and some of the midamble symbols can also be used as the pilot symbols to compensate for fading. Although the modulation level of the data sections (D_1 and D_2 in Figure 17.26) is changing according to the channel conditions, BPSK is employed for symbols in the preamble, midamble, and postamble regardless of the modulation parameters for the data sections. As a result, CE and MC are more robust to noise and fading than the data sections.

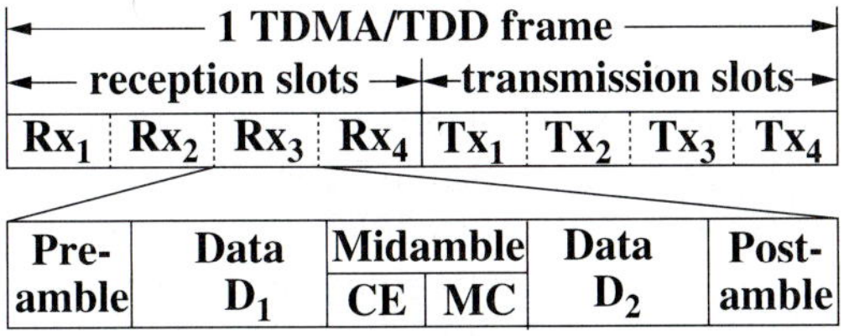

Fig. 17.26 An example of the frame and slot formats for fast adaptive modulation/TDMA/TDD systems.

17.7.3 Configuration of the Transmitter and Receiver

Figure 17.27 shows the configuration of the transmitter and receiver of fast adaptive modulation systems, where the solid lines show the transmitted or received signal flow and the dashed lines show control and timing signal flow. This configuration is almost the same as that of the M-ary QAM systems except for the modulation parameter estimator, the fading monitor and predictor, and the modulation parameter controller.

At the transmitter, the transmitted serial data is converted to parallel data, and the complex baseband signal for the selected modulation parameters is generated at the baseband signal generator (BSG). At this stage, Gray coding is employed. Then the TDMA/TDD slot is formatted with the selected modulation parameters. After that, the TDMA/TDD slot signal band-limited by the LPF modulates a carrier, and the signal is transmitted.

At the receiver, after the desired signal is picked up by a BPF, the signal is quasi-coherently detected using a local oscillator and stored in the burst memory. The stored data are then band-limited by an LPF. As the transmitter and receiver LPF, a root Nyquist filter with a roll-off factor of 0.5 is employed. After the symbol timing and

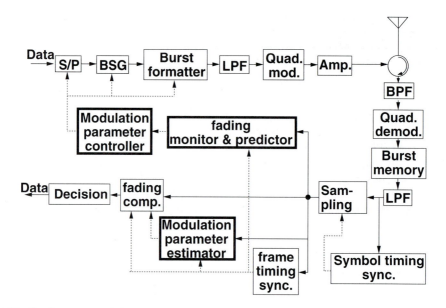

Fig. 17.27 Configuration of the transmitter and receiver of fast adaptive modulation systems.

frame timing are taken and the modulation parameters are detected using the band-limited data, fading distortion in the received baseband signal is compensated for, and the transmitted data sequence is regenerated.

At the same time, the instantaneous delay profile is measured using the CE in each burst, and the instantaneous delay profile for the next transmission time slot is estimated by extrapolating the delay profile sequence of the received signal. Using the estimated delay profile, the optimum modulation parameter of the next transmission time slot is selected, and the result is transferred to the modulation parameter controller.

17.7.4 Modulation Parameter Estimator

In fast adaptive modulation systems, accuracy of the modulation parameter estimator is essential because we cannot correctly demodulate the received signal unless the modulation parameters are correctly decoded. In other words, the MC should be more robust to noise and fading than the other data. For this purpose, we will employ a BPSK-modulated Walsh function as the MC, where $A_m\exp(\pi/4)$ and $A_m\exp(-3\pi/4)$ (A_m is the maximum amplitude of the modulation scheme for m-th option of the modulation parameters) are used as the BPSK symbols.

Figure 17.28 shows the concept of the modulation parameter decoder using the Walsh function. In this figure, a 4-symbol Walsh code is shown as an example, where the selectable modulation schemes of QPSK, 16QAM, 64QAM, and 256QAM are assumed.

Let's assume that the embedded Walsh code corresponds to m-th option of the modulation parameters is given by

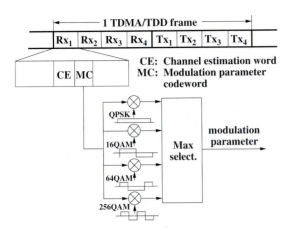

Fig. 17.28 Concept of the modulation parameter decoder using a Walsh function.

$$s_m(t) = \sum_{k=1}^{N_{MC}} A_m w_{mk} e^{j\pi/4} \delta(t - kT_s) \qquad (17.38)$$

where N_{MC} is the number of symbols in the MC, w_{mk} (= 1 or –1) is k-th Walsh symbol for m-th option of the modulation parameters, T_s is a symbol duration of the MC, and $t = T_s$ is assumed to be the timing of the first symbol of the MC. To simplify discussion, let's disregard distortion due to transmitter and receiver filters. When $s_m(t)$ is received, i-th sample of the received codeword is given by

$$y_m(i) = c(i)s_m(i) + n(i) \qquad (17.39)$$

where $c(t)$ and $n(t)$ are the complex envelope variation due to flat Rayleigh fading and noise component expressed as an equivalent low-pass system and $y_m(i)$, $c(i)$, $s_m(i)$, and $n(i)$ mean $y_m(iT_s)$, $c(iT_s)$, $s_m(iT_s)$, and $n(iT_s)$, respectively.

When a slot signal is received, we will take correlation between the received MC and each codeword of MC and calculate its power. When we take correlation between the received signal and the Walsh codeword for m-th option, the result is given by

$$V_m = |\lambda_m|^2 \qquad (17.40)$$

where

$$\lambda_m = \sum_{i=1}^{N_{MC}} y(i)s_m^*(i) \qquad (17.41)$$

When the received codeword and the multiplied codeword are the same, V_m has a certain value. On the other hand, when they are different, V_m becomes 0 because Walsh codes are orthogonal with each other. As a result, we can find the transmitted MC by searching for a Walsh code with the largest correlation value.

Figure 17.29 shows computer-simulated results of the frame error rate due to misdetection of MC versus E_s/N_0 with a parameter of N_{MC} under flat Rayleigh fading conditions with $f_d T_s = 3.125 \times 10^{-4}$. This figure shows that the decision error decreases with increasing N_{MC} because energy per codeword is increased. From this viewpoint, larger N_{MC} is preferable. However, larger N_{MC} degrades the frame efficiency. Therefore, we have to select the minimum length of N_{MC} that does not degrade the BER performance too much.

Figure 17.30 shows the BER performance of QPSK/TDMA and 16QAM/TDMA systems with a parameter of the modulation parameter estimation scheme under flat Rayleigh fading conditions. In this simulation, the modulation scheme is fixed at any time, and the corresponding MC is embedded in each burst. Therefore, the decision error means that the embedded MC indicates a code other than the used modulation scheme. This figure shows that there is no performance difference between $N_{MC} = 4$ and $N_{MC} = 4$. The performance degradation from the theoretical value is due to the pilot symbol-aided fading compensation as discussed in chapter 4. Therefore, the results show that $N_{MC} = 16$ is sufficient as the modulation parameter codeword.

Next let's discuss the robustness of MC to the delay spread. As we have discussed before, the adaptive modulation has to cope with time-varying delay spread. Therefore, the modulation parameter estimation should also be robust to delay spread. Figure 17.31 shows irreducible BER versus normalized delay spread with MC estimation and perfect modulation parameter estimation systems, where $N_{MC} =$

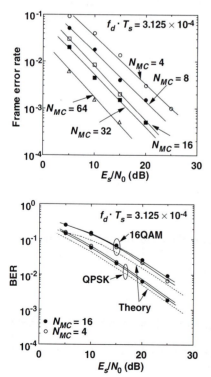

Fig. 17.29 Computer-simulated results of the decision error rate of MC versus E_s/N_0 with a parameter of N_{MC} under flat Rayleigh fading conditions with $f_d T_s = 3.125 \times 10^{-4}$.

Fig. 17.30 BER performance of QPSK/TDMA and 16QAM/TDMA systems with the modulation parameter estimation scheme under flat Rayleigh fading conditions.

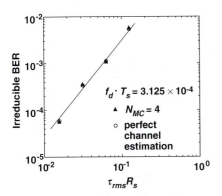

Fig. 17.31 Irreducible BER versus normalized delay spread with MC estimation and perfect modulation parameter estimation systems, where $N_{MC} = 4$.

4. As shown in this figure, there is no difference between the performance with or without the MC estimation. As a result, we can conclude that $N_{MC} = 4$ is sufficient as the modulation parameter codeword under both flat Rayleigh and frequency-selective fading conditions.

17.7.5 Selection Rule for the Modulation Parameters

In fast adaptive modulation systems, the modulation parameters are selected according to the instantaneous C/N_0 (C/I_0) and delay spread. Therefore, the threshold C/N_0 and delay spread are determined by the BER performance under AWGN conditions. Figure 17.32 shows an example of the modulation parameter selection chart, where QPSK, 16QAM, 64QAM, and 256QAM are employed as the selectable modulation parameters.

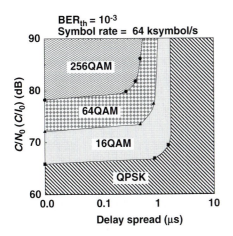

Fig. 17.32 An example of modulation parameter selection chart, where QPSK, 16QAM, 64QAM, and 256QAM are employed as the selectable modulation parameters.

17.7.6 Theoretical Performance Analysis of Fast Adaptive Modulation Systems

When instantaneous C/N_0 and its average value are x and x_{av}, respectively, the p.d.f. of x is given by

$$p(x) = \frac{1}{x_{av}} e^{-x/x_{av}} \tag{17.42}$$

When we define x_1, x_2, and x_3 as the threshold values of C/N_0 for 16QAM, 64QAM, and 256QAM, respectively, that satisfy BER $\leq$ BER$_{th}$, the selection probabilities of QPSK, 16QAM, 64QAM, and 256QAM are given by

$$\Pr(\text{QPSK}) = \int_0^{x_1} \frac{1}{x_{av}} e^{-x/x_{av}} dx \tag{17.43}$$

$$\Pr(\text{16QAM}) = \int_{x_1}^{x_2} \frac{1}{x_{av}} e^{-x/x_{av}} dx \tag{17.44}$$

$$\Pr(\text{64QAM}) = \int_{x_2}^{x_3} \frac{1}{x_{av}} e^{-x/x_{av}} dx \tag{17.45}$$

$$\Pr(\text{256QAM}) = \int_{x_3}^{\infty} \frac{1}{x_{av}} e^{-x/x_{av}} dx \tag{17.46}$$

When we employ QPSK, we can transmit 2 bits per each symbol. Similarly, we can transmit 4 bits, 6 bits, and 8 bits using 16QAM, 64QAM, and 256QAM, respectively. Therefore, we can transmit data with the average bit rate of

$$R_{bav} = \sum_{i=1}^{4} 2iR_s \cdot \int_{x_{i-1}}^{x_i} \frac{1}{x_{av}} e^{-x/x_{av}} dx \tag{17.47}$$

where $x_0 = 0$ and $x_4 = \infty$. The average BER is defined as the number of error bits divided by the number of transmitted bits. Therefore, it is given by [17-23]

$$P_b(x_{av}) = \frac{\displaystyle\sum_{i=1}^{4} 2iR_s \int_{x_{i-1}}^{x_i} \frac{1}{x_{av}} e^{x/x_{av}} \cdot \alpha_i erfc\left(\sqrt{\beta_i x}\right) dx}{\displaystyle\sum_{i=1}^{4} 2iR_s \int_{x_{i-1}}^{x_i} \frac{1}{x_{av}} e^{-x/x_{av}} dx} \tag{17.48}$$

where α_i and β_i, which include degradation due to pilot symbol-aided fading compensation, are shown in Table 17.3 [17-23].

Figure 17.33 shows theoretical BER and average-bit-rate performance under flat Rayleigh fading conditions, where BER$_{th}$ = 10^{-3} is applied; the performance of the QPSK system is also shown. As shown in this figure, the average bit rate is drastically improved with increasing C/N_0, although the BER performance is almost the same as the BER performance of QPSK systems. Therefore, if the transmitted power is assumed to be the same, the fast adaptive modulation scheme is considered to be effective in achieving higher transmission quality than that of QPSK systems.

Table 17.3 Coefficients α_i and β_i for each modulation.

i	α_i	β_i
1	1/2	$1/(2.58R_s)$
2	3/8	$0.4/(5.4R_s)$
3	7/24	$1/(60.9R_s)$
4	15/64	$1/(255R_s)$

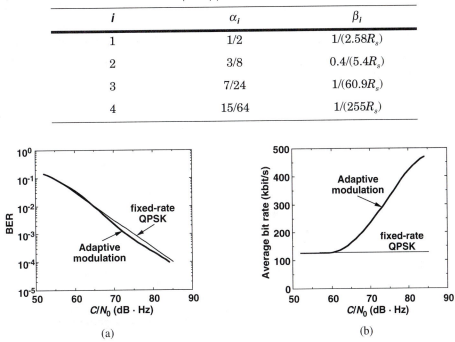

Fig. 17.33 Theoretical BER and average-bit-rate performance under flat Rayleigh fading conditions, where $\text{BER}_{th} = 10^{-3}$ is applied.

17.7.7 Computer Simulation Results of Fast Adaptive Modulation Systems

Figure 17.34 shows the computer-simulated results of BER and average-bit-rate performance of fast adaptive modulation systems under flat Rayleigh fading conditions, where $f_d = 1.563 \times 10^{-5}$. The results show that the computer-simulated results are almost the same as the theoretical performance results.

Next we will evaluate the effect of channel variation speed. Figure 17.35 shows the simulated results of the BER versus $f_d T_F$ performance under flat Rayleigh fading conditions, where f_d is normalized by a frame length T_F. When $f_d T_F \geq 8.0 \times 10^{-2}$, BER performance is degraded with f_d because the channel condition estimator cannot sufficiently estimate the channel variation as well due to fast channel variation. When we want to achieve an average BER of less than 10^{-3}, we have to satisfy $f_d T_F = 8.0 \times 10^{-2}$ in the case of $C/N_0 = 78$ dB·Hz, and $f_d T_F = 1.5 \times 10^{-1}$ in the case of $C/N_0 = 88$ dB·Hz. When we assume a frame length of 5 ms, these requirements correspond to $f_d = 16$ Hz and 30 Hz, respectively.

Delay spread immunity is important when we apply fast adaptive modulation systems. Figure 17.36 shows the irreducible BER versus normalized delay spread performance of QPSK and fast adaptive modulation systems. Because the average bit rate of fast adaptive modulation (R_{bav}) is subject to channel conditions, the inverse of the average bit rate is defined as the average bit duration, and the delay spread is normalized by this average bit duration. We can find from this figure that, when we employ fast adaptive modulation systems, we can accept approximately larger delay

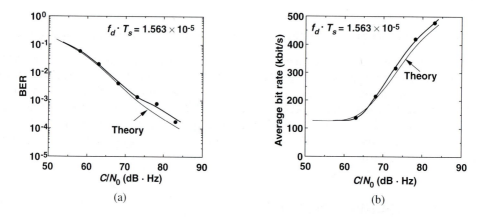

Fig. 17.34 Computer-simulated results of BER and average-bit-rate performance of fast adaptive modulation systems under flat Rayleigh fading conditions, where $f_d = 1.563 \times 10^{-5}$.

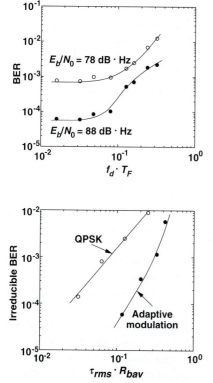

Fig. 17.35 Simulated results of the BER versus $f_d T_F$ performance under flat Rayleigh fading conditions, where f_d is normalized by a frame length T_F.

Fig. 17.36 Irreducible BER versus normalized delay spread performance of QPSK and fast adaptive modualtion systems.

spread that is approximately four times the size as that for the QPSK system. This means that, when the average delay spread is 5 µs, the maximum bit rate that satisfies BER = 10^{-3} is 15.4 kbit/s in the case of QPSK, whereas we can achieve 62 kbit/s when we employ fast adaptive modulation systems.

17.7.8 Other Ongoing Research Topics on Adaptive Modulation Techniques

There is other research covering the various possible configurations of fast adaptive modulation techniques, such as the symbol rate and modulation-level-controlled systems [17-25], coding-rate-controlled adaptive modulation [17-26], a combination of the coding and adaptive modulation scheme [17-27], and improvement of channel variation tracking ability using directive antennas [17-28]. Because adaptive modulation is a very attractive technique, the results of other investigation into this topic will be reported in the future.

17.8 RADIO HIGHWAY NETWORK

When we consider the whole wireless communication systems, we can divide it into two portions—the information transfer network, which is in charge of transferring information using switching, routing, and some other functions; and the radio link, which is in charge of establishing a wireless link between the BS and the terminal. Using this classification, the intelligent radio communication technologies discussed in this chapter are techniques that offer flexibility and intelligence to the radio link. On the other hand, Komaki et al. proposed the *radio highway network* concept, which offers flexibility and intelligence to the information transfer network [17-29].

Figure 17.37 shows the proposed concept of the radio highway network. This network is composed of an optical fiber link and optical switching nodes. The optical switching node consists of optical switches to route a signal to another optical switch-

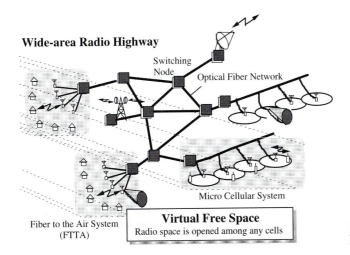

Wide-area Radio Highway

Switching Node

Optical Fiber Network

Micro Cellular System

Fiber to the Air System (FTTA)

Virtual Free Space
Radio space is opened among any cells

Fig. 17.37 Concept of the radio highway network.

ing node or the BS near the destination terminal. In this system, each BS consists of only an electric-to-optic converter (E/O) and an optic-to-electric converter (O/E). When the signal from a terminal is received at a BS, the signal is encapsulated into the envelope of the optical signal by the direct modulation of the optical carrier and transferred to an appropriate BS via several optical routing nodes. Because the optical carrier is directly modulated by the received RF signal from a terminal, this system is called the *fiber and radio extension link* (FREx link). As a result, this system can open the radio-free space of a wireless communication service at any location simply by installing a simple BS that consists of only the E/O converter and the O/E converter and connecting it to the switching office by the optical fiber, even if the BS is located far away from the switching office. Therefore, this radio-free space is called *virtual free space*, or the radio highway network. Because we can employ any infrastructure with optical fiber cables, such as the fiber-optic micro/picocell radio, cable television (CATV), and fiber to the air (FTTA) systems, and can cope with any type of radio signals, the radio highway network can flexibly cope with the extension of the network configuration as well as the modification of the radio interface of wireless communication systems.

Moreover, Komaki and his associates propose various application fields of the radio highway network, such as the radio wide area network (WAN) [17-29], and many key technologies for the radio highway network, such as the intercell connection bus link (ICBL) using TDM [17-30] and CDM [17-31] techniques.

Although the concept of the radio highway network is new, and there are many problems to be solved, it is one of the most promising networking concepts for flexible and intelligent wireless personal multimedia communication systems of the future.

REFERENCES

17-1. Morinaga, N., Yokoyama, M. and Sampei, S., "Intelligent radio communication techniques for advanced wireless communication systems," IEICE Trans. Comm., Vol. E79-B, No. 3, pp. 214–21, March 1996.

17-2. Hayes, J. K., "Adaptive feedback communications," IEEE Trans. Commun. Tech., Vol. COM-16, pp. 29–34, February 1968.

17-3. Kosaka, K., Suzuki, Y., Nishiyama, I., Kohri, T. and Egami, S., "Communication experiments: Experiments on measures against rain attenuation," IEEE Trans. Aerosp. Electron. Syst., Vol. AES-22, No. 3, pp. 302–9, May 1986.

17-4. Ariyavisitakul, S., "Autonomous SIR-based power control for a TDMA radio system," IEEE GLOBECOM'93 (Houston, Texas), pp. 307–10, November 1993.

17-5. Viterbi, A. J., *CDMA, principle of spread spectrum communications*, Addison-Wesley Publishing Company, 1995.

17-6. Dohi, T., Okumura, Y., Higashi, A., Ohno, K. and Adachi, F., "Experiments on coherent multicode DS-CDMA," 46th IEEE Veh. Tech. Conf. (Atlanta, Georgia), pp. 889–93, April 1996.

17-7. Hayashi, M., Miya, K., Kato, O. and Homma, K., "CDMA/TDD cellular systems utilizing a base-station-based diversity scheme," 45th IEEE Veh. Tech. Conf. (Chicago, Illinois), pp. 799–803, August 1995.

17-8. Abeta, S., Sampei, S. and Morinaga, N., "Adaptive coding rate and processing gain control for cellular DS/CDMA systems," 4th IEEE ICUPC, pp. 241–45, November 1995.

17-9. Cavers, J. K., "Variable rate transmission for Rayleigh fading channels," IEEE Trans. Commun., Vol. COM-20, pp. 15–22, February 1972.

17-10. Bello, P. A. and Cowan, W. M., "Theoretical study of on/off transmission over Gaussian multiplicative circuit," Proc. IRE 8th National Commun. Symp. (Utica, New York), October 1962.

17-11. Webb, W. T. and Hanzo, L., *Quadrature amplitude modulation–Principle and application for fixed and wireless communications*, Pentech Press, 1994.

17-12. Komaki, S., "Theoretical analysis of a capacity controlled digital microwave radio," Trans. IEICE, Vol. J73-B-II, No. 10, pp. 498–503, October 1990.

17-13. Lee, H. J., Omae, T., Komaki, S. and Morinaga, N., "Performance analysis of the capacity controlled system with adaptive equalizer," IEICE Trans. Commun., Vol. J76-B, No. 2, pp. 148–54, February 1993.

17-14. Ouchi, M., Lee, H. J., Komaki, S. and Morinaga, N., "Proposal for modulation level controlled radio system applied to ATM networks," 4th European Conference on Radio Relay Systems (Edinburgh, Scotland), pp. 322–27, October 1993.

17-15. Sampei, S., Komaki, S. and Morinaga, N., "Adaptive modulation/TDMA scheme for large capacity personal multi-media communication systems," IEICE Trans. Commun., Vol. E77-9, No. 9, pp. 1096–1103, September 1994.

17-16. Kamio, Y. and Sampei, S., "Performance of a trellis-coded 16QAM/TDMA system for land mobile communications," IEEE Trans. Veh. Technol., Vol. 43, No. 3, pp. 528–36, August 1994.

17-17. Sumiyoshi, H., Tanimoto, M. and Komai, M., "Theoretical study on synchronized spread spectrum systems," Technical Report of IECE, CS81-11, April 1981.

17-18. Sampei, S., Leung, P. and Feher, K., "High capacity cell configuration strategy in ACI and CCI conditions," 43rd IEEE Veh. Tech. Conf. (Secaucus, New Jersey), pp. 185–88, May 1993.

17-19. Kanai, T., "Channel assignment for sector cell layout," Trans. IEICE (B-II), Vol. J73-B-II, No. 11, pp. 595–601, November 1990.

17-20. Tanaka, T., Akeyama, A. and Kozono, S., "Urban multipath propagation delay characteristics in mobile communications," Trans. IEICE (B-II), Vol. J73-B-II, No. 11, pp. 772–78, November 1990.

17-21. Sampei, S., Morinaga, N. and Kamio, Y., "Adaptive modulation/TDMA with a BDDFE for 2 Mbit/s multi-media wireless communication systems," IEEE 45th Veh. Tech. Conf. (Chicago, Illinois), pp. 311–15, July 1995.

17-22. Driessen, P. F., "Multipath delay characteristics in mountainous terrain at 900 MHz," 42d IEEE Veh. Tech. Conf. (Denver, Colorado), pp. 520–23, May 1992.

17-23. Otsuki, S., Sampei, S. and Morinaga, N., "Performance of modulation level controlled adaptive modulation systems," Trans. IEICE (B-II), Vol. J78-B-II, No. 6, pp. 435–44, June 1995.

17-24. Alamouti, S. M. and Kallel, S., "Adaptive trellis-coded multiple-phase-shift keying for Rayleigh fading channels," IEEE Trans. Commun., Vol. 42, No. 6, pp. 2305–14, June 1994.

17-25. Ue, T., Sampei, S. and Morinaga, N., "Symbol rate and modulation level controlled adaptive modulation/TDMA/TDD for personal communication systems," 45th IEEE Veh. Tech. Conf. (Chicago, Illinois), pp. 306–10, July 1995.

17-26. Saifuddin, A. and Kohno, R., "Adaptive multilevel coding and multicarrier interference cancellation for CDMA in time varying channel," Technical Report of IEICE, SST95-7, May 1995.

17-27. Matsuoka, H., Sampei, S., Morinaga, N. and Kamio, Y., "Adaptive modulation system with punctured convolutional code for high quality personal communication systems," 4th IEEE ICUPC (Tokyo, Japan), pp. 22–26, November 1995.

17-28. Suzuki, T., Sampei, S. and Morinaga, N., "Directive antennas diversity reception for an adaptive modulation system in land mobile communications," 4th IEEE ICUPC (Tokyo, Japan), pp. 595–99, November 1995.

17-29. Komaki, S., Tsukamoto, K., Okada, M. and Harada, H., "Proposal of radio high-way networks for future multi-media-personal wireless communications," 1994 IEEE Int. Conf. on Personal Wireless Commun. (Bangalore, India), pp. 204–8, August 1994.

17-30. Tsukamoto, K., Harada, H., Kajiya, S., Komaki, S. and Morinaga, N., "TDM intercell connection fiber-optic bus link for personal radio communication system," 1994 APMC (Tokyo, Japan), pp. 1039–42, December 1994.

17-31. Kajiya, S., Harada, H., Tsukamoto, K. and Komaki, S., "A consideration on radio-highway networks with CDMA optical fiber link," Technical Report of IEICE, RCS94-96, October 1994.

A

Abbreviations and Acronyms

A

ACCH	Associated control channel
ACI	Adjacent channel interference
ACK	Acknowledgment
ACT	Adaptive carrier tracking
ACTS	Advanced communications technologies and services
A/D	Analog-to-digital
ADPCM	Adaptive differential pulse code modulation
AFC	Automatic frequency controller
AGC	Automatic gain controller
AMPS	Advanced mobile phone system
ARIB	Association of Radio Industries and Businesses
ARQ	Automatic repeat request
ATM	Asynchronous transfer mode
ATT	Attenuator
AWGN	Additive white Gaussian noise

B

BBC	Binary block coding
BCCH	Broadcast control channel
BDDFE	Bidirectional decision feedback equalizer
BEF	Band elimination filter
BER	Bit error rate
BPF	Band-pass filter

BS	Base station
BSG	Baseband signal generator

C

CAC	Common access channel
CC	Call control
CCCH	Common control channel
CCH	Control channel
CCI	Co-channel interference
CCIR	Consultative Committee for International Radiocommunications
CCITT	Consultative Committee for International Telegraph and Telephone
CDM	Code division multiplexing
CDMA	Code division multiple access
CDR	Call detail record
CEPT	Conference of European Posts and Telecommunications
C/I (also CIR)	Carrier-to-interference power ratio
CI	Channel indicator
CMA	Constant modulus algorithm
CPU	Central processing unit
CQPSK	Coherent quaternary phase shift keying
CRC	Cyclic redundancy code
CRSW	Carrier synchronization word
CS	Cell station
CT-2	Second-generation cordless telephone
CW	Carrier wave or continuous wave

D

D/A	Digital-to-analog
dB	Decibel(s)
dBi	Decibles relative to an isotropic radiator
dBm	Decibels relative to 1 milliwatt
dBW	Decibels relative to 1 watt
DC	Direct current
DCA	Dynamic channel assignment
DCR	Dual-mode carrier recovery
DDFSE	Delayed decision feedback sequence estimation
DFBPSK	Differentially encoded binary phase shift keying
DFE	Decision feedback equalizer
D-FF	D-type flip flop
DPSK	Differential phase shift keying
DPSK/PDI	Differential phase shift keying with post-detection integrator
DQPSK	Differential quaternary phase shift keying
DS	Direct-sequence
DS/CDMA	Direct-sequence/code division multiple access

DSP	Digital signal processor
DTCT	Dual-tone calibration technique

E

EIRP	Effective isoroeopic radiated power
EMC	Electric magnetic compatibility
ESMR	Extended Specialized Mobile Radio
ETSI	European Telecommunications Standards Institute

F

FACCH	Fast associated control channel
FB	Feedback
FCA	Fixed channel assignment
FCC	Federal Communications Commission
FDD	Frequency division duplex
FDM	Frequency division multiplexing
FEC	Forward error correction
FER	Frame error rate
FF	Feed forward
FFH	Fast frequency hopping
FFT	Fast Fourier Transform
FH	Frequency hopping
FM	Frequency modulation
FPLMTS	Future Public Land Mobile Telecommunications Systems
FQAM	Feher's quadrature amplitude modulation
FQPSK	Feher's quaternary phase shift keying

G

GBN	Go-back-N
GEO	Geostationary earth orbit
GLR	Gate location register
GMSK	Gaussian-filtered minimum shift keying
GP-IB	General-purpose interface bus
GSM	Group Special Mobile. Global Systems for Mobile Commuications

H

HDLC	High-level data link control
HLR	Home location register

I

I&D	Integrate and dump
IC	Integrated circuit

ICMA	Idle-signal casting multiple access
ICMA-PE	Idle-signal casting multiple access with partial echo
IF	Intermediate frequency
IGS	Interconnection gateway switch
ILC	Instantaneous likelihood calculator
IMTS	Improved Mobile Telephone System
I-Q	In-phase and quadrature-phase
IS-54	Interim standard 54
IS-95	Interim standard 95
ISDN	Integrated services digital network
ISI	Intersymbol interference
ISM	Industrial, scientific, and medical
ITU	International Telecommunication Union
ITU-R	International Telecommunication Union Radio Section
ITU-T	International Telecommunication Union Telecommunication Section
IWP 8/13	Interim Working Party 8/13

J

JDC	Japanese digital cellular
JSMR	Japan shared mobile radio

L

LAN	Local area network
LAPDC	Link access procedure for digital cordless
LAPDM	Link access procedure for digital mobile channel
LCCH	Link control channel
LCH	Link channel
LEO	Low earth orbit
LINC	Linear amplification with nonlinear components
LM	Level monitoring period
LMS	Least mean square
LOS	Line-of-sight
LPF	Low-pass filter
LS	Local switch
LSI	Large-scale integration

M

MAM	Maximum amplitude method
MCA	Multichannel access
MCC	Mobile communication control center
MCPSK	M-ary coherent phase shift keying
MDPSK	M-ary differential phase shift keying

MES	Mobile earth station
MGC	Mobile gateway switching center
ML	Maximum likelihood
MLSE	Maximum likelihood sequence estimation
MM	Mobility management
MPT	Ministry of Posts and Telecommunications
M-QAM	Multicarrier quadrature amplitude modulation
MRC	Mobile Radio Center
MS	Mobile station
MSK	Minimum shift keying
MTS	Mobile Telephone System
MTSO	Mobile telephone switching office

N

NACK	Nonacknowledgment
NCC	New common carrier
NF	Noise figure
NLOS	Non line-of-sight
NMT	Nordic Mobile Telephone
NRZ	Non return-to-zero
NTT	Nippon Telegram and Telephone

O

OQPSK	Offset quaternary phase shift keying
OSI	Open system interconnect

P

p.d.f.	Probability density function
PBS	Paging base station
PBX	Private branch exchange
PC	Personal computer
PDC	Personal digital cellular
PCH	Paging channel
PCS	Personal communication system
PD	Phase detector
PDA	Personal digital assistance
PES	Personal earth station
PHS	Personal handyphone system
PLL	Phase-locked loop
PMR	Private mobile radio
PN	Pseudo-noise
POI	Point of interface
PRBS	Pseudorandom binary sequence

PS   Personal station
PSTN   Public switched telecommunication network

Q

QAM   Quadrature amplitude modulation
QMF   Quadrature mirror filter
QoS   Quality of service
QPSK   Quaternary phase shift keying

R

R&D   Research and development
RACE   R&D in advanced communications technologies in Europe
RCCH   Radio control channel
RCH   Radio-channel housekeeping channel
RCR   Research and Development Center for Radio Systems (presently ARIB)
RDS   Running digital sum
RF   Radio frequency
RLS   Recursive least squares
rms   Root mean square
ROM   Read-only memory
RRC   Radio Regulation Council
RT   Radio transmission management

S

SACCH   Slow associated control channel
SAW   Stop-and-wait
SCCH   Signaling control channel
SCH   Service channel
SCP   Service control point
SER   Symbol error rate
SF   Steal flag
SLSW   Slot synchronization word
SMG   Special Mobile Group
SMR   Shared mobile radio
SNR   Signal-to-noise power ratio
S/P   Serial-to-parallel
SR   Selective repeat
SS   Spread spectrum
SS7   Signaling system No. 7
SSC   Superframe synchronization counter
STC   Sub-technical committee
STSW   Symbol timing synchronization word
SW   Synchronization word. (also switch)

T

TA	Time alignment
TACS	Total access communication system
TCH	Traffic channel
TCT	Tone calibration technique
TDD	Time division duplex
TDM	Time division multiplex
TDMA	Time division multiple access
TG-8/1	Task group 8/1
TIA	Telecommunications Industry Association
TPC	Transmitter power control
TTC	Telecommunication Technology Council
TTIB	Transparent tone-in band

U

UHF	Ultra-high frequency
UMTS	Universal Mobile Telecommunication System
UPCH	User packet channel
USC	User-specific channel
USCCH	User-specific control channel
USPCH	User-specific packet channel

V

VCO	Voltage-controlled oscillator
VLSI	Very large-scale integration
VSELP	Vector sum excited linear predictor

W

WARC	World Administrative Radio Conference
WDM	Wave differential method
WP	Wide area pager
WRC	World Radio Communication Conference

X

xM	times M

Z

ZCM	Zero-crossing method